Advanced Methodology and Analysis in Coal Mine Gas Control

Advanced Methodology and Analysis in Coal Mine Gas Control

Guest Editors

Haijun Guo
Jian Chen
Yingfeng Sun
Hao Zhang

MDPI

Basel • Beijing • Wuhan • Barcelona • Belgrade • Novi Sad • Cluj • Manchester

Guest Editors
Haijun Guo
School of Emergency Management and Safety Engineering
China University of Mining and Technology (Beijing)
Beijing
China

Jian Chen
School of Emergency Management and Safety Engineering
North China University of Science and Technology
Tangshan
China

Yingfeng Sun
Research Institute of Macro-Safety Science
University of Science and Technology Beijing
Beijing
China

Hao Zhang
College of Safety & Emergency Management Engineering
Taiyuan University of Technology
Taiyuan
China

Editorial Office
MDPI AG
Grosspeteranlage 5
4052 Basel, Switzerland

This is a reprint of the Special Issue, published open access by the journal *Applied Sciences* (ISSN 2076-3417), freely accessible at: https://www.mdpi.com/journal/applsci/special_issues/A2NV3MWTSK.

For citation purposes, cite each article independently as indicated on the article page online and as indicated below:

Lastname, A.A.; Lastname, B.B. Article Title. *Journal Name* **Year**, *Volume Number*, Page Range.

ISBN 978-3-7258-5863-7 (Hbk)
ISBN 978-3-7258-5864-4 (PDF)
https://doi.org/10.3390/books978-3-7258-5864-4

Contents

applied sciences

MDPI

Editorial

Special Issue: Advanced Methodology and Analysis in Coal Mine Gas Control

Haijun Guo [1,*], Jian Chen [2], Yingfeng Sun [3] and Hao Zhang [4]

1 School of Emergency Management and Safety Engineering, China University of Mining and Technology (Beijing), Beijing 100083, China
2 School of Emergency Management and Safety Engineering, North China University of Science and Technology, Tangshan 063210, China; chenjian@ncst.edu.cn
3 Research Institute of Macro-Safety Science, University of Science and Technology Beijing, Beijing 100083, China; yingfengsun@ustb.edu.cn
4 College of Safety & Emergency Management Engineering, Taiyuan University of Technology, Taiyuan 030024, China; haozhang@cumt.edu.cn
* Correspondence: Haijun Guo: navyguohj@cumtb.edu.cn

1. Introduction

The significance of coal mine gas control in the field of engineering cannot be underestimated. Gas disasters are one of the most common and destructive types that occur in coal mines [1]. Due to the depletion of shallow resources in recent years, mining depths in coal mines have increased year by year [2,3]. With the presence of micro-pores and the low permeability and high adsorption of coal seams being common issues, the problem of gas in mining has become one of the main issues that restricts their safe development [4,5]. Coal mine gas control is an important requirement for safe and efficient coal production. However, gas control and utilization face a series of challenges in coal and gas outburst prevention, intelligent gas disaster warnings, gas explosions, and efficient coal bed methane extraction. This Special Issue, "Advanced Methodology and Analysis in Coal Mine Gas Control", aims to report the latest research in coal mine gas control in terms of basic theories, key technology, and equipment development to improve gas disaster prevention in coal mining processes.

2. Overview of Contributions

This Special Issue's contributions can be categorized into three key themes:

1. Mechanism and control technology of gas disasters

Ma et al. (contribution 7) established normal and tangential analysis models of ore rock impacting shaft walls. Du et al. (contribution 9) described a gas explosion simulation that combines the chemical reaction mechanism and tunnel structure. Liu et al. (contribution 1) suggested that coal mine gas accidents could be prevented to a large extent by encouraging good management culture in enterprise management and discourage illegal production behaviors. Xia et al. (contribution 11) reported the quantitative characterization of the degree of damage caused by ore rock flow and suggested a mechanism of damage to the shaft wall in the storage section of the ore pass. Jin et al. (contribution 6) took coal and rock samples subjected to the coupled effects of water, temperature, and dynamic loads as the research object and discussed their mechanical properties. Fu et al. (contribution 12) studied the roof movement and ground pressure evolution characteristics of an ultralong working face in a shallow coal seam with a high mining height. Gao et al. (contribution 2) developed a deformation and damage mechanism for the surrounding rock of auxiliary retracement channels and a method for determining the reasonable spacing of two retracement channels at the end of the mining period. Pu et al. (contribution 4) established two sets of active support schemes for controlling the surrounding rock in thick coal seams with ultra-large span open-off cuts and proposed an optimized support scheme through numerical simulation.

2. Gas disaster monitoring and early warning technology

Lutyński et al. (contribution 17) presented the results of an analysis of the hazards posed by coal mine dust and methane in coal preparation plants of hard coal mines in Poland. Zhang et al. (contribution 16) determined the upper limit of time to ensure safe operation under different failure modes; they found that the acquired temperature distribution characteristics under failure modes can be used as a reference for the design, inspection, and status warning of flameproof cable boxes for coal mines. Guo et al. (contribution 3) established a XGBoost–GR–stacking gas outburst early warning model to demonstrate high accuracy and practical performance.

3. Gas extraction and unloading

Hao et al. (contribution 14) developed a novel borehole sealed quality detection device for the extraction of gas in a coal mine. Yu et al. (contribution 13) employed a model of the equivalent fissure to reveal the heat transfer mechanism between the rock matrix and CO_2. Xie et al. (contribution 10) proposed a dual method of hydraulic fracturing of key layers of overlying rock layers combined with the pre-extraction of gas via large-diameter caving boreholes. Shu (contribution 8) examined gas extraction technology in the goaf of an L-shaped borehole in the mining fissure zone of a short-distance coal seam group. Ma et al. (contribution 15) showed that stress directly affects the gas desorption of coal and plays a decisive role in the gas desorption and emission characteristics of water-bearing coal in the stress-affected zone. Zhang et al. (contribution 5) constructed a molecular structure model of anthracite and coking coal–rock compound models containing different calcite and kaolinite contents.

3. Conclusions

The articles collected in our Special Issue highlight the great progress being made in gas storage, disaster mechanisms, monitoring and warnings, precise prevention and control, and efficient extraction, and provide a valuable frontier perspective and technical guidance for mine gas control. At present, mine gas control still faces many challenges. With continuous improvements to the intelligent construction of coal mines, future research will continue to improve the level of comprehensive gas control and development with more scientific technology, more advanced equipment, and enhanced management techniques. This will enable the coal industry to develop in a more efficient, green, and low-carbon direction, and contribute further to the realization of the "double carbon" goal and the protection of energy security.

Funding: This research received no external funding.

Acknowledgments: Thanks to all the authors and peer reviewers for their valuable contributions to this Special Issue 'Advanced Methodology and Analysis in Coal Mine Gas Control'. I would also like to express my gratitude to all the staff and people involved in this Special Issue.

Conflicts of Interest: The author declares no conflicts of interest.

List of Contributions:

1. Liu, Y.; Liang, Y.; Li, Q. Cause Analysis of Coal Mine Gas Accidents in China Based on Association Rules. *Appl. Sci.* **2023**, *13*, 9266. https://doi.org/10.3390/app13169266.
2. Gao, X.; Liu, C.; Zhang, H.; Yang, K.; Hu, Y.; Guo, X. The Change in the Shape Characteristics of the Plastic Zone in the Surrounding Rock of an Auxiliary Retracement Channel and a Reasonable Channel Spacing Determination Method. *Appl. Sci.* **2023**, *13*, 10543. https://doi.org/10.3390/app131810543.
3. Guo, Y.; Liu, H.; Zhou, X.; Chen, J.; Guo, L. Research on Coal and Gas Outburst Risk Warning Based on Multiple Algorithm Fusion. *Appl. Sci.* **2023**, *13*, 12283. https://doi.org/10.3390/app132212283.
4. Pu, L.; Liu, Y.; Cai, Y.; Sun, Z.; Zhou, X. Study on Active Support Parameters for Surrounding Rock with Ultra-Large Span Open-Off Cut in Thick Coal Seam. *Appl. Sci.* **2023**, *13*, 12804. https://doi.org/10.3390/app132312804.
5. Zhang, X.; Liu, Y.; Zhou, T.; Cai, Y.; Zhang, B. Molecular Simulation Study on the Effect of Co-Associated Minerals on Methane Adsorption and Mechanical Properties of Coal. *Appl. Sci.* **2023**, *13*, 12975. https://doi.org/10.3390/app132412975.
6. Jin, T.; Sun, X.; Liu, K.; Lin, S.; Yang, S.; Xie, J. Experimental Study of the Multiple Fractalisation of Coal and Rock Failure Subjected to the Coupled Effects of Water, Temperature and Dynamic Loads. *Appl. Sci.* **2023**, *13*, 13004. https://doi.org/10.3390/app132413004.
7. Ma, S.; Cao, J.; Zhang, Q.; Xue, S. Study on the Influence of Gas Desorption Characteristics under High-Pressure Fluid Fracturing of Deep Coal. *Appl. Sci.* **2023**, *13*, 13327. https://doi.org/10.3390/app132413327.
8. Shu, L. Study on Gas Extraction Technology for Goaf Using L-Shaped Borehole on the Ground. *Appl. Sci.* **2024**, *14*, 1594. https://doi.org/10.3390/app14041594.
9. Du, J.; Chen, J.; Zhu, L.; Guo, L.; Wang, F.; Hu, X. Numerical Study on the Dynamic Response of Gas Explosion in Uneven Coal Mine Tunnels Using CESE Reaction Dynamics Model. *Appl. Sci.* **2024**, *14*, 2372. https://doi.org/10.3390/app14062372.
10. Xie, J.; Li, F.; Yan, Z.; Huo, J. Double Unloading Gas Control Technology for Fracturing Soft Coal Seams in Overlying Key Strata. *Appl. Sci.* **2024**, *14*, 3202. https://doi.org/10.3390/app14083202.
11. Xia, Z.; Deng, Z.; Lu, Z.; Ma, C. Study on the Macro-Fine Mechanical Behavior of Ore Flow Based on the Discrete Element Method. *Appl. Sci.* **2024**, *14*, 3457. https://doi.org/10.3390/app14083457.
12. Fu, Y.; Li, C.; He, Y. Research on the Movement of Overlying Strata in Shallow Coal Seams with High Mining Heights and Ultralong Working Faces. *Appl. Sci.* **2024**, *14*, 4685. https://doi.org/10.3390/app14114685.
13. Yu, F.; Deng, G. The Thermodynamic Change Laws of CO_2-Coupled Fractured Rock. *Appl. Sci.* **2024**, *14*, 5122. https://doi.org/10.3390/app14125122.
14. Hao, Q.; Chen, H.; Xu, G.; Yang, Y.; Chen, X.; Wang, Z.; An, K. Development and Application of Unsealed Borehole Leakage Detection Device Based on Flow Method. *Appl. Sci.* **2024**, *14*, 6684. https://doi.org/10.3390/app14156684.
15. Ma, Q.; Ma, C.; Li, J.; Lu, Z.; Xia, Z. Theoretical Analysis of Shaft Wall Damage and Failure Under Impacting of Ore-Rock Falling in Vertical Ore Pass. *Appl. Sci.* **2024**, *14*, 10695. https://doi.org/10.3390/app142210695.
16. Zhang, H.; Cui, Z.; Lan, Y. Magnetic–Thermal Coupling-Based Study on the Temperature Characteristics of Flameproof Cable Boxes for Coal Mines at Various Failure Modes. *Appl. Sci.* **2024**, *14*, 11209. https://doi.org/10.3390/app142311209.
17. Lutyński, A.; Lutyński, M. Coal Dust and Methane as a Hazard in Coal Preparation Plants. *Appl. Sci.* **2024**, *14*, 11433. https://doi.org/10.3390/app142311433.

References

1. Wang, K.; Du, F. Coal-gas compound dynamic disasters in China: A review. *Process Saf. Environ. Prot.* **2020**, *133*, 1–17. [CrossRef]
2. Zhao, P.; Chang, Z.; Li, S.; Zhuo, R.; Jia, Y.; Shao, Q.; Lei, W.; Guo, X. Study of the velocity effect on the energy dissipation structure of overlying rock at a fully mechanized mining face. *Energy* **2025**, *320*, 135315. [CrossRef]

3. Yuan, L.; Wang, E.; Ma, Y.; Liu, Y.; Li, X. Progress of research on coal-rock power disasters in China and the scientific an technological difficulties faced. *J. China Coal Soc.* **2023**, *48*, 1825–1845.
4. Zhou, B.; Xu, J.; Peng, S.; Geng, J.; Yan, F. Test system for the visualization of dynamic disasters and its application to coal and ga outburst. *Int. J. Rock Mech. Min.* **2019**, *122*, 104083. [CrossRef]
5. Lu, G.; Wei, C.; Wang, J.; Meng, R.; Tamehe, L. Influence of pore structure and surface free energy on the contents of adsorbe and free methane in tectonically deformed coal. *Fuel* **2021**, *285*, 119087. [CrossRef]

applied sciences

MDPI

rticle

Cause Analysis of Coal Mine Gas Accidents in China Based on Association Rules

ing Liu [1,2], Yunpei Liang [1,2,*] and Quangui Li [1,2,*]

1 State Key Laboratory of Coal Mine Disaster Dynamics and Control, Chongqing University, Chongqing 400044, China; cquliuying@163.com
2 School of Resources and Safety Engineering, Chongqing University, Chongqing 400044, China
* Correspondence: liangyunpei@cqu.edu.cn (Y.L.); liqg@cqu.edu.cn (Q.L.); Tel.: +86-135-9460-8029 (Y.L.); +86-185-2359-4231 (Q.L.)

Abstract: Coal mine gas accidents will cause great economic losses and casualties. It is of great significance to find out the essential causes of coal mine gas accidents and put forward measures to prevent them. In this paper, 110 coal mine gas accidents which occurred in China from 2001 to 2022 are selected to analyze the causes of the accidents by extracting the keywords of human factors, equipment factors, environment factors, and management factors from the accident investigation reports. Firstly, the accident statistical analysis is carried out from three dimensions of factor frequency, accident type, and accident grade. Secondly, the Apriori algorithm is used for data mining to obtain frequent item sets and association rules of coal mine gas accident factors. Finally, the coal mine gas accident cause chains which are obtained by using the association rule. The frequent terms of 9 factors, 23 association rules, and 3 coal mine gas accident cause chains are obtained. The results show that the production of coal mine enterprises by illegal organizations is an important reason for the occurrence of coal mine gas accidents. The lack of good management culture easily leads to habitual violations of personnel and decision-making errors, and then causes equipment problems, reflected in the ventilation system which is not perfect, resulting in gas accumulation. The occurrence of coal mine gas accidents can be prevented to a large extent by preventing the absence of good management culture in enterprise management and the occurrence of illegal production behaviors.

Keywords: coal mine gas accident; cause factors; cause analysis; association rules; Apriori algorithm

1. Introduction

A comparative analysis of the fatality rate per million tons of various coal mine accidents in China shows that coal mine gas accidents often cause large economic losses and casualties [1,2] In recent years, although coal mine safety production technology has made great progress, the issue of how to prevent the occurrence of coal mine gas accidents is still an important subject of the current research. A large number of historical accident statistics can better reflect the mechanisms of and reasons for the accident. Gao et al. [3] sorted out nearly 170 gas explosion accidents in China and nearly 50 gas explosion accidents abroad, analyzed the accidents from the three aspects of accident location, occurrence time, and types of ignition sources, and obtained ten types of ignition sources that are most likely to cause accidents. Zhao et al. [3] analyzed the tremendous and devastating coal mine gas explosion accidents from 1981 to 2010 from the three dimensions of accident occurrence time, prone provinces, and accident types based on accident statistical analysis and the Granger causality test. Chen et al. [4] analyzed 377 tremendous and devastating accidents which occurred in China's coal mines from 2002 to 2011 based on accident inducements, and concluded that more than 90% of these accidents were caused by unsafe behaviors of personnel. Montewka et al. [5] obtained the potential relationship between accident causes of different types of marine ship accidents based on the MDTC model through statistical analysis. Wang et al. [6] analyzed the data of coal mine gas accidents in

China from 2006 to 2010, and found that the number of accidents and deaths in low-gas mines were more than those in high-gas mines, and the occurrence time of coal mine gas accidents presented exponential distribution. Zhang et al. [7] analyzed serious coal mine gas explosion accidents in China from 1950 to 2015 by using statistical methods, and found that improper ventilation management was the most common cause of gas accumulation.

Some progress has also been made in the analysis of the cause of accidents from a large number of historical accident statistics. Kim et al. [8] proposed the causality model of railway accidents by analyzing the investigation report of railway accidents. Manu et al. [9] proposed a new systematic research method to analyze the causes of construction accidents and put forward corresponding countermeasures. Maiti et al. [10] developed an EEA event assessment method to assess the safety level of coal production by studying coal mine death accidents in India in the past 100 years. Zhang [11] studied and constructed a causal model between life events and human errors of coal miners by using the structural equation method. Niu et al. [12] analyzed the causes of coal mine accidents in recent years and found that most coal mine gas explosion accidents occurred in low-gas mines. Liu et al. [13] used the statistical data of 362 major coal mine accidents in China from 2000 to 2016 to establish the Chinese Coal Mine Human Factor Analysis and Classification System (HFACS-CM), and conducted qualitative and quantitative analysis on the causes of accidents.

However, when determining the cause of accidents, most studies rely on expert experience and previous literature as the main sources, resulting in a lack of objectivity. As an important basis for the cause of coal mine gas accidents, the accident investigation report contains the cause, process, and result of the accident, which cannot be ignored. It should be analyzed as an important report to prevent coal mine gas accidents. At the same time, the occurrence of coal mine gas accidents is not the result of a single factor, but the joint action of multiple factors, and there is usually a potential relationship between these factors. The association rule is one of the data mining methods to reflect the dependence and correlation between things, with good generalization ability, and has become a hotspot of application research in many fields [14]. However, in the field of coal mining, there is little application of association rules and few historical statistics for accident investigation reports. Based on this, this paper combines data mining with theory. The keywords of unsafe factors are extracted from a large number of accident reports and the HFACS framework is modified according to the characteristics of the coal mining industry on the basis of the previous research. The potential relationship between factors is revealed by the association rule method, which provides a better theoretical framework for accident analysis and prevention.

2. Methods

2.1. Data Source

In order to ensure the reliability and integrity of the accident data, this paper takes 2001 to 2022 as the research period. The data sources are the website of The National Mine Safety Administration and the investigation reports of coal mine gas accidents of the sub-stations of the mine safety supervision bureaus of all provinces (autonomous regions and municipalities).

After statistical sorting, a total of 110 coal mine gas accidents with complete accident investigation reports were selected, covering 20 provinces in southeast, southwest, and northeast China, excluding the possible geographical specificities of the causes of the accidents and ensuring universality. The specific number and location distribution of accident areas are shown in Figure 1.

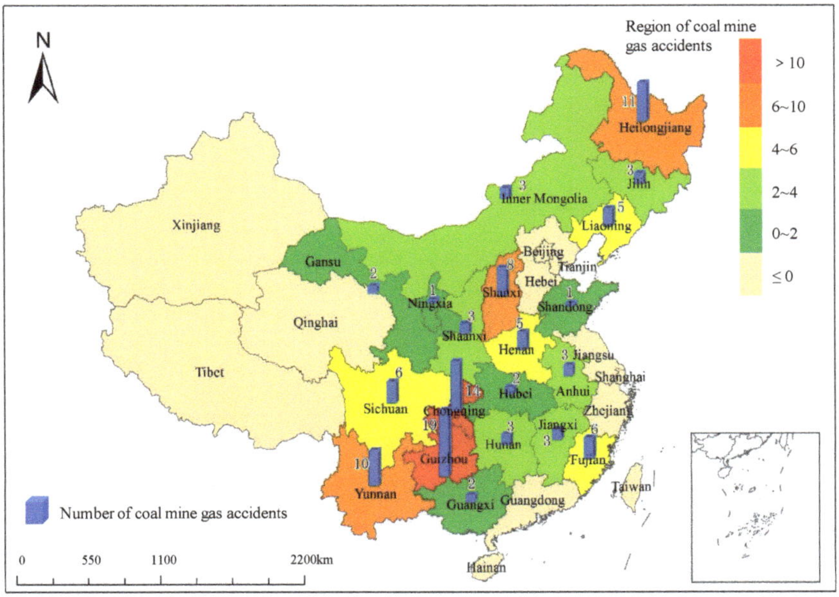

Figure 1. Distribution area and quantity of coal mine gas accident samples.

2.2. Data Preprocessing

The causes of coal mine gas accidents are complex and changeable, and the identification of important factors leading to the occurrence of accidents can effectively prevent the occurrence of accidents [15]. This paper adopts the idea of data-driven methods, extracts the key factors in the coal mine gas accident investigation report, and counts the frequency of each keyword. This paper also summarizes the causes comprehensively from the four aspects of human, equipment, environment, and management. Based on the statistical results of the frequency of the occurrence of keywords, combined with the existing literature and objective facts, 35 keywords with relatively high frequency are selected as research elements. As shown in Figure 2.

(1) Human factors

The HFACS causation theory model divides human factors into four levels: unsafe behaviors, preconditions of unsafe behaviors, unsafe supervision, and organizational influence [16]. According to the specific accident causes in the coal mine gas accident investigation report, this paper classifies human factors into three unsafe behaviors including skills error, decision-making error, and violation operation, as well as the preconditions of unsafe behaviors, and then divides them into different specific manifestations according to keywords. In order to avoid subjective factors affecting the classification process, the classification of factors in Figure 2 is defined and explained as follows:

A. Skills error

This refers to the "Missing procedures steps", "Using inadequate construction technology", "Using wrong methods/instruments", etc. The missing procedures steps include failure to take anti-outburst measures, failure to test the effect of anti-outburst, blind operation without eliminating the risk of outburst, failure to measure the gas content in advance, etc. The construction technology is not in place, including the construction strength which is not in place, is not in accordance with the provisions of the design and construction, and equipment (facilities) installation is not reasonable. The use of wrong methods/instruments includes the use of wrong measuring tools, coal mining methods, gas extraction methods, etc.

B. Decision-making error

This refers to the situation of incomplete understanding and wrong judgment of emergency situations due to the lack of on-site experience or training of personnel, including "Illegally order", "Inadequate emergency handling", and "Illegal entry into blind lane and other closed areas".

C. Violation operation

This refers to the behavior that the personnel do not abide by the rules and regulations subjectively, which is divided into habitual violation and accidental violation. Habitual violations often occur underground and are often condoned by managers, while operators are accustomed to natural violations, such as failure to carry or wear safety protection equipment, illegal shooting, illegal modification/use of equipment, falsified data parameters, and cross-operation of multiple types of underground work.

Figure 2. Classification of factor keywords.

(2) Equipment factor

The objects in underground coal mine are mainly all kinds of working equipment and machines. With the progress and development of coal mining technology in China, the types and quantities of equipment and machines in underground coal mines are gradually increasing, and the unsafe state factors of objects are also increasing. In the process of sorting out and classifying all the equipment factors in the accident investigation report, it was found that there are various equipment factors involved in the specific content of the coal mine gas accident investigation report, and there are many fault types. The machine factors leading to accidents can be divided into missing or damaged parts of the equipment itself, working equipment is not in normal operation state, etc. This article is divided into two categories: "Equipment missing/imperfect" and "Equipment not running/failure".

A. Security monitoring system missing/imperfect

This includes insufficient sensor types or quantity, insufficient gas monitor, and damaged safety monitoring equipment.

B. Gas extraction system missing/imperfect

Including: the ground gas extraction system has not been established, gas extraction drilling construction is not in place, and gas extraction time is not enough. The gas extraction system here includes equipment factors before and during the work, among which, insufficient gas extraction time can be a design defect, but from the overall situation, it should be classified as an imperfect gas extraction system.

C. Equipment failure

This refers to the long-term and occasional faults of all downhole equipment, such as short circuit, device detonation, cable damage, protection device does not move in time, gas data cannot be uploaded, etc.

(3) Environmental factors

Environmental factors include: coal mining geology is complicated, geological structure properties are different and so on. Environmental factors will indirectly lead to coal mine gas accidents [17–19]. For example, if the spontaneous combustion of coal is strong, coal mines are prone to gas combustion accidents. This is because there is air leakage in the goaf or closed area, and the coal seam will come into contact with oxygen, resulting in spontaneous combustion [20,21]. It is easy to oxidize the coal body and cause gas combustion accidents [22].

Secondly, the working environment of underground coal mine is another influencing factor causing coal mine gas accidents [23,24]. For example, the high noise in underground coal mines can easily affect the psychological state of operators, which may lead to operators making operation errors and to coal mine gas accidents [25]. However, according to the specific content of the coal mine gas accident investigation report, there are few records of accidents caused by the poor working environment. Therefore, this paper mainly analyzes the natural geological conditions.

(4) Management factors

Management factors refer to the factors that cause safety accidents due to wrong or inadequate decision-making at the management level of coal mine safety production enterprises, which is the essential cause of the occurrence of coal mine safety accidents. The article classifies the management factors that appear in the accident investigation report as follows:

A. Poor safety technology management Including no safety technical measures and safety technical measures are not perfect.
B. Flawed area design The ventilation partition design is insufficient, the regional anti-penetration design is unreasonable, and the mining and excavation deployment is unreasonable.
C. Inadequate implementation of measures

The implementation of gas control measures is not in place, the implementation of blasting system is not in place, and the implementation of anti-intrusion measures is not in place.

2.3. Data Analysis Method

2.3.1. Accident Statistical Method

The samples of 110 coal mine gas accidents are comprehensively counted from three aspects of regional distribution, accident type, and accident grade, and the occurrence regularity of accidents is preliminarily explored.

2.3.2. Association Rule Method

The factors in coal mine gas accident are discrete data, and the occurrence frequency and frequency of factors are not enough to obtain the potential information from them. In order to get the relationship between each factor, the association rule method was selected as a data mining method for analysis.

(1) Association rule principle

The association rule data mining method is a means to discover potential connections hidden in a large amount of data. These potential connections can reveal the relationship between data, which is more conducive to discovering the occurrence mechanism of events [26].

I for the item set, $I = \{i_1, i_2, \ldots i_n\}$, where i_n called the item, each transaction T is a subset of item set I. All T term sets, like $X \rightarrow Y$ ($X,Y \subseteq I$, $X \cap Y = \varnothing$) is the implied expression of the association rule.

A. Support:

This refers to the probability that transaction X and transaction Y occur simultaneously in the item set. In the application of the association rule method, the most important thing is to select the support threshold. The support threshold needs to be set reasonably according to the data situation. If the calculated support is lower than the threshold, it indicates that the occurrence times of the item set $\{X, Y\}$ in the total item set are too few. Otherwise, subsequent analysis can be carried out. Support is expressed as:

$$Support\ (X \rightarrow Y) = P\ (X,Y)/P\ (I) \tag{1}$$

B. Confidence:

This refers to the probability of transaction Y occurring if transaction X has already occurred. More important is the choice of confidence threshold, which reflects the ability of the occurrence of transaction X to affect the occurrence of transaction Y. Confidence is expressed as:

$$Confidence\ (X \rightarrow Y) = P\ (Y|X) = P\ (X,Y)/P\ (X) \tag{2}$$

C. Frequent item sets:

The frequent item set refers to the set of all transactions whose support is greater than or equal to the support threshold. If the support of item set I meets the selected minimum support threshold, item set I is the frequent item set.

D. Association rules:

The association rules that meet the minimum support threshold and minimum confidence threshold are those selected to be mined.

(2) Apriori algorithm

Apriori algorithm is a data mining algorithm for mining frequent item sets and association rules. Its purpose is to carry out the two steps of association and mining continuously to find the association rules that meet the threshold of support and confidence.

Firstly, by scanning the database, the count of each item is accumulated and the items that meet the minimum support threshold are collected to find the set of frequent item set 1, denoted as I_1. Then, use I_1 to find I_2 of frequent item set 2, use I to find I_3, and so on until frequent k item set can no longer be found. A full scan of the database is required for each I_k found.

(3) Analysis steps

The frequent item set in this paper is a combination of frequently occurring factors that lead to coal mine gas accidents. The mining results of frequent item sets are closely related to the selection of support threshold. If the support threshold is set too large, there will be too few causes in frequent item sets, which is not conducive to the generation of association rule results. However, if the support threshold is set too small, the number of frequent item sets generated will be too large, and it is not easy to determine the root cause of coal mine gas accidents. In order to prevent improper selection of the support threshold from affecting the conclusion of association rules, this paper determines the most appropriate support degree by comparing frequent item sets obtained from different support degrees. Four support degrees of 0.2, 0.15, 0.1, and 0.05 were selected, and 0.05 was selected as the frequent item set of minimum support mining after comparison of results.

After the frequent item set is obtained, the Apriori algorithm is used to mine association rules. According to the comparison of experience and results, the confidence degree is set as 0.4. One factor in the frequent item set is taken as the front term of association rules, and the other factors are taken as the back term, respectively.

2.3.3. Accident Cause Chains Method

The occurrence of coal mine gas accidents is the result of the interaction and joint action of various factors. According to association rules, a more intuitive cause chain of coal mine gas accidents is constructed, from which three cause chains are extracted for analysis and countermeasures and suggestions are put forward.

3. Results and Discussion

3.1. The Accident Statistics

3.1.1. Factor Frequency Statistics

First of all, the analysis of the occurrence frequency of these accident factors can provide a better basis for the production of association rules. The frequency of occurrence of factor keywords is analyzed and the proportion of these factors in accidents is analyzed. The factor names in Figures 3 and 4 correspond to the label in Figure 2.

The results of statistical analysis for human and equipment factors are as follows:

It can be seen from Figure 3a that "Habitual violations" occurred 121 times in total, accounting for 32.3%, which is also the factor with the most frequency among the four aspects of the factors, followed by "Missing procedures steps" in skills errors, which occurred 70 times, accounting for 18.7%. "Poor personnel safety awareness" appeared 43 times, which shows that the safety education and training of personnel in most coal mining enterprises, as well as the establishment of personnel safety awareness, are not in place.

In Figure 3b, the factors of "equipment missing/imperfect" is mainly reflected in the safety monitoring system and ventilation system, 43 times and 38 times respectively, accounting for 30.7% and 27.1% of the machine factors; security monitoring. For example, gas accumulation in the roadway is caused by unreasonable ventilation design or failure to open local ventilators, and abnormal concentration of gas is not detected by safety monitoring equipment, so coal mine gas accidents easily occur [27,28]. "Explosion-proof measures not set" and "Gas extraction system missing/imperfect" only accounted for 5.7% of the total equipment factors, which therefor had little reference significance. In addition, "Normal equipment but not running" and "Equipment failure" occurred 20 times and 31 times, respectively, accounting for 14.3% and 22.1% of the total equipment factors, indicating that compared with equipment missing/imperfect, the factor of "equipment failure" is relatively rare.

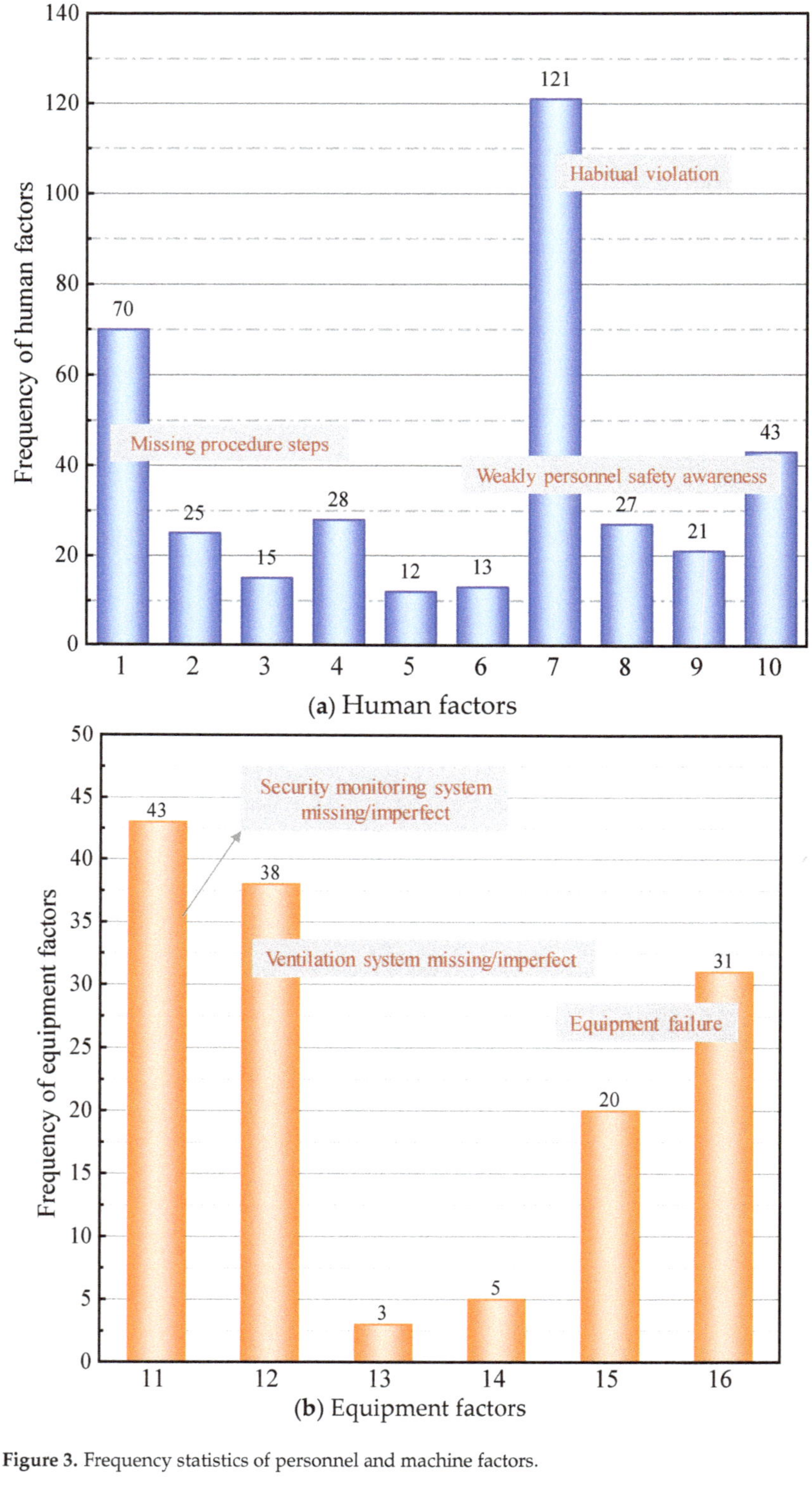

Figure 3. Frequency statistics of personnel and machine factors.

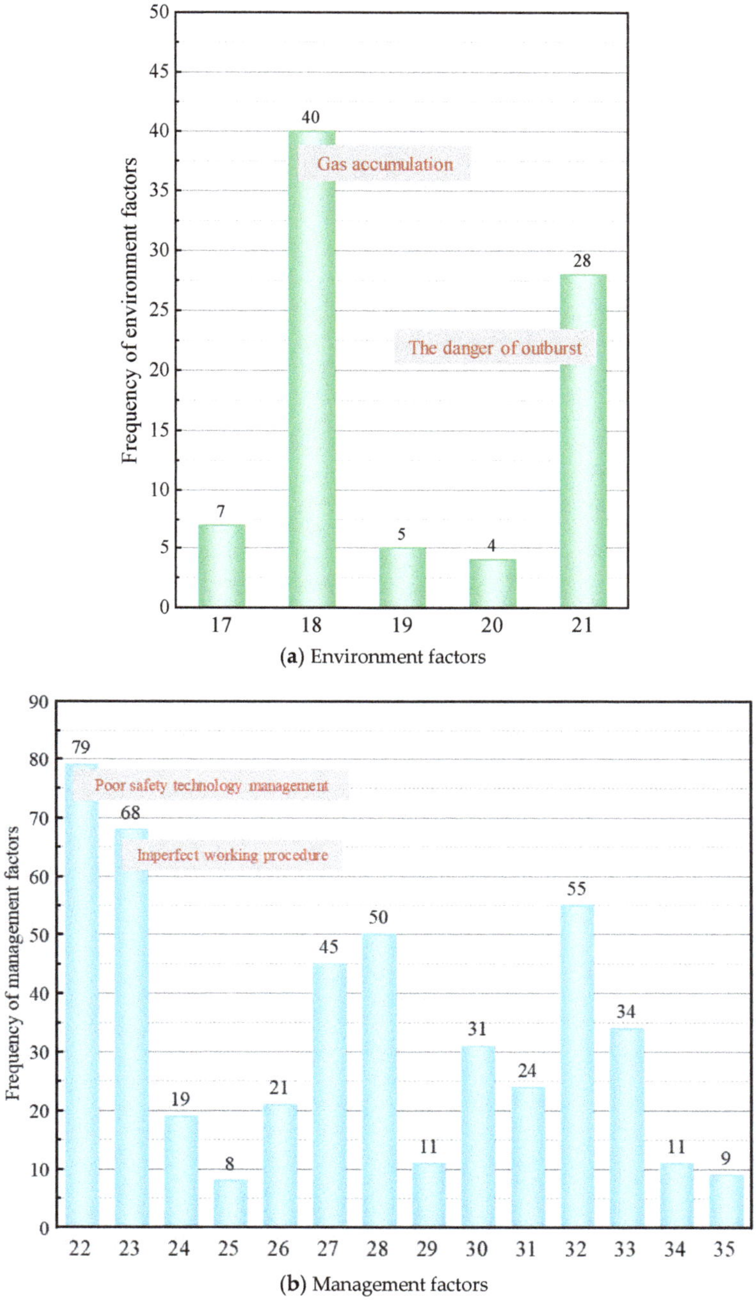

Figure 4. Frequency statistics of environmental and management factors.

The results of statistical analysis for environment and management factors are as follows:

Figure 4a shows that in terms of environment, which is the main cause of accidents, gas accumulation accounted for 45.5% of the environmental cause factors, followed by coal seams with outburst danger which accounted for 31.8%. In the process of sorting out the accident causes of the accident report, it is also found that most of the coal mines where coal and gas outburst accidents occurred are identified as having outstanding risks. The occurrence of "Strong spontaneous combustion tendency", "Abnormal gas emission", and "Soft coal seam" is less common.

It can be seen from Figure 4b that among all factors, management factors appear the most frequently, indicating that management factors are more important in ensuring coal mine production safety. The hidden dangers of management factors mainly exist in "illegal supervision", "lack of good management culture", and "management process loopholes." Among them, the number of times that ineffective supervision by the government/safety supervision bureau of illegal supervision occurred was 79, accounting for 16.9%; the number of times illegal production organization factors occurred was 68, accounting for 14.6%; the lack of good management culture was reflected in the failure to implement the main responsibility of safety production and the lack of safety education and training work were 9.6% and 10.7%, respectively. The loopholes in the management process are mainly reflected in the inadequate implementation of measures.

3.1.2. Accident Type Statistics

Further statistical analysis of accident types, accident grades, and number of deaths is helpful to better explore the nature of accident causes.

The statistical results for the types of accident and the number of deaths are as follows:

As can be seen from Figure 5, gas explosion (combustion) accidents are the most frequent in terms of the number of occurrences and deaths, with 52 cases occurring in total, accounting for 47.3%, and 984 deaths, accounting for 73.7% of the total. This is mainly because gas explosion is powerful, and easily causes a series of other disasters. If aroused, the dust of the ground can produce a two-dust explosion. Therefore, once a gas explosion accident occurs, there are often more casualties. Preventing the occurrence of gas explosion accidents is of great significance to control the occurrence of casualties and accidents.

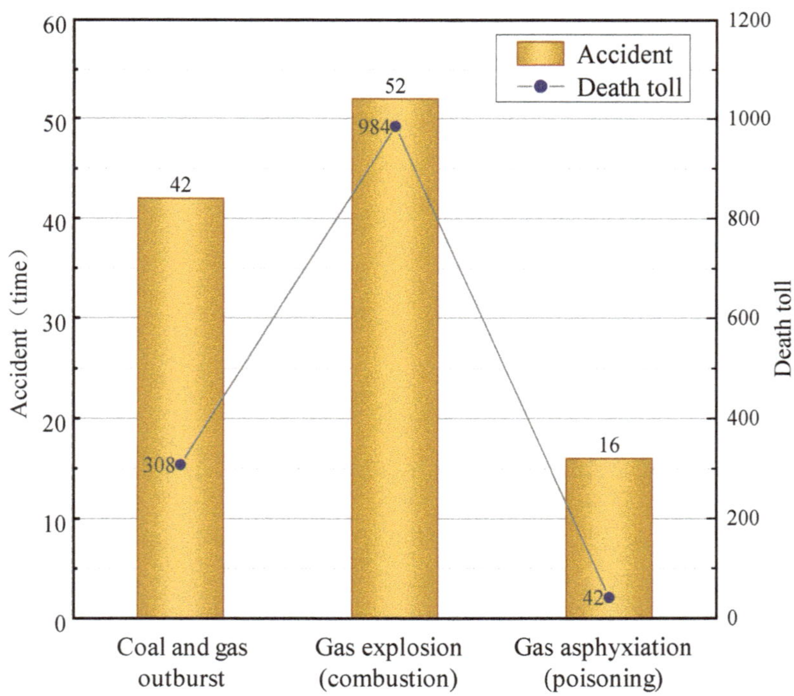

Figure 5. Accident types and death statistics.

There are 42 coal and gas outburst accidents, with 308 deaths, accounting for 38.2% and 23.1% of the total, respectively. Gas asphyxiation (poisoning) accidents are relatively rare, with 16 cases accounting for 14.5% of the total, and 42 deaths accounting for only 3.2% of the total. Among all the coal mine gas accident investigation reports, only one is on a gas poisoning accident, which is a gas poisoning accident in the Xudong Coal industry in the Gansu province in 2015, and the other 15 are gas asphyxiation accidents.

3.1.3. Accident Grade Statistics

The statistical results of the accident grade are as follows:

It can be seen from Figure 6 that coal and gas outburst accidents and gas explosion (combustion) accidents are often categorized as considerable and serious accidents. Gas explosion (combustion) accidents are also tremendous and devastating accidents, which further supports the finding in Figure 5 that gas explosion accidents cause the highest number of deaths. Gas asphyxiation (poisoning) accidents resulted in relatively light consequences, no serious accidents, and especially serious accidents.

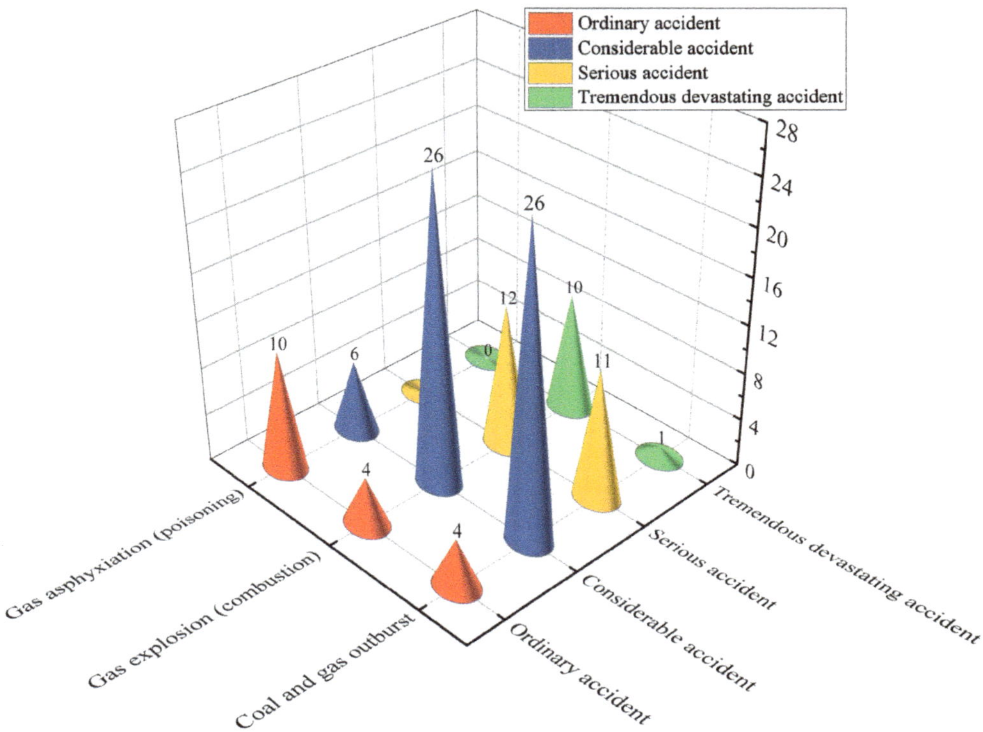

Figure 6. Statistics of accident types and accident grades.

As can be seen from Figure 7, the total number of ordinary accidents, considerable accidents, serious accidents, and tremendous, devastating accidents is 18, 58, 23, and 11, respectively, accounting for 16.4%, 52.7%, 20.9%, and 10.0% of the total number of accidents; resulting in 27, 260, 474, and 573 deaths respectively, accounting for 2.0%, 19.5%, 35.5%, and 43.0% of the total number of deaths. Although the number of tremendous, devastating accidents is relatively small, the death toll accounts for nearly half of the total, such as the gas explosion in the Liaoning Sunjiawan Coal Mine in 2005, which killed 214 people and caused heavy casualties and economic losses. It can be seen that coal mine gas accidents in particular need to control the occurrence of tremendous, devastating accidents.

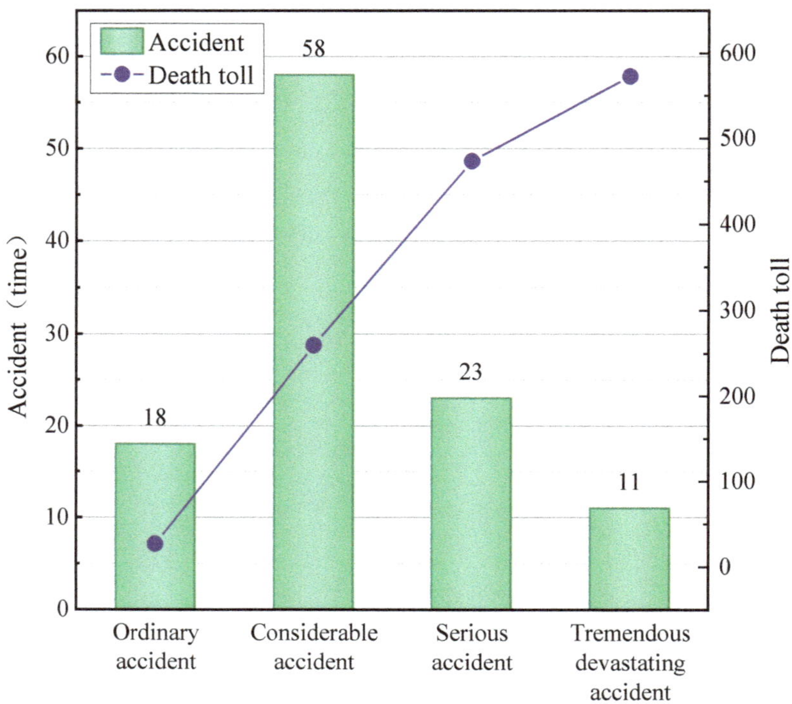

Figure 7. Statistics of accident grades and deaths.

3.2. Association Rule Statistics

3.2.1. Frequent Item Sets

As mentioned above, the setting of the support threshold is crucial when calculating association rules. In order to ensure the reasonable setting of support, the paper first adjusts the calculation of the support threshold. When trying to set it to 0.5, 0.4, and 0.3, there is no frequent return of item set results, which will lose the value of subsequent research. When the support threshold is set to 0.2, two items are generated, so the paper conducts subsequent research from the support threshold of 0.2.

According to Table 1, when the support threshold is set to 0.2, the mining result of the frequent item set of coal mine gas accident factors is "habitual violation", "poor supervision of the safety supervision bureau". This indicates that the support threshold is set too high and the frequent items are too few. It is not conducive to the generation of association rules.

Table 1. Results of frequent item sets.

Frequent Item Sets	Support
{Habitual violation, Poor supervision of the safety supervision bureau}	0.24
{Habitual violation, Poor supervision of the safety supervision bureau, Illegal production}	0.14
{Habitual violation, Poor supervision of the safety supervision bureau, Gas accumulation}	0.12
{Habitual violation, Gas accumulation, Illegal production}	0.11
{Habitual violation, Poor supervision of the safety supervision bureau, Gas accumulation Illegally production, Ventilation system missing/imperfect}	0.05

When the support threshold is set to 0.1, two factors, "Illegally production" and "Gas accumulation", are increased compared with the support threshold of 0.2, indicating that environmental factors will be returned after the support threshold is lowered.

The support threshold is set to 0.05, meaning that the results of the frequent item in the equipment aspect is more reasonable. If the support threshold is lowered further, there will be too many causative factors, which means the coverage of the frequent item set in the transaction set is very small, therefore the occurrence of factors in frequent item set is very accidental, and the final result lacks feasibility. After synthesis, the support threshold is selected as 0.05 for subsequent analysis. The frequent item sets of coal mine gas accident factors are summarized as "habitual violation", "poor supervision of the safety supervision bureau", "gas accumulation", "illegal production", and "ventilation system missing/imperfect".

In the analysis of frequent item set results, it can be found that it is difficult to return the results of environment factors and equipment factors of accidents and the number is small. The reasons may be as follows: Coal mine gas accidents are mostly liability accidents. At present, the investigation reports of coal mine gas accidents mainly focus on liability identification. The direct and indirect causes are generally analyzed for human and management reasons, and the analysis of equipment and environment is less commonly completed, and the records are not detailed enough.

3.2.2. Association Rule

"Habitual violation" is used as the preceding term to obtain the association rule:

As the result of Article 5 in Table 2, the support degree of "Habitual violation" and "Weakly supervision of the safety supervision bureau" is 0.23, indicating that the probability of accidents caused by both factors is 23.5%. The confidence degree is 0.61, indicating that the probability of ineffective supervision by the safety supervision bureau is 61.1% when habitual violations occur, and the reliability of "Weakly supervision of the safety supervision bureau" is 61%. Similarly, in the case of habitual violations, the probability of "Illegally production" and "Inadequate implementation of measures" is 45.8%, the probability of "Lack of safety education" is 44.4%, and the probability of "Gas accumulation" is 43.1%.

Table 2. The association rule obtained as "Habitual violation".

Number	Preceding Term	Consequent	Support	Confidence
1	Habitual violation	Gas accumulation	0.165775401	0.430555556
2	Habitual violation	Lack of safety education	0.171122995	0.444444444
3	Habitual violation	Inadequate implementation of measures	0.176470588	0.458333333
4	Habitual violation	Illegal production	0.176470588	0.458333333
5	Habitual violation	Poor supervision of the safety supervision bureau	0.235294118	0.611111111

"Weakly supervision of the safety supervision bureau" is used as the preceding term to obtain the association rule:

As shown in Table 3, the probability of "Habitual violations" is 72.1%, "illegally production" is 59.0%, "Lack of safety education" is 44.3%, and "Gas accumulation" is 42.6% when there is "Weakly supervision of the safety supervision bureau".

Table 3. The association rule obtained as "Weakly supervision of the safety supervision bureau".

Number	Preceding Term	Consequent	Support	Confidence
1	Poor supervision of the safety supervision bureau	Gas accumulation	0.139037433	0.426229508
2	Poor supervision of the safety supervision bureau	Lack of safety education	0.144385027	0.442622951
3	Poor supervision of the safety supervision bureau	Illegal production	0.192513369	0.590163934
4	Poor supervision of the safety supervision bureau	Habitual violation	0.235294118	0.721311475

"Gas accumulation" is used as the preceding term to obtain the association rule:

As shown in Table 4, in the case of gas accumulation in the mine working area, the probability of "Habitual violations" is 81.6%, the probability of "Weakly supervision of the safety supervision bureau" is 68.4%, the probability of "Illegally production" is 63.2%, and the probability of "Ventilation system missing/imperfect" is 42.1%.

Table 4. The association rule obtained as "Gas accumulation".

Number	Preceding Term	Consequent	Support	Confidence
1	Gas accumulation	Ventilation system missing/imperfect	0.085561497	0.421052632
2	Gas accumulation	Illegal production	0.128342246	0.631578947
3	Gas accumulation	Poor supervision of the safety supervision bureau	0.139037433	0.684210526
4	Gas accumulation	Habitual violation	0.165775401	0.815789474

"Illegally production" is used as the preceding term to obtain the association rule:

As shown in Table 5, when there is illegal production in the mine, the probability of "Weakly supervision of the safety supervision bureau" is 73.5%, the probability of "Habitual violation" is 67.3%, the probability of "Gas accumulation" is 49.0%, and the probability of "Ventilation system missing/imperfect" is 44.9%. The probability of "Lack of safety education" is 42.9%, the probability of "Missing procedure steps" is 40.8%, and the probability of "The primary responsibility is not implemented" is 40.8%.

Table 5. The association rule obtained as "Illegally production".

Number	Preceding Term	Consequent	Support	Confidence
1	Illegal production	The primary responsibility is not implemented	0.106951872	0.408163265
2	Illegal production	Missing procedure steps	0.106951872	0.408163265
3	Illegal production	Lack of safety education	0.112299465	0.428571429
4	Illegal production	Ventilation system missing/imperfect	0.117647059	0.448979592
5	Illegal production	Gas accumulation	0.128342246	0.489795918
6	Illegal production	Habitual violation	0.176470588	0.673469388
7	Illegal production	Poor supervision of the safety supervision bureau	0.192513369	0.734693878

"Ventilation system missing/imperfect" is used as the preceding term to obtain the association rule:

As shown in Table 6, under the condition of "Ventilation system missing/imperfect", the probability of "Habitual violation" is 69.4%, the probability of "Weakly supervision of the safety supervision bureau" is 63.9%, the probability of "Illegally production" is 61.1%, and the probability of "The primary responsibility is not implemented" is 47.2%. The probability of underground "Gas accumulation" is 44%, and the probability of "Lack of safety education" is 44%.

Table 6. The association rule obtained as "Ventilation system missing/imperfect".

Number	Preceding Term	Consequent	Support	Confidence
1	Ventilation system missing/imperfect	Lack of safety education	0.085561497	0.444444444
2	Ventilation system missing/imperfect	Gas accumulation	0.085561497	0.444444444
3	Ventilation system missing/imperfect	The primary responsibility is not implemented	0.09090909	0.472222222
4	Ventilation system missing/imperfect	Illegal production	0.117647059	0.611111111
5	Ventilation system missing/imperfect	Poor supervision of the safety supervision bureau	0.122994652	0.638888889
6	Ventilation system missing/imperfect	Habitual violation	0.13368984	0.694444444

From the above, 26 association rules of 110 coal mine gas accidents were obtained, involving 9 factors in total. Where, 01 is "Illegally production", 02 is "Weakly supervision of the safety supervision bureau", 03 is "Lack of safety education", 04 is "The primary responsibility is not implemented", 05 is "Inadequate implementation of measures", 06 is "Habitual violation", 07 is "Missing procedure steps", 08 is "Ventilation system missing/imperfect", and 09 is "Gas accumulation".

The association rules obtained from the above results meet the support threshold and confidence threshold, but not all the association rules that meet the requirements have application value. Human factors and management factors in coal mine gas accident factors refer to the HFACS causation model, so the causal relationship between them should follow the principle of sequential occurrence in the HFACS model, rather than the reverse order. The hierarchical classification is: organizational factors, unsafe supervision, the premise of unsafe behavior, and unsafe behavior. Among them, 01 and 02 belong to first level, 03, 04, and 05 belong to second level, and 06 and 07 belong to fourth level. At the same time, equipment factors and environmental factors cannot lead to the occurrence of management factors and human factors.

Based on these two principles, the confidence degree that does not meet the requirements is represented by 0, and the confidence degree that meets the requirements is represented by 1, and the final matrix with a modified confidence degree is obtained. The confidence degree in the matrix is the confidence degree of row to column. The matrix is shown in Table 7.

Table 7. Confidence modified matrix.

	01	02	03	04	05	06	07	08	09
01	0	0.59	0	0	0	0	0	0	0
02	0.73	0	0	0	0	0	0	0	0
03	0.43	0.44	0	0.45	0	0	0	0	0
04	0.41	0	0	0	0	0	0	0	0
05	0	0	0	0.45	0	0	0	0	0
06	0.67	0.72	0.64	0.66	0.73	0	0.65	0	0
07	0.41	0	0.44	0.47	0.58	0	0	0	0
08	0.45	0	0	0.45	0	0.43	0	0	0
09	0.49	0.43	0	0	0	0	0	0.44	0

After confidence degree modification, 23 association rules are obtained, and the results are shown in Table 8.

Table 8. Association rule result table.

Association Rule	Confidence
Illegal production => Poor supervision of the safety supervision bureau	0.73
Illegal production => Lack of safety education	0.43
Illegal production => The primary responsibility is not implemented	0.41
Illegal production => Habitual violation	0.67
Illegal production => Missing procedure steps	0.41
Illegal production => Ventilation system missing/imperfect	0.45
Illegal production => Gas accumulation	0.49
Poor supervision of the safety supervision bureau => Illegal production	0.59
Poor supervision of the safety supervision bureau => Lack of safety education	0.44
Poor supervision of the safety supervision bureau => Habitual violation	0.72
Poor supervision of the safety supervision bureau => Gas accumulation	0.43
Lack of safety education => Habitual violation	0.64
Lack of safety education => Missing procedure steps	0.44
The primary responsibility is not implemented => Lack of safety education	0.45
The primary responsibility is not implemented => Inadequate implementation of measures	0.45
The primary responsibility is not implemented => Habitual violation	0.66
The primary responsibility is not implemented => Missing procedure steps	0.47
The primary responsibility is not implemented => Ventilation system missing/imperfect	0.45
Inadequate implementation of measures => Habitual violation	0.73
Inadequate implementation of measures => Missing procedure steps	0.58
Habitual violation => Ventilation system missing/imperfect	0.43
Missing procedure steps => Habitual violation	0.65
Ventilation system missing/imperfect => Gas accumulation	0.44

3.3. Accident Cause Chains

The cause chains of coal mine gas accidents can be obtained by connecting the factors in the association rules. Three cause chains are obtained and further reflect the deep relationship between the causes of coal mine gas accidents.

3.3.1. Management Factors Cause Chain

In coal mine gas accident cause chain 1 (Figure 8), the implementation of measure is the direct cause of the accident. The main responsibility of work safety is not implemented and this is the main reason leading to the implementation of measures not being in place.

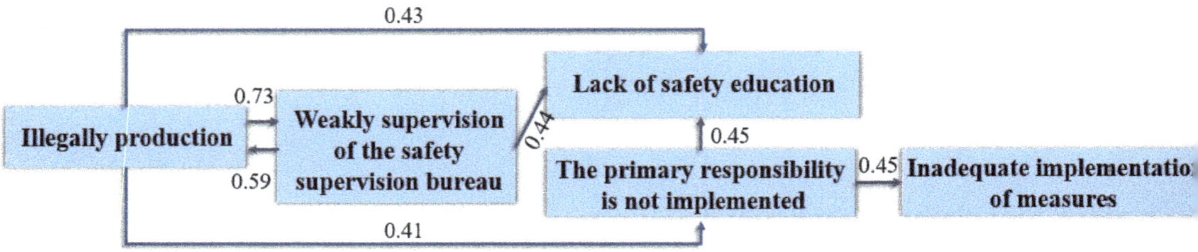

Figure 8. Coal mine gas accident cause chain 1 (The numbers on the arrows indicate the degree to which the two are related).

The primary responsibility is not implemented and this is the main reason leading to the inadequate implementation of measures, and at the same time, it will also lead to the lack of safety education. This is because the primary safety responsibility contains "security system formulation" and "safety in production inputs", so if it is not implemented, the safety production system of coal mine enterprises is not perfect or the safety production investment is insufficient, meaning the theoretical provisions and material support of prevention are not enough.

When coal mining enterprises appear in illegal organization production, there is a 73% possibility that the safety supervision bureau has ineffective supervision; when the safety bureau's supervision of coal mining enterprises is lax, coupled with the management of coal mining enterprises' "Focus on production, Light safety", the wrong concept easily appears in cross-border mining, production without licenses, and other illegal production behavior. Production by illegal organizations, ineffective supervision by the safety supervision bureau, and failure to implement the responsibility of the main body for work safety easily lead to the failure of safety education and training, but the failure to implement the responsibility of work safety accounts for 45%, indicating that this is the most reliable reason. The production of illegal organizations and the safety supervision bureau's poor supervision will lead to each other, which shows that the management and accident prevention of coal mine enterprises are inseparable from the local government, safety supervision bureau, and the enterprise itself. Therefore, in terms of management in general, illegal supervision will lead to a lack of good management culture, and ultimately lead to loopholes in the management process.

3.3.2. Management and Human Factors Cause Chain

When coal mine gas accidents cause chain 2 (Figure 9), management factors and human factors are closely related to accident causes, and the cause chain is complex. It can be seen that when the safety education is not in place, the primary responsibility of the safety production is not implemented, and the implementation of measures are not in place, which will lead to habitual violations and omissions of procedures. Among them, improper implementation of measures is most likely to lead to habitual violations and missing procedure steps, with a confidence rate of 73% and 58%, respectively. The inadequate implementation of measures means that the theoretical knowledge or practical experience of safety measures is too little, the overall cognitive level of personnel is low, and decision-making errors are easy to occur, so violations and omissions of operation steps easily occur. At the same time, omission of procedural steps may lead to habitual violations by 65%. This is because when operators are not familiar with operational procedures and technical operations, the omission of procedural steps will occur. In the long run, such omission will form habits, produce fluke psychology, and eventually develop into habitual violations.

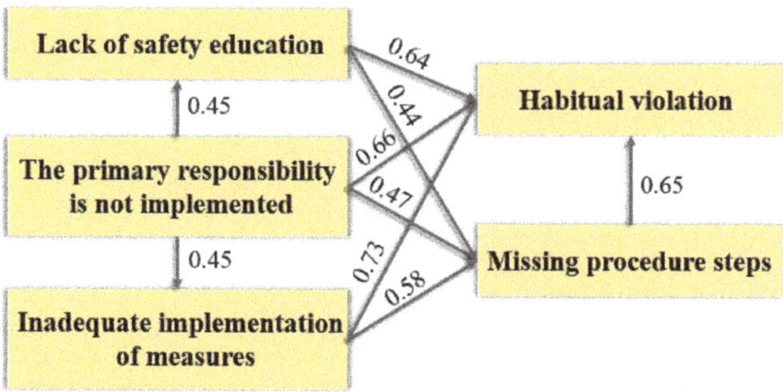

Figure 9. Coal mine gas accident causes chain 2 (The numbers on the arrows indicate the degree to which the two are related).

3.3.3. Comprehensive Factors Cause Chain

When coal mine gas accidents cause chain 3 (Figure 10), gas accumulation is an important cause of coal mine gas accidents, and the cause of gas accumulation is the ventilation system being missing/imperfect. Due to the imperfection or non-operation of the ventilation system, the gas in the working area cannot be discharged in time and thus accumulates. People's habitual violation and omission of procedural steps will cause problems in the ventilation system, and the reason for the habitual violation and omission of procedural steps is that the measures are not implemented in place. Therefore, in terms of comprehensive factors, loopholes in the management process will lead to decision-making errors and illegal operations, which will lead to the "Ventilation system missing/imperfect" in equipment factors, leading to the occurrence of gas accumulation.

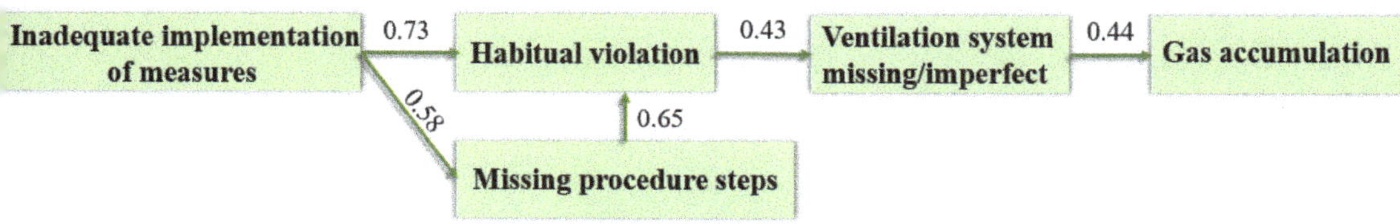

Figure 10. Coal mine gas accident causes chain 3.

3.4. Countermeasures and Suggestions

According to the coal mine gas accident cause chains combined with the actual coal mine gas accident case, the following countermeasures and suggestions are put forward:

Strictly perform the duties of coal mine safety supervision and strengthen the inspection and guidance of coal mine production enterprises in the area under their jurisdiction.

For example, on 10 May 2017, a serious gas combustion accident occurred in the Yandefang Coal Mine in Shanziwei, Longyan City, Fujian Province. The expiration date of the safety production license of this coal mine was 2016, so the mine was suspended from 2016. The absence of safety protection measures and the illegal organization of the workers into mining operations resulted in the accident.

When the coal mine gas accident causes chain 1, poor supervision by the safety supervision bureau and illegal production by coal mining enterprises are the essential causes of accidents. As a result of this, the safety supervision bureau should take responsibility and strengthen safety management. At the same time, the safety supervision should strictly investigate and punish illegal mining behavior and cross-border mining behavior of coal mine production enterprises operating without a production license, and ask the coal mine to stop production and rectify their enterprise within a time limit, so as to eliminate hidden safety risks in the embryonic stage. The management of coal mine production enterprises should strictly abide by the relevant laws and regulations of the state, set up correct values, put safety in production first, and should not blindly pursue output and interests at the expense of safety protection.

Earnestly implement the main responsibility of enterprise safety production and strengthen safety education.

For example, there was a gas explosion in the Goutou Coal Mine in the Yunnan province on 22 September 2019, a gas explosion in the Kongjiagou Coal Mine in the Guizhou Province on 13 December 2017, and a large coal and gas outburst accident in the Zhushanchong Coal Mine in the Hunan Province on 8 December 2017. In view of such situations, coal mining enterprises should earnestly implement the provisions of the "Regulations on Reporting and Investigation of Production Safety Accidents" and report coal mine accidents truthfully and in a timely manner.

Illegal supervision will lead to a lack of good management culture in the daily operation of coal mine production enterprises. First of all, good management culture must: have a perfect safety management organization; be equipped with enough professionals; not have personnel without a certificate or a certificate that does not match the job; not illegally hire personnel; formulate practical emergency rescue plans for accidents, and conduct regular emergency training for personnel; ensure adequate investment in production safety, material, and financial security; and strictly follow incident reporting procedures when accidents occur. In addition, coal enterprises need to, on the one hand: strengthen safety education and training of personnel, to pay attention to workers' safety and in turn pay special attention to the worker's coal mine gas accident prevention education. On the other hand, lectures, competitions, and other forms of training should be held regularly to popularize and assess the theoretical knowledge of coal mine gas accidents, and safety publicity and education should be carried out in the form of radio, television, and short videos, so that employees can consciously abide by safety rules and regulations and carry out production operations in strict accordance with the operation rules.

Ensure that the relevant measures of operation are put in place to improve the gas control level.

For example, a large coal and gas outburst accident occurred in the Liaoyuan Coal Mine in the Shaanxi province on 22 June 2021. The direct cause of the accident was that the underground outburst prevention measures were not implemented in place and the gas extraction time was not enough. The coal mining operation was carried out without eliminating the danger of outburst at the accident site.

Before mining, protruding coal seam, "four-in-one" outburst prevention measures must be implemented. For gas control measures, it is necessary to carry out targeted scheme design according to different geological conditions of different coal mines. Unreasonable control measures cannot achieve the expected effect of gas extraction, and it is likely to lead to accidents.

4. Conclusions

This paper selects 110 coal mine gas accidents in China from 2001 to 2022. Firstly, the regional distribution, accident type, and accident grade are analyzed statistically. Secondly, the keywords of human factors, equipment factors, environment factors, and management factors are extracted from the accident investigation reports. Finally, the Apriori algorithm is used for data mining to obtain frequent item sets and association rules of coal mine gas accident factors, and the cause chains of coal mine gas accidents are obtained by association rules. The following conclusions were drawn:

(1) From the perspective of accident statistical analysis, the keyword analysis of factors in 110 accident reports shows that "Habitual violation" is the most frequent factor in human factors, "Security monitoring system missing/imperfect" is the most frequent factor in equipment factors, and "Gas accumulation" is the most frequent factor in environmental factors. Among the management factors, the most frequent factor is "Weakly supervision of the safety supervision bureau". Among the accident types, gas explosion (combustion) accounted for the largest number of deaths and accidents. From the grade of accidents, especially gas explosion (combustion).

(2) Frequent item sets and association rules of coal mine gas accident factors are generated. The support degree is set to 0.05 and the confidence degree is set to 0.4. The frequent item set is obtained after mining the keywords of factors, which contained 9 factors in total, including "habitual violation", "poor supervision of the safety supervision bureau", "gas accumulation", 'illegal production", and "ventilation system missing/imperfect". Mining the association relation between the factors and establishing a matrix to exclude the association rules that do not conform to reality, 23 coal mine gas accident association rules are obtained.

(3) The causes chain of coal mine gas accidents are obtained. By analyzing the relationship of association rules, three causes chains are extracted and generated, and the main causes of coal mine gas accidents are obtained by analyzing the cause chains, and relevant prevention, control measures and, countermeasures are put forward based on accident cases.

Author Contributions: Resources, Q.L.; Writing—original draft, Y.L. (Ying Liu); Supervision, Y.L. (Yunpei Liang). All authors have read and agreed to the published version of the manuscript.

Funding: National Natural Science Foundation of China (Grant No. 52174166).

Institutional Review Board Statement: Not applicable.

Informed Consent Statement: Not applicable.

Data Availability Statement: Not applicable.

Acknowledgments: This work is financially supported by the National Natural Science Foundation of China (Grant No. 52174166), which are gratefully acknowledged.

Conflicts of Interest: The authors declare no conflict of interest.

References

1. Yin, W.T.; Fu, G.; Yang, C.; Jiang, Z.G.; Zhu, K.; Gao, Y. Fatal gas explosion accidents on Chinese coal mines and the characteristics of unsafe behaviors: 2000–2014. *Saf. Sci.* **2017**, *92*, 173–179. [CrossRef]
2. Ran, Q.C.; Liang, Y.P.; Zou, Q.L.; Hong, Y.; Zhang, B.C.; Liu, H.; Kong, F.J. Experimental investigation on me-chanical characteristics of red sandstone under graded cyclic loading and its inspirations for stability of overlying strata. *Geomech. Geophys. Geo-Energy Geo-Resour.* **2023**, *9*, 11. [CrossRef]
3. Gao, Y.; Fu, G.; Nieto, A. A comparative study of gas explosion occurrences and causes in China and the United States. *Int. J. Min. Reclam. Environ.* **2016**, *30*, 269–278. [CrossRef]
4. Zhao, D.Y.; Nie, B.S. Statistical Analysis of China's Coal Mine Particularly Serious Accidents. *Procedia Eng.* **2011**, *26*, 2213–2221. [CrossRef]
5. Chen, H.; Qi, H.; Long, R.Y.; Zhang, M.L. Research on 10-year tendency of China coal mine accidents and the characteristics of human factors. *Saf. Sci.* **2012**, *50*, 745–750. [CrossRef]
6. Montewka, J.; Goerlandt, F.; Kujala, P. Determination of collision criteria and causation factors appropriate to a model for estimating the probability of maritime accidents. *Ocean Eng.* **2012**, *40*, 50–61. [CrossRef]
7. Wang, L.; Cheng, Y.P.; Liu, H.Y. An analysis of fatal coal mine gas accident in Chinese coal mines. *Saf. Sci.* **2014**, *62*, 107–113. [CrossRef]
8. Zhang, J.J.; Cliff, D.; Xu, K.L.; You, G. Focusing on the patterns and characteristics of extraordinarily severe gas explosion accidents in Chinese coal mines. *Process Saf. Environ. Prot.* **2018**, *117*, 390–398. [CrossRef]
9. Kim, D.S.; Yoon, W.C. An accident causation model for the railway industry: Application of the model to 80 rail accident investigation reports from the UK. *Saf. Sci.* **2013**, *60*, 57–68. [CrossRef]
10. Manu, P.; Ankrah, N.; Proverbs, D.; Suresh, S. An approach for determining the extent of contribution of construction project features to accident causation. *Saf. Sci.* **2010**, *48*, 687–692. [CrossRef]

11. Maiti, J.; Khanzode, V.V.; Ray, P.K. Severity analysis of Indian coal mine accidents-A retrospective study for 100 years. *Saf. Sc* **2009**, *47*, 1033–1042. [CrossRef]
12. Zhang, W. Causation mechanism of coal miners' human errors in the perspective of life events. *Int. J. Min. Sci. Technol.* **2014**, 2 581–586. [CrossRef]
13. Niu, H.Y.; Deng, J.; Zhou, X.Q.; Wang, H.Q. Association Analysis of Emergency Rescue and Accident Prevention in Coal Min *Int. Symp. Saf. Sci. Eng. China* **2012**, *43*, 71–75. [CrossRef]
14. Liu, R.L.; Cheng, W.M.; Yu, Y.B.; Xu, Q.F. Human factors analysis of major coal mine accidents in China based on the HFACS-CM model and AHP method. *Int. J. Ind. Ergon.* **2018**, *68*, 270–279. [CrossRef]
15. Fa, Z.W.; Li, X.C.; Qiu, Z.X.; Liu, Q.L.; Zhai, Z.Y. From correlation to causality: Path analysis of accident-causing factors in coa mines from the perspective of human, machinery, environment and management. *Resour. Policy* **2021**, *73*, 102157. [CrossRef]
16. Wiegmann, D.A.; Shappell, S.A. Human error analysis of commercial aviation accidents: Application of the human factor analysis and classification system (HFACS). *Aviat. Space Environ. Med.* **2001**, *72*, 1006.
17. Mahdevari, S.; Shahriar, K.; Esfahanipour, A. Human health and safety risks management in underground coal mines usin fuzzy TOPSIS. *Sci. Total Environ.* **2014**, *488*, 85–99. [CrossRef] [PubMed]
18. Zou, Q.L.; Liu, H.; Jiang, Z.B.; Wu, X. Gas flow laws in coal subjected to hydraulic slotting and a prediction model for it permeability-enhancing effect. *Energy Sources Part A Recovery Util. Environ. Effects* **2021**, *1556–7036*, 1–15. [CrossRef]
19. Zou, Q.L.; Zhang, T.C.; Cheng, Z.H.; Jiang, Z.B.; Tian, S.X. A method for selection rationality evaluation of the first-mining sear in multi-seam mining. *Geomech. Geophys. Geo-Energy Geo-Resour.* **2022**, *8*, 17. [CrossRef]
20. Ran, Q.C.; Liang, Y.P.; Zou, Q.L.; Zhang, B.C.; Li, R.F.; Chen, Z.H.; Ma, T.F.; Kong, F.J.; Liu, H. Characteristics of mining-induce fractures under inclined coal seam group multiple mining and implications for gas migration. *Nat. Resour. Res.* **2023**, *3* 1481–1501. [CrossRef]
21. Gao, M.Z.; Xie, J.; Gao, Y.A.; Wang, W.Y.; Li, C.; Yang, B.G.; Liu, J.J.; Xie, H.P. Mechanical behavior of coal under different minin rates: A case study from laboratory experiments to field testing. *Int. J. Min. Sci. Technol.* **2021**, *31*, 825–841. [CrossRef]
22. Gao, R.Z.; Wang, P.F.; Li, Y.J.; Liu, R.H. Determination of optimal blowing-to-suction flow ratio in mechanized excavation fac with wall-mounted swirling ventilation using numerical simulations. *Int. J. Coal Sci. Technol.* **2021**, *8*, 248–264. [CrossRef]
23. Nadudvari, A.; Abramowicz, A.; Fabianska, M.; Misz-Kennan, M.; Ciesielczuk, J. Classification of fires in coal waste dumps base on Landsat, Aster thermal bands and thermal camera in Polish and Ukrainian mining regions. *Int. J. Coal Sci. Technol.* **2021**, 441–456. [CrossRef]
24. Tu, Q.Y.; Cheng, Y.P.; Xue, S.; Ren, T.; Cheng, X. Energy-limiting factor for coal and gas outburst occurrence in intact coal sean *Int. J. Min. Sci. Technol.* **2021**, *31*, 729–742. [CrossRef]
25. Yin, S.; Li, Z.H.; Song, D.Z.; He, X.Q.; Qiu, L.M.; Lou, Q.; Tian, H. Experimental study on the infrared precursor characteristics c gas-bearing coal failure under loading. *Int. J. Min. Sci. Technol.* **2016**, *31*, 901–912. [CrossRef]
26. Liu, Q.L.; Meng, X.F.; Hassall, M.; Li, X.C. Accident-causing mechanism in coal mines based on hazards and polarized manag ment. *Saf. Sci.* **2016**, *85*, 276–281. [CrossRef]
27. Koteeswaran, S.; Malarvizhi, N.; Kannan, E.; Sasikala, S.; Geetha, S. Data mining application on aviation accident data fo predicting topmost causes for accidents. *Clust. Comput. J. Netw. Softw. Tools Appl.* **2019**, *22*, 11379–11399. [CrossRef]
28. Linghu, J.S.; Zhao, W.; Zhou, J.B.; Yan, Z.M.; Wang, K.; Xu, C.; Sun, C.W. Influence of deep magma-induced thermal effects on th regional gas outburst risk of coal seams. *Int. J. Coal Sci. Technol.* **2021**, *8*, 1411–1422. [CrossRef]

applied sciences

MDPI

Article

The Change in the Shape Characteristics of the Plastic Zone in the Surrounding Rock of an Auxiliary Retracement Channel and a Reasonable Channel Spacing Determination Method

u Gao [1,*], Chenyi Liu [2], Hongkai Zhang [2], Kunlin Yang [2], Yingjie Hu [2] and Xiaofei Guo [2,3]

1 Railway Engineering Research Institute, China Academy of Railway Sciences Corporation Limited, Beijing 100081, China
2 School of Energy and Mining Engineering, China University of Mining and Technology (Beijing), Beijing 100083, China
3 Research Center of Roadway Support and Disaster Prevention Engineering in Coal Industry, Beijing 100083, China
* Correspondence: cumtb_gx@163.com

Abstract: In underground coal mines, the stability of the retracement channel in the surrounding rock is crucial for the safe and efficient retracement of the equipment and to guarantee the continuity of the retracement work. To reveal the deformation and damage mechanism of the surrounding rock of an auxiliary retracement channel (ARC) and the determination method for the reasonable spacing of two retracement channels during the end of the mining period, the deviatoric stress field in front of the working face and the change in the shape characteristics of the plastic zone in the ARC are investigated in this paper. The formation of ultimate stress equilibrium, high deviatoric stress, decreasing deviatoric stress, and low deviatoric stress environments in front of the working face during the end of mining occur successively, and the different deviatoric stress environments are the main reasons for the different shape characteristics of the plastic zone in the surrounding rock. The changes in the shape characteristics of the plastic zone correspond to the changes in the shape characteristics in the zone with deviatoric stress and exhibit the following order: full plastic deformation zone, butterfly-shaped zone, elliptical zone, and circular plastic zone. A reasonable spacing determination method for the two retracement channels is proposed: the ARC is arranged in the decreasing deviatoric stress environment, where the surrounding rock plastic zone shape is elliptical, and the ARC is relatively stable. Based on this research result, the spacing of the double retracement channels at the Lijiahao 22-116 working face was determined to be 25 m, which achieved a positive application effect and allowed the safe and efficient retracement of the working face equipment.

Keywords: auxiliary retracement channel; plastic zone; mining stress; surrounding rock

1. Introduction

In underground coal mining, heavy equipment such as hydraulic supports are retraced to the next working face after mining has been completed in a comprehensive working face, and the speed of the retracement is a significant factor in the efficiency of coal production [1]. Pre-excavation technology for double retracement channels arranges two channels parallel to the working face in advance near the stopping line, which allows quick retracement of the equipment through the auxiliary retracement channel (ARC) after the working face has been cut through the main retracement channel (MRC). The stability of the retracement channel surrounding rock is crucial for the safe and efficient retracement of the equipment and to guarantee the continuity of the retracement work [2]. The spacing of two retracement channels is the key factor in controlling the stability of the ARC surrounding rock. Therefore, it is beneficial in practical engineering to improve the understanding of the stress environment of the ARC under different spacings, clarify the surrounding rock deformation damage mechanism, and propose a reasonable spacing setting method.

The retracement channel is strongly influenced by mining; the surrounding rock can be seriously deformed and cause disasters, such as roof falling and coal wall spalling, at the end of the mining period [3–5]. It has been determined that the stress concentration from exploitation is the main reason for roadway destruction [6–8]. Field measurements, numerical simulations, and other means have revealed that the horizontal and vertical stresses rapidly increased after mining, and the stress environment was complex at the working face [9–11]. Using elastic-plastic theory, the appearance of different plastic zone shapes and the expansion laws of the roadway surrounding rock in the nonuniform stress field were analyzed, which proposed that the plastic zones in the roadway surrounding rock were butterfly-shaped, elliptical, and circular [12,13]. The mechanism of the analyzed damage to the roadway surrounding rock indicated that different surrounding rock plastic zone shapes would contribute to different degrees of deformation in the roadway [14,15].

These research results have shown the morphological variability and mechanistic complexity of the damage and deformation to roadway surrounding rock. However, the stress conditions, where the pre-excavation of the two retracement channels is located, are more complex due to the specificity of being arranged in advance and the influence from continuous mining [16,17]. Therefore, many scholars currently focus on controlling the stability of the retracement channel. The stress vector distribution in the retracement channel region is obtained through numerical simulation and field investigation [18]. LV [19] discovered the stress transfer phenomenon in the coal pillars on two sides of the retracement channel during the end of the mining period. HE et al. [20] proposed that the load on the upper coal pillars and mining disturbances caused plastic zone stress asymmetries in the retracement channel. GUO and YU [21] presented a stress release and roof cutting method to control the retracement channel stability. GU [22] investigated the mechanism of the coal pillar load transfer from the working face to the MRC, with a method to determine the widths of the coal pillars. Nevertheless, there are few studies on the stressful environment and surrounding rock deformation damage in the ARC during mining, which makes it more challenging to control the surrounding rock stability in the ARC.

The purpose of this work is to reveal the deviatoric stress field in front of the working face during the end of the mining period, analyze the change in the shape characteristics of the plastic zone in the ARC in different stress environments, propose a method to determine the reasonable spacing of the pre-excavation for double retracement channels based on the stability of the ARC, and reduce the width of the protective coal pillars while ensuring the ARC surrounding rock stability. The objective is to avoid the waste of mine resources and complete equipment retracement from the working face, safely and efficiently.

2. Engineering Background

2.1. Engineering Situations

Lijiahao Mine is in Hohhot, in the Inner Mongolia Autonomous Region of China. The 22-116 comprehensive mining working face is mainly used to mine the 2-2 coal seam. A schematic diagram of the working face layout is shown in Figure 1. To quickly transfer the hydraulic supports and other heavy equipment to the next working face after the mining has been completed, double retracement channels are prearranged near the stopping line at the back of the working face. The retracement channel near the working face is the MRC, the other is the ARC.

When the working face cuts through to the MRC at the end of the mining period, the equipment is transferred to the ARC via the connection roadway and, subsequently, to the next working face. The pre-excavation technology used for the two retracement channels saves equipment retracement time and improves coal mining efficiency.

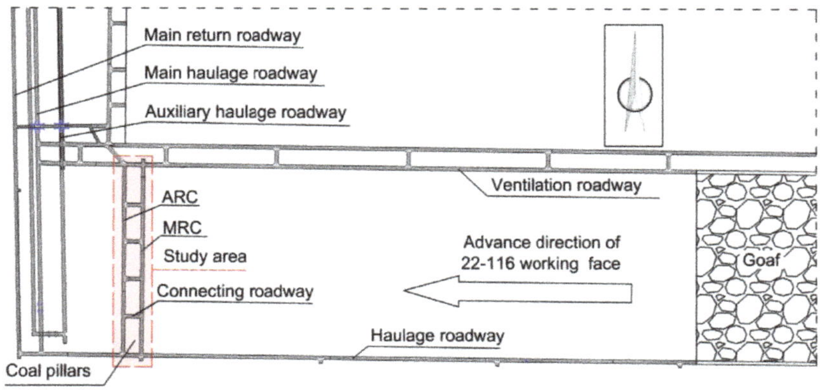

Figure 1. Schematic diagram of the 22-116 working face layout.

2.2. Problems in Production

The problem faced by the pre-excavation technology used for the two retracement channels is that the ARC is influenced by mining during the end of the mining period. This influence results in severe deformation and damage to the ARC and challenges the equipment's safe and efficient reinforcement. The MRC in the 22-116 working face is supported by a stacking support, and the ARC is supported by a bolt and a cable anchor.

During the end of mining, if the two retracement channels are too closely arranged, the ARC is strongly influenced by the mining and causes serious deformation and damage to the channel, and the equipment cannot achieve smooth retracement. This outcome will directly decrease the retracement efficiency and cause economic loss when the ARC must be repaired, which will also impact the layout of the next working face. If the spacing of two retracement channels is too wide, many of the coal pillars left behind cannot be recovered, which causes a waste of coal resources.

2.3. Geological Conditions

The 22-116 working face in the Lijiahao Mine is mainly used for mining the 2-2 medium coal seam, with a tendency length of 300 m, a strike length of 2406 m, and an average burial depth of 185 m. In the working face area, the average thickness of the coal seam is 2.01 m. The seam is relatively stable with a more significant thickness variation, and it is thicker in the southern part of the survey area. The coal seam dips from 0~3°, which is a near-horizontal coal seam, and contains 0~2 layers of dirt band, part of which contains one layer of dirt band. The coal seam structure is simple and stable. The geological columnar section of the 22-116 working face is shown in Figure 2.

The floor lithology is mainly composed of sandy mudstone and siltstone, which have low mechanical strength and easily soften and deform when encountering water. The roof lithology is mainly composed of sandy mudstone and coarse-grained sandstone. From the mechanical test results, the compressive strength is very low, mainly less than 30 MPa, while the shear and tensile strengths are even lower. The rocks in the coal seam floor and roof are mostly soft, and some of the rocks are semihard or hard rocks. The equipment retracement adopts the pre-excavated double retracement channel technology. The ARC section dimension is 5.4 m × 2.6 m, and the MRC section dimension is 5.2 m × 2.4 m.

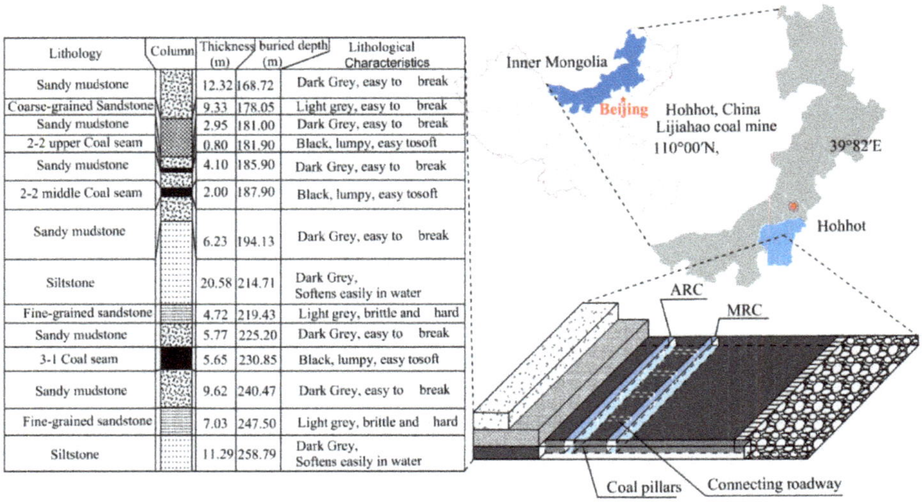

Figure 2. Stratigraphic column of the 22-116 working face.

3. Analysis of the Plastic Zone Characteristics in the Rock Surrounding the Auxiliary Retracement Channel

3.1. Evolution of the Shape of the Plastic Zone in the Nonuniform Stress Field in the Roadway Surrounding Rock

3.1.1. Mechanical Model of a Nonuniform Stress Field in a Circular Roadway

Based on the current research on the elastic-plastic deformation of the hole, the damage range of the roadway is quantified by mechanical-mathematical methods, which adopt the circular hole plane from the strain model. The stress field under the mining influence is simplified to a planar axisymmetric, nonuniform hydrostatic pressure condition. A bidirectional, nonuniform pressure circular roadway model is established, as shown in Figure 3.

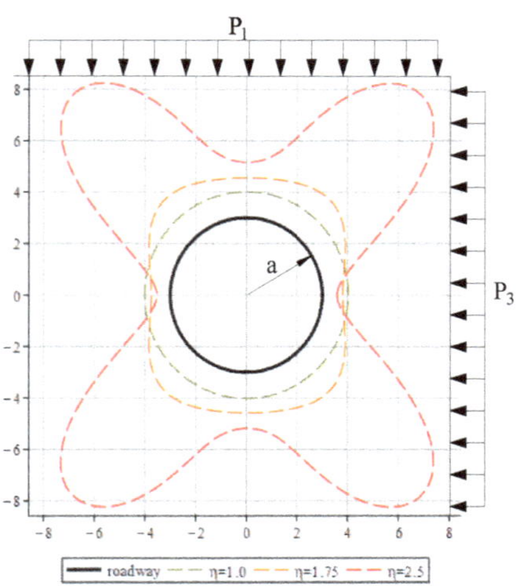

Figure 3. Computational model of the surrounding rock plastic zone in a bidirectional, nonuniform stress circular roadway.

In the model, the radius of the circular roadway is a, the coordinates of any point in the polar coordinate system are (r, θ), the nonuniform stress field around the circular roadway is represented using the regional principal stress field, P_1 is the regional maximum principal stress, and P_3 is the regional minimum principal stress.

A bidirectional, nonuniform stress condition is called a deviatoric stress environment, which uses the ratio η of P_1 to P_3 to reflect the degree of nonuniform stress distribution, as shown in Equation (1).

$$\eta = \frac{P_1}{P_3} > 1, \tag{1}$$

3.1.2. Evolution of the Shape of the Surrounding Rock Plastic Zone in the Deviatoric Stress Environment

In the literature, the bidirectional, nonuniform stress circular roadway model is analyzed, which combines the stress expression at a point in the roadway surrounding rock using polar coordinates in elastic mechanics and the Mohr–Coulomb damage criterion to derive the implicit equation for the boundary of the plastic zone in the circular roadway in a nonuniform stress field [23]. The plastic zone boundary location in the roadway can be calculated for the given roadway specifics and surrounding rock stress conditions, as shown in Equation (2).

$$f(R,\theta) = a \times F(P_1, P_3, \eta, C, \varphi), \tag{2}$$

Equation (2) shows that the surrounding rock plastic zone boundary in a circular roadway in a nonuniform stress field is mainly related to the roadway radius a, the regional principal stress environment (P_1, P_3, and η), and the rock lithology (φ: internal friction angle, C: cohesion). For the same lithology, the degree of nonuniformity in the stress field distribution (η) is the critical factor for determining the damage form and size of the roadway surrounding rock in a partial stress environment [24].

The following parameters are used in building the FLAC^{3D} numerical simulation model: the circular roadway has a radius of 3 m, the mechanical properties of the surrounding rock are C = 3 MPa and φ = 25°, and the minimum principal stress P_3 is 20 MPa. Under these conditions, Equation (2) was prepared using the MAPLE2020 mathematical software to calculate the roadway plastic zone boundary with a change in the degree of the nonuniform stress distribution (η). The evolution of the size and shape of the plastic zone in the surrounding rock in the roadway is shown in Figure 4.

With an increase in η, the surrounding rock plastic zone shape from the FLAC^{3D} simulation and MAPLE calculation indicates the development pattern of the circular–elliptical–butterfly shaped, and the maximum boundary range of the plastic zone increases with the change in shape. When η = 1, the roadway is in a bidirectional equal stress condition, the plastic zone is not large, it is uniformly distributed in the surroundings of the roadway, and the form is a regular circular shape. With an increase in η, the plastic zone begins to expand above the roof and below the floor in the roadway. When η increases, the plastic zone shape develops from circular to elliptical, where the ellipse's long axis is in the direction of P_3. When η > 1.8, the growth rate of the plastic zone in the four corners of the roadway is larger than the two sides of the roadway, which is expanded by a significant amount in the four corners of the roadway, and the shape develops from elliptical to butterfly shaped. When η > 2.3, the shape is like a butterfly, and any further minor increases in η will significantly expand the plastic zone near the angular parallels in the P_1 and P_3 directions.

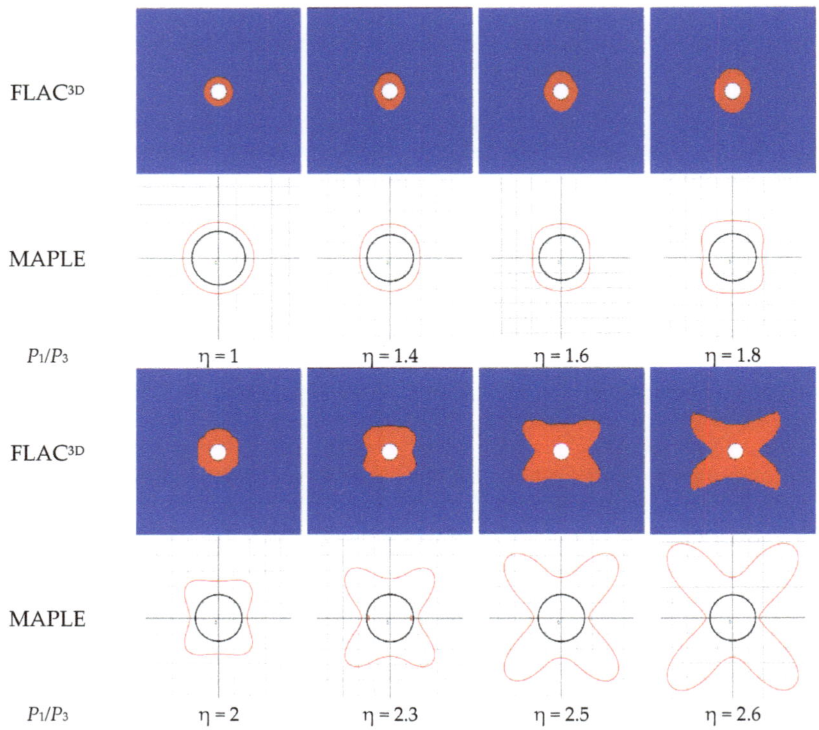

Figure 4. Evolution of the plastic zone calculated by simulation.

3.1.3. Relationship between the Principal Stress Direction and the Shape Characteristics of the Plastic Zone

When the roadway is in a deviatoric stress environment, the plastic zone shape changes mainly due to the nonuniform stress distribution. With increasing η, the plastic zone evolves in the order of circular–elliptical–butterfly shaped. During the evolution of the shape, the plastic zone extension is mainly near the angle bisector between the directions P_1 and P_3. Therefore, when the directions of the maximum and minimum principal stress change, the expansion of the plastic zone in the surrounding rock will also change location.

A butterfly-shaped plastic zone in the roadway surrounding rock in a deviatoric stress environment is shown in Figure 5a. When P_1 is in the vertical direction and P_3 is in the horizontal direction, the butterfly-shaped plastic zone significantly expands at the roadway's four corners. If a certain deflection angle occurs in the direction when applying P_1 and P_3, the location where the expansion of the butterfly-shaped plastic zone occurs also has the same deflection angle.

The plastic zone expansion location in the roadway is near both sides and the floor and roof plates after the principal stress has been deflected. The plastic zone near the roof is deeper than the effective range of the available anchor bolts and cable supports. Figure 5b shows that the roadway disaster manifests as a large nonuniform deformation of the surrounding rock and a large amount of roof subsidence, and the anchor bolts and anchor cable support failure appear to result in roof caving.

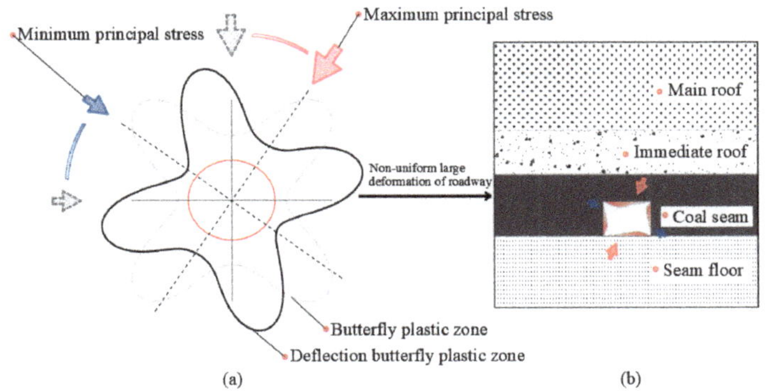

Figure 5. Deflection of the butterfly-shaped plastic zone in the roadway surrounding rock.

3.2. Migration Law of the Mining Stress Field in front of the Working Face

After the working face has been mined, the bearing capacity of [6,9,25] the waste rock collapsed in the goaf is reduced. The self-weight stress of the overlying strata is transferred to the coal body in front of the working face, which results in the stress concentration phenomenon after the stress redistribution. The vertical stress starts from less than the original rock stress (γH) and rapidly increases to KγH (K is the abutment pressure coefficient, 3~5), which reaches the maximum value and decreases as it moves away from the working face until it decreases near γH. The horizontal stress increases from 0 to close to γH and, subsequently, remains unchanged. At the location far from the working face, the vertical and horizontal stress are γH, where the roadway is in the original rock stress state and is almost unaffected by mining.

Figure 6 shows the stress field migration law of mining in front of the working face. The retracement channels are prearranged near the stop line when the working face is mined, away from the working face unaffected by the mining stress. The mining stress field migrates with the working face as the working face is mined. The coal body in front of the working face is always under the influence of concentrated stresses. At the end of mining, the retracement channel gradually enters the influence range of the mining stress field, which gradually increases the stress on the retracement channel. After the working face cuts through the MRC, the spacing of the two retracement channels will determine the degree to which the mining stress field influences the ARC.

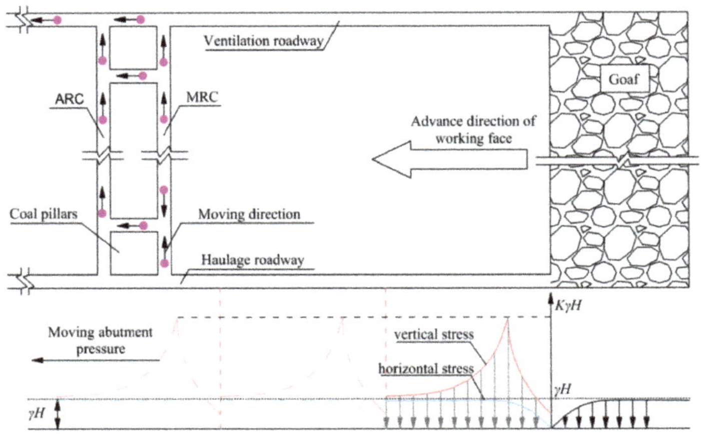

Figure 6. Migration law of the mining stress field in front of the working face.

3.3. *Shape Characteristics of the Plastic Zone in the Auxiliary Retracement Channel during the End of the Mining Period*

Under the influence of mining stress, the maximum principal stress near the working face is similar to the vertical stress size change law, and the minimum principal stress is similar to the horizontal stress [26]. The current results show that P_1 in front of the working face and the vertical stress change with the same trend [27,28]. At the end of the mining period, when the working face cuts through the MRC, the self-weight stress of the overburden above the goaf will be transferred to the retracement channel to the protective coal pillars. The mining stress concentration phenomenon means that the ARC is in a deviatoric stress environment. When the concentrated stress exceeds the rock strength, the surrounding rock will undergo plastic zone deformation. The plastic zone is the damage zone obtained by theoretical calculation, which reflects the damage characteristics of the roadway surrounding rock.

As previously mentioned, the spacing of the two retracement channels is crucial for determining the stress environment where the ARC is located. In the deviatoric stress region with no uniform distribution of P_1 and P_3, the surrounding rock of the ARC resembles plastic zones of different shapes, as shown in Figure 7. Different widths of the protective coal pillars mean that the ARC is located in different stress environments.

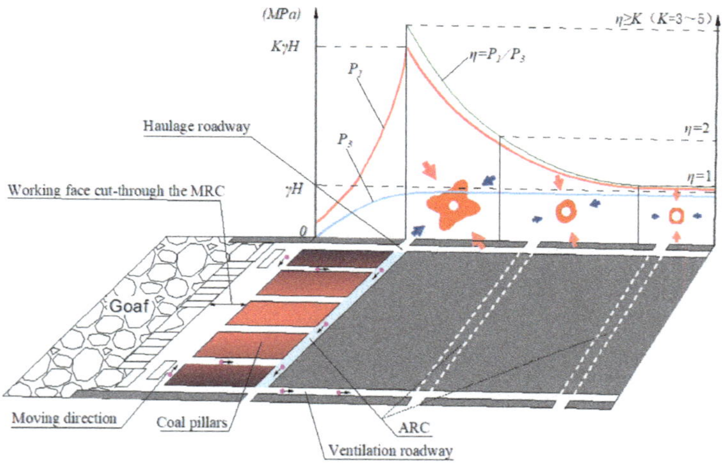

Figure 7. Morphological characteristics of the plastic zone in the auxiliary retracement channel with different spacings.

If the protective coal pillars are narrow, the ARC is located in a high deviatoric stress environment, deflects the principal stress at a certain angle and produces a butterfly-shaped plastic zone in the ARC surrounding the rock. With an increase in the width of the protective coal pillars, the mining influence on the ARC gradually diminishes, which decreases the nonuniformity in the main stress distribution. The plastic zone in the ARC is reduced in range with weakened asymmetry as a result of an elliptical-shaped plastic zone. When the protective coal pillars are wide, the ARC is slightly influenced by mining, the principal stress distribution is uniform, and the plastic zone shape is circular.

4. Numerical Simulation Analysis

4.1. *Numerical Simulation Model*

According to the mechanical parameters of each coal and rock stratum in the Lijiahao 22-116 working face [29], combined with the site geological conditions, the FLAC^{3D} 6.0 finite element software was used to establish a three-dimensional numerical model, as shown in Figure 8. The deformation and damage to the ARC are simulated after the working face has been cut through the MRC.

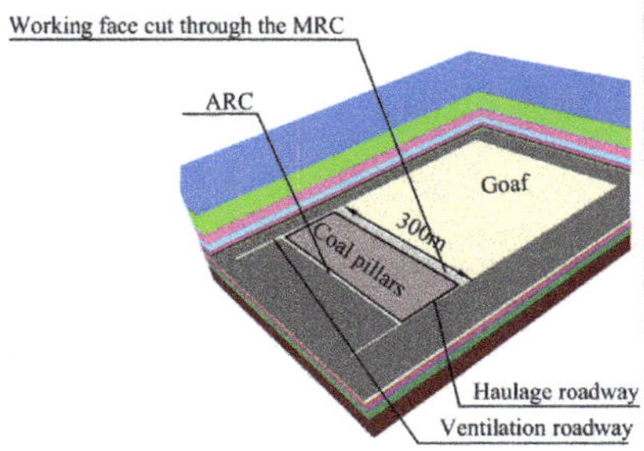

Lithology	Density (kg/m^3)	Bulk modulus (GPa)	Shear modulus (GPa)	Tensile strength(MPa)	Cohesion (MPa)	Internal friction angle (°)
2-2 middle Coal seam	1470	2.30	1.98	0.60	1.80	27
2-2 upper Coal seam	1350	2.50	1.20	0.60	0.80	28
Fine-grained sandstone1	2690	4.98	6.58	1.58	2.70	33
Fine-grained sandstone2	2400	3.60	5.78	1.75	2.30	31
Fine-grained sandstone3	2560	4.90	7.58	1.35	1.60	35
Overlying strata	2500	5.50	3.30	2.60	6.20	32
Sandy mudstone1	2600	3.50	2.80	2.10	3.60	31
Sandy mudstone2	2500	2.70	1.80	1.10	2.50	28
Sandy mudstone3	2400	3.80	1.80	1.20	1.60	28
Sandy mudstone4	2300	2.70	1.60	1.06	1.60	29
Sandy mudstone5	2600	3.60	2.80	2.30	1.60	30
Sandy mudstone6	2630	4.20	2.20	1.50	1.50	31
Silstone1	2600	4.50	2.80	2.10	3.60	31
Silstone2	2500	3.20	2.60	1.50	2.40	30

Figure 8. Numerical simulation model and mechanical parameters of each coal and rock stratum.

The model adopts the Mohr–Coulomb damage criterion based on the elastic-plastic theory, with fixed constraints around and at the floor surface and a free boundary at the top surface. The overlying rock was 120 m from the surface, and 3MPa of stress was applied on the top surface to compensate for the self-weight stress of the overlying rock. The coal seam with its roof and floor cell size was 0.5 m. The ARC was a rectangular section (3 m high; 5 m wide). The model excavated the haulage roadway and ventilation roadway, excavated the MRC and ARC, and gradually excavated the coal seam until it cut through with the MRC.

4.2. Distribution of the Mining Stress Field in Front of the Working Face

Working face mining disrupts the original rock stress equilibrium state and redistributes stresses. Due to the strong influence of mining, stress concentration occurs in the coal pillars in front of the working face. The stress field changes after the stress redistribution. The changes in P_1 and P_3 in the range of 0~70 m for the front coal pillars were tested to obtain the stress field distribution change law, as shown in Figure 9.

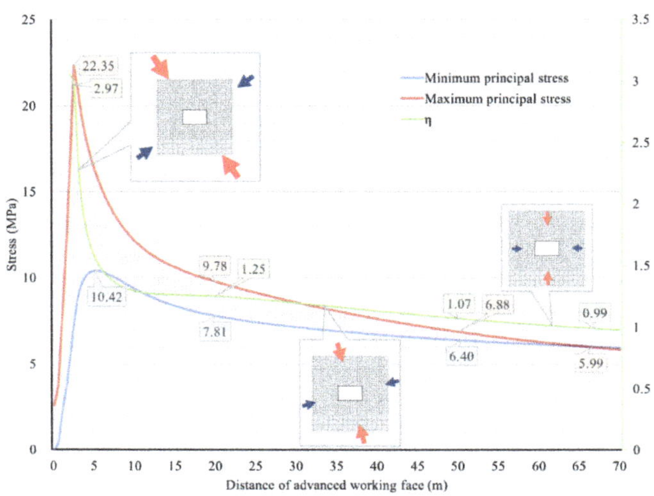

Figure 9. Distribution of the 0~70 m mining stress field in front of the working face.

The original rock stress (γH) at the site without mining disturbance is 5 MPa. P_1, in the coal pillars in front of the working face, sharply increases from less than γH to a peak of 22.35 MPa at 2.5 m, which is 4.47 times the original rock stress; then, it decreases to $2\gamma H$ at a faster rate and slowly returns to γH when the stress magnitude moves further from the working face. P_3 relatively quickly increases from 0 to a peak of 10.42 MPa at 5 m in front, slowly decreases to γH and, subsequently, stabilizes.

Based on the above analysis, the numerical simulation results for the stress field distribution in front of the working surface are consistent with the theoretical analysis. The stress concentration phenomenon caused by mining leads to a nonuniform distribution of P_1 and P_3, and forms a deviatoric stress environment. Moreover, while the stress is concentrated, the directions of P_1 and P_3 will simultaneously be deflected at a certain angle. Protective coal pillars with different widths will mean that the ARC is located under different degrees of deviatoric stress, and the roadway surrounding rock will be damaged in different forms.

4.3. Shape Characteristics of the Plastic Zone in the Surrounding Rock of the Auxiliary Retracement Channel in the Deviatoric Stress Environment

With the changes in the width of the protective coal pillars between the two retracement channels, the evolution characteristics of the shape of the plastic zone in the ARC surrounding rocks in different deviatoric stress environments are obtained after the working face has been cut through the MRC. With the increasing width of the protective coal pillars, the plastic zones in the ARC are butterfly shaped, elliptical and circular, as shown in Figure 10.

For example, Figure 10 shows that the surrounding rock plastic zone is uniformly distributed along the centerline of the rectangular roadway when the retracement channel is not influenced by mining. The plastic zone in the two sides and floor has a maximum depth of 0.5 m, while that of the roof is 1 m.

When the working face is cut through the MRC, the self-weight stress of the overlying rock is transferred to the protective coal pillars between the two retracement channels. When the protective coal pillars are narrow, the ARC is in a high deviatoric stress environment with stress concentration and deflection, which forms a butterfly-shaped plastic zone. Deflection of the butterfly-shaped plastic zone occurs. The deepest part of plasticity occurs above the roof, the plastic zone at the roof expands to the secondary side, and the plastic zone at the floor expands to the primary side (it is defined as the primary side near the working face, and the far side is the secondary side). Asymmetric, large deformation occurs in the ARC (Figure 10b–d). When the width of the protective coal pillars increases, the ARC is located far from the working face and the mining stress influence is reduced. The concentration degree, deflection angle, and uneven distribution of stresses are weakened, which gradually reduces the range of the plastic zone. As shown in Figure 10e–h, the plastic zone is concentrated in the corner of the rectangular roadway, and the deepest part is in the roof near the secondary side, which has an elliptical shape. When the width of the protective coal pillars continues to increase, the ARC is less influenced by mining, and the plastic zone form is similar to a circular shape, which is more uniformly distributed around the roadway (Figure 10i).

At the end of the mining period, the numerical simulation calculated the shape characteristics of the plastic zone in the ARC, which were consistent with the shape characteristics of the nonuniform stress field plastic zone derived from the theoretical calculations. With the nonuniform degree of distribution of P_1 and P_3 in the stress field where the roadway is located, the plastic zones in the surrounding rock are round, elliptical, and butterfly shaped. Because the numerically simulated roadway is a rectangular section and the redistributed stress field caused by mining is very complex, the size range of the numerically simulated plastic zone shape is slightly different from the shape characteristics calculated according to the theory. However, the shape characteristics of the plastic zone change law remain consistent.

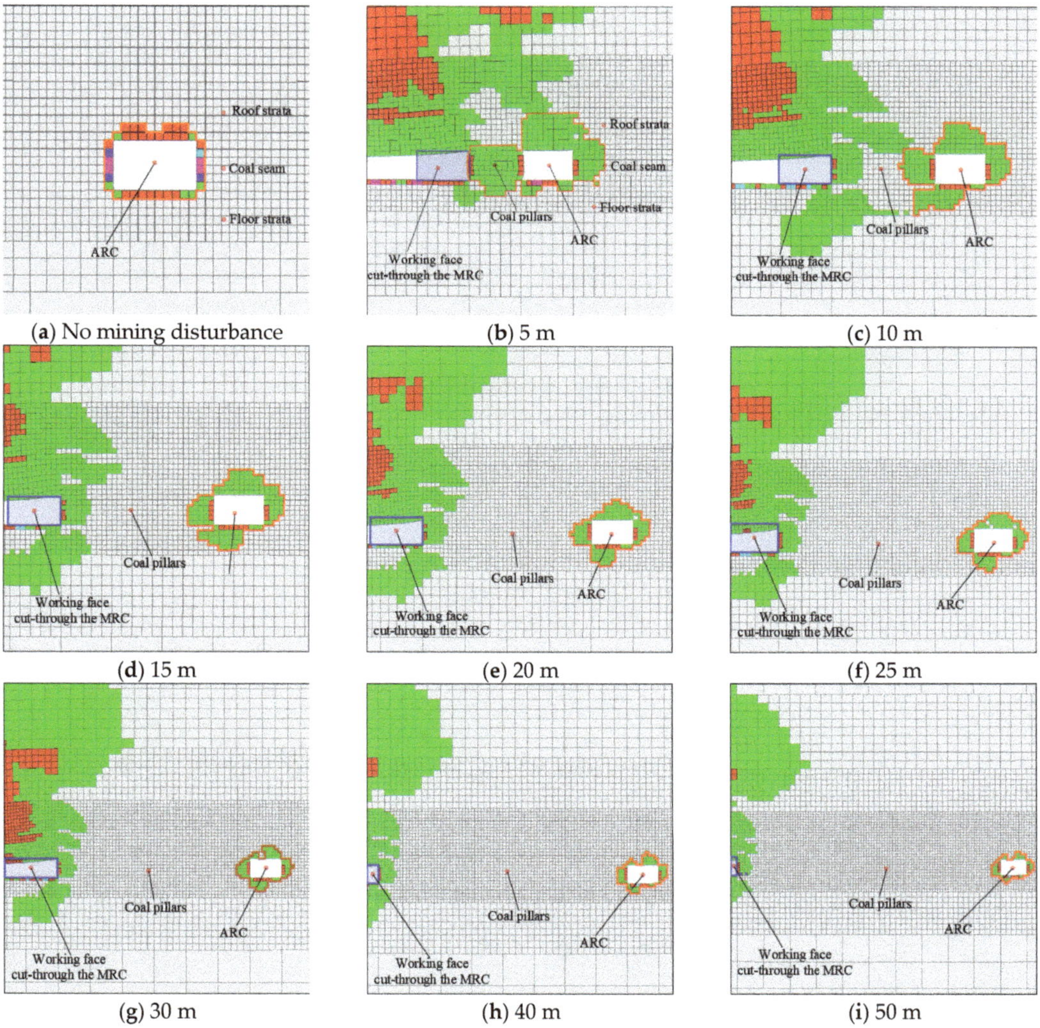

Figure 10. Shape characteristics of the plastic zone in the auxiliary retracement channel with different protective coal pillar widths.

5. Results and Discussion

5.1. Influence of the Plastic Zone Shape on the Stability of the Auxiliary Retracement Channel

Theoretical and numerical simulations verified that mining will cause stress concentrations in a certain range of the coal pillar in front of the working face and form a high deviatoric stress environment with a large difference between P_1 and P_3. The plastic zone in the ARC surrounding rock in a high deviatoric stress environment is butterfly shaped. When the ARC is far from the working face, the mining influence is weakened, and the plastic zone shape of the roadway becomes elliptical and circular.

The butterfly-shaped plastic zone is not uniformly distributed, easily making the ARC surrounding rock undergo serious asymmetric deformation, and is in an unstable state. In addition, influenced by the concentrated and deflection stress, the plastic zone with the butterfly shape expands, a large amount, above the roof. The plastic zone depth is beyond the effective range of the available anchor bolts and cables, which makes the support body fail, the roof sinking is large, and even the roof falls. The elliptical plastic zone is more uniformly distributed than the butterfly-shaped plastic zone, and the maximum depth is reduced. The anchor bolts and cables can provide effective support, and the ARC is relatively stable. The circular plastic zone is barely influenced by mining, the extent of the plastic zone is the smallest, and the surrounding rock has the best stability.

5.2. Method for Determining the Reasonable Spacing of Two Retracement Channels

At the end of the mining period, after the working face has been cut through the MRC, the self-weight stress of the overlying rock is mainly transferred to the protective coal pillar between the two retracement channels, so the spacing of the two retracement channels becomes critical to protect the stability of the ARC. Determining the reasonable spacing of the two retracement channels optimizes the channel layout under the premise of ARC stability, minimizes the size of the protective coal pillars, and improves the productivity of coal mining enterprises, while reducing the waste of coal resources.

The key to reasonable spacing is to ensure the stability of the surrounding rock in the ARC during the retracement of the equipment. Based on the above analysis, Figure 11 illustrates the ARC with different protective coal pillar widths in the stress environment, the plastic zone shape characteristics, and the surrounding rock stability state.

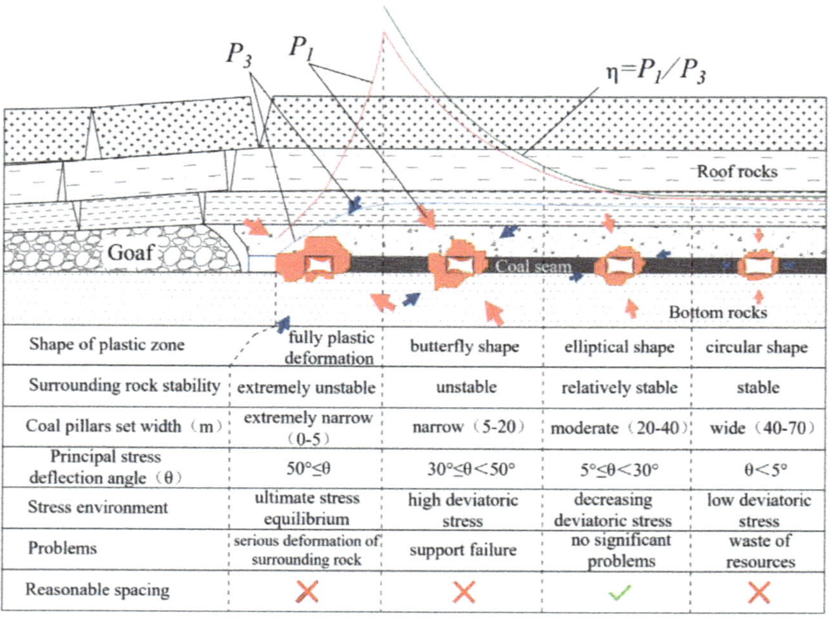

Shape of plastic zone	fully plastic deformation	butterfly shape	elliptical shape	circular shape
Surrounding rock stability	extremely unstable	unstable	relatively stable	stable
Coal pillars set width (m)	extremely narrow (0-5)	narrow (5-20)	moderate (20-40)	wide (40-70)
Principal stress deflection angle (θ)	50°≤θ	30°≤θ<50°	5°≤θ<30°	θ<5°
Stress environment	ultimate stress equilibrium	high deviatoric stress	decreasing deviatoric stress	low deviatoric stress
Problems	serious deformation of surrounding rock	support failure	no significant problems	waste of resources
Reasonable spacing	✗	✗	✓	✗

Figure 11. Reasonable spacing determination method.

Mining makes P_1 and P_3 nonuniformly distributed in magnitude and direction in front of the working face, which results in ultimate stress equilibrium, high deviatoric stress, decreasing deviatoric stress, and low deviatoric stress environments. With an increase in the double retracement channel spacing, the ARC is in different stress environments during the end of the mining period, which produces different shape characteristics of the plastic zone.

The ARC is in the ultimate stress equilibrium environment when the coal pillars are set extremely narrow (0~5 m). Because of the stress field migration, the protective coal pillars have experienced the action of multiple stress fields, which have entirely undergone plastic deformation and lost their bearing capacity. The surrounding rock in the ARC is severely non-symmetrically deformed and extremely unstable. When narrow coal pillars

(5~20 m) are set, the ARC is in a high deviatoric stress environment, the plastic zone has a butterfly shape, and the surrounding rock is asymmetrically deformed with poor stability and challenging support control. When moderate-width coal pillars (2~40 m) are set, the ARC is in the decreasing deviatoric stress environment. The shape of the plastic zone is elliptical and more uniformly distributed, which can control the stability of the surrounding rock through the support of anchor bolts and anchor cables. When wide coal pillars (40~70 m) are set, the ARC is less influenced by mining and is in a low deviatoric stress environment. The shape of the plastic zone is circular, which has stable surrounding rock, but the coal pillars cannot be recovered, and additional resources are wasted.

The width of the protective coal pillars that correspond to the different shape characteristics of the previously mentioned plastic zones is not an invariant quantity. In general, greater ground stress and weaker mechanical strength of the rock correspond to wider protective coal pillars and greater stability in the ARC.

The above analysis shows that the reasonable spacing of the two retracement channels is 20~40 m; however, this range is not invariant and can be adjusted according to actual environmental changes. The key to reasonable spacing is to ensure that the protective coal pillars are sufficiently wide to mean that the ARC is in a decreasing deviatoric stress environment, and the plastic zone should exhibit a stable elliptical shape.

5.3. Engineering Application Effect Analysis

Based on the above reasonable spacing determination method, the spacing of the pre-excavated double retracement channels for the 22-116 working face at the Lijiahao Mine is determined. Through the results from the numerical simulation analysis, when the protective coal pillar is wider than 20 m, the ARC is in the decreasing deviatoric stress environment, the plastic zone shape is elliptical, and the plastic zone size effect on the surrounding rock stability is considered. The spacing between the two retracement channels, set to 25 m, is determined during the application of the actual project.

To obtain the surrounding rock deformation in the ARC at the end of the mining period, 17 surface displacement monitoring stations were uniformly arranged in the ARC. The displacement between the floor and roof and between the two sides are shown in Figure 12. With a 25 m protective coal pillar set, the surface displacement of the ARC is small overall: the average is 22.9 mm between the roof and floor and 17.1 mm between the two sides. The deformation and damage in the surrounding rock in the ARC are not significant, with a small amount of sinking in the roof and an insignificant coal wall spalling phenomenon, which does not affect the normal use of the ARC.

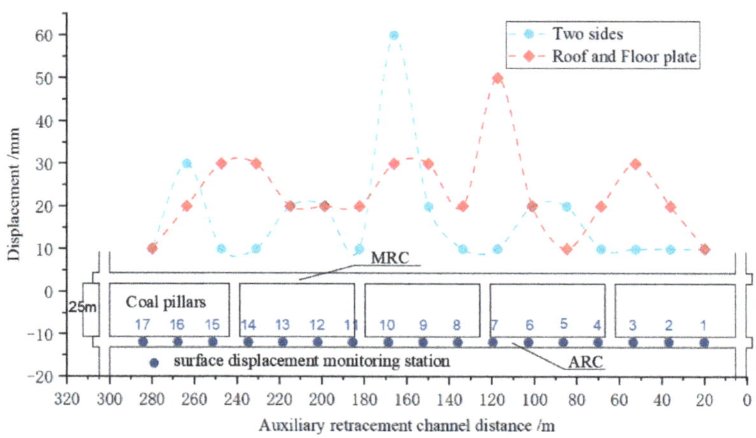

Figure 12. Surface displacement changes at 17 monitoring points in the auxiliary retracement channel.

6. Conclusions

(1) The deviatoric stress environment formed by the nonuniform distribution of the maximum principal stress and minimum principal stress caused by mining is revealed, which reflects the formation of ultimate stress equilibrium, high deviatoric stress, decreasing deviatoric stress, and low deviatoric stress environments, away from the front of the working face.

The relationship between the shape characteristics of the plastic zone in the auxiliary retracement channel surrounding rock and the channel stability is revealed. With an increase in the width of the protective coal pillars, the shape of the surrounding rock plastic zones experience full deformation, butterfly shaped, elliptical, and circular, which indicates extremely unstable, unstable, relatively stable, and stable channels, respectively.

(2) The mechanical essence of the surrounding rock in the auxiliary retracement channel forming the plastic zones with different shape characteristics is that the channel is experiencing different degrees of deviatoric stress. When the auxiliary retracement channel is in the high deviatoric stress environment, the plastic zone in the surrounding rock is butterfly shaped; in the decreasing deviatoric stress environment, the plastic zone is elliptical; and in the low deviatoric stress environment, the plastic zone is circular.

(3) A method to determine the reasonable spacing of the pre-excavated double retracement channels is proposed. The necessary condition for reasonable spacing is to arrange the auxiliary retracement channel in the decreasing deviatoric stress environment, which can effectively control the stabilization of the surrounding rock and avoid a waste of resources caused by being unable to recover numerous coal pillars. The retracement efficiency is effectively improved through engineering practice verification, setting the pre-excavation spacing for the two retracement channels at the Lijiahao Mine 22-116 working face to 25 m, placing the auxiliary retracement channel in the deviatoric stress environment, which ensures the surrounding rock stability at the end of the mining period, and the safe and efficient completion of the equipment retracement.

Author Contributions: Conceptualization, X.G. (Xu Gao), H.Z. and X.G. (Xiaofei Guo); Methodology, H.Z.; Software, K.Y. and Y.H.; Formal analysis, C.L.; Investigation, Y.H.; Resources, H.Z.; Data curation, C.L. and K.Y.; Writing—original draft, X.G. (Xu Gao) and C.L.; Supervision, X.G. (Xiaofei Guo); Project administration, X.G. (Xu Gao) and X.G. (Xiaofei Guo); Funding acquisition, X.G. (Xu Gao). All authors have read and agreed to the published version of the manuscript.

Funding: China Academy of Railway Sciences Youth Fund Project: 2023YJ187; China State Railway Group Co., Ltd. Focused Fund: N2021G002; National Natural Science Foundation of China High Speed Rail Joint Fund: U2268217.

Institutional Review Board Statement: Not applicable.

Informed Consent Statement: Not applicable.

Data Availability Statement: The data can be provided if necessary.

Conflicts of Interest: The authors declare that they have no conflict of interest.

References

1. Ju, Y.; Wang, Y.; Su, C.; Zhang, D.; Ren, Z. Numerical analysis of the dynamic evolution of mining-induced stresses and fractur in multilayered rock strata using continuum-based discrete element methods. *Int. J. Rock Mech. Min. Sci.* **2019**, *113*, 191–2 [CrossRef]
2. Fei, L.; Jiang, Z. Research on Deformation Mechanism of Retracement Channel during Fully Mechanized Caving Mining Superhigh Seam. *Adv. Civ. Eng.* **2018**, *2018*, 1368965. [CrossRef]
3. Zhang, F.; Wang, X.; Bai, J.; Wu, W.; Wu, B.; Wang, G. Fixed-length roof cutting with vertical hydraulic fracture based on the stre shadow effect: A case study. *Int. J. Min. Sci. Technol.* **2022**, *32*, 295–308. [CrossRef]
4. Malashkevych, D.; Petlovanyi, M.; Sai, K.; Zubko, S. Research into the coal quality with a new selective mining technology of t waste rock accumulation in the mined-out area. *Min. Miner. Depos.* **2022**, *16*, 103–114. [CrossRef]
5. Vu, T.T. Solutions to prevent face spall and roof falling in fully mechanized longwall at underground mines, Vietnam. *Min. Min Depos.* **2022**, *16*, 127–134. [CrossRef]

. Lou, J.; Gao, F.; Yang, J.; Ren, Y.; Li, J.; Wang, X.; Yang, L. Characteristics of evolution of mining-induced stress field in the longwall panel: Insights from physical modeling. *Int. J. Coal Sci. Technol.* **2021**, *8*, 938–955. [CrossRef]

. Zhao, H.; Cheng, H.; Ji, D.; Wang, T.; Wang, Y.; Jiang, C. Study of the mechanism and evolution law of unsymmetrical failure of the mining roadway in close distance coal seam. *Int. J. Min. Sci. Technol.* **2021**, *50*, 1029–1040+1050. [CrossRef]

. Li, W.; Tu, S.; Tu, H.; Li, Y.; Liu, X.; Miao, K. Failure characteristics and control techniques for mining roadway affected by stress accumulation of residual pillars in contiguous coal seams. *Eng. Fail. Anal.* **2022**, *141*, 106646. [CrossRef]

. Wu, R.; Zhang, P.; Kulatilake, P.H.; Luo, H.; He, Q. Stress and deformation analysis of gob-side pre-backfill driving procedure of longwall mining: A case study. *Int. J. Coal Sci. Technol.* **2021**, *8*, 1351–1370. [CrossRef]

0. Suchowerska, A.; Merifield, R.; Carter, J. Vertical stress changes in multi-seam mining under supercritical longwall panels. *Int. J. Rock Mech. Min. Sci.* **2013**, *61*, 306–320. [CrossRef]

1. Babets, D.; Sdvyzhkova, O.; Hapieiev, S.; Shashenko, O.; Vasyl, V. Multifactorial analysis of a gateroad stability at goaf interface during longwall coal mining—A case study. *Min. Miner. Depos.* **2023**, *17*, 9–19. [CrossRef]

2. Zhao, Z.Q.; Ma, N.J.; Liu, H.; Guo, X.F. A butterfly failure theory of rock mass around roadway and its application prospect. *J. China Univ. Min. Technol.* **2018**, *47*, 969–978. [CrossRef]

3. Guo, X.; Ma, N.; Zhao, X.; Zhao, Z.; Li, Y. General shapes and criterion for surrounding rock mass plastic zone of round roadway. *J. China Coal Soc.* **2016**, *41*, 1871–1877. [CrossRef]

4. Ma, N.; Zhao, X.; Zhao, Z.; Li, J.; Guo, X. Stability analysis and control technology of mine roadway roof in deep mining. *J. China Coal Soc.* **2015**, *40*, 2287–2295. [CrossRef]

5. Smoliński, A.; Malashkevych, D.; Petlovanyi, M.; Rysbekov, K.; Lozynskyi, V.; Sai, K. Research into impact of leaving waste rocks in the mined-out space on the geomechanical state of the rock mass surrounding the longwall face. *Energies* **2022**, *15*, 9522. [CrossRef]

6. Wang, B.; Dang, F.; Chao, W.; Miao, Y.; Li, J.; Chen, F. Surrounding rock deformation and stress evolution in pre-driven longwall recovery rooms at the end of mining stage. *Int. J. Coal Sci. Technol.* **2019**, *6*, 536–546. [CrossRef]

7. He, Y.; Song, Y.; Shi, Z.; Li, J.; Chen, K.; Li, Z. Study on failure mechanism of withdrawal channel's surrounding rock during last mining period. *China Saf. Sci. J.* **2022**, *32*, 158. [CrossRef]

8. Li, C.; Guo, X.; Huo, T.; Liu, R. Coal pillar design of pre-excavated double equipment withdrawal channel and its surrounding rock stability control. *J. Huazhong Univ. Sci. Technol.* **2021**, *49*, 1–7. [CrossRef]

9. LV, H.-W. The mechanism of stability of predriven rooms and the practical techniques. *J. China Coal Soc.* **2014**, *39*, 50–56. [CrossRef]

0. He, F.; Lv, K.; Li, X.; Qin, B.; Li, L. Failure mechanism and control of lower retracement channel in close-distance double-thick coal seams. *Shock. Vib.* **2021**, *2021*, 1–19. [CrossRef]

1. Guo, Y.; Yu, H. Research and application of roof cutting and pressure release technology of retracement roadway in large height mining working face. *Geotech. Geol. Eng.* **2021**, *39*, 1289–1298. [CrossRef]

2. Gu, S.; Wang, B.; Huang, R.; Miao, Y. Method for determining the load on and width of coal pillar at the recovery room end of fully-mechanized longwall mining. *J. China Univ. Min. Technol.* **2015**, *44*, 990–995. [CrossRef]

3. Zhao, Z. *Study on Mechanism and Control Method of Deformation and Failure of Surrounding Rock in Large Deformation Mining Roadway*; China University of Mining and Technology: Beijing, China, 2014; pp. 26–32.

4. Guo, X.; Zhao, Z.; Gao, X.; Wu, X.; Ma, N. Analytical solutions for characteristic radii of circular roadway surrounding rock plastic zone and their application. *Int. J. Min. Sci. Technol.* **2019**, *29*, 263–272. [CrossRef]

5. Qian, M.; Shi, P.; Xu, J. *Mining Pressure and Strata Control*; China University of Mining and Technology Press: Xuzhou, China, 2010; pp. 98–101.

6. Han, Y.; Wang, Z.; Tang, Y. Principal stress rotation in longwall top-coal caving face adjacent to the gob. *J. China Coal Soc.* **2020**, *45* (Suppl. S1), 12–22. [CrossRef]

7. Wang, J.C.; Wang, Z.H.; Yang, S.L. Stress analysis of longwall top-coal caving face adjacent to the gob. *Int. J. Min. Reclam. Environ.* **2020**, *34*, 476–497. [CrossRef]

8. Tao, Z.; Song, Z.; He, M.; Meng, Z.; Pang, S. Principles of the roof cut short-arm beam mining method (110 method) and its mining-induced stress distribution. *Int. J. Min. Sci. Technol.* **2018**, *28*, 391–396. [CrossRef]

9. LI, F. *Research on the Technology of Upper and Lower Simultaneous Mining in Shallow Thin Bedrock Coal Seam*; China University of Mining and Technology: Beijing, China, 2013.

applied sciences

MDPI

Article

Research on Coal and Gas Outburst Risk Warning Based on Multiple Algorithm Fusion

Yanlei Guo [1,2], Haibin Liu [1,*], Xu Zhou [2], Jian Chen [2] and Liwen Guo [2]

[1] School of Management, China University of Mining and Technology-Beijing, Beijing 100083, China; bqt1900502012@student.cumtb.edu.cn

[2] School of Emergency Management and Safety Engineering, North China University of Science and Technology, Tangshan 063000, China; sxzhouxu@126.com (X.Z.); chenjian@ncst.edu.cn (J.C.); guoliwen64@163.com (L.G.)

* Correspondence: hbliu@cumtb.edu.cn

Featured Application: The XGBoost–GR–stacking gas outburst early warning model established in this article demonstrates high accuracy and practical performance, making it suitable for gas outburst risk warning in mining safety.

Abstract: To improve the accuracy of gas outburst early warning, this paper proposes a gas outburst risk warning model based on XGBoost–GR–stacking. The statistic is based on gas outburst data from 26 mines and establishes a data generation model based on XGBoost. The obtained virtual datasets are analyzed through visualization analysis and ROC curve analysis with respect to the original data. If the augmented data has an ROC area under the curve of 1, it indicates good predictive performance of the augmented data. Grey correlation analysis is used to calculate the grey correlation degrees between each indicator and the "gas emission". The indicator groups with correlation degrees greater than 0.670 are selected as the main control factor groups based on the sorting of correlation degrees. In this study, SVM, RF, XGBoost, and GBDT are selected as the original models for stacking. The original data and virtual data with correlation degrees greater than 0.670 are used as inputs for SVM, RF, XGBoost, GBDT, and stacking fusion models. The results show that the stacking fusion model has an MAE, MSE, and R2 of 0.031, 0.031, and 0.981. Comparing the actual and predicted values for each model, the stacking fusion model achieves the highest accuracy in gas outburst prediction and the best model fitting effect.

Keywords: directional splitting; damage region; coalbed methane mining; coalbed permeability enhancement

1. Introduction

Coal and gas outbursts are a frequent and dangerous occurrence in the coal mining industry. These accidents pose a significant threat to both the equipment used in coal mining and the safety of miners. As mining operations continue to deepen and intensify, the frequency of coal and gas outburst accidents has been on the rise, greatly impacting the safety of coal mine production. In fact, China has experienced a significant number of these accidents, accounting for over 40% of the global total. As of 2022, approximately 33.6% [1] of Chinese coal mines are classified as high-risk mines prone to coal and gas outbursts. To address these challenges and support the goals outlined in the national "14th Five-Year Plan", experts and scholars have been conducting research and analysis to understand the mechanisms and risks involved in coal and gas outbursts. Their aim is to predict and prevent such accidents in order to ensure the safe production of coal mines and contribute to the sustainable development of China's energy sector.

Coal and gas outburst accidents are the result of complex phenomena, driven by the dynamic interaction of coal and gas within the mine. This involves the uncontrolled evolution of various nonlinear factors, making it a highly destructive gas dynamic phenomenon [2]. In studying the mechanisms behind these outbursts, scholars have examined multiple factors such as ground stress, gas properties, and coal mechanical properties. They have constructed theoretical models based on traditional algorithms, although traditional algorithms often struggle to analyze the nonlinear relationship between these factors. With advancements in intelligent algorithms and machine learning, more and more experts and scholars are utilizing these technologies to analyze coal and gas outburst risks, yielding promising results. The challenge lies in uncovering the nonlinear relationships between various influencing factors, accurately predicting the risk and severity of coal seam gas outbursts, and implementing early warning systems to prevent or mitigate the disasters caused by these outbursts. This has become a critical task in ensuring the safe production of mines [3].

Over the past century, experts and scholars from both domestic and international institutions have conducted extensive research on the mechanism of coal and gas outbursts and have proposed numerous hypotheses. However, there is still no unified theory that can completely reveal the development mechanism of coal and gas outbursts. The theory proposed by Pingping Ye [4] analyzed the mechanism of pore pressure on coal and conducted deformation tests of coal rock under cyclic loading and unloading of pore pressure. Norbert [5] derived a relationship model between coal porosity and mechanical strength and used porosity and gas pressure to classify and predict the level of outburst hazards. Zhao [6] through theoretical analysis, pointed out that structurally weak coal has a lower bearing capacity, and the fine coal particles formed after fragmentation have an extremely fast gas desorption rate. This can sustain the development of outbursts, making structurally weak coal seams prone to coal and gas outbursts. Wold [7] suggested that coal and gas outbursts are influenced by factors such as gas pressure, gas composition, coal permeability, and adsorption desorption characteristics, and analyzed the relationship between these influencing factors and the control of outbursts. Dazhao Li [8] proposed a coal and gas outburst support model, and analyzed the mechanism of non synchronous deformation induced coal and gas outburst in soft and hard layer. Chaolin Zhang [9] systematically summarized the research progress on coal and gas outburst mechanisms in China from three aspects: theoretical hypothesis, physical simulation, and numerical simulation. Hu Qianting [10] described in detail the entire process of occurrence and development of outbursts based on numerical simulations and theoretical analysis. The process was ultimately divided into four stages: preparation, initiation, development, and termination. Lijun Zhao [11] summarized the research progress on the mechanism of coal and gas outburst, and analyzed the shortcomings of existing theoretical models. Liangcheng Wang [12] and Shoujian Peng [13], based on theoretical exploration combined with a large number of numerical simulation analyses, discussed the evolution process of coal and gas outbursts. Guo Pinkun [14], combining experimental research, established a model for the development of layer fractures during the outburst process and explored the mechanism of layer fracture development. Xu Mangui [15] and others constructed a microelement model of coal-rock mass and believed that the destruction of coal and gas microelements is the primary cause of outbursts.

With the increasing popularity of data analysis and data science theory, numerous experts and scholars have extensively analyzed various factors related to coal and gas outbursts using traditional algorithm and mathematical analysis methods. Dan Dakuo [16] combined mathematical and statistical analysis methods to determine the prediction indicators and critical values for coal and gas outbursts, with gas as the dominant factor. This achieved the prediction of the risk level of coal and gas outbursts. Si Hu [17] extracted 27 factors that influence the occurrence of coal and gas outburst accidents. By using association analysis and cross-coupling analysis, they conducted statistical analysis and in-depth exploration of coal and gas outburst accidents of average and above average severity that had occurred in the last 15 years. Wen Changping [18] constructed attribute measurement functions to calculate single-index attribute measurements and comprehensive sample attribute measurements. By applying confidence criteria, they conducted attribute recognition of gas outbursts in tunnel samples and established an attribute recognition model for gas outburst evaluation in the tunnel survey and design stage. Wang Gang [19] analyzed the energy relationship in the process of coal and gas outbursts using the energy method. They obtained the relationship between the energy conditions of coal and gas outbursts, coal seam geostress, cohesion coefficient, coal seam thickness, and the risk of gas outburst accidents. Cao Shugang [20] conducted experiments on the adsorption desorption deformation process of outburst-prone coal under different gas pressure conditions. They found a good power-function relationship and quadratic function relationship between the desorption shrinkage deformation of coal samples and the original gas pressure. Dingding Y [21] studied the influence of temperature on the "energy-mass" characteristics of gas and discovered the function relationship between the initial gas expansion energy released and temperature under different conditions, improving the prediction indicators for outburst hazards. Li Yunbo [22] studied the initial gas desorption velocity and amount of gas-prone coal and structurally weak coal using a self-made gas desorption experimental apparatus. They analyzed and established mathematical models for the influencing factors during the initial gas desorption period of structurally weak coal. They concluded that the initial desorption velocity of gas exhibits a power-law relationship with adsorption equilibrium pressure, and that the initial desorption curve of structurally weak coal conforms to the Vent formula.

Based on mathematical theory and machine learning, predictive methods have shown a high degree of adaptability to the complex problem of coal and gas outbursts, which involve non-linear relationships among various factors. An increasing number of experts and scholars are adopting intelligent algorithms to predict coal and gas outbursts and they have achieved a certain amount of success. Xiang Zeng Du [23] used a grey comprehensive correlation analysis model to quantitatively analyze six predictive indicators of coal and gas outbursts and determine the optimal prediction indicators. This provides a quantitative basis for the selection of prediction indicators for coal and gas outbursts. Zhou Xihua [24] predicted the intensity of coal and gas outbursts using an RBF neural network model and principal component analysis, ultimately achieving high prediction accuracy. Liu Xiaoyue [25], Cao Bo [26], Ren Shaowei [27], and others used the BP neural network to predict coal and gas outbursts and optimized the dimensions of the influencing factors through principal component analysis. By reducing the correlation among variables and selecting the main control factors, they improved the prediction efficiency of the entire model. The optimized models also achieved high prediction accuracy. Zhao Huatian [28], Zeng Weishun [29], Zhang Wenjuan [30], and others made full use of support vector machines (SVMs) to address the advantages of solving small sample problems and combined them with other optimization algorithms to predict coal and gas outbursts. Zhao Huatian and Zeng Weishun used particle swarm optimization to optimize SVM, searching for global optimal solutions from a global perspective and greatly reducing the probability of local optimal solutions. Zhang Wenjuan utilized the least squares method to optimize SVM, effectively removing noise from gas data and improving prediction accuracy. Wu Yaqin [31] and others combined genetic algorithms with simulated annealing algorithms to propose a genetic simulated annealing algorithm. They introduced adaptive learning rates into the BP neural network and further optimized the BP network using the GASA algorithm. They ultimately established an improved GASA—BP neural network model for outburst prediction. The accuracy of the predicted results of this model was verified through practical application in coal mines.Xuning Liu [32] proposed a hybrid prediction model that combines feature extraction and pattern classification for coal and gas outbursts. Experimental results on a coal and gas outburst dataset showed that, compared to other models from the current coal and gas outburst prediction models, this method significantly influenced various indicators. In the field of machine learning, the improvement in model accuracy through dataset optimization often surpasses the improvement achieved through algorithm enhancements [33]. However, in practical production activities, the occurrence of coal and gas outburst accidents may result in limited and missing accident data due to the damage of monitoring devices. This leads to reduced model accuracy, overfitting issues, and other problems.

In view of this, this paper analyzes the correlation between different indicators and the risk level of coal and gas outbursts using the grey relational algorithm, aiming to select feature indicators and perform attribute reduction based on the mechanism of coal and gas outbursts. A data generation algorithm based on machine learning and data reconstruction is constructed to generate virtual data from the original data. XGBoost, SVM, and GBDT are selected as primary learners, and random forest is used as a secondary learner to construct a predictive model for the risk level of coal and gas outbursts based on stacking ensemble learning. This model predicts the magnitude of the risk level of coal and gas outbursts, and the results are compared. This work aims to reduce personnel casualties and related economic losses caused by coal mine accidents from coal and gas outbursts, and promote the construction of a smart mine safety system.

This paper introduces the basic theoretical part of the gas outburst risk warning model, including the process and principles of grey relational analysis and the XGBoost, SVM, GBDT, and random forest algorithms. Then, a gas outburst risk warning model is constructed based on stacking ensemble learning. The construction process for this model consists of three parts. The first step is to build a data generation model based on XGBoost-Regressor to expand the capacity of the original dataset. The second step is the selection of main control factors based on grey relational analysis. The factors with the highest grey correlation coefficients, indicating the greatest relevance to coal and gas outburst risks, are selected from a set of 20 indicators. The third step is to construct a warning model based on the stacking fusion algorithm framework. Through the implementation of XGBoost, SVM, GBDT, and random forest as the four learners, an effective warning for coal and gas outburst risks is achieved. Furthermore, sufficient experimental analysis is conducted, and the model's performance is analyzed to select the factors that most influence gas outburst risks. Finally, the experimental results are summarized and analyzed.

2. Basic Theory

2.1. Grey Relationship Analysis

Grey system theory can utilize limited small sample data to solve uncertainty problems. Grey relational analysis, a prominent technique in the theory, is used to measure the degree of correlation between influencing factors and the research issue. It finds wide application in various domains for system diagnosis and analysis. The main principle is to assess the correlation between factors based on the similarity or dissimilarity of their development trends. By conducting grey relational analysis, one can analyze the impact of sub-factors on the main factor, aiming to optimize the dimensions of the system.

Grey relational analysis typically begins with selecting a reference sequence. Due to the varying dimensions among different influencing factors, direct comparison is not feasible. Therefore, commonly used approaches such as mean normalization, initial value normalization, standardization, or extreme value normalization are employed to eliminate the dimensional differences. These techniques help convert the indicators into dimensionless values, facilitating further analysis and comparison.

For grey relational analysis on processed data, the grey relational coefficient is calculated using Formula (1).

$$y(x_0(k), x_i(k)) = \frac{a + \rho b}{|x_0(k), x_i(k)| + \rho b} \tag{1}$$

The a is the minimum value, b is the maximum value of the two levels, and ρ is the discrimination coefficient (generally 0.5) [34]. $y(x_0(k), x_i(k))$ is the grey correlation coefficient, representing the correlation degree value between the sequence and the reference sequence.

The grey correlation degree of each subsequence is calculated by Formula (2), and the grey correlation degree is set as r_i:

$$r_i = \frac{1}{n} y(x_0(k), x_i(k)) \quad (2)$$

According to the obtained grey correlation degree r_i, it is sorted according to the size of r_i. If $r_1 > r_2$, it means that r_1 and the mother sequence are more related to r_0 and more correlated. According to the sorting results, the factor with the largest grey correlation degree value is selected as the main control factor. As shown in Figure 1.

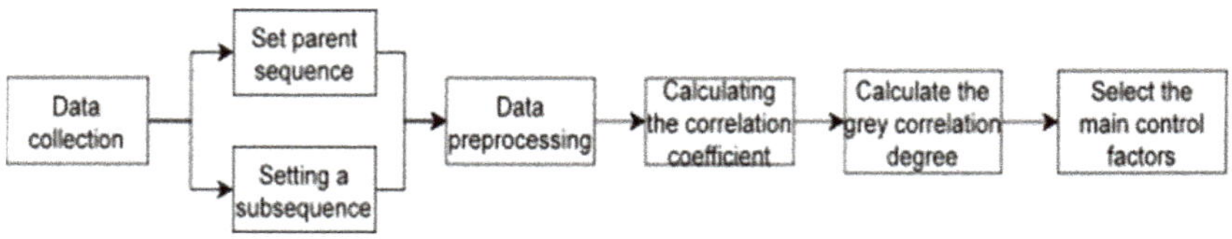

Figure 1. Grey correlation analysis flowchart.

2.2. Data Generation Algorithm

In machine learning, gas outburst is a prominent issue related to small sample problems. One important approach to improve model accuracy, prediction, and risk identification is through data augmentation and reconstruction. This paper constructs a gas outburst data generation model based on collected data to generate synthetic samples. This method can generate data samples that are consistent with the real data distribution, enhancing data and improving the effectiveness and quality of the model. The specific process of the data generation strategy consists of three steps: (1) training feature models, (2) sampling features to generate virtual data, and (3) generating the final synthesized data.

In the data generation algorithm, first, in the known data samples with certain features, a feature in the data sample is designated as the label, while the remaining features are treated as elements of the new feature vector to form new training samples. Based on the new training samples, data are randomly sampled and recombined by analyzing the correlations between various features, thus obtaining feature values for n features. Finally, by selecting individual features, using Feature 1 as input to the model for training and obtaining output F1, and using Feature 2 as input to the model for training and obtaining output F2, after completing the sampling process for all indicators, a temporary data sample can be obtained, thereby obtaining a complete dataset. Its data distribution characteristics are similar to the original data and have good representativeness. As shown in Figure 2.

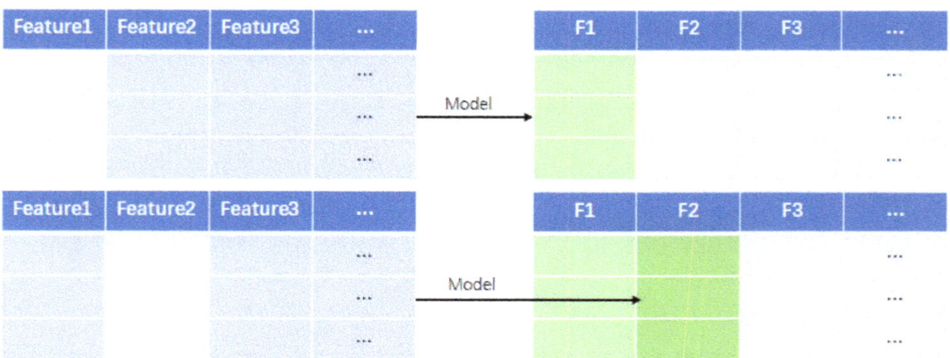

Figure 2. Virtual data generation process.

2.3. XGBoost

XGBoost (eXtreme Gradient Boosting) is a powerful supervised multi-parameter model that operates within the gradient boosting framework. It is an implementation of the boosting algorithm, designed to handle both classification and regression problems. The fundamental concept behind XGBoost involves combining multiple weak learners to form a robust learner using specific techniques. This method utilizes multiple classification and regression trees in a collaborative manner, resulting in improved model performance. The following steps outline the working process of XGBoost:

Step 1: Calculate the predicted results of the model on the samples after t iterations, and then define a function that incorporates the model's loss function and a regularization term to suppress the complexity of the model.

$$\hat{y}_i^{(t)} = \sum_{k=1}^{t} f_k(x_i) = \hat{y}_i^{(t-1)} + f_t(x_i) \tag{3}$$

Here, x_i represents the i-th feature in the feature vector, $\hat{y}_i^{(t)}$ represents the predicted value of sample i after t iterations, k is the number of base models, $f_k(x_i)$ represents the kth base model, and $\hat{y}_i^{(t-1)}$ represents the predicted value of sample after $t-1$ iterations. $f_k(x_i)$ is the model of the t th tree.

Step 2: Calculate the loss function of the model and the objective function consisting of the regularization term that inhibits the complexity of the model:

$$Obj^{(t)} = \sum_{i=1}^{n} l\left(y_i, \hat{y}_i^{(t-1)} + f_t(x_i)\right) + \Omega(f_t) + \sum_{t=1}^{T-1} \Omega(f_t) \tag{4}$$

Here, y_i represents true value, $\sum_{i=1}^{n} l(y_i, \hat{y}_i)$ represents the loss function L, and Ω represents a regular term that suppresses the complexity of the model.

Step 3: Simplify the calculation to obtain the final objective function and solve the model. The objective function is close to the Taylor expression, so the objective function can eventually be simplified as follows.

$$Obj^{(t)} \simeq \sum_{i=1}^{n} \left[g_i f_t(x_i) + \frac{1}{2} h_i f_t^2(x_i) \right] + \Omega(f_t) \tag{5}$$

Here, g_i, h_i are the first derivative and the second derivative of the loss function L, $g_i = \partial_{\hat{y}^{(t-1)}} l\left(y_i, \hat{y}^{(t-1)}\right)$, $h_i = \partial^2_{\hat{y}^{(t-1)}} l\left(y_i, \hat{y}^{(t-1)}\right)$. Therefore, as long as the loss function is determined, then g_i and h_i are determined, and the objective function is determined.

2.4. SVM

Support vector machine (SVM) is a powerful generalized linear classifier primarily employed for binary classification in supervised learning. The underlying principle involves mapping the original data into a high-dimensional feature space through a series of transformations, enabling efficient classification within this transformed space. SVM exhibits strong generalization and self-learning capabilities, ensuring effective performance even with limited statistical sample datasets. The following steps outline the working process of SVM:

Step 1: Given the training set $T = \{(x1, y1), (x2, y2), \ldots, (xn, yn)\}$.

Step 2: Solve the quadratic programming problem, which is solved by $\alpha* = (\alpha_1^*, \ldots, \alpha_n^*)^T$.

$$\min \frac{1}{2} \sum_{i=1}^{n} \sum_{j=1}^{n} a_i a_j y_i y_j (x_i . x_j) - \sum_{i=1}^{n} a_i \tag{6}$$

$$s.t. \sum_{i} a_i y_i = 0, a_i \geq 0 \tag{7}$$

Here, x_i represents the i-th feature in the feature vector, and y_i represents true value.

Step 3: Calculate the parameter w and take a positive component α_i^*, and calculate the b.

$$w^* = \sum \alpha_i^* y_i x_i \tag{8}$$

$$b^* = y_j - \sum a_i^* y_i (x_i \cdot x_j) \tag{9}$$

Step 4: Structural decision boundary: $g(x) = (w^* \cdot x) + b^* = 0$, the decision function is thus obtained:

$$f(x) = \text{sgn}(g(x)) \tag{10}$$

2.5. GBDT

Gradient boosted decision trees (GBDT) is an iterative algorithm that combines the concepts of boosting and gradient descent. It leverages the forward distribution algorithm to train multiple weak learners, with each weak learner being constructed using a CART regression tree. By combining these weak learners through an additive model, it forms a powerful strong learner. The training process of each weak classifier is guided by the negative gradient of the loss from the previous weak classifier. This iterative optimization approach gradually reduces the loss, leading to the convergence towards the optimal solution. The principles of GBDT can be summarized as follows:

The model constants are initially given:

$$F_0(\vec{x}) = \underset{\gamma}{\text{argmin}} \sum_{i=1}^{n} L(y_i, \gamma) \tag{11}$$

Here, y_i represents true value, and γ represents the prior probability of a class. $\sum\limits_{i=1}^{n} L(y_i, \gamma)$ represents the loss function.

For m = 1 to M, compute the pseudo-residual:

$$r_{im} = -\left[\frac{\partial L(y_i, F_{m-1}(\vec{x_i}))}{\partial F_{m-1}(\vec{x_i})}\right] (i = 1, 2, \ldots, n) \tag{12}$$

Using data to calculate the basis function for fitting residuals, in gradient enhancement, a decision tree is calculated based on pseudo residuals. The input space is divided into disjoint regions, and the decision tree can provide a certain type of deterministic prediction in each region.

Update the current model to:

$$F_m(x) = F_{m-1}(x) + \gamma t_m(x) \tag{13}$$

$$F_m(x) = F_{m-1}(x) + \sum_{j=1}^{J} \gamma_{jm} I(x \in R_{jm}) \tag{14}$$

$$\gamma_{jm} = \underset{w}{\text{argmin}} \sum_{\vec{x_i} \in R_{jm}}^{n} L(y_i, F_{m-1}(x_i) + w) \tag{15}$$

Here, x_i represents the i-th feature in the feature vector, and $t_m(x)$ is the basis function that fits the residual. $I(x)$ represents indication mark.

The final model as:

$$F_M(x) = F_0(x) + \sum_{m=1}^{M} \sum_{j=1}^{J} \gamma_{jm} I(x \in R_{jm}) \tag{16}$$

2.6. Random Forest

Random forest is an ensemble learning algorithm in the field of machine learning. It consists of multiple decision trees as classifiers. Each decision tree independently outputs a class, and the final prediction is determined by taking the majority class among these decision tree outputs. Random forest integrates the predictions of multiple trees using the idea of ensemble learning. By combining the effects of random forest, we can obtain more robust and accurate classification results. Random forest is widely used in practical applications, especially for handling large-scale datasets and high-dimensional features. It can not only effectively handle classification problems but also be used for regression and feature selection tasks.

The root of the decision tree algorithm is information. The basic concepts of entropy and information gain can be understood through these three concepts to determine the order of feature selection in the decision tree.

$$H(x) = \sum_{i=1}^{n} p(x_i)I(x_i) = -\sum_{i=1}^{n} p(x_i)\log_b p(x_i) \tag{17}$$

Here, $H(x)$ indicates entropy, which depends on x. The distribution, but with x It doesn't matter. In the decision tree. In the middle, the greater the entropy. The greater the category uncertainty, the smaller the reverse. Information gain is used to select feature indicators in the decision tree. The greater the information gain, the better the selectivity of features. A feature corresponds to multiple categories, and conditional entropy needs to be introduced in the calculation. The formula is as follows:

$$H(Y|X) = \sum p(x)H(Y|X = x) \tag{18}$$

In fact, the information gain also expresses the difference between the entropy of the set to be classified and the conditional entropy of selecting a feature. Therefore, the information gain formula is introduced:

$$IG(Y|X) = H(Y) - H(Y|X) \tag{19}$$

The prediction effect of ensemble learning is judged by the error rate. According to the Hoeffding inequality, the error rate of ensemble learning is:(where T is the number of decision trees):

$$P(H(X) \neq f(x)) = \sum_{\mathrm{k}=0}^{T/2} \binom{T}{k} (1-\varepsilon)^k \varepsilon^{T-k} \leq \exp(-\frac{1}{2}T(1-2\varepsilon)^2) \tag{20}$$

The algorithm steps are as follows:

Step 1: Using the Bagging method, randomly select N samples from N sample sets, and use these N samples to train a decision tree as samples for the root node of the decision tree.

Step 2: From the sample M Random selection among the features m Characteristics, satisfied $m << M$, Select a feature from m features using a certain strategy (information gain, Gini index, etc.) as the splitting feature of the node.

Step 3: Repeat step 2 to split the node until it cannot be split, forming a decision tree.

Step 4: Follow steps 1~2. Establish a large number of decision trees to form random forests.

3. An Account of Stacking-Framed Coal and Gas Outburst Risk Warning Model

All data undergo pre-training data cleaning, data filtering, and data sorting. The influence of dimensionality differences among different indicators is removed through standardization processing. Missing values are filled in using data cleansing techniques, and certain data points are identified or removed. Once data processing is complete, model training can commence.

Step 1: Training feature model. In the known presence ofIn the data sample of features, each dataset has its own characteristics. One feature of the given data sample is used as a label, and the rest of the feature are used as elements of the new feature vector to form a new training sample.

Step 2: Sampling features to generate virtual data. From the collection, replace a value of the sample: the The value is the characteristic of the temporarily synthesized data sample. 1. The first eigenvalue, and then the remaining features are taken in turn, and finally you can getthe eigenvalues of the feature.

Step 3: The features to be generated in the temporary data sample 1. As an output, characteristics 2 and other features are used as input respectively... Put in the feature matrix and put The output value of the model is characteristic.1The virtual generation value of In the same sense, all virtual values are obtained.

Step 4: Based on the original data and the newly generated data, calculate the grey correlation coefficient between the degree of gas outburst and various factors.

$$y(x_0(k), x_i(k)) = \frac{a + \rho b}{|x_0(k), x_i(k)| + \rho b} \tag{21}$$

Step 5: Calculate the grey correlation between each subsequence, and sort each indicator according to the grey correlation.

$$r_i = \frac{1}{n} y(x_0(k), x_i(k)) \tag{22}$$

Step 6: Divide indicator groups based on the size of grey correlation degree, and use XGBoost, RF, GBDT, and SVM training data to comprehensively compare MAE, MSE, and R2 in each indicator. Select the data group with the smallest error and the best fit degree as the main control factor.

Step 7: Based on the main control factors of coal and gas outburst, SVM, XGBoost, GBDT, and RF training models are selected to predict the situation of gas outburst.

Step 8: Using the above predictions on the training set as the training set and the predictions on the test set as the test set, retrain the data to obtain new prediction results. Stacking combines steps 7 and 8 together, using the output of step 7 as the input to step 8 to obtain the final output result.

Step 9: Compare the training effects of SVM, XGBoost, RF, GBDT, and Stacking fusion models under the main control factors, and select the Stacking fusion model with the best prediction effect to construct a gas outburst risk warning model.

The occurrence of coal and gas outbursts is comparatively intricate. It is influenced by multiple factors and exhibits a certain degree of nonlinearity. This article employs grey relational analysis to explore the nonlinear relationship between coal and gas outburst issues. It aims to unveil the correlation between influencing factors and the occurrence of outbursts. A set of key controlling factors that impact gas outburst situations is selected. By doing so, the prediction work is streamlined, resulting in improved model forecasting performance.

Due to variations in gas outburst factors among different mines, there is a certain relationship between these factors and the mining environment and coal occurrence conditions. Therefore, when constructing a gas outburst risk warning model, the model's generalization ability should be considered. Integrated models have better generalization ability compared to single machine learning models, which can improve the accuracy of the model. Classic methods of integration include bagging, boosting, and stacking. The main idea of stacking ensemble learning is to combine multiple models and fully utilize their respective advantages to make final predictions and achieve the best results. As shown in Figure 3.

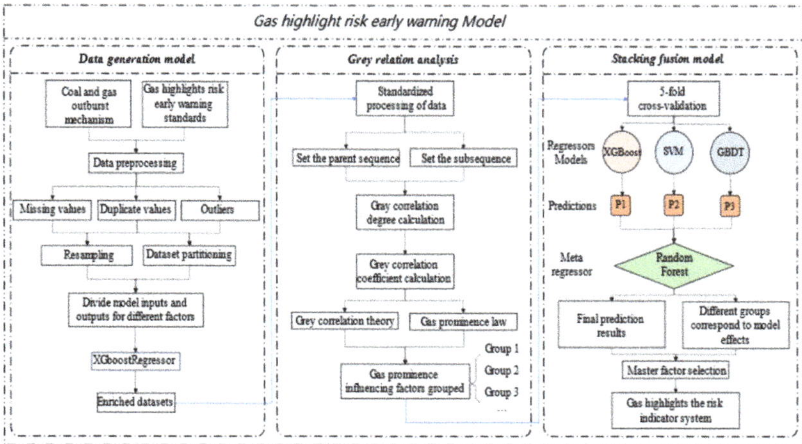

Figure 3. XGBoost–GR–stacking flowchart of gas outburst warning model.

4. Experimental Simulation

4.1. Data Source

The influential factors of gas outburst can include the static data of the mine and the dynamic data of real-time monitoring. The static parameters of the outburst mine partially reveal the potential patterns of gas outbursts. To study the laws of gas outbursts, this paper collected relevant data on gas outbursts from 26 mines in the Southwest China, North China, and Central China regions, as well as numerical simulation indicators. The related data indicators included mining depth, coal solidification coefficient, ash content, volatile matter, initial gas emission velocity, gas adsorption constants a and b, Δh_2, absolute gas outburst (m^3/min), relative gas outburst (m^3/t), K_1, Smax (kg/m), q (L/min), and ejection amount. At the same time, this paper used numerical simulation experiments to simulate the dynamic slope of gas pressure, geostress, goaf thickness, fault height, and fitting curve of pre-gas outburst data with similar ejection amounts. Part of the raw data is shown in Table 1.

Table 1. Part of the original data.

Mining Depth	Coefficient of Coal Firmness	Ash Content	Volatile Matter	Seam Thickness	Initial Velocity of Gas Release	Gas Adsorption Constant a	Gas Adsorption Constant b	Δh_2	Absolute Gas Emission	Relative Gas Emission
800	0.34	0.18	0.12	1.25	14.84	38.59	0.787	170	32.01	13.94
452.7	0.31	0.15	0.1	5.69	12.16	37.32	0.723	140	6.58	4.21
850	0.39	0.23	0.12	4	18.6	18.13	2.3445	180	34.8	15.09
600	0.49	0.12	0.08	1.7	28	30.303	1.3346	175	32.04	13.95
500	0.24	0.18	0.12	1.25	31	33.4832	1.6166	170	32.01	13.94
515	0.17	0.17	0.11	4.6	38	31.08	1.13	172	16.01	10.88
500	0.23	0.17	0.11	3.2	26.3	26.3459	1.2572	160	9.78	10.45
…	…	…	…	…	…	…	…	…	…	…

K_1	Smax	q	Initial gas pres-sure	Numerical simulation of gas pressure	Numerical simulation of ground stress	Numerical simulation of stone gate thickness	Numerical simulation of fault height	Dynamic slope
0.36	3.5	57	0.75	28	4	5	15	0.75
0.32	3.4	26.05	0.75	35	4	1	2	0.75
0.34	3.7	34	2.4	16	0.2	1	5	2.4
0.41	2.1	38	0.75	10	3	5	7.5	0.75
0.413	2.2	42	2.4	35	4	1	1.1	2.4
0.36	2.2	83	0.75	10	0.2	5	20	0.75
0.47	2	36	0.75	22	1	1	1.5	0.75
…	…	…	…	…	…	…	…	…

4.2. Data Generation

Part of the virtual data is shown in Table 2. The data samples contained in this table were part of the virtual samples generated based on the above methods, and their data distribution characteristics were similar to the original data, which had a good representation. This study tested the effect of the model based on the following data.

Table 2. Partial virtual data tables.

Mining Depth	Coefficient of Coal Firmness	Ash Content	Volatile Matter	Seam Thickness	Initial Velocity of Gas Release	Gas Adsorption Constant a	Gas Adsorption Constant b	Δh_2	Absolute Gas Emission	Relative Gas Emission
681.0	0.474	0.124	0.299	1.277	5.118	27.82	0.949	152.6	30.85	681.0
400.0	1.499	0.310	0.210	1.320	16.39	38.58	0.790	150.0	2.490	400.0
400.0	1.500	0.310	0.210	1.320	16.39	38.59	0.790	150.0	2.490	400.0
399.9	1.499	0.310	0.210	1.320	16.39	38.59	0.790	150.0	2.490	399.9
400.0	1.499	0.310	0.209	1.321	16.39	38.58	0.790	150.0	2.490	400.0
450.0	0.490	0.131	0.100	1.550	18.35	33.09	1.120	179.9	27.58	450.0
450.0	0.490	0.130	0.100	1.550	18.35	33.09	1.120	180.0	27.59	450.0
...	...	...	...	...	...	...	...	...	...	...

K_1	Smax	q	Initial gas pressure	Numerical simulation of gas pressure	Numerical simulation of ground stress	Numerical simulation of stone gate thickness	Numerical simulation of fault height	Dynamic slope
0.403	2.949	42.681	0.962	1.766	21.832	2.222	2.868	3.967
0.389	2.800	43.000	1.500	1.300	16.000	0.200	1.001	5.000
0.390	2.800	43.000	1.500	1.300	16.000	0.200	1.000	5.000
0.389	2.800	43.000	1.499	1.300	16.000	0.200	1.000	5.000
0.390	2.800	43.000	1.500	1.300	16.000	0.200	1.001	5.000
0.340	2.900	38.000	0.650	2.399	16.000	0.200	1.000	4.500
0.340	2.900	38.000	0.650	2.400	16.000	0.200	1.000	4.500
...	...	...	...	...	...	...	...	...

The diagram directly reflects the distribution of the original data and extended data. It also standardizes the original data and extended data of the 20 index groups to remove the dimensional influence. The experimental results are shown in Figure 4.

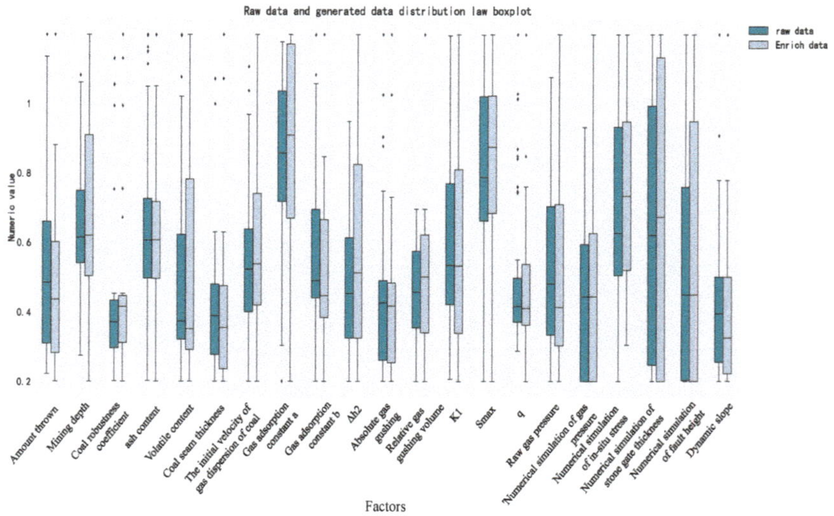

Figure 4. Box diagram of distribution of original data sample and expanded data.

By comparing the distribution of the original data and the extended data, it was found that the distribution interval of the 20 indicators of the extended data was basically consistent with the distribution interval of the original data, indicating that the extended data was consistent with the distribution of gas outburst indicators in the mine, had strong credibility, and could be used for the construction of the gas outburst warning model.

In order to verify whether the generated virtual data could effectively improve the training effect of the model in this paper, the original data samples and the expanded data samples were respectively placed in the stacking fusion model for training. Then, we analyzed the ROC curve of the stacking fusion model to prove the feasibility of the above methods. The experimental results are shown in Figure 5.

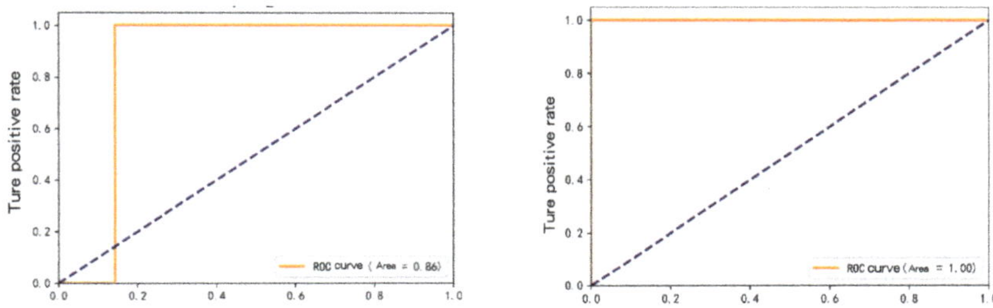

Figure 5. Comparison of the effects of the stacking model between original data samples and expanded data samples.

For comparing the ROC curve of the original data and the extended data, it is generally believed that the closer the area under the ROC curve is to 1, the better the classification prediction of this model is. When the value is greater than 0.5, the model is better than random guessing. If the model sets a reasonable threshold, the classifier can have predictive value. For the expanded data, the ROC curve area value of the stacking model was equal to 1, so the prediction results were perfect. This model belongs to a good predictor. For the raw data, where the ROC curve area of the stacking fusion model was equal to 0.86, the prediction effect was acceptable. The ROC curve shows that the expanded data model was better than that for the original data.

In addition, this paper trained XGBoost, GBDT, SVM, RF, and stacking prediction models based on the original and extended data. After visual comparison of the difference between the real value and the predicted value (Figures 6 and 7), it was found that the error between the real value and the predicted value of the expanded data was significantly smaller than that between the real value and the predicted value of the original data. The error of SVM-predicted value and true value was the largest for raw data and extended data, but the prediction error was larger for the raw data. The errors between the predicted values and the true values of the stacking model were the minimum for the original data and the expanded data, and the prediction effect was better for the expanded data. The actual values and errors of the five models showed that the extended data model was better than that for the original data.

All in all, the ROC curve comparison between the original data and the extended data and the errors between the real and predicted values of the XGBoost, GBDT, RF, SVM, and stacking models trained with the original data and the extended data, respectively, indicated that the model effect after data expansion was significantly improved compared with the model effect with the original data. This paper used extended data to train the model.

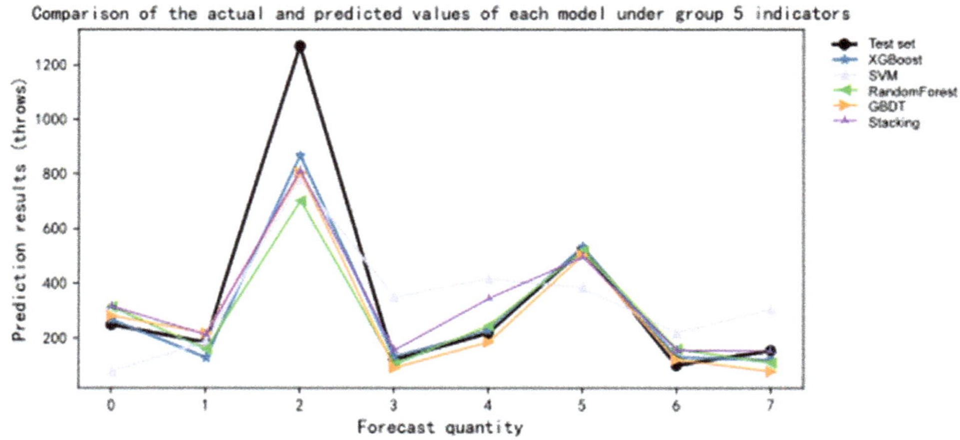

Figure 6. Comparison between the predicted value and the true value of each model for the original data.

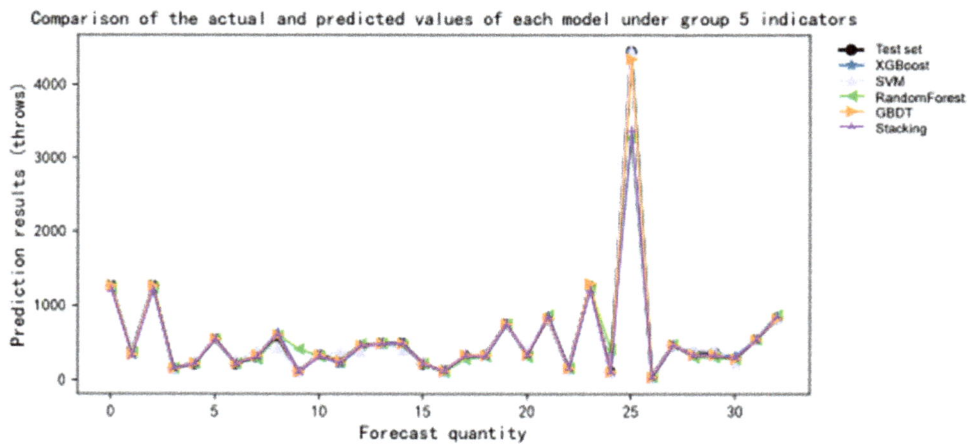

Figure 7. Comparison between the predicted value and the true value of each model for the extended data.

4.3. Analysis of Correlation Degree of Risk Factors of Gas Outburst

This paper explored the concept of throw-out quantity as the main factor, with other indicators as subsidiary factors. The dimensional unit of each data group was eliminated through standardization processing. By calculating the grey correlation coefficient using Formula (1), combined with Formula (2), the magnitude of the grey correlation degree between the grey subsequence and the main sequence was obtained. The results were then sorted based on the grey correlation degree, as shown in Table 3:

Table 3. Grey correlation degree ranking of subsequence and parent sequence.

Name	Grey Correlation Degree Value	Rank
Dynamic slope	0.830	1
Initial velocity of coal gas release	0.745	2
Initial velocity of coal gas release	0.742	3
Δh_2	0.728	4

Table 3. *Cont.*

Name	Grey Correlation Degree Value	Rank
q	0.727	5
Relative abundance of methane	0.709	6
Numerical simulation of fault height	0.699	7
Absolute gas emission rate	0.691	8
Coal firmness coefficient	0.678	9
Ash content	0.670	10
Thickness of coal seam	0.668	11
Exploitation depth	0.667	12
Numerical simulation of gas pressure	0.666	13
K_1	0.649	14
Initial gas pressure	0.646	15
Volatiles	0.638	16
Numerical simulation of ground stress	0.635	17
Gas adsorption constant a	0.593	18
Numerical simulation of stone gate thickness	0.578	19
Smax	0.572	20

The results in the table indicate a clear distinction in the grey correlation of all 20 indicators, ranging from 0.572 to 0.830. Among the 20 indicators, the grey correlation value between the dynamic slope and gas emission is the highest at 0.830. This suggests that there is a strong and close relationship between the dynamic slope and gas emission. Following that, the grey correlation values decrease significantly for the initial gas emission velocity of coal and the gas adsorption constant b, reaching 0.745 and 0.742, respectively, indicating a weakening correlation. Among the twenty indicators, the grey correlation between Smax and gas emission is the lowest, at 0.572, implying the weakest and least significant relationship between Smax and grey correlation. The value of 0.578 for the simulated thickness of the gate in numerical simulations indicates a loose relationship between the simulated gate thickness and gas emission.

According to the results of grey correlation degree calculated in Table 2, we set five conditions respectively for prediction, and then selected the main influencing factors according to the error of the predicted results. We set up five groups of experiments to study the correlation and influencing factors. The correlation degrees were, respectively, $r_i > 0.700$, $r_i > 0.670$, $r_i > 0.650$, and $r_i > 0.630$. The results are shown in Table 4:

Table 4. Classification of influencing factors based on correlation degree.

Group	Grey Relational Degree	Influence Factors
1	$r_i > 0.700$	Dynamic slope, initial velocity of coal gas release, gas adsorption constant, q, relative abundance of methane.
2	$r_i > 0.670$	Dynamic slope, the initial speed of gas dispersion of coal, gas adsorption constant, q, relative gas outflow, numerical simulation fault height, absolute gas outflow, coal toughness coefficient, ash content.
3	$r_i > 0.650$	Dynamic slope, initial velocity of coal gas release, gas adsorption constant, q, relative gas emission, numerical simulated fault height, absolute gas emission, coal firmness coefficient, ash content, coal seam thickness, mining depth, numerical simulated gas pressure.
4	$r_i > 0.630$	Dynamic slope, initial velocity of coal gas release, gas adsorption constant, q, relative gas emission amount, numerical simulated fault height, absolute gas emission amount, coal firmness coefficient, ash content, seam thickness, mining depth, numerical simulated gas pressure, K1, original gas pressure, volatile content, numerical simulated ground stress.
5	$r_i > 0.570$	All factors.

The impact of various factors on the level of gas outburst hazard differed based on the grouping results shown in the above table. Furthermore, there existed either strong or weak relationships among these factors. By conducting error analysis and analyzing the fitting degree of the models for each indicator group, the key controlling factor indicator group was determined. The comparative effects of each model are shown in Table 5.

Table 5. Comparison of model results.

Model	XGBoost			SVM			RF		
Index	MAE	MSE	R^2	MAE	MSE	R^2	MAE	MSE	R^2
Group 1	0.091	0.091	0.933	0.152	0.152	0.888	0.061	0.061	0.955
Group 2	0.031	0.031	0.981	0.094	0.094	0.944	0.031	0.031	0.978
Group 3	0.091	0.091	0.933	0.242	0.303	0.776	0.091	0.091	0.933
Group 4	0.152	0.152	0.900	0.424	1.152	0.244	0.091	0.091	0.940
Group 5	0.121	0.121	0.920	0.333	0.333	0.781	0.091	0.091	0.940

Model	GBDT			Stacking		
Index	MAE	MSE	R^2	MAE	MSE	R^2
Group 1	0.061	0.061	0.955	0.061	0.061	0.955
Group 2	0.031	0.031	0.981	0.031	0.031	0.981
Group 3	0.091	0.091	0.933	0.091	0.091	0.932
Group 4	0.061	0.061	0.960	0.121	0.121	0.920
Group 5	0.091	0.091	0.940	0.152	0.152	0.900

For the five models (XGBoost, GBDT, RF, SVM, and stacking), the table above presents metrics such as MAE, MSE, and R^2, which reflect the performance of the models. MAE and MSE represent the errors of the models, and a lower value indicates higher accuracy in the predictions. The R^2 coefficient represents the goodness of fit of the models. Within a certain range, a higher value indicates a better fit and a more effective prediction. Based on the data in the table:

For the first set of indicators, the XGBoost predictive model yields an MAE, MSE, and R^2 of 0.091, 0.091, and 0.933, respectively. The SVM predictive model scores an MAE, MSE, and R^2 of 0.152, 0.152, and 0.888, respectively. The RF predictive model, on the other hand, performs with an MAE, MSE, and R^2 of 0.061, 0.061, and 0.955. As for the GBDT predictive model, it achieves an MAE, MSE, and R^2 of 0.061, 0.061, and 0.955. Lastly, the stacking predictive model exhibits an MAE, MSE, and R^2 of 0.061, 0.061, and 0.955.

For the second index group, the MAE, MSE and R^2 of the XGBoost prediction model are 0.031, 0.031 and 0.981, respectively, while the MAE, MSE and R^2 of the SVM prediction model are 0.094, 0.094 and 0.944. The MAE, MSE and R^2 of the RF prediction model are 0.031, 0.031 and 0.978, respectively, and the MAE, MSE and R^2 of the GBDT prediction model are 0.031, 0.031 and 0.981, respectively. The MAE, MSE, and R^2 of the stacking prediction model are 0.031, 0.031, and 0.981, respectively.

For the third group of indicators, the XGBoost predictive model exhibits an MAE, MSE, and R^2 of 0.091, 0.091, and 0.933, respectively. The SVM predictive model shows an MAE, MSE, and R^2 of 0.242, 0.303, and 0.776, respectively. For the RF predictive model, the MAE, MSE, and R^2 are 0.091, 0.091, and 0.933, respectively. The GBDT predictive model demonstrates an MAE, MSE, and R^2 of 0.091, 0.091, and 0.933, respectively. Lastly, the stacking predictive model yields an MAE, MSE, and R^2 of 0.091, 0.091, and 0.932, respectively.

For the fourth group of indicators, the XGBoost predictive model achieves an MAE, MSE, and R^2 of 0.152, 0.152, and 0.9, respectively. The SVM predictive model presents an MAE, MSE, and R^2 of 0.424, 1.152, and 0.244, respectively. For the RF predictive model, the MAE, MSE, and R^2 are 0.091, 0.091, and 0.94, respectively. The GBDT predictive model demonstrates an MAE, MSE, and R^2 of 0.061, 0.061, and 0.96, respectively. Lastly, the stacking predictive model yields an MAE, MSE, and R^2 of 0.121, 0.121, and 0.92, respectively.

For the fifth group of indicators, the XGBoost predictive model exhibits an MAE, MSE, and R^2 of 0.121, 0.121, and 0.92, respectively. The SVM predictive model shows an MAE, MSE, and R^2 of 0.333, 0.333, and 0.781, respectively. For the RF predictive model, the

MAE, MSE, and R^2 are 0.091, 0.091, and 0.94, respectively. The GBDT predictive model demonstrates an MAE, MSE, and R^2 of 0.091, 0.091, and 0.94, respectively. Lastly, the stacking predictive model yields an MAE, MSE, and R^2 of 0.152, 0.152, and 0.9, respectively. As shown in Figure 8.

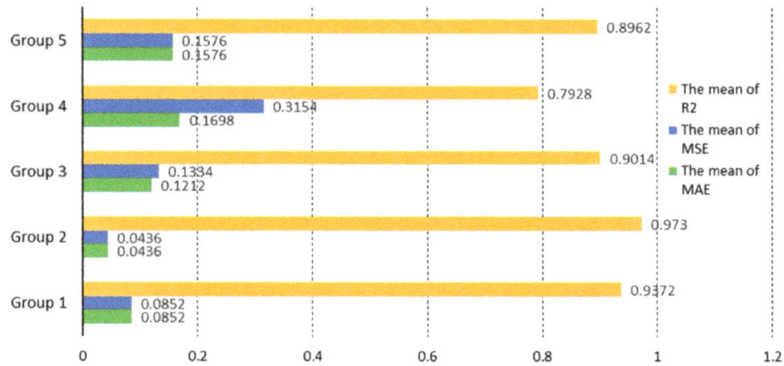

Figure 8. Average indicators of each group.

By plotting bar charts for the average MAE, MSE, and R^2 of the five datasets, it is visually evident that the second group of metrics performs optimally. Specifically, the MAE, MSE, and R^2 for this group are 0.0436, 0.0436, and 0.973, respectively. On the other hand, the fifth group exhibits the poorest performance with MAE, MSE, and R^2 values of 0.1576, 0.1576, and 0.8962, respectively. As shown in Figure 9.

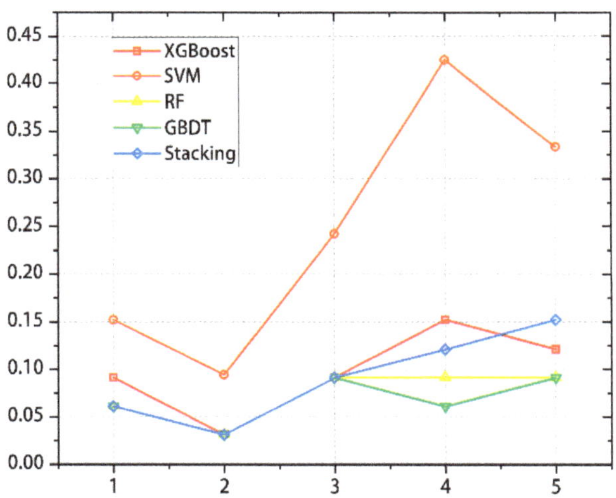

Figure 9. MAE index diagram for each mode.

By drawing five models, it was concluded that the change in MAE for the different data groups directly reflected the prediction effects of the different indicator groups in each model. By comparing the five points on the same line, it was found that the MAE value of the second indicator group was the smallest, while the MAE value of the fourth indicator group was the largest in SVM and XGBoost. The fifth indicator group had the largest MAE value in stacking, RF, and GBDT. As shown in Figure 10.

By drawing five models, it was concluded that the change about MSE values for the different data groups directly reflected the prediction effects of the different indicator groups in each model. By comparing the five points on the same line, it was found that the MSE value of the second indicator group was the smallest, while the MSE value of the fourth indicator group was the largest in SVM and XGBoost. The MSE value of the fifth indicator group in stacking, RF, and GBDT was the largest. As shown in Figure 11.

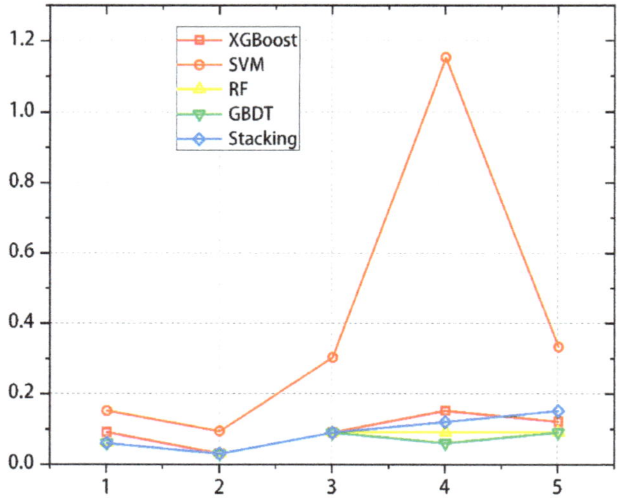

Figure 10. MSE index diagram for each model.

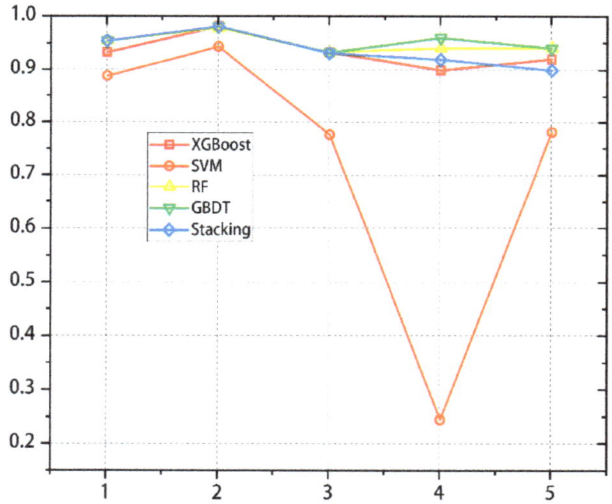

Figure 11. R^2 indicators for each model.

By drawing five models, it was concluded that the change in R^2 values for the different data groups directly reflected the prediction effects of the different indicator groups in each model. By comparing the five points on the same line, it was found that the R^2 value of the second indicator group was the largest, while the R^2 value of the fourth indicator group was the largest in SVM and XGBoost. The R^2 value of the fifth indicator group in stacking, RF, and GBDT was the largest.

After a comprehensive analysis of the aforementioned results, it was observed that the average MSE and MRE of the five models, with the second set of indicators as the main controlling factor, were the lowest, while the average R^2 was the highest. This indicated that, when the second set of indicators was used as the main controlling factor, the models had the lowest errors and the best fitting performance. Therefore, it was possible to select the datasets with a grey correlation degree greater than 0.670 as the main controlling factors, namely, "dynamic slope", "q", "absolute gas emission volume", "coal seam thickness", "coal hardness coefficient", "relative gas emission volume", "volatile matter", and "gas adsorption constant b". Furthermore, among the first three sets of indicators, the stacking fusion model showed the best performance. As for the fourth and fifth sets of indicators, GBDT and RF demonstrated relatively good performance, but the overall performance was inferior to that of the stacking fusion model of the first three sets of indicators.

4.4. Analysis of Gas Outburst Risk Early Warning Model

The present study conducted a series of experiments to investigate the training performance of different models when input with gas outburst factor data under various degrees of relevance. Specifically, the gas outburst data were categorized into several groups based on their grey relevance, including factors with relevance degrees $r_i > 0.700$, $r_i > 0.670$, $r_i > 0.630$, $r_i > 0.570$, and all factors combined. The GBDT, RF, SVM, XGBoost, and stacking ensemble models were employed for training. In order to visually compare the differences between the actual values and predicted values, line graphs were plotted as shown in Figures 12–19.

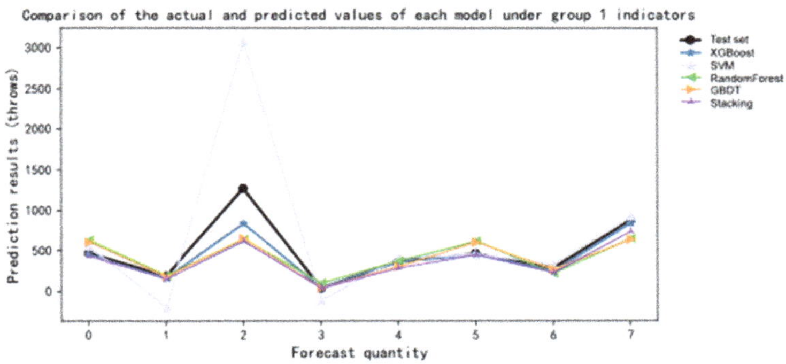

Figure 12. Comparison between the predicted value and the true value of each model under the first set of indicators in the original data.

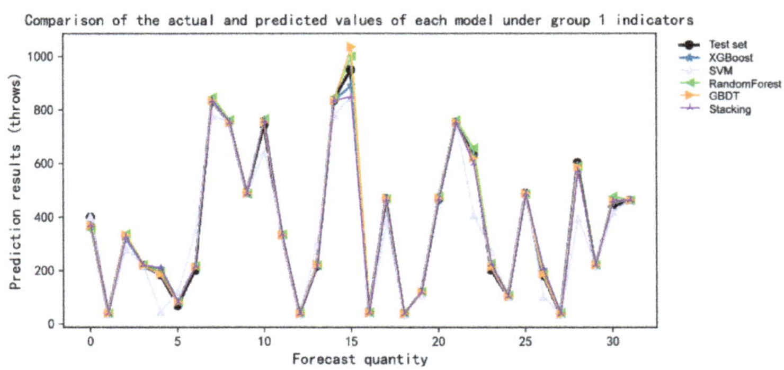

Figure 13. Comparison between the predicted value and the true value of each model of the original data under the first set of indicators in the extended data.

For the comparison of errors between actual values and predicted values, Figures 6 and 7 represent the complete set of indicators, which is the fifth group of indicators. The agreement between the actual values and predicted values for the expanded data in all five indicator groups was higher than that for the original data. This indicates that the predictive performance of the models was better for the expanded dataset compared to the original dataset. Among the original and expanded datasets, the second group of indicators exhibited the highest agreement between the actual values and predicted values among the XGBoost, SVM, RF, GBDT, and stacking prediction models. This suggests that selecting the second group of indicators as the main control factor leads to the best predictive performance. Among the XGBoost, RF, GBDT, SVM, and stacking models, it is observed that, particularly for models trained on the second group of indicators from the expanded data, the stacking fusion model shows a closer fit between predicted values and actual values. The higher the accuracy of the model, the better the predictive performance.

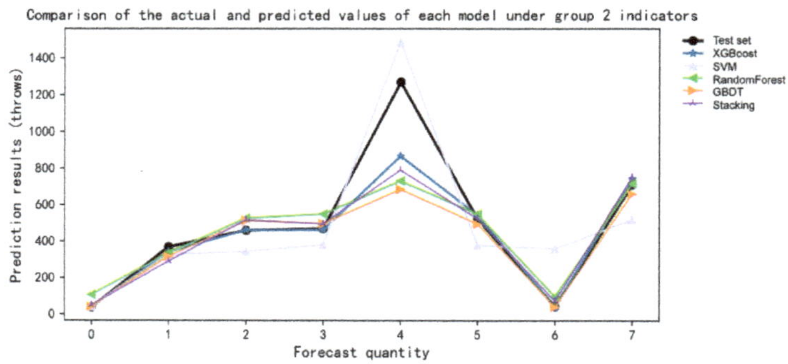

Figure 14. Comparison between the predicted value and the true value of each model of the original data under the second group of indicators in the original data.

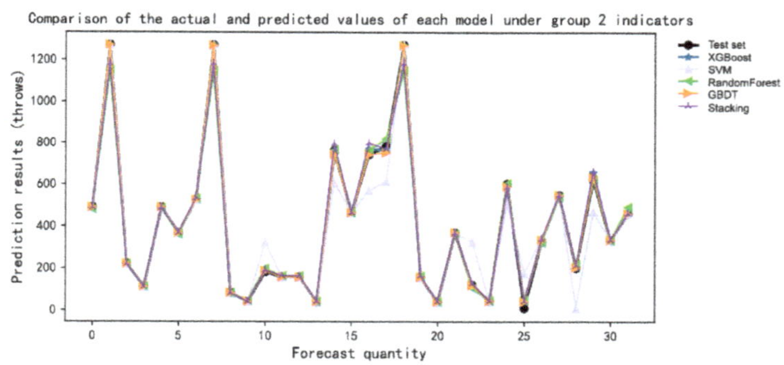

Figure 15. Comparison between the predicted value and the true value of each model of the original data under the second group of indicators in the extended data.

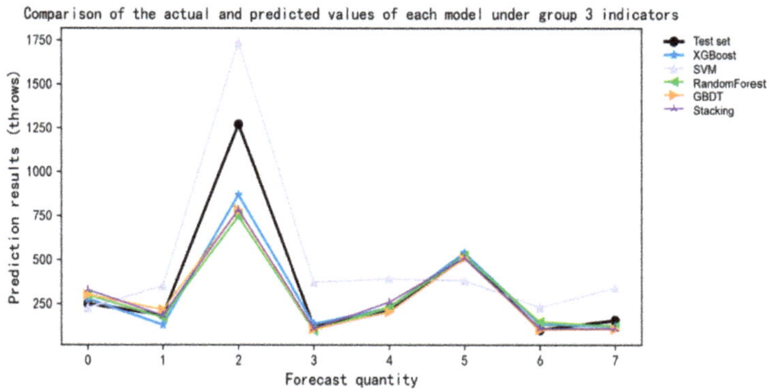

Figure 16. Comparison between the predicted value and the real value of each model of the original data under the third group of indicators in the original data.

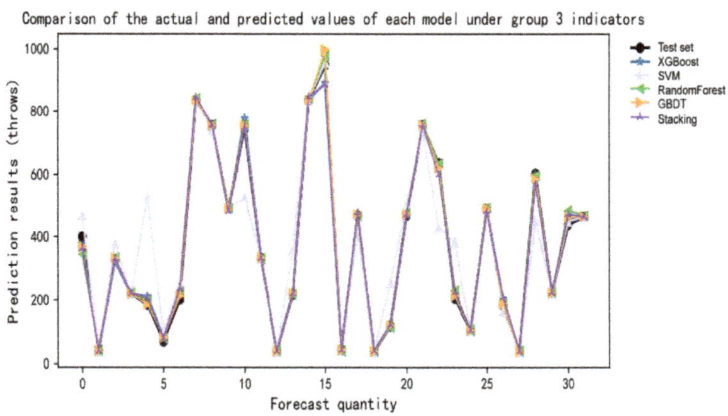

Figure 17. Comparison between the predicted value and the true value of each model of the original data under the third group of indicators in the extended data.

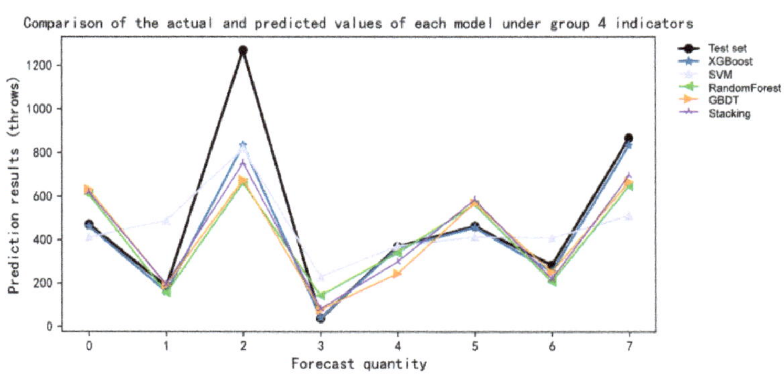

Figure 18. Comparison between the predicted value and the true value of each model of the original data under the fourth group of indicators in the original data.

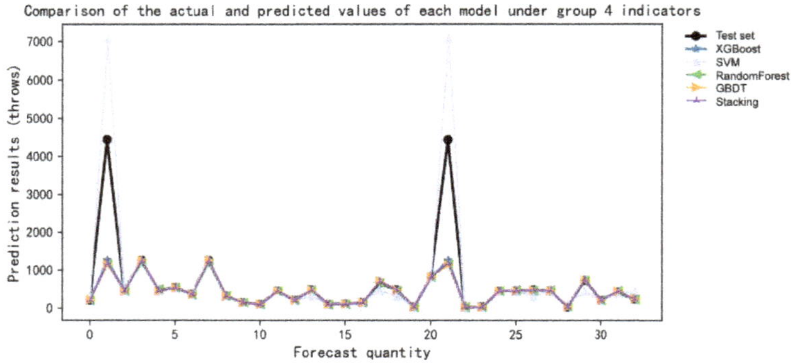

Figure 19. Comparison between the predicted value and the true value of each model of the original data under the fourth group of indicators in the extended data.

In the course of the experiment, we utilized various evaluation metrics to comprehensively assess the performance and effectiveness of the model. These metrics included mean squared error (MSE), mean absolute error (MAE), and the coefficient of determination (R^2), among others.

Based on the comparison and analysis of the experimental results, stacking and RF exhibited the best predictive performance among the first set of metrics, while SVM performed the poorest. In terms of the second set of metrics, RF and stacking showed the most favorable outcomes with the selected controlling factors, while SVM performed the worst. However, overall, the predictive efficacy with the second set of metrics as controlling factors surpassed that of the first set. Regarding the third set of metrics, stacking demonstrated relatively good predictive performance. However, for models other than stacking, there were significant errors when comparing the actual values with the predicted values, making them unsuitable for forecasting purposes. As for the fourth and fifth sets of metrics, SVM again displayed the worst predictive performance, and stacking also exhibited inferior results compared to the previous sets. Additionally, by examining the data in Tables 4 and 5, we found that when selecting data with $r > 0.700$ as input for the model after augmentation, all models performed relatively well except for SVM. However, the comparison between actual values and predicted values in the analysis table indicated that stacking and RF achieved superior results with minimal error. Conversely, SVM performed the worst in terms of the comparison. Therefore, we concluded that stacking demonstrates greater stability and reliability compared to the RF model. Taking into consideration all the aforementioned factors, we chose stacking as the final gas outburst risk warning model.

5. Conclusions

1. This paper proposes a data generation model based on XGBoost to address the issue of coal and gas outburst risk warning. Virtual samples are generated to expand the dataset. By comparing the original data with the expanded data, it is found that the expanded samples outperform the original data in multiple methods. After data expansion, the model's predicted ROC curve area (AUC) value increased from 0.86 to 1.00, indicating a significant improvement in prediction effectiveness using the expanded data model.
2. The process of gas emission is influenced by various factors, and the relationship between each factor and the emission rate is nonlinear. This paper proposes the use of grey correlation analysis to select the main controlling factors based on the magnitude of the grey correlation degree. The experiments conducted indicate that the model with a grey correlation degree ranging from 0.67 to 0.70 achieves the best predictive performance, with average MSE, MRE, and R2 values of 0.0436, 0.0436, and 0.973, respectively. Therefore, the model with the least prediction error and the optimum

model fit is identified. The group of factors includes "Dynamic Slope Indicator", "q", "Absolute Gas Emission Rate", "Coal Thickness", "Coal Firmness Coefficient", "Relative Gas Emission Rate", "Volatile Matter", and "Gas Adsorption Constant b".

3. This paper proposes the XGBoost–GR–stacking model for addressing the issue of coal and gas outburst risk warning. The XGBoost algorithm is utilized to generate data, while the GR algorithm is employed for feature selection. Furthermore, a prediction model based on the stacking fusion algorithm is established. The results show that the MSE, MRE, and R2 predictions of this model are 0.031, 0.031, and 0.981, respectively, which are superior to those of the XGBoost, GBDT, RF, and SVM models. This indicates that the proposed model exhibits lower prediction errors and a higher fitting degree, making it highly applicable in the domain of gas outburst warning.

Author Contributions: Conceptualization, methodology, visualization, Y.G.; software, validation, formal analysis, investigation, writing—original draft preparation, Y.G.; resources, data curation, supervision, project administration, funding acquisition, J.C. and H.L.; writing—review and editing, L.G. and X.Z. All authors have read and agreed to the published version of the manuscript.

Funding: This research was funded by the National Natural Science Foundation of China (52174182, 52274202), the Central Guidance on Local Development Science and Technology Funds of Hebei Projects (226Z4601G), the Youth Foundation of Hebei Province (E2022209051), and the North China University of Science and Technology Youth Support Program.

Institutional Review Board Statement: Not applicable.

Informed Consent Statement: Not applicable.

Data Availability Statement: Data are contained within the article.

Acknowledgments: The first author would like to thank Zhishen Li, Qingze He, Yingxin Li, and Zien Yang for their technical help.

Conflicts of Interest: The authors declare no conflict of interest.

eferences

Zhang, C.; Wang, P.; Wang, E.; Peng, S.; Yin, S.; Jiang, Q.; Liu, M. Development history and prospect of the mechanism of coal and gas outburst in China over 70 years. *Coal Geol. Explor.* **2023**, *51*, 59–94.

Chen, L.; Chen, J. Study on the prediction effect of coal and gas outburst based on data imputation-machine learning. *China Saf. Sci. J.* **2022**, *18*, 69–74.

Lin, H.; Zhou, J.; Jin, H.; Li, S.; Zhao, P.; Liu, S. A collaborative prediction method for coal and gas outburst hazard level based on feature selection and machine learning. *J. Min. Saf. Eng.* **2023**, *40*, 361–370.

Ping, Y.; Bo, L.; Xue, W.; Hao, S.; Qiao, C.; Jiang, X. Study on the Evolution Mechanism of Coal and Rock Seepage under Cyclic Loading and Unloading of Pore Pressure. *J. China Univ. Min. Technol.* **2023**, *52*.

Skoczylas, N. Laboratory study of the phenomenon of methane and coal outburst. *Int. J. Rock Mech. Min. Sci.* **2012**, *55*, 102–107. [CrossRef]

Zhao, W.; Cheng, Y.; Jiang, H.; Jin, K.; Wang, H.; Wang, L. Role of the rapid gas desorption of coal powders in the development stage of outbursts. *J. Nat. Gas Sci. Eng.* **2016**, *28*, 491–501. [CrossRef]

Wold, M.B.; Connell, L.D.; Choi, S.K. The role of spatial variability in coal seam parameters on gas outburst behaviour during coal mining. *Int. J. Coal Geol.* **2008**, *75*, 1–14. [CrossRef]

Dazhao, S.; Minggong, G.; Gang, Y.; Liming, Q. Mechanism of Coal and Gas Outburst Induced by Non synchronous Deformation of Soft and Hard Coal Stratification. *Coal Sci. Technol. Mag.* **2023**, *2023*, 44.

Zhang, C.; Wang, P.; Wang, E.; Xu, J.; Li, Z.; Liu, X.; Peng, S. 70 year development process and prospects of coal and gas outburst mechanisms in China. *Coal Geol. Explor.* **2023**, *2023*, 51.

Hu, Q. Mechanical action mechanism of coal and gas outburst process. *J. Coal Sci.* **2008**, *3*, 1368–1372.

Lijun, Z.; Wentuan, Q.; Wei, Y. Research progress on coal and gas outburst mechanism based on geological structure perspective. *Saf. Coal Mines* **2022**, *2022*, 53.

Liang, W.; Bi, S.; Qing, T.; Yi, L. Experimental Study on Strata Fracture Evolution Characteristics of middling coal in the Process of Coal and Gas Outburst. *J. Saf. Sci. Technol.* **2023**, *2023*, 19.

Peng, S.; Yang, W.; Zhou, B.; Xu, J.; Cheng, L.; Yang, X. Study on the spatiotemporal evolution law of gas pressure during coal and gas outburst under true three-dimensional stress state. *Chin. J. Rock Mech. Eng.* **2020**, *39*, 1762–1772.

14. Guo, P. Study on the Development Mechanism of Coal and Gas Outburst Induced by Layer Fracture. Ph.D. Thesis, Chi University of Mining and Technology, Xuzhou, China, 2014.
15. Xu, M.; Dong, K.; Dong, Y. Microelement destruction and outburst mechanism of coal-rock and gas. *J. Xi'an Univ. Sci. Techn* **2014**, *34*, 249–254.
16. Shan, D. Investigation of sensitive indicators and critical values for coal seam outbursts dominated by gas as the main factor. C **2021**, *30*, 20–22+93.
17. Si, H.; Zhao, J.; Hu, Q. Analysis of coal and gas outburst accidents under the theory of big data. *J. Xi'an Univ. Sci. Technol.* **20** *38*, 515–522+537.
18. Wen, C. Attribute recognition model and examples for risk assessment of tunnel gas outburst. *J. China Coal Soc.* **2011**, 1322–1328.
19. Wang, G.; Cheng, W.; Xie, J.; Zhou, G. Analysis of the role of gas content in the outburst process. *J. China Coal Soc.* **2011**, 429–434.
20. Cao, S.; Zhang, Z.; Li, Y.; Ping, G.; Yan, B. Experimental study on adsorption and desorption gas deformation characteristics dangerous coal prone to outbursts. *J. China Coal Soc.* **2013**, *38*, 1792–1799.
21. Yang, D.; Peng, K.; Zheng, Y.; Chen, Y.; Zheng, J.; Wang, M.; Chen, S. Study on the characteristics of coal and gas outburst haza under the influence of high formation temperature in deep mines. *Energy* **2023**, *268*, 126645. [CrossRef]
22. Li, Y.; Zhang, Y.; Zhang, Z.; Jiang, B. Experimental study on initial characteristics of coalbed methane desorption. *J. China Coal S* **2013**, *38*, 15–20.
23. Xiang, Z. Optimization for Prediction Indexes of Coal and Gas Outburst in Coal Roadway Excavation Working Face Based Grey Synthetically Relational Degree. *Adv. Mater. Res.* **2012**, *1792*, 524–527.
24. Zhou, X.; Sun, J. Gas Outburst Quantity Prediction Based on Improved BP Neural Network Using Principal Component Analy *Min. Saf. Environ. Prot.* **2018**, *45*, 43–47.
25. Liu, X.; Li, P. Soft Measurement Modeling of Coal and Gas Outburst Based on PCA and IFOA-BP. *Min. Res. Dev.* **2018**, *38*, 109–1
26. Cao, B.; Bai, G.; Li, H. Prediction Analysis of Gas Content Based on PCA-GA-BP Neural Network. *China Saf. Sci. J.* **2015**, *11*, 84–
27. Ren, S. Application Research of PSO-DE Hybrid Algorithm Optimizing BP Neural Network in Coal and Gas Outburst Predicti Ph.D. Thesis, Taiyuan University of Technology, Taiyuan, China, 2015.
28. Zhao, H. Application Research on Mine Gas Concentration Prediction Based on Support Vector Machine. *Coal Sci. Technol.* **20** *4*, 41–44.
29. Zeng, W.; Han, X.; Liu, J. Study on Gas Outburst Layer Based on Particle Swarm Optimization Support Vector Machine. *Coal G China* **2018**, *30*, 44–46.
30. Zhang, W. Research on Gas Concentration Prediction Based on Support Vector Machine and Immune Genetic BP. Master's The Xi'an University of Science and Technology, Xi'an, China, 2017.
31. Wu, Y.; Li, H.; Xu, D. Coal and Gas Outburst Prediction Algorithm Based on IPSO-Powell Optimized SVM. *Ind. Autom. M Metall.* **2020**, *46*, 46–53.
32. Liu, X.; Li, Z.; Zhang, Z.; Zhang, G. Coal and gas outbursts prediction based on combination of hybrid feature extracti DWT+FICA–LDA and optimized QPSO-DELM classifier. *J. Supercomput.* **2022**, *78*, 2909–2936. [CrossRef]
33. Lu, G.; Guo, W.; Hu, H. Prediction of coal and gas outburst intensity based on KPCA-CS-ELM coupling model. *J. Appl. Fu Anal.* **2018**, *20*, 218–226.
34. Qian, Y.; Qiu, L. Quantitative study on the value of resolution coefficient in grey correlation analysis. *Stat. Decis.* **2019**, *35*, 10–

applied sciences

MDPI

rticle

Study on the Influence of Gas Desorption Characteristics under High-Pressure Fluid Fracturing of Deep Coal

huyin Ma [1,2,3,4], Jianjun Cao [3,4], Qinghua Zhang [3,4] and Sheng Xue [1,5,*]

1 School of Safety Science and Engineering, Anhui University of Science and Technology, Huainan 232001, China
2 State Key Laboratory of Mining Response and Disaster Prevention and Control in Deep Coal Mines, Anhui University of Science and Technology, Huainan 232001, China
3 Chongqing Research Institute Co., Ltd., China Coal Technology and Engineering Group Corp., Chongqing 400037, China
4 State Key Laboratory of the Gas Disaster Detecting, Preventing and Emergency Controlling, Chongqing 400037, China
5 Joint National-Local Engineering Research Centre for Safe and Precise Coal Mining, Anhui University of Science and Technology, Huainan 232001, China
* Correspondence: sheng.xue@aust.edu.cn

Abstract: In order to study the influence law of gas desorption accumulation and emission characteristics under hydraulic fracturing, this experiment uses coal-rock adsorption–desorption test equipment to carry out isothermal desorption tests of water-bearing coal under various stress paths. The experimental object is anthracite from Four Seasons Chun coal mine in Guizhou Province. In this experiment, the influence law of the desorption emission characteristics of coal under different stresses is analyzed. Research shows that the stress directly affects the gas desorption of coal and plays a decisive role in the gas desorption and emission characteristics of water-bearing coal in the stress-affected zone. Under equivalent gas adsorption of water-bearing coal, the total accumulated gas desorption displayed by coal increases with the increase in stress under certain conditions and the increase rate slows down with the time; coal samples differing in moisture content are subjected to various stress paths, leading to the difference in the total gas desorption. The total accumulated gas desorption displayed by coal with higher moisture content is generally smaller than that with lower moisture content. Through field observation, a zone with high accumulated gas desorption is formed in the stress-affected zone beyond the radius of effective fracture influence, generating an imbalance of gas desorption and emission. The study results are of theoretical and practical engineering significance for the prevention and control of stress-induced disasters and gas disasters in deep coal seams.

Keywords: stress action; water-bearing coal; gas desorption; gas disaster prevention and control

1. Introduction

The supply of coal resources still occupies an important position in China's energy system. The horizontal depth of coal mining has been increasing due to the massive exploitation of shallow coal resources over the years. In recent years, the output ratio of deep coal resources has increased, accompanied by more complicated deep stress environments, geological conditions, and gas occurrence. As the mining depth increases, the risk of dynamic disasters of coal and gas outburst is aggravated, and the gas pressure and gas content in coal body increase accordingly. Given the complicated deep stress environment and geological conditions, the proportions of soft coal and broken coal increase, and it is more difficult to drill holes and extract gas from outburst coal seams, which directly leads to more frequent gas ultra-limits, becoming the main controlling factor restricting the safe production of coal mines.

Underground coal is in a state of stress balance before mining. After mining, the stress imbalance superimposed by the geological structure and the comprehensive stress of the mining disturbance exerts a significant impact on the process of gas adsorption–desorption–diffusion–seepage in coal seams [1,2]. If the coal body is severely disturbed by continuous high stress, a composite dynamic disaster of coal and gas outbursts and gas impacts will be triggered. To manage the main controlling factors of comprehensive stress-dominated gas disasters in deep coal bodies, pressure relief in deep coal has been realized using high-pressure hydraulic fracturing measures, and the characteristics of gas desorption and emission under fracturing have been studied to provide theoretical and practical experience for gas disaster prevention in deep coal.

For many years, the gas absorption/desorption laws of coal have been extensively examined by Chinese experts from such perspectives as coal moisture content [3,4], damage condition [5,6], stress environment [7–11], temperature [12,13], and coal particle structure [14,15]. Among these studies, gas absorption/desorption models and diffusion formulas have been proposed by some scholars, laying a theoretical foundation for characterizing the mechanism of coal and gas outburst risk. Stress exhibits certain influencing characteristics on gas adsorption/desorption. Li X. C. et al. [16] found that effective stress evidently affects the methane adsorption capacity. Lv X. F. et al. [17] obtained the relationship between pore pressure and the desorption amount through simulation. In order to clarify the influencing mechanism of high-pressure fluid fracturing stress on the desorption characteristics of water-bearing gas coal, the stress state of anthracite in Sijichun Coal Mine, which was taken as the study object, was approximated and simulated using different stress combinations under the action of high-pressure fluid. Then, water-bearing gas coal samples were subjected to isothermal desorption tests under different stress states, so as to expound the influencing mechanism of the zone of stress influence on the desorption and emission characteristics of water-bearing gas coal under high-pressure fluid fracturing. This may help us to understand the phenomenon of gas emission after fracturing, and also provide a theoretical basis for gas disaster management.

2. Test Samples and Methods

In the experiment, the self-improved triaxial coal-rock adsorption/desorption testing apparatus was used to carry out isothermal adsorption and desorption tests of water-bearing coal under various stress paths [18,19], as shown in Figure 1.

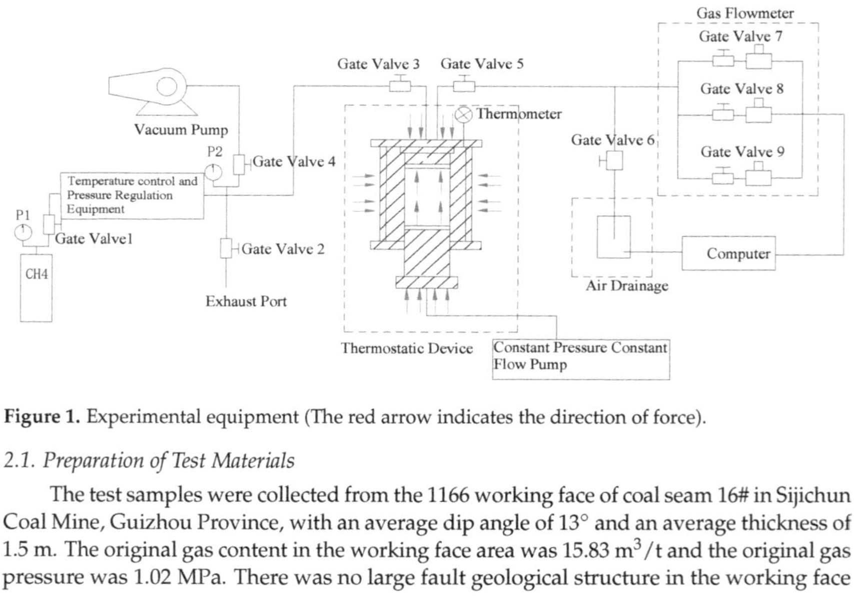

Figure 1. Experimental equipment (The red arrow indicates the direction of force).

2.1. *Preparation of Test Materials*

The test samples were collected from the 1166 working face of coal seam 16# in Sijichun Coal Mine, Guizhou Province, with an average dip angle of 13° and an average thickness of 1.5 m. The original gas content in the working face area was 15.83 m^3/t and the original gas pressure was 1.02 MPa. There was no large fault geological structure in the working face area of this coal seam, and the coal body strike structure was stable. The coal body of this working face was selected as the material for preparing Φ50 mm × 100 mm standard coal samples. Considering the stress balance between the upper and lower cylinders during sample loading, the upper and lower cylinders of the coal samples were polished to make their surfaces smooth. The coal samples are shown in Figure 2.

Figure 2. Processed coal samples.

The fabrication process of samples differing in moisture content was as follows [20,21]:

(1) The processed samples in Figure 2 were dehydrated and dried in a drying box and weighed (m_0).
(2) The dry coal samples were completely soaked in a closed container filled with distilled water for 3 h.
(3) The coal samples were taken out and their soaking mass m_s was measured. The moisture content w_s of the coal samples is fixed as follows:

$$w_s = \frac{m_s - m_0}{m_0} \times 100\% \tag{1}$$

(4) The soaked samples were placed in the drying box and weighed every 15 min. The weighing time could be adjusted according to the specific needs until their mass reached the required value m_p. The coal samples were taken out, placed in a constant-temperature and -humidity box, and weighed at regular intervals until the mass m_u of the samples remained unchanged. The moisture content w_u of the coal samples is expressed as follows:

$$w_u = \frac{m_u - m_0}{m_0} \times 100\% \tag{2}$$

(5) Steps (2) and (4) were repeated to acquire the moisture content of different coal samples.
(6) The coal samples were divided into two groups in the experiment, with a moisture content of about 1.15% and 3.24%, respectively.

2.2. Experimental Method

To achieve the purpose of isothermal desorption tests of coal with different water-bearing states (gas-bearing coal) under various stress paths, a gas adsorption equilibrium was achieved in the experiment using a pressure of 2 MPa. Specifically, the coal samples were tested under a confining pressure of 3 MPa and an axial pressure of 0, 4, 8, 12 and 16 MPa, respectively. The gas desorption of coal samples with different water-bearing states under multiple stress paths was measured, and the desorption laws of loaded coal samples under different conditions were analyzed. The specific test steps were as follows: (1) the valves and switches of the system equipment were all closed, the coal samples were fixed in a stress chamber and subjected to the confining pressure of 3 MPa, and then the vacuum pump, valve 3, and valve 4 were opened for vacuum degassing of the coal samples. (2) After degassing was completed, valve 1 of the CH_4 gas tank was opened to charge the gas into the temperature-controlling and pressure-regulating device, and the temperature was regulated at 20 °C. (3) At this time, the stress chamber was kept at a constant temperature of 20 °C, and the temperature-controlling and pressure-regulating device and valve 3 were opened to charge the gas into the stress chamber. When the coal samples adsorbed gas, a gas pressure change was recorded. (4) After the reading of the pressure gauge became stable and adsorption was completed, the set axial pressure was applied to the coal samples, valve 5 was opened so that the free gas in the stress chamber was discharged, and when the pressure reading of the instrument was 0, the pressure was considered to identical to atmospheric pressure. In this case, the water-bearing gas coal samples under different stress states were subjected to isothermal desorption tests using a gas metering device, and the gas content was recorded.

3. Gas Desorption Laws of Coal with Different Water-Bearing States under Various Stress Paths

Under the stress produced by high-pressure fluid presplitting and permeability enhancement, water-bearing coal in the gas atmosphere was subjected to an ever-changing stress load, generating a significant impact on coal adsorption capacity, inevitably leading to a change in the adsorption–desorption state of coal after being subjected to different stresses. To study the desorption and emission laws of water-bearing coal under different stresses in the gas atmosphere, the gas desorption amount was analyzed by combining the above test methods.

3.1. Influencing Analysis of Various Stress Paths on Gas Desorption from Water-Bearing Coal Samples

During the test, the influence of stress on the gas desorption displayed by water-bearing coal could be analyzed using the accumulated gas desorption. The isothermal adsorption characteristics of coal samples could be described using the Langmuir equation, namely:

$$Q = \frac{abP}{1+bP} \tag{3}$$

where Q is the gas adsorption capacity with the unit of mL/g; a is the maximum adsorption capacity with the unit of mL/g; b is the Langmuir constant with the unit of Pa^{-1}; and P is the pore pressure with the unit of Pa.

Combined with the above formula, the accumulated gas desorption from the water-bearing coal body under various stress paths was analyzed by using a Langmuir-type relational expression of gas desorption with time [22]:

$$Q_t = \frac{AB\sqrt{t}}{1+B\sqrt{t}} \tag{4}$$

where Q is the accumulated gas desorption at time t with the unit of mL/g; t is desorption time with the unit of s; A is the limiting gas desorption with the unit of mL/g; and B is the desorption rate constant.

Under constant temperature, the accumulated gas desorption displayed by water-bearing coal under multiple stress paths was measured and fitted as per Formula (4) to acquire the limiting gas desorption under different test conditions. The water-bearing coal desorption curves and the change curves of the limiting gas desorption under multiple stress paths were drawn, as displayed in Figures 3–6.

By analyzing Figures 3 and 4, it can be seen that (1) the coal samples differing in moisture content were subjected to different stress paths, which also led to the difference in the corresponding gas desorption process. Within 0–3 min in the initial test stage, the stress borne by the coal samples gradually increased, accompanied by the overlapping or crossing of desorption curves under each stress path, but under equivalent gas adsorption by coal samples, the total desorption displayed by coal samples grew with the stress within 60 min. (2) The coal samples with different moisture contents were under the action of multiple stress paths, and the total gas adsorption produced by each stress path was different, but the variation trend of gas desorption was largely the same. In addition, the gas desorption displayed by the coal sample with higher moisture content was generally smaller than that with a lower moisture content under different stress paths.

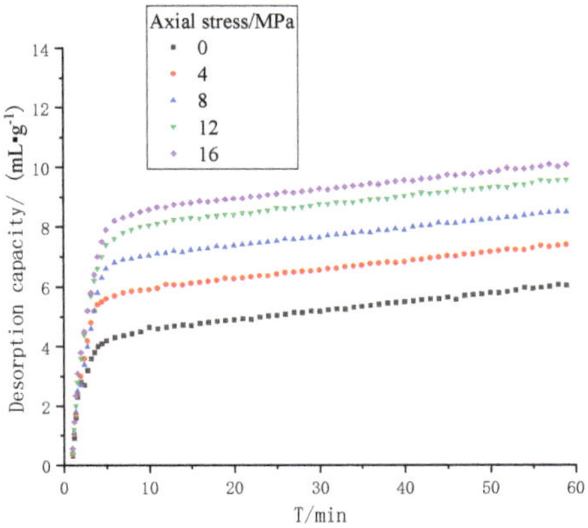

Figure 3. Gas desorption curves of a coal body with a moisture content of 1.15% under various stress path states.

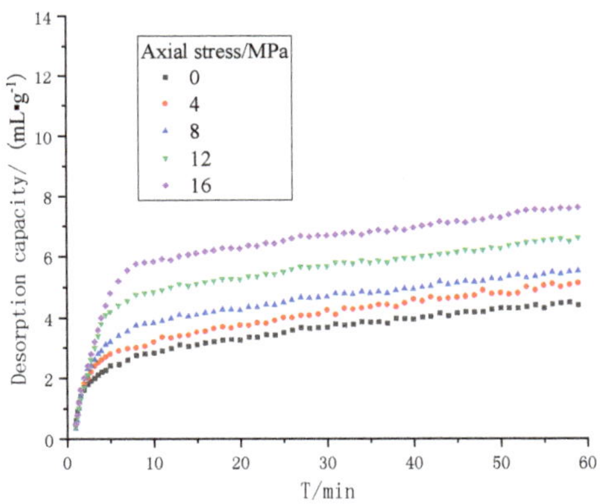

Figure 4. Gas desorption curves of coal with a moisture content of 3.24% under various stress path states.

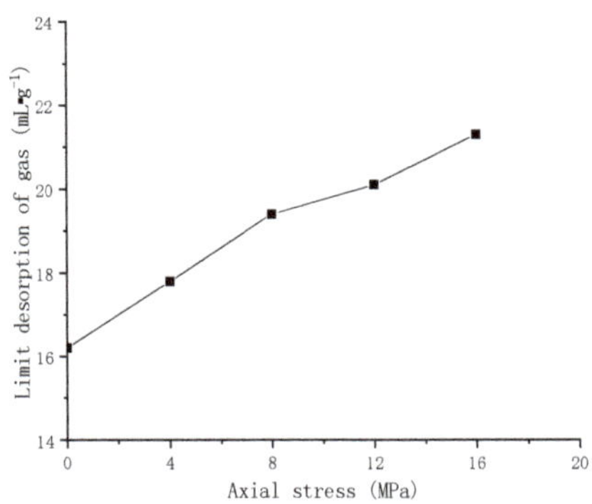

Figure 5. Variation of limiting desorption amount provided by coal samples with a moisture content of 1.15% under various stress path states.

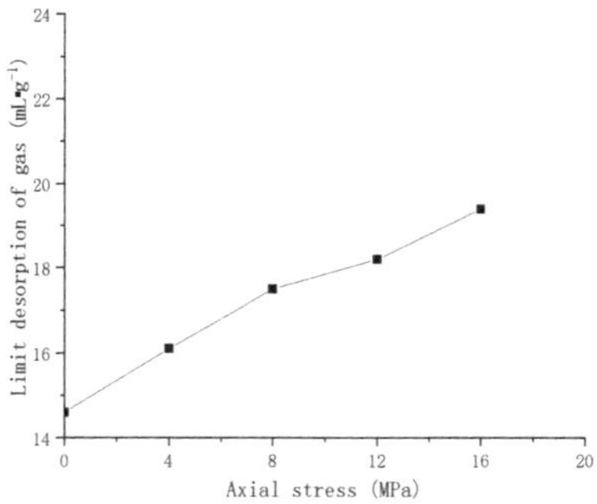

Figure 6. Variation of limiting desorption amount provided by coal samples with a moisture content of 3.24% under various stress path states.

It can be seen in Figures 5 and 6 that when the axial stress borne by the coal samples with two different moisture contents showed hierarchical growth (0, 4, 8, 12, and 16 MPa), the limiting desorption displayed by coal samples was fitted, and then the limiting desorption displayed by the coal sample with a moisture content of 1.15% was about 16.12–21.33 mL/g, with an increase of 5.21 mL/g. The limiting desorption of the coal sample with a moisture content of 3.24% was about 14.81–19.34 mL/g, increasing by 4.53 mL/g. The desorption displayed by coal samples was obviously promoted by the hierarchical growth of the axial stress borne by coal samples. However, the increase amplitude of the limiting desorption declined due to the increase in the moisture content of coal samples.

3.2. Influencing Analysis of Stress on Gas Desorption and Emission from Water-Bearing Coal Samples

As a heterogeneous porous solid medium, coal can be characterized as a solid particle skeleton structure, and the forced compression of coal can be regarded as the combination of pore volume reduction and solid particle skeleton volume reduction [23]. The high-pressure fluid presplitting and permeability enhancement process is adopted for deep coal. In addition to ground stress, coal is also affected by the change in the high-pressure fluid stress, which directly influences the gas occurrence in coal and the gas desorption and emission.

Through the above tests, it can be observed that the stress promoted the gas desorption of water-bearing coal, and the total accumulated gas desorption observed in water-bearing coal also increased under various stress paths when the stress grew. The influence of gas desorption and emission could be further analyzed as follows:

(1) There is a multi-stage pore system in the coal body, in which micro-pores filled with gas are characterized as "a gas storage molecular system", representing a strong gas adsorption zone; under the stress action, the particle skeleton structure and the pore volume of coal are compressed, the gas adsorption equilibrium is destroyed, and a large amount of free gas is desorbed by the pore diffusion system, representing a weak gas adsorption zone; a connective channel exists between the pore system and the seepage system, and the compressed gas molecules are quickly discharged, which leads to the increase in the total accumulated gas desorption from coal.

(2) The test coal is anthracite, which has a high degree of micropore development, and the larger the specific surface area of coal, the greater the gas adsorption capacity [24,25]. The coal is infiltrated by high-pressure fluid stress, which results in the competitive adsorption and desorption of water and gas molecules in the pore diffusion system in the weak pore adsorption zone. Under the high-pressure fluid stress, however, the gas desorption capacity is weakened, a large number of water molecules infiltrate the pore diffusion system to displace gas molecules in a specific stage, and water

molecules block the connective channel between the pore diffusion system and the seepage system, resulting in the difference in gas emission in this zone.

(3) The stress on coal reduces the connective channel between the pore diffusion system and the seepage system, and changes the gas occurrence and gas migration resistance, as shown in Figure 7 (water molecules are blue, gas molecules are colorless). Pore channel A of the water-bearing coal is the channel size before the stress is applied (solid line part), and A_1 and A_2 are the channel sizes after the stress is applied (dotted line part). Due to the blockage of water molecules and the narrowing of gas migration channels, the pore pressure increases, which increases the pressure difference between gas channels.

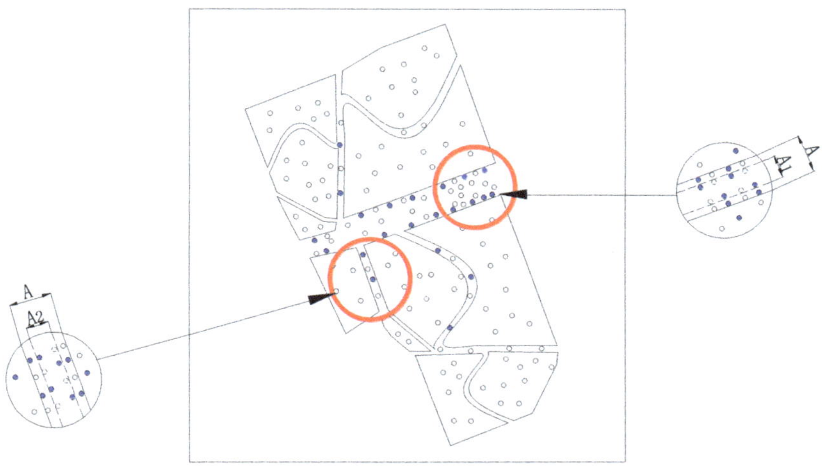

Figure 7. Changes to gas channels under high-pressure fluid stress.

4. Gas Measurement and Analysis under Field Stress Action

To investigate the influence of high-pressure fluid stress borne by coal on gas desorption, the hydraulic presplitting process test was implemented in the haulage roadway of the 1166 working face in Sijichun Coal Mine, Guizhou Province. There was no fault geological structure in the test area, and the coal seam was stable. Fracturing holes and observation holes (1#–5#) were drilled to investigate the gas pressure, gas content, gas concentration, and single-hole pure gas extraction, as displayed in Figure 8. The observation holes were used to measure the gas pressure.

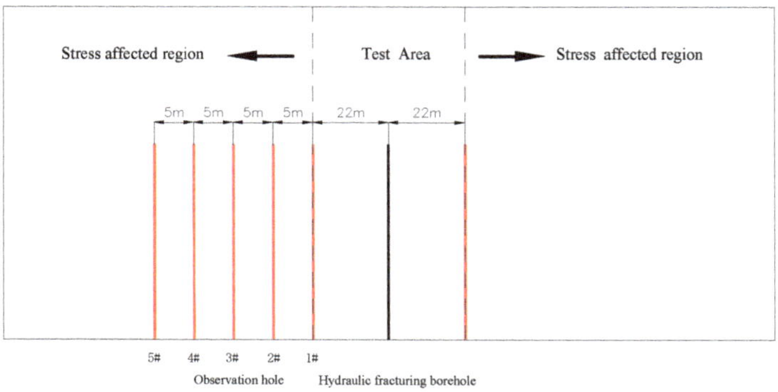

Figure 8. Layout plan of drill holes in hydraulic presplitting test area.

Before the hydraulic presplitting process was implemented, the fracturing holes and observation holes (1#–5#) were constructed first, and the observation holes were simultaneously used for the measurement of the gas pressure and gas extraction. During this period, the actual pressure range of hydraulic fracturing was 0–34 MPa, and the pressure was stabilized at 32–34 MPa after pressure was initially applied. When the process measure was completed, the total amount of water injected was 198.6 m^3.

According to the previous hydraulic presplitting test in this mine, the radius of the effective fracture influence was 22 m. The previous test area belonged to the same geological unit and coal seam with the existing test area, and the coal seam occurrence was steady without geological structures, so the radius of effective fracture influence was set to 22 m during hydraulic presplitting in this test area. Beyond this distance was a zone where the extension of the hydraulic fracture ended and the high-pressure fluid stress exerted its action. The gas measurement in the stress-affected zone was analyzed as follows.

(1) Analysis of gas pressure in the stress-affected zone

The observation holes (1#–5#) were constructed simultaneously with fracturing holes. After that, the gas pressure changes before and after hydraulic presplitting were observed, as shown in Figure 9.

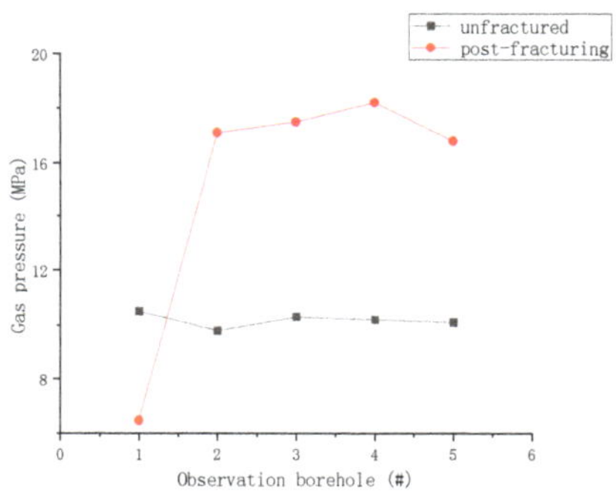

Figure 9. Variation curves of gas pressure before and after hydraulic presplitting.

Because observation hole 1# was located at the radius of the effective fracture influence of 22 m and on the edge of the effective fracture influence, a big gap was observed between observation hole 1# and observation holes 2#–5# in terms of gas pressure. As can be seen from the change curve in Figure 9, the gas pressure in the observation holes (1#–5#) before the stress affected the test area was 9.8–10.5 MPa, and the gas pressure changed only by 0.7 MPa. Under the stress action, the gas pressure was 6.46–18.2 MPa, and the maximum gas pressure exceeded 7.7 MPa before the stress action.

(2) Concentration analysis of gas extracted in the stress-affected zone

The gas concentration changes in the observation holes (1#–5#) in the stress-affected zone are exhibited in Figure 10.

Observation hole 1# was located on the edge of the radius of the effective fracture influence, and the gas concentration ranged from 35.1% to 40.6%. The gas concentration in observation holes 2#–5# was 77.4–90.4%. Therein, the highest gas concentration in observation hole 2# was 86.6–90.4%, always being higher than that in observation hole 1#.

(3) Analysis of pure gas extraction in the stress-affected zone

The changes in the pure gas extraction in the observation holes (1#–5#) in the stress-affected zone are displayed in Figure 11.

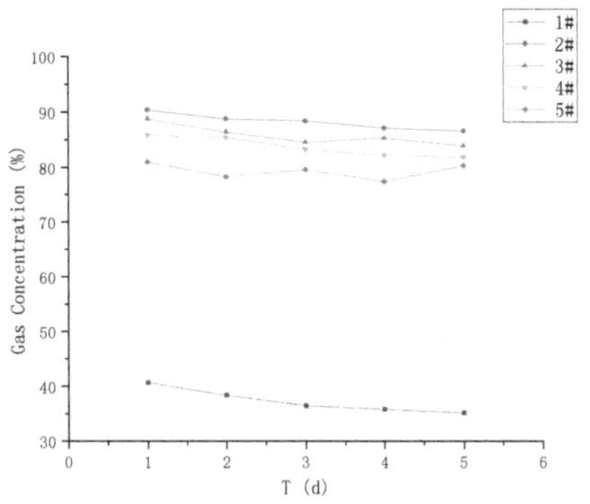

Figure 10. Curves of gas concentration under stress action.

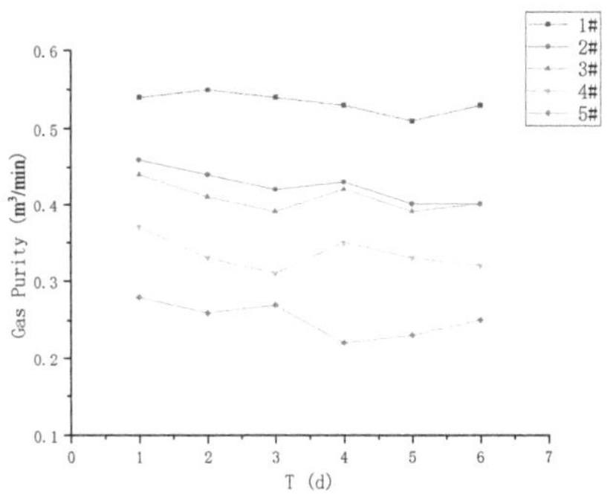

Figure 11. Variation curve of pure gas extraction in the stress-affected zone.

The pure gas extraction from observation hole 1# was 0.51–0.55 m^3/min, which was higher than that from the other observation holes. As the distance from the observation hole to the fracturing hole increases, the pure gas extraction presents a progressive declining trend.

To sum up, after the implementation of the hydraulic presplitting and permeability enhancement process, the coal body was in the stress-affected zone, and under certain conditions, the total accumulated desorption displayed by the coal body infiltrated by high-pressure water increases with the increase in stress, and the increase rate slows down with the time. This is because the adsorption capacity of gas molecules is reduced and the pore volume is compressed due to the competitive adsorption of water molecules and gas molecules in the coal body, and the multi-level gas connective channels in the coal body are also compressed, which inhibits the seepage and migration of gas and increases the gas pressure. Within 1 week after gas extraction, the gas concentration was high, the gas extraction was considerable, and the pure gas extraction was much greater than the pure gas amount without the implementation of the hydraulic fracturing process. It can be found by analyzing the process that the gas desorption and emission from the coal body were affected by multiple factors. The gas desorption displayed by coal could be promoted by the hydraulic presplitting process. Since the permeability of the coal body in the stress-affected zone was not enhanced, most gas connective channels were not connected, and a zone with high accumulated gas desorption was formed within a certain range beyond the radius of the effective fracture influence during hydraulic fracturing.

5. Conclusions

(1) The ground stress and high-pressure water stress directly affect the desorption ability of gas-bearing coal during gas control in the deep coal body, which determines the gas emission characteristics of the coal seam in the stress-affected zone.

(2) Because coal samples with different moisture contents are affected by various stress paths, the total amount of gas desorption produced by each stress path is different, but the change trend of gas desorption is roughly the same.

(3) Most of the gas flow channels in the stress-affected area are not connected, and a "high-volume area" of gas desorption accumulation is formed in a certain area outside the influence radius of the effective crack of water pre-escalation, resulting in the imbalance of gas desorption and emission.

Author Contributions: Software, Q.Z.; Validation, S.M. and S.X.; Formal analysis, S.M.; Resources, J.C., Q.Z. and S.X.; Writing—original draft, S.M.; Writing—review & editing, S.M.; Supervision, J.C.; Project administration, J.C.; Funding acquisition, S.X. All authors have read and agreed to the published version of the manuscript.

Funding: This research is supported by Open Research fund of the Joint National-Local Engineering Research Centre for Safe and Precise Coal Mining (EC2021018).

Institutional Review Board Statement: Not applicable.

Informed Consent Statement: Not applicable.

Data Availability Statement: The data, figures and tables used to support the findings of this study are included in the article.

Acknowledgments: The authors would like to express their sincere thanks to those technicians who have contributed to this research.

Conflicts of Interest: Author Shuyin Ma, Jianjun Cao and Qinghua Zhang was employed by the company Chongqing Research Institute Co., Ltd., China Coal Technology and Engineering Group Corp. The remaining authors declare that the research was conducted in the absence of any commercial or financial relationships that could be construed as a potential conflict of interest.

References

1. Zhang, Z.; Liang, W. Experimental study about the effect of adsorbed methane on coal permeabil-ity. *J. Taiyuan Univ. Techno* **2013**, *44*, 356–360.
2. Jiang, D.Y.; Yuan, X.; Chen, J.; Jiang, X.; Fan, J.Y.; Ren, S.; Li, L. Effect of adsorptional gases on permeability of outburst coal. *China Coal Soc.* **2015**, *40*, 2091–2096.
3. Gao, Z.; Ma, D.; Chen, Y.; Zhang, H.; Zhang, L. Effect of water content on adsorption/desorption of methane of differer macroscopic lithotypes. *Coal Sci. Technol.* **2020**, *48*, 97–105.
4. Mou, J.; Cheng, Y.; Liu, H. Research on gas dif- fusion characteristics of injected water coal. *J. Min. Saf. Eng.* **2012**, *29*, 746–749.
5. Chen, L.; Li, X.; Bi, J.; Liu, Y. Study on characteristics of initial desorption of tectonic coal and primary structural coal in norther Guizhou. *Coal Sci. Technol.* **2019**, *47*, 107–113.
6. He, M.C.; Wang, C.G.; Li, D.J.; Liu, J.; Zhang, X.H. Desorption characteristics of adsorbed gas in coal samples under couplin tem perature and uniaxial compression. *Chin. J. Rock Mech. Eng.* **2010**, *29*, 865–872.
7. Li, C.; Lei, D. Experimental research of the influence of static electric field on the characteristics of coal gas emission. *J. China Co Soc.* **2012**, *37*, 962–966.
8. Nie, B.; He, X.; Wang, E. Research on the mechanises of the influence of electromagnetic field on coalbed methane adsorptio *Nat. Gas Ind.* **2004**, *24*, 32–34.
9. Jiang, Y.D.; Song, X.; Liu, H.; Zheng, Q.; Liang, Y.Q.; Cui, Y.Z.; Zhou, J.P. Adsorption model and law of methane under the effe of high-power acoustic wave. *J. China Coal Soc.* **2014**, *39*, 152–157.
10. Li, S.; Zhao, Y.; Zhang, T. Test systems of the coal sample adsorption/desorption characteristics based on lowfrequency vibratio *J. China Coal Soc.* **2010**, *35*, 1142–1146.
11. Hu, G.; Zhu, Y.; Li, Z. Experimental study on desorption enhancing of methane in coal mass using a controlled microwave fiel *Chin. J. Rock Mech. En Gineering* **2017**, *36*, 874–880.
12. Qin, L.; Lin, H.F.; Lan, S.R.; Zhao, Y.P.; Yan, M. Experimental study on coal gas adsorption characteristics under action of lo temperature liquid nitrogen. *Coal Sci. Technol.* **2020**, *48*, 105–112.

Rong, T.; Zhou, H.; Wang, L.; Ren, W.; Wang, Z.; Su, T. Study on coal permeability model in front of working face under the influence of mining disturbance and temperature coupling. *Rock Soil Mech.* **2019**, *40*, 4289–4298.

Liu, Y.; Liu, M. Effect of particle size on difference of gas desorption and diffusion between soft coal and hard coal. *J. China Coal Soc.* **2015**, *40*, 579–587.

Liu, Y.; Han, G.; Wang, W. Evolution mechanism of fracture structure of loaded coal under different unloading rates. *J. China Coal Soc.* **2020**, *45*, 3806–3816.

Li, X.; Fu, X.; Fang, Z. Experimental study of influence of effective stress on coal adsorption performance. *Rock Soil Mech.* **2013**, *34*, 1247–1252.

Lü, X.F.; Pan, Y.S.; Liu, J.J.; Tang, J.P.; Di, J.Z. Experimental study of effect of pore pressure on desorption deformation of coal matrix. *Rock Soil Mech.* **2010**, *31*, 3447–3451.

Wei, J.; Wen, Z.; Yuan, Y.; Wang, Q.; Si, L. Study on influence of stress on desorption characteristics of coal containing gas. *Coal Sci. Technol.* **2021**, *49*, 35–43.

Zhang, S.; Liu, Y.; Meng, T. Experimental study on influence of water with different salinity on methane desorption per- formance of coal seam. *Coal Sci. Technol.* **2021**, *49*, 110–117.

Wei, J.; Wei, L.; Wang, D. Experimental study of moisture content influences on permeability of coal containing gas. *J. China Coal Soc.* **2014**, *39*, 97–103.

Qin, Y.; Wang, Y.; Yang, X.; Liu, W.; Luo, W. Experimental study on dynamic gas adsorption. *Int. J. Min. Sci. Technol.* **2012**, *22*, 761–765. [CrossRef]

Busch, A.; Gensterblum, Y. CBM and CO_2-ECBM related sorption processes in coal: A review. *Int. J. Coal Geol.* **2011**, *87*, 49–71. [CrossRef]

Wei, J.; Wang, H.; Si, L.; Xi, Y. Characteristics of coal-water solid-liquid contact in gas atmosphere. *J. China Coal Soc.* **2022**, *47*, 323–332.

Zhang, J.; Li, H.; Liu, Y.; Li, X.; Xie, J.; Dai, Z. Micro-wet- ting characteristics of coal dust in Pingdingshan mining area. *J. China Coal Soc.* **2021**, *46*, 812–825.

Cheng, Y.; Hu, B. Main occurrence form of methane in coal: Micropore filling. *J. China Coal Soc.* **2021**, *46*, 2933–2948.

applied sciences

MDPI

Article

Study on Active Support Parameters for Surrounding Rock with Ultra-Large Span Open-Off Cut in Thick Coal Seam

Lin Pu [1,2], Yingjie Liu [3], Yongbo Cai [2,3,*], Zuo Sun [2,3] and Xin Zhou [2]

1 Baode Coal Mine of Shendong Coal Group, Xinzhou 036600, China; 10040656@ceic.com
2 School of Emergency Management and Safety Engineering, China University of Mining and Technology, Beijing 100083, China; sunzuo@mail.ccri.ccteg.cn (Z.S.); sqt2210103042@student.cumtb.edu.cn (X.Z.)
3 Emergency Science Research Institute, Chinese Institute of Coal Science, Beijing 100013, China; liuyingjie@mail.ccri.ccteg.cn
* Correspondence: caiyongbo@mail.ccri.ccteg.cn

Abstract: In order to effectively control the stability of surrounding rock in ultra-large span open-off cuts by employing the techniques of support strength theory calculations and analogical application methods, two sets of rational support schemes were proposed, and the optimal design of active support parameters in thick coal seams with ultra-large span open-off cuts was explored by using theoretical analysis, numerical simulation, and field experiments. The results demonstrated that the span is one of the key factors influencing the stability of the roadway roof, exhibiting an inverse quadratic relationship with the peak stress borne by the roadway roof. By utilizing the pre-stressing force of anchor cables and support strength formulas, two sets of active support schemes for controlling the surrounding rock in thick coal seams with ultra-large span open-off cuts were established, and an optimized support scheme was obtained through numerical simulation. These findings provide references and guidance for related mining engineering under actual conditions in mines.

Keywords: surrounding rock control; thick coal seams; open-off cuts; support parameter

1. Introduction

With the development and utilization of ultra-large-scale coal mining machinery, the shearer and hydraulic supports are becoming increasingly larger. The underground coal mining is moving towards mechanization and unmanned operation [1]. In order to accommodate large modern machinery, the size of the entry section of the working face is gradually increasing, and the size of some open-off cuts exceeds 10 m [2–4]. Therefore, it is of great significance to study the surrounding rock failure status and control techniques of cuts in thick coal seams with ultra-large spans.

In thick coal seam conditions, cutting areas are typically located within the coal seam itself, and their width is much larger than conventional roadway dimensions. Inadequate support parameters can jeopardize the stability of the surrounding strata in the cutting areas, severely impacting the normal progress of mining operations and the safety of underground personnel and mining equipment [5,6].

Regarding the support of ultra-large span open-off cuts, many researchers have made significant progress and achieved preliminary results. Chi G.M. et al. [7] investigated the control methods for the surrounding rock in ultra-large cutting areas in the Bu Lianta coal mine. They demonstrated that segmented support and hydraulic pre-splitting for roof control contribute to maintaining the stability of the roadway, providing valuable insights into deformation control for cutting areas. Ping Z. et al. [8] researched the impact of support structure instability on roadways in high-stress fractured strata and proposed a combined support technology based on the failure principles of anchor rod support, addressing the support challenges in high-stress fractured strata roadways. Wang H.W. et al. [9] studied the "anchor rod + short anchor cable + long anchor cable + steel shed" active-passive combined support method for a 5.5 m thick ultra-large span open-off cutting area in Zhaojiazhai, providing guidance for support in roadways with weak interlayers. He F.L. et al. [10], to address the control challenges in thick coal seam roofs and ultra-large span open-off cutting areas, used numerical simulation to analyze the influence of roof thickness and cutting areas width on surrounding rock failure characteristics and proposed a "cross-step" joint control technology. Xu Z.J. et al. [11] conducted theoretical analysis, numerical simulation, and field measurements to analyze stress distribution characteristics in roadways and deformation failure characteristics of surrounding rock, proposing a high pre-stressed roof beam high-strength double-anchor truss support system. Chai J. et al. [12], through theoretical analysis, numerical simulation, and field practice, analyzed the deformation and failure characteristics of large span open-off cutting areas, proposing and implementing a support method of "short, dense anchor cables + suspended anchor rods" that achieved good support results. Xie Z.Z. et al. [13,14] conducted comprehensive experiments using a pneumatic loading system to compare and study the control effects of three support systems on coal–rock composite roof roadways, offering insights into rapid excavation stability in deep mines and proposing a new support system. Liu Jianyu et al. [15] studied the comprehensive control technology for ultra-large span open-off cutting areas under complex technical conditions. The results showed that the application of this technology is safe and reliable, ensuring the stability of surrounding strata in cutting areas. Hu Y.R. et al. [16] addressed the challenges of large-span open-off cutting area support by conducting numerical simulation analysis of cutting area stress and plastic zone distribution, proposing a new long-short cable joint support scheme. Zhou Bo et al. [17] established a three-directional load-bearing beam structural analysis model for roadway roofs, determined stress analysis methods for beam structures under three-directional stress, and obtained the failure patterns under three-directional stress for open-off cutting areas.

While these studies have primarily focused on ultra-large spans or thick coal seams, research on support in ultra-large span open-off cutting areas in thick coal seams is relatively limited. Moreover, due to the diverse and complex mining environments, support solutions developed for one mine may not be entirely applicable to others. Therefore, this paper, based on the context of open-off cutting areas in the 81309 working face of Baode Coal Mine, explores strategies for controlling the surrounding rock in open-off cutting areas with thick coal seams and large spans. By analyzing the relationship between span and roof stability in thick coal seams, studying the influence of different active support parameters on roadway strata stability, and designing active support parameter schemes for cutting areas in Baode Coal Mine, this research aims to guide practical production work and holds significant engineering significance.

2. Stability Analysis of Ultra-Large Span Open-Off Cutting Areas

2.1. Engineering Background

The Baode Coal Mine is located within the jurisdiction of Qiaotou Town, Baode County, Shanxi Province. Figure 1 shows the schematic diagram of the traffic location of Baode Coal Mine, where the situ investigation was performed. The 81309 working face is situated west of the 81308 working face, with the southern boundary being the mining field boundary and the western boundary representing unmined coal seams. The coal rock layers exhibit a predominantly north–south orientation and are characterized by a westward dipping monocline structure, with coal seam dip angles averaging around 5°.

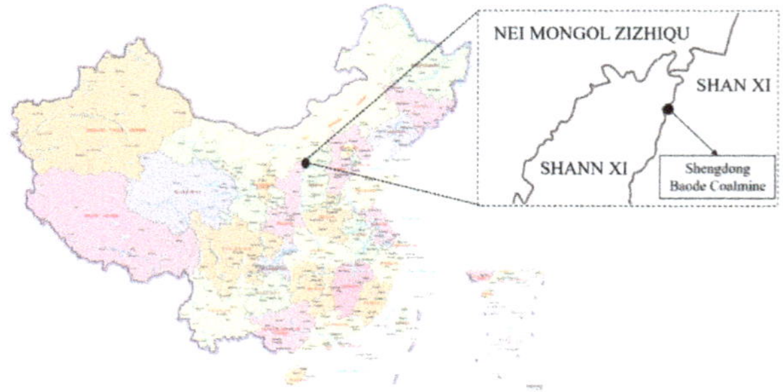

Figure 1. The location of Baode Coal Mine.

The coal seam thickness in the cutting section of the 81309 working face ranges from 5 to 8 m, with an average thickness of 7.1 m. The coal seam exhibits a complex structure with multiple interlayers of gangue, including 2–3 stable gangue layers. The rock types present include mudstone and carbonaceous mudstone. The immediate roof of the cutting section coal seam is composed of coarse-grained sandstone, with a thickness of 3.7 m, while the overlying stratum is characterized by medium-grained sandstone, with a thickness of 9.8 m. The immediate floor directly beneath the coal seam consists of fine-grained sandstone with an approximate thickness of 2.1 m. The coal seam pillar configuration is illustrated in Figure 2.

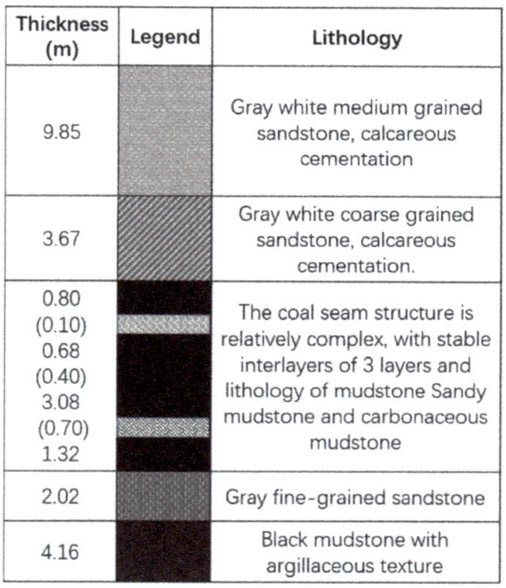

Figure 2. Coal seam histogram.

The width of the cutting section in the 81308 working face is approximately 8.8 m. However, for the upcoming excavation of the 81309 working face, the cutting section width will be increased to 9.8 m, with a height of 3.8 m. The maximum span of the roadway at the location where the chamber is installed in the cutting area is 11.8 m. The substantial increase in the cutting width for the 81309 working face necessitates a reevaluation of the support parameters for cutting section support, as adhering to the original support plan designed for the 81308 working face would pose challenges in ensuring effective cutting section support for the 81309 working face.

2.2. Analysis of Factors Influencing Stability in the Open-Off Cutting Areas

The 81309 working face at Baode Coal Mine is characterized as a large-scale, high-cut, comprehensive mining operation with pronounced susceptibility to the influence of substantial mechanical equipment. The design parameters specify a net width of 9.8 m at the cutting location, and an exceptional feature of this working face is the maximum span of the roadway, which reaches 11.8 m at the cutting installation chamber. The instability phenomenon is easy to occur in the middle of the roof of the roadway with such a large span. The thickness of the coal seam in the position of cutting holes fluctuates greatly, the roof structure is complicated, the coal seam contains many dirt layers, and the surrounding rock of the roadway is broken and difficult to support. Baode Coal Mine mining alternating work is relatively perfect; during the 81308 mining face, 81309 working roadway has completed the excavation work, and the service cycle of cutting the hole is as long as two years, further aggravating the difficulty of roadway support.

In light of the comprehensive analysis provided, it is evident that the stability control of the cutting area in the 81309 working face is influenced by multiple factors, with the significant impact of the expansive open-off cutting span being particularly prominent. Notably, certain sections of the excavation exhibit span up to 11.8 m, thereby presenting formidable challenges in terms of maintenance. The pronounced subsidence of the mid-section of the roadway roof, coupled with the inadequacy of the existing support strength, underscores the imperative need to explore effective methodologies for actively supporting the surrounding rock in ultra-large-span cutting areas to ensure the safety and integrity of mining operations within the coal mine.

2.3. Cutting Span and Stability of Working Face

(1) Mechanical Model

The relationship between the span of open-off cuts and the stability of the roof is a significant factor. To investigate this relationship, a theoretical model for analyzing the stress characteristics of open-off cuts was established. In this model, the open-off cut is treated as infinitely long, with no influence on the stability of the roof. Based on principles from elasticity theory, a theoretical model with fixed supports at both ends was developed, as shown in Figure 3.

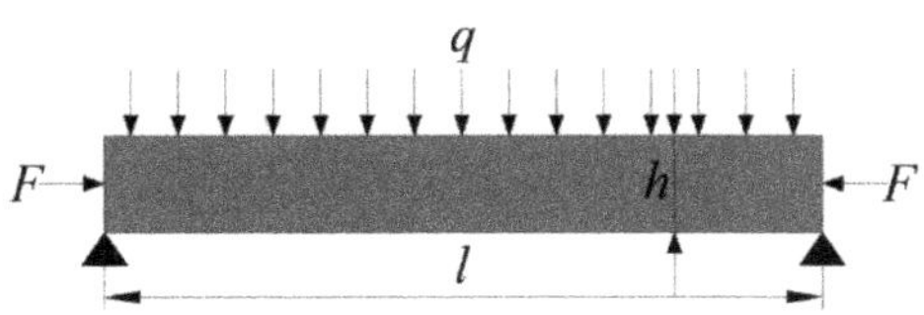

Figure 3. Theoretical model of stress on open-off cuts.

In the figure, h (m) corresponds to the height of the roadway, l (m) is the span, the upper portion of the roof is subjected to a uniformly distributed load q (Pa), and the horizontal support force is F (Pa).

According to the mechanical model, the relationship between the span of open-off cuts and the stability of the working face was analyzed. The critical stress when both ends are fixed is:

$$\sigma_s = \frac{\pi^2 E}{3(1-\mu^2)} \cdot \frac{1}{i^2} \quad (1)$$

where i represents the ratio of the open-off cut span to the thickness of the roof rock layer, known as the span-to-thickness ratio. From the equation, it can be observed that the critical stress for the model's stability is inversely proportional to the square of the span-to-thickness ratio. It is directly related to the elastic modulus and dependent on the Poisson's ratio but unrelated to the rock strength. This suggests that, under constant roof rock layer thickness, the span of the open-off cut and the mechanical properties of the rock layer are the determining factors for roof stability. As the span of open-off cuts in roadway excavation is twice that of conventional mining tunnels, the ratio of the critical stress-bearing capacity of the roof in open-off cuts to conventional tunnels is only one-fourth, indicating reduced roof stability in open-off cuts.

(2) Conditions for fracture instability

Due to the lower tensile strength of the rock layer, under significant in situ stress, it is highly likely that the rock beam may crack at the upper ends of the supports and then fracture near the middle or bottom of the beam. As the rock blocks rotate, horizontal compressive forces are formed, resulting in a three-hinge arch-shaped equilibrium. However, it is important to note that rock beams do not necessarily collapse as soon as they reach the fracture limit. The arching conditions primarily depend on the original rock stress and the magnitude of horizontal compressive forces formed during rock block rotation. When the horizontal compressive forces are sufficiently high, multiple rock blocks can still remain compressed together in a suspended state.

The conditions for sliding instability of the structure involve the magnitude of the frictional force at the interlocking point, which is the product of the horizontal compressive force and the coefficient of friction at that point. This force acts in the opposite direction to the direction of rock block sliding, serving to prevent the mutual sliding of rock blocks.

When the shear force exceeds the frictional force, the structure experiences instability, thus the conditions of instability, as shown in Figure 4.

$$i \geq \frac{1}{4} tg\phi \tag{2}$$

$$\sigma_t = K \cdot q \cdot \frac{l^2}{h^2/6} = 6Ki^2 q \tag{3}$$

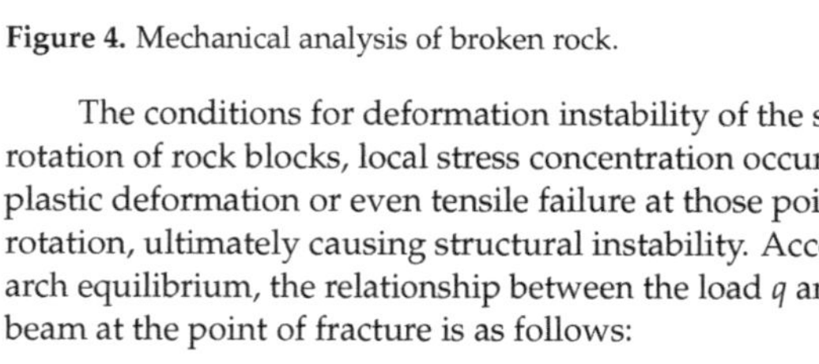

Figure 4. Mechanical analysis of broken rock.

The conditions for deformation instability of the structure are as follows: During the rotation of rock blocks, local stress concentration occurs at compression points, leading to plastic deformation or even tensile failure at those points. This can exacerbate rock block rotation, ultimately causing structural instability. According to the theory of three-hinge arch equilibrium, the relationship between the load q and the tensile strength σ_t of the rock beam at the point of fracture is as follows:

$$\sigma_t = K \cdot q \cdot \frac{l^2}{h^2/6} = 6Ki^2 q \tag{4}$$

The condition for structural deformation instability is that σ_t is greater than σ_c. The instability and failure of fractured rock layers are directly proportional to the square of the span-to-thickness ratio, positively correlated with the load, and inversely correlated with the compressive strength of the rock layer. This highlights the span-to-thickness ratio as a critical factor influencing the instability of rock beams.

Therefore, strengthening the roof support strength of ultra-large span open-off cuts and selecting appropriate support parameters become essential tasks in ensuring mine safety and production efficiency.

3. Design of Active Support Parameters for Surrounding Rock with Ultra-Large Span Open-Off Cut in Thick Coal Seam

3.1. Theoretical Analysis of Support Parameter Design

Ultra-large span open-off cuts differ from conventional roadways, as the critical stress-to-span ratio that the cutting area roof can withstand is only one-fourth that of the roadway. At the installation area of the underground chamber within the open-off cuts, the span reaches a maximum of 11.8 m. Numerical simulation results indicate a significant stress concentration at the expanding openings and the central roof area, which poses considerable challenges for the support of these galleries. To design and optimize the support parameters for these ultra-large span open-off cuts, Proctor's theory and the natural equilibrium arch theory were employed to calculate the unstable rock layer thickness suspended in the open-off cuts [15].

The expression for the natural equilibrium arch height b can be represented as follows:

$$b = \frac{\left[a + h \cdot ctg\left(45^\circ + \frac{\theta}{2}\right)\right]}{f} \tag{5}$$

where a is half of the open-off cut span, taken as 4.9 m; f is the direct roof Proctor's coefficient, with a value of 3 for coarse sandstone; h is the gallery height, set at 3.8 m; θ is the friction angle, with a value of 25° for the coal layer where the open-off cuts are located; f is the Proctor's coefficient for coal, set at 1.13.

The formula for calculating the length of the rock bolts is given by:

$$l = l_1 + l_2 + l_3 \tag{6}$$

where l_1 is the exposed length of the rock bolt and is taken as 0.1 m; l_2 is the effective length of the rock bolt, ranging from 1.0 to 1.2 m; l_3 is the length of the rock bolt deeply embedded into the basic roof, taken as 0.3–0.4 m based on field experience.

The formula for calculating the rock bolt diameter is:

$$d = \sqrt{\frac{4Q}{\pi \sigma_t}} \tag{7}$$

where d is the diameter of the rock bolt in meters; Q is the anchoring force of the rock bolt, obtained from on-site tests, measured at 150 MPa; σ_t is the tensile strength of the rock bolt, which is a high-strength left-handed threaded steel bolt without longitudinal ribs.

The designed length of the anchor cable should satisfy the following equation:

$$L = L_a + L_b + L_c + L_d \tag{8}$$

where L represents the design length of the anchor cable. L_a denotes the anchoring length of the anchor cable into the relatively stable rock layer.

$$L_a \geq K \times \frac{d_1 f_a}{4 f_c} \tag{9}$$

where K is the safety factor, commonly set at 2 based on empirical data; d_1 is the diameter of the cable; f_a is the tensile strength of the cable itself; f_c is the anchoring strength, set at 400 MPa; L_b is the thickness of the unstable rock layer to be suspended, equivalent to the height of the natural equilibrium arch [18]; L_c is the thickness of the tray and anchors; L_d is the exposed lifting length.

By combining Equations (4) to (8), the calculations for the length of the bolts, bolt diameter, cable length, and the spacing between bolts were obtained, as shown in Table 1.

Table 1. Rock bolt, cable calculations, and design values.

Bolt Cable Design	Theoretical Calculated Value (m)	Design Value (m)
Bolt length	≥1.925	2.0
Bolt diameter	≥19.0	20.0
Cable length	≥7.365	8.0

Based on the calculated support parameters and in conjunction with commonly used materials in mines, rational design values for the support parameters were derived. The designed support cross-sections are illustrated in Figure 5 (roadway surrounding rock support diagram). The initial design for the support of the open-off cuts in the 81309 working face, design Scheme 1 is as follows:

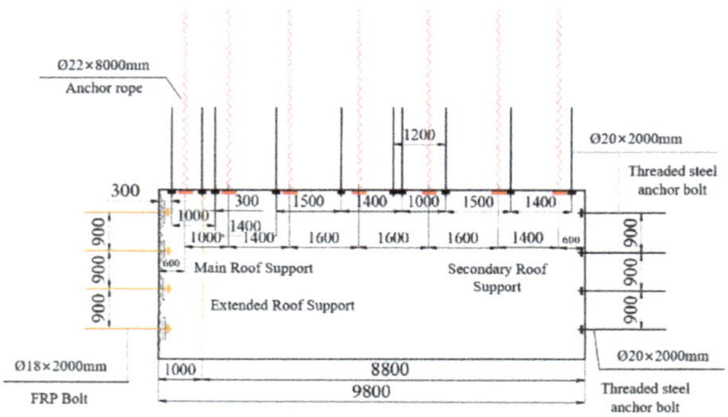

(**a**) Support design Scheme 1 profile.

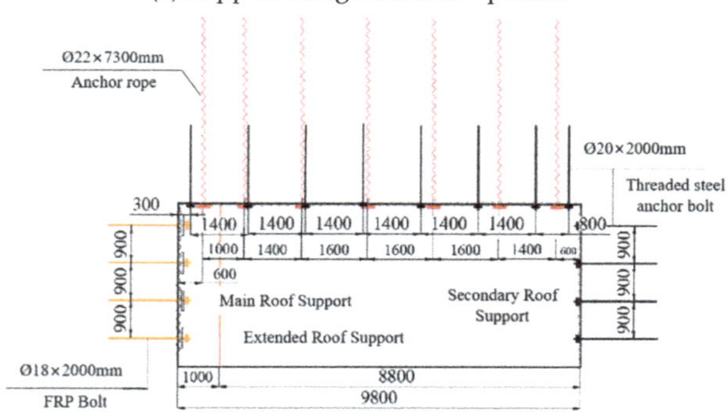

(**b**) Support design Scheme 2 profile.

Figure 5. Roadway surrounding rock support diagram.

Roof: Ø20 mm × 2000 mm high-strength left-handed threaded steel bolts, spaced at 1000 mm, 1400 mm, and 1500 mm, with intervals of 900 mm; Ø22 mm × 8000 mm anchor cables, spaced at 1400 mm × 900 mm and 1600 mm × 900 mm.

Rib: Ø20 mm × 2000 mm high-strength left-handed threaded steel bolts; Ø18 mm × 2000 mm fiberglass reinforced plastic (FRP) bolts used for cut-off ribs, spaced at 900 mm × 900 mm. Two additional rock bolts were added at the expanding openings of the open-off cuts to enhance the support strength.

The support parameters for the cutting areas in the 81309 working face were obtained using the analogy method. The support scheme was applied by comparing the lengths and spacing of the rock bolts and cables from the 81308 working face. The support parameters for the open-off cuts in the 81309 working face, design Scheme 2, are as follows:

Roof: Ø20 mm × 2000 mm high-strength left-handed threaded steel bolts, spaced at 1400 mm × 1000 mm; Ø22 mm × 7300 mm anchor cables, spaced at 1400 mm × 1000 mm and 1600 mm × 1000 mm.

Rib: Ø20 mm × 2000 mm high-strength left-handed threaded steel bolts; Ø18 mm × 2000 mm fiberglass reinforced plastic (FRP) bolts used for cut-off ribs, spaced at 900 mm × 1000 mm. The rib bolts are installed horizontally, and on one side of the cut-off rib, fiberglass-reinforced plastic (FRP) bolts are used.

3.2. Numerical Model Establishment

Based on the geological data background of the Baode Coal Mine and the mechanical parameters of the roof and floor of the 81309 working face, a geological model was established using the FLAC3D 5.00 numerical simulation software, with numerical dimensions of 100 m × 100 m × 80 m. Displacement was constrained in the X, Y, and Z-zero directions of the model, and a uniform load of 15 MPa was applied to the top surface, with gravity acceleration set at 10 m/s^2. The Mohr–Coulomb constitutive model was adopted, and the model's bottom and sides were fixed. Different excavation steps were set up to conduct separate simulations for design Scheme 1 and design Scheme 2. The aim was to investigate the stress, deformation, and failure characteristics of thick coal seam surrounding rock with an exceptionally large span under different support conditions and to perform a comparative analysis of the overall control effectiveness of the surrounding rock. The mechanical parameters of the surrounding rock are shown in Table 2, and the numerical simulation model is depicted in Figure 6.

Table 2. Rock mechanical parameters.

Lithology	Density/(kg·m^{-3})	K/GPa	C/MPa	φ/(°)
Coarse sandstone	2390	8.3	4.0	30
Coal	1390	4.2	1.0	28
Mudstone	2250	5.0	2.0	30
Sandy mudstone	1990	6.0	2.5	30
Fine sandstone	2550	10.4	5.4	33
Medium-grained sandstone	2050	9.4	4.4	31

Figure 6. Numerical model diagram.

3.3. Numerical Analysis of Support Parameter Design

Monitoring of displacements, stress states, and plastic deformation zones at 30 m and 60 m positions in the roadway for coal mining revealed the following simulated results:

(1) Displacement field analysis of surrounding rock

As shown in Figure 7, numerical simulation results indicate that compared to support design Scheme 2, both the bottom heave in the floor and floor heave in Scheme 1 remain within 40 mm. However, the peak subsidence of the roof in Scheme 2 reaches 81.3 mm, while in Scheme 1, the peak subsidence of the roof is 67.9 mm. Under the conditions of Scheme 1, compared to Scheme 2, the sidewall displacements are reduced by 16.6%, and the roof subsidence is reduced by 16.3%. Therefore, Scheme 1 exhibits superior support effectiveness in controlling the roof layer displacement in the roadway compared to Scheme 2.

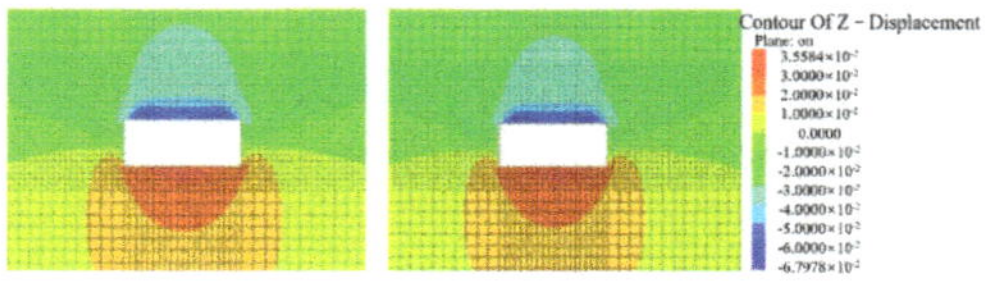

(a)

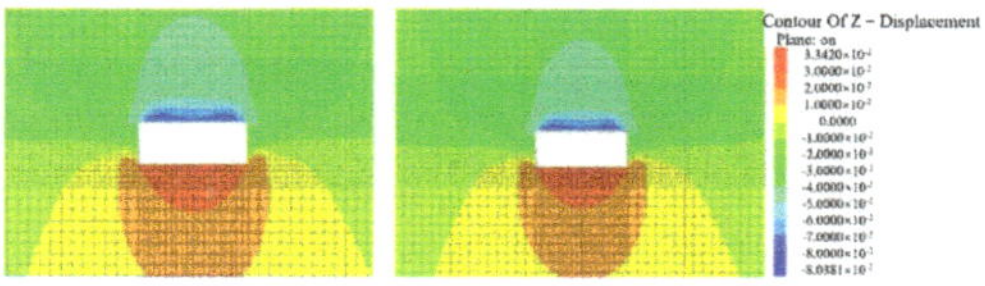

(b)

Figure 7. Displacement field of surrounding rock: (**a**) rock displacement in the surrounding area at 30 m and 60 m for Scheme 1; (**b**) rock displacement in the surrounding area at 30 m and 60 m for Scheme 2.

(2) Stress field analysis of surrounding rock

When comparing the stress fields of surrounding rock for the two support design schemes, it is observed that under Scheme 1, the load-bearing capacity of the surrounding rock is increased, resulting in a reduction in peak compressive stress by 3 MPa compared to Scheme 2. The stress concentration areas on both sidewalls are reduced for both schemes. Both schemes exhibit a bimodal stress relief area on the roof, with peak tensile stress concentrated in the middle of the roof. In Scheme 1, the peak tensile stress is reduced by 1 MPa compared to Scheme 2. The stress relief range in the sidewall is reduced by 5% in Scheme 1 compared to Scheme 2, and the high-stress concentration range is reduced by 21%. Therefore, Scheme 1 exhibits a significant effect in controlling the degree of stress concentration in the surrounding rock of the roadway compared to Scheme 2.

(3) Plastic zone analysis of surrounding rock

Through numerical simulation calculations, the distribution of plastic zones in the surrounding rock of the roadway is obtained for both support design schemes, as shown in Figure 8.

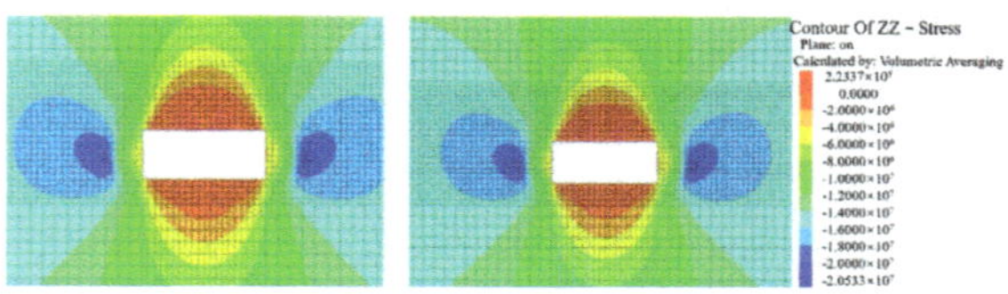

(a)

Figure 8. *Cont.*

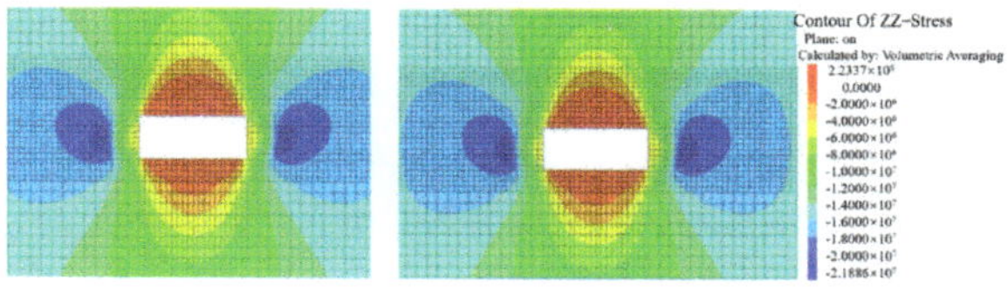

(b)

Figure 8. Surrounding rock stress field: (**a**) vertical stress at 30 m and 60 m for Scheme 1. (**b**) Vertical stress at 30 m and 60 m for Scheme 2.

From Figure 9, it is evident that under support design Scheme 1, the distribution range of the plastic zone in the surrounding rock is significantly reduced compared to Scheme 2. In Scheme 2, there is a tension–shear failure zone near the roadway, whereas in Scheme 1, shear failure is observed around the tunnel, with no tension–shear failure. The depth of floor failure in Scheme 2 is greater than that in Scheme 1, with a relatively larger range of deformation failure. Numerical simulation demonstrates that under Scheme 1, the support enhances the stability of the roadway, significantly reducing plastic failure in the top corners of the tunnel and resulting in a 26% decrease in the plastic failure area.

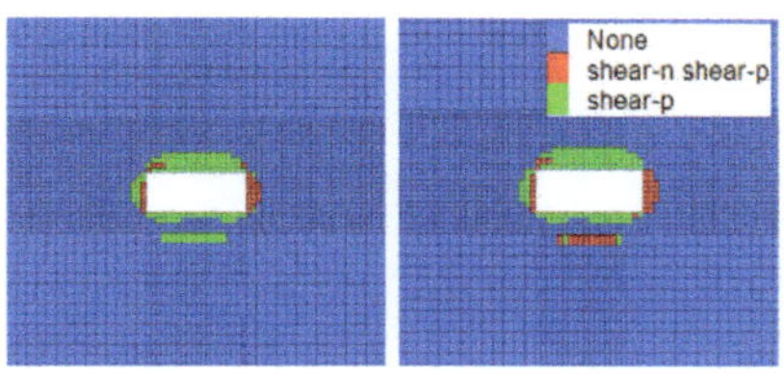

(a)

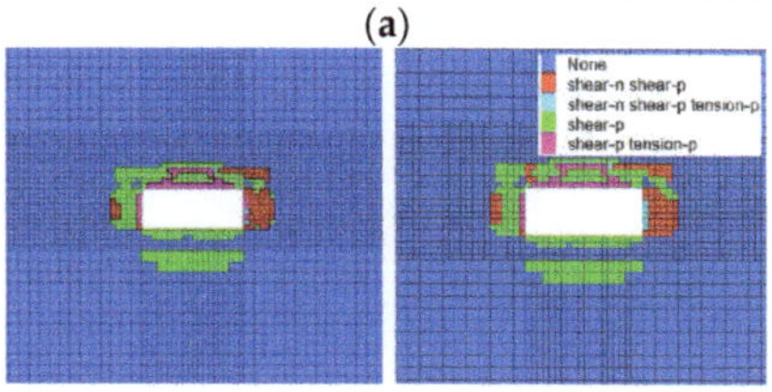

(b)

Figure 9. Plastic zone of surrounding rock: (**a**) plastic failure at 30 m and 60 m for Scheme 1; (**b**) plastic failure at 30 m and 60 m for Scheme 2.

Through a numerical simulation and analysis using FLAC3D, a comparison of displacement variations, stress states, and plastic failure zones of the surrounding rock in the roadway was conducted. It is evident that the support effectiveness of Support Scheme 1 surpasses that of Scheme 2 obtained through analogous methods. Scheme 1 effectively maintains the stability of the surrounding rock in ultra-large span roadway excavation, achieving the required support effect. Therefore, Scheme 1 should be applied for the roadway support in the 81309 working face during excavation in the coal mining industry.

4. Field Application

4.1. Test Site

In order to assess the effectiveness of the support design scheme for large-span cutting in thick coal seams, displacement monitoring of the surrounding rocks in the excavated area was conducted. Within the cutting area of 81309 working face, three sets of roadway surrounding rock displacement monitoring stations were installed. The first station was positioned 30 m from the end of the cutting, and subsequent stations were placed at 30 m intervals, named 1# measuring point, 2# measuring point, and 3# measuring point. Deep-seated displacement monitoring devices were buried to measure the deformation of the surrounding rocks, aiming to validate the rationality of the support optimization scheme. The specific layout of monitoring stations is illustrated in Figure 10.

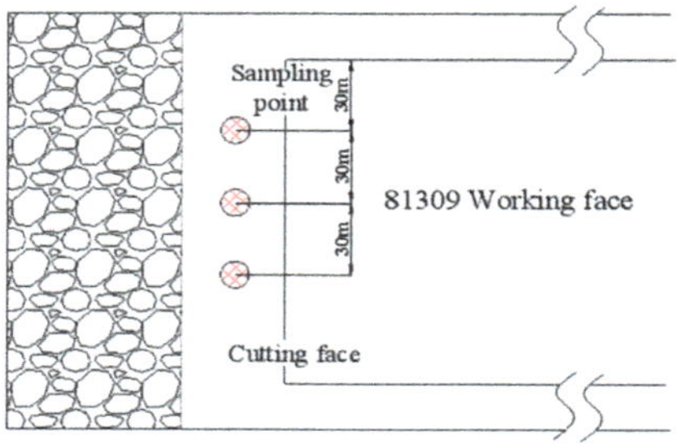

Figure 10. Layout diagram of roadway surface displacement measuring station.

4.2. Monitoring of Cutting Surrounding Rock Deformation

Accurate measurement of the displacement of the cutting surrounding rocks was performed, and irrelevant data from abandoned monitoring points were excluded. The relationship between the number of monitoring days and displacement is illustrated in Figure 11.

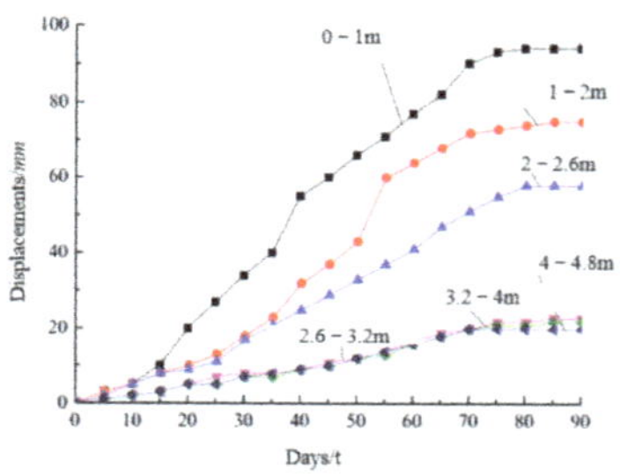

(a)

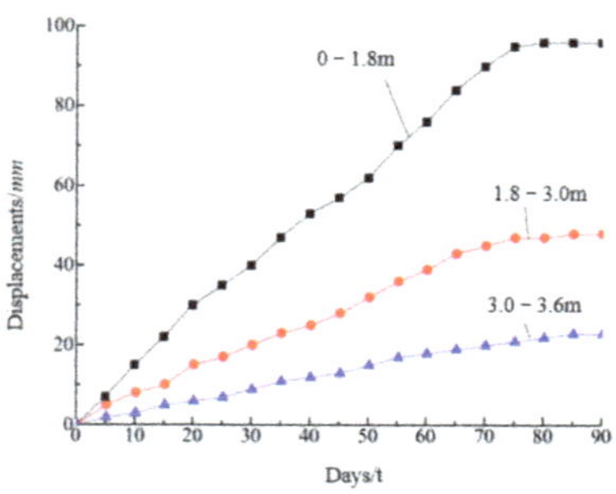

(b)

Figure 11. Approach amount of roadway roof and two sides in 81309 working face cutting: (**a**) approach amount of both sides of the roadway in the 81309 working face cutting: (**b**) changes in the roof and floor of the 81309 working face cutting.

From the observation data in Figure 10, it is observed that with the passage of time, the displacement of the roadway roof, floor, and sides generally increased within the first 60 days and then gradually stabilized.

Both the roof and the sides of the roadway exhibited maximum displacement at shallow depths. The maximum approach amount of the roof reached 96 mm. A comprehensive analysis of the monitoring results indicated that the displacement of the roadway sides and the roof was less than 100 mm, and the displacement at a depth of more than 2 m from the cutting surface was less than 60 mm. These values met the requirements of the design regulations. The on-site experimental results were consistent with the results obtained from numerical simulations.

5. Conclusions

(1) In order to effectively control the stability of the surrounding rock of the large-span cutting roadway, combined with the geological conditions of Baode Coal Mine, the optimal supporting scheme parameters are proposed by using theoretical analysis, numerical simulation, and field tests.
(2) The ultra-large span of the open-off cutting in thick coal seams is the primary factor affecting the stability of the cutting. The span is inversely proportional to the stress borne by the cutting roof. The ratio of span to thickness is a key factor influencing the stability of the roof layer and the stability of rock beams.
(3) Through comprehensive comparative analysis of changes in surrounding rock stress, displacement, and plastic zone through numerical simulation, Scheme 1 proved significantly superior to Scheme 2. Under Scheme 1, the approach amount of the roof, floor, and sides decreased by approximately 16%, and the extent of plastic zone damage reduced by 26%.
(4) Under Scheme 1, roadway surrounding rock displacement monitoring was conducted after the cutting excavation. The monitoring data showed that the maximum deformation of the surrounding rocks was less than 100 mm, complying with the design regulations. This indicates that the support parameters of this scheme effectively maintain the stability of the cutting area.

Author Contributions: Methodology, investigation, and writing—original draft, L.P.; formal analysis, conceptualization, and writing—review and editing, Y.L. and Z.S.; writing—review and editing, Y.C. and X.Z. All authors have read and agreed to the published version of the manuscript.

Funding: This study was supported by the Science and Technology Innovation and Entrepreneurship project of TDTEC (No. 2022-QN001, No. 2022-2-MS003).

Institutional Review Board Statement: Not applicable.

Informed Consent Statement: Not applicable.

Data Availability Statement: The data presented in this study are available on request from the corresponding author. The data are not publicly available due to the containing information that could compromise the privacy of research participants.

Conflicts of Interest: The authors declare that this study received funding from the Science and Technology Innovation and Entrepreneurship project of TDTEC. The funder was not involved in the study design, collection, analysis, interpretation of data, the writing of this article or the decision to submit it for publication.

·eferences

Wang, X.; Tang, J.; Li, Y. The Failure Law and Combined Support Technology of Roadways with Weak Surrounding Rock in Deep Wells. *Appl. Sci.* **2023**, *13*, 9738. [CrossRef]
Gu, S.; Wang, X.; Huang, R.; He, H. Analysis of High-Stress Roadway Floor Deformation Mechanism and Control Technology. *J. Saf. Sci. Technol.* **2020**, *16*, 57–63.
Chai, Z.; Kang, T.; Li, Y. The Role of Large Section Cutting Hole Anchorage Support in Extra Thick Coal Seam. *J. China Coal Soc.* **2008**, *7*, 732–737.

4. Jia, H.; Ma, N.; Zhao, X. Collapse Law of Large-Span Cutting Roof in Deep Buried Thin Bedrock. *J. Min. Saf. Eng.* **2014**, *5*, 702–7
5. Li, C.; Cao, Y.; Cheng, Z. Research on Roof Collapse and Bending Law of Large-Span Roadway and Support Technology. *J. M Saf. Eng.* **2015**, *6*, 978–983+988.
6. He, F.; Xu, L.; Wu, H.; Wang, Y. Evolution of Fracture Field in Thick Coal Roof Large Section Cutting Hole and Surrounding Ro Stability Analysis. *J. China Coal Soc.* **2014**, *2*, 336–346.
7. Chi, G.; Zhang, L.; Tang, H. Comprehensive Control Technology for Extra Large Section Cutting Hole Surrounding Rock. *Coal S Technol.* **2017**, *11*, 27–31.
8. Ping, Z.; Ma, N.; Wei, J. Combined Support Technology of Large Section Roadway in High-stress Fractured Surrounding Ro *Procedia Eng.* **2011**, *26*, 1270–1278. [CrossRef]
9. Wang, H. Active and Passive Combined Support Technology for Large Span Cutting Hole in Thick Coal Seam. *Coal Sci. Techn* **2018**, *46*, 33–38.
10. He, F.; Bo, Y.; Xu, Z. Study on Surrounding Rock Stability and Reasonable Anchor Configuration of Thick Coal Seam Cutti Hole. *J. Min. Saf. Eng.* **2015**, *2*, 233–239.
11. Xu, Z.; Li, C.; Cao, Y. Study on the High Strength Cable Truss System Control of the Surrounding Coal Roadway Excavated Regions Left Intensively Mining-Induced Stresses. *Appl. Sci.* **2023**, *13*, 10504. [CrossRef]
12. Chai, J.; Han, Z.; Qiao, Y. Stability Research of Top Coal in Large Span Cutting Hole under Stratified Mining of Goaf. *J. Min. S Eng.* **2022**, *2*, 282–291.
13. Xie, Z.; Zhang, N.; Qian, D. Rapid Excavation and Stability Control of Deep Roadways for an Underground Coal Mine with Hi Production in Inner Mongolia. *Sustainability* **2018**, *10*, 1160. [CrossRef]
14. Xie, Z.; Li, Y.; Zhang, N. Model experiment research on HPTL anchoring technology for coal-rock composite roof in deep roadw *Sci. Rep.* **2023**, *13*, 2381. [CrossRef] [PubMed]
15. Liu, J.; Zhang, L.; Xie, L. Comprehensive Control Technology Research of Extra Large Section Cutting Hole under Compl Technical Conditions. *Coal Sci. Technol.* **2021**, *49*, 55–59.
16. Hu, Y.; Liu, Y.; Shi, L. Research on Supporting Technology for Surrounding Rock of Inclined Large-Span Open-Off Cut Roadw *Geotech. Geol. Eng.* **2020**, *38*, 1873–1884. [CrossRef]
17. Zhou, B.; Yuan, L.; Xue, S. Analysis of Weakening the Bearing Capacity of Roof Strata in Coal Roadways. *J. Saf. Sci. Technol.* **20** *9*, 122–128.
18. Luo, X.; Gao, X.; Wang, W. Optimization and Application of Bolt and Cable Support Parameters in Roadway Based on Natu Balance Arch. *Coal Technol.* **2022**, *41*, 56–59.

applied sciences

MDPI

rticle

Molecular Simulation Study on the Effect of Co-Associated Minerals on Methane Adsorption and Mechanical Properties of Coal

iaoyu Zhang [1], Yingjie Liu [1], Tianbai Zhou [1], Yongbo Cai [1,*] and Bin Zhang [2]

1 Chinese Institute of Coal Science, Beijing 100013, China; zhangxy@ccteg-bigdata.com (X.Z.); liuyingjie@mail.ccri.ccteg.cn (Y.L.); zhoutb@ccteg-bigdata.com (T.Z.)

2 Key Laboratory of In-Situ Property Improving Mining of Ministry of Education, Taiyuan University of Technology, Taiyuan 030024, China; zhangbin@tyut.edu.cn

* Correspondence: caiyongbo@mail.ccri.ccteg.cn

Abstract: When rockbursts and coal and gas outbursts simultaneously occur in a coal mine, changes in gas adsorption (concentration of ambient methane) and displacement of coal and rock must occur. The co-associated minerals in coal reservoirs can affect the mechanical properties and methane adsorption capacity, which are commonly disregarded. It is important to construct compound molecular structure models of coal and rock and conduct molecular dynamic simulations to gain a microscopic understanding of underground disasters. In this work, the molecular structure models of anthracite and coking coal–rock compound models containing different contents of calcite and kaolinite were constructed, and the methane adsorption amount and mechanical properties considering temperature, pressure, and mineral contents were simulated and analysed. The results showed that the methane adsorption amount of the compound models increased rapidly, then increased moderately, and stabilized eventually with increasing adsorption pressure, and the Langmuir fitting findings were good. The saturation adsorption amount of methane in the coal models linearly decreased with increasing temperature, while the methane adsorption heat increased. The presence of minerals adsorbed a certain amount of methane, and the methane adsorption amount increased with increasing mineral contents. The mechanical properties of coal molecules changed when mineral molecules such as calcite and kaolinite were present, which had opposite contribution effects. The addition of kaolinite minerals to the coal molecular model always increased the bulk modulus and shear modulus, while the addition of calcite decreased the bulk modulus of the anthracite, causing an increase in the brittleness of the models. The results of the study further explain the adsorption behaviour and mechanical properties of methane in coal and minerals.

Keywords: coal–rock compound molecular structure models; methane adsorption; mechanical properties; coal; mineral matter; coal and gas outburst

1. Introduction

Since the 21st century, coal mining technology has been greatly upgraded, integrated mining and caving mining methods have been promoted, and the number of rockburst coal mines in China increased to 121 by 2008. With the increasing coal mining intensity and depth in China, the number of mines with rockbursts rapidly increased from 142 in 2012 to 177 in 2017 [1]. The accident investigations revealed that two disasters, a rock burst and a coal and gas outburst, simultaneously occurred. For example, the average mining depth of the Pingdingshan No. 12 Coal Mine reached 1100 m, and the rockburst and coal and gas outbursts occurred on 29 June 2005 and 19 March 2006 [2]. The occurrence mechanism of compound disasters is complex, and the prediction index for a single disaster may be invalid [3]. Research on the mechanism of compound disasters is one of the most fundamental research contents in the study of compound disasters and provides a theoretical basis for the evaluation, prediction, and prevention of compound disasters [4]. Just like the human gene sequence, the molecular structures determine the thermodynamic, optical, electromagnetic, and surface properties of coal. The mechanism of methane adsorption and mechanical properties can be reflected through molecular dynamic simulation, so varied coal molecular structure models under different geological conditions need to be built urgently, and the effects of pressure, temperature and mineral contents on methane adsorption need to be clarified. To clarify the mechanism of rockbursts, coal and gas outbursts, and the relationship between the two disasters, this study constructs coal–rock compound structure models, which are intended to provide a reference for the dynamic mechanism of compound disasters.

Analysis and interpretation of the occurrence mechanism of rockburst include strength theory [5], stiffness theory [6], energy theory [7], impact propensity theory [8], "three criteria" theory [9], deformation system instability theory [10], and "three factors" theory [11]. There are three important factors in the occurrence of a rockburst hazard, the propensity to impact is the intrinsic factor, the high-stress concentration and dynamic disturbance is the source factor, and the presence of soft layers in the coal rock is the structural factor. Qi et al. pointed out that a coal rock structure with impact tendency deforms under the action of high stress, forming a high-stress concentration and locally gathering energy. Under the disturbance of mining stress, stick–slip and the release of a large amount of energy along the weak surface or contact surface of the coal rock structure will cause a rockburst disaster [12]. Analysis and interpretation of the occurrence mechanism of coal and gas outbursts include the comprehensive action hypothesis [13], rheological hypothesis [14], spherical shell instability hypothesis [15], and solid–fluid coupling instability theory [16]. Wang et al. reported that coal and gas outbursts are the result of the comprehensive action of stress, the gas contained in the coal seam, and the physical and mechanical properties of the coal seam itself [17]. Field monitoring data, laboratory tests, and molecular dynamics simulation results confirmed that a large amount of gases such as CH_4 and CO burst out when the shear toughness zone of the coal seam was damaged [18–21]. Therefore, the construction of the atomic representation of coal–rock compound structure models and the adsorption characteristics and mechanical properties of methane in the model need further development.

Molecular dynamics simulation was first proposed by Alder and Wainwright in 1957 [22]. To date, molecular simulation technology has been successfully applied in the chemical [23], materials [24], and energy [25,26] industries. In 1942, the University of Pennsylvania built the first coal molecular structure model [27]. Hundreds of coal molecular structure models have since been successfully constructed worldwide. Recently, molecular dynamics simulation in coal and rock models has become an important tool for analysing the adsorption and desorption behaviour of CH_4, CO_2, and other gases in extreme environments [28–32]. In addition, Wang et al. used molecular dynamics to simulate the change in coal molecular structures and the type of gas generated under shear stress and pointed out that shear gas production is the main source of excessive gas emissions when coal and gas outburst disasters occur [20,21]. To explore the mineralization mechanism of coal-bearing graphite, Ma et al. constructed a high-rank coal molecular structure model of Fengxian in Shaanxi [33]. The CH_4 adsorption capacity of coal is influenced by different dynamic and static factors. The dynamic factors generated by the redistribution of the initial rock stress by coal mining or methane extraction, static factors such as coal ranks, moisture contents, temperatures, and minerals of methane adsorption in coal also have attracted extensive attention. Molecular dynamics simulation of coal or rock reflects the influence mechanism of temperature, pressure, water content, and other parameters on the molecular structures and the adsorbed gas therein through volume expansion, interaction energy, and radial distribution function. Zhang et al. constructed an anthracite molecular structure model of the Qinshui Basin and simulated the methane adsorption capacity under different adsorption pressure conditions and the changes in the mechanical properties of anthracite after saturated adsorption [30].

In this work, atomistic representations of a coal–rock compound structure of anthracite in the Qinshui Basin and coking coal in the Ordos Basin were constructed, and molecular dynamic simulations were performed to characterize methane adsorption and mechanical properties considering temperature, pressure, and mineral content. The response characteristics of different associated mineral types and contents to the adsorption characteristics and mechanical properties of methane in coal from the molecular level were explained, and a micro mechanism for the occurrence mechanism of rockburst and coal and gas outburst combined disasters was provided.

2. Molecular Structure Models

Combined with elemental analysis, Fourier transform infrared (FTIR) spectrometry, ^{13}C NMR spectroscopy, X-ray photoelectron spectroscopy (XPS), and transmission electron microscopy of coal samples, the molecular structures of anthracite and coking coal samples, are constructed [34], so the detailed construction process has been overlapped. According to the test results, the 3D models of the molecular structure of the anthracite and coking coal are constructed in Materials Studio software 2020 (as shown in Figure 1). The chemical formulas of the anthracite and coking coal are $C_{220}H_{82}N_2O_{13}S_2$ and $C_{213}H_{114}O_{16}N_2S$, respectively, and the dimensions of their 3D models are 1.53 × 1.53 × 1.53 nm and 1.56 × 1.56 × 1.56 nm, respectively. The minerals on the coal surface are mainly kaolinite and calcite [35], thus, calcite and kaolinite molecular are added to the molecular structure models of the anthracite and coking coal to build the coal–rock compound molecular structure model with different mineral contents in Materials Studio software 2020 [36]. It is noted that the methane adsorption on the edges of kaolinite and calcite should be studied before the adsorption simulations in the compound models; the omission of this section is due to the methane isothermal adsorption on a single pure mineral surface (0 0 1) and coal surfaces having been presented and validated in our previous work [37].

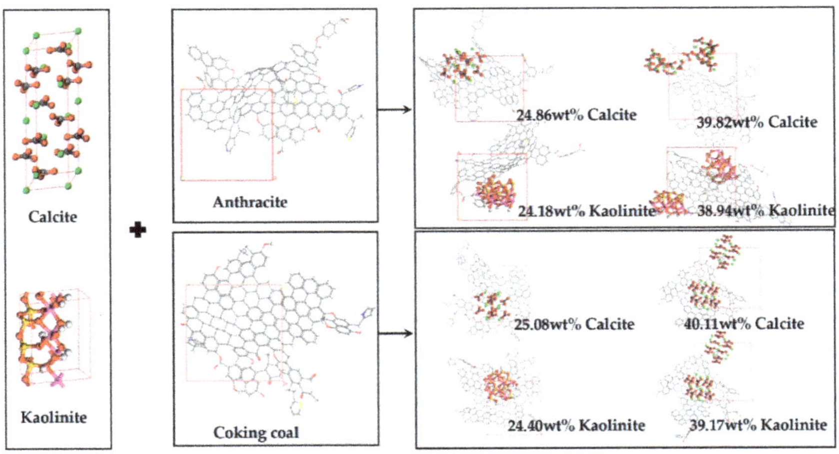

Figure 1. Schematic of coal–rock compound structure models.

In Figure 1, carbon, hydrogen, nitrogen, oxygen, sulphur, calcium, silicon, and aluminium atoms are represented by grey, white, blue, red, yellow, green, dark yellow, and magenta, respectively. Eight coal–rock compound molecular structure models with different kaolinite and calcite mineral contents were constructed. When adding calcite with mass fractions of 25.08w% and 40.11 wt% to the anthracite coal model, the chemical formulas of the model are $C_{230}H_{82}N_2O_{43}S_2Ca_{10}$ and $C_{240}H_{82}N_2O_{73}S_2Ca_{20}$, and the model sizes are 1.68 × 1.68 × 1.68 nm and 1.81 × 1.81 × 1.81 nm. When kaolinite with mass fractions of 24.18w% and 38.94 wt% is added to the anthracite coal model, the chemical formulas of the model are $C_{220}H_{90}N_2O_{45}S_2Al_8Si_8$ and $C_{220}H_{98}N_2O_{77}S_2Al_{16}Si_{16}$, with model sizes of 1.68 × 1.68 × 1.68 nm and 1.80 × 1.80 × 1.80 nm. When adding calcite with mass fractions of 25.08w% and 40.11 wt% to the coking coal model, the chemical formulas of the model are $C_{230}H_{82}N_2O_{43}S_2Ca_{10}$ and $C_{240}H_{82}N_2O_{73}S_2Ca_{20}$, and the model sizes are 1.72 × 1.72 × 1.72 nm and 1.85 × 1.85 × 1.85 nm. When kaolinite with mass fractions of 24.40w% and 39.17 wt% is added to the coking coal model, the chemical formulas of the model are $C_{213}H_{122}O_{48}N_2SAl_8Si_8$ and $C_{213}H_{130}O_{80}N_2SAl_{16}Si_{16}$, with model sizes of 1.72 × 1.72 × 1.72 nm and 1.84 × 1.84 × 1.84 nm. Note that the size of the coal models created in the manuscript was indeed small (<2 nm), and readers should critically refer to refer to the simulation results.

3. Computational Methods

3.1. Simulation Scheme

As mentioned above, the minerals on the coal surface are mainly kaolinite and calcite; thus, calcite and kaolinite molecules were added to the molecular structure models of anthracite and coking coal to build the coal–rock compound molecular structure model with different mineral contents. The molecular dynamics simulation schemes include two coal molecular structure models and eight coal–rock compound models. For each coal molecular model, five adsorption temperatures (273.15, 283.15, 293.15, 303.15, and 313.15 K) and thirteen adsorption pressures (0.1, 0.3, 0.5, 1, 2, 3, 4, 5, ..., 10 MPa) were selected and calculated, Refs. [30,32], as showed in Table 1. Under each constant temperature, methane adsorption simulations with different adsorption pressures were carried out in each coal molecular model. Each coal–rock compound model was used to simulate different methane adsorption pressures at 273.15 K.

Table 1. Simulation Schemes.

Scheme	Molecular Models	Adsorption Pressure (MPa)	Adsorption Temperature (K)
1	Anthracite coal	0.1, 0.3, 0.5, 1, 2, 3...10	273.15, 283.15, 293.15, 303.15, 313.15
2	Anthracite + 24.86 wt% Calcite	0.1, 0.3, 0.5, 1, 2, 3...10	273.15
3	Anthracite + 39.82 wt% Calcite	0.1, 0.3, 0.5, 1, 2, 3...10	273.15
4	Anthracite + 24.18 wt% Kaolinite	0.1, 0.3, 0.5, 1, 2, 3...10	273.15
5	Anthracite + 38.94 wt% Kaolinite	0.1, 0.3, 0.5, 1, 2, 3...10	273.15
6	Coking coal	0.1, 0.3, 0.5, 1, 2, 3...10	273.15, 283.15, 293.15, 303.15, 313.15
7	Coking coal + 25.08 wt% Calcite	0.1, 0.3, 0.5, 1, 2, 3...10	273.15
8	Coking coal + 40.11 wt% Calcite	0.1, 0.3, 0.5, 1, 2, 3...10	273.15
9	Coking coal + 24.40 wt% Kaolinite	0.1, 0.3, 0.5, 1, 2, 3...10	273.15
10	Coking coal + 39.17 wt% Kaolinite	0.1, 0.3, 0.5, 1, 2, 3...10	273.15

3.2. Implementation of Molecular Simulations

The simulation details of molecular dynamics have been elaborated in detail in our previous work [30,37]. The force field utilized in the model construction and methane adsorption simulations was the Dreiding force field, and the simulations were completed by the Amorphous cell module, Sorption module, and Force module in the Materials Studio software 2020. The mechanical property simulation was completed in Mechanical Properties in the Force module, also using the Dreiding force field [38]. Note that a suitable force field selection is the key to the accuracy of molecular simulation results. The force field form selected in the simulation was only one of many force fields [38–45], and the simulation parameters and results were expected to be critically referenced by researchers. The cutoff distance for methane adsorption is 12.5 Å, and the amount of absolute adsorption was collected. To calculate the saturation sorption amount of methane, V_L, on modified coal, the Langmuir isothermal sorption equation was fitted to the test results.

$$V_{ab} = \frac{V_L P}{P_L + P} \quad (1)$$

where P is the methane gas pressure, MPa; V_{ab} is the methane adsorption volume of coal at pressure P, mL/g; V_L is the methane saturation adsorption volume of coal at a certain temperature, mL/g, and P_L is the methane gas pressure at half of the saturation adsorption volume, MPa. The simulation results of mechanical properties were mainly presented in the form of a stiffness matrix, and then the mechanical parameters were obtained.

4. Results and Discussion

4.1. Methane Adsorption in Coal Models

To verify the correctness and applicability of the models and the applied force field, the adsorption of methane on the coal was fitted by the Langmuir equation. Figure 2 shows the methane isotherm adsorption curves and Langmuir fitting parameter, a, of the anthracite and coking coal at 273.15–313.15 K. When the adsorption temperature is constant, the adsorption capacity of methane molecules in the coal molecular structure model rapidly increases and then stabilises with increasing adsorption pressure, which is consistent with the Langmuir adsorption equation. When the adsorption pressure is constant, the amount of methane adsorption decreases with increasing temperature. In the molecular structure model of anthracite, the saturated adsorption amount of methane is 5.80 N/uc at 273.15 K; it decreases to 4.85 N/uc at 313.15 K. In the molecular structure model of coking coal, the methane adsorption amount is 5.23 N/uc at 273.15 K; it decreases to 4.78 N/uc at 313.15 K. Regarding the methane adsorption amount in the molecular structure model of coking coal under different temperature and pressure conditions, Zhu et al. built a coking coal model of the Chiyu coal mine and also found that the methane adsorption amount decreased with the increase in temperature [46]. A probe with a diameter equal to the methane molecular diameter (0.38 nm) was used to detect and count the micropores of the two coal molecular structure models. The results show that the free volumes of methane adsorbed by the anthracite and coking coal are 16.25×10^{-3} and 8.98×10^{-3} nm^3/uc, respectively, and that the accessible surface areas of methane are 0.45 and 0.56 nm^2/uc, respectively. The simulation results are consistent with Ref [47], and Cheng et al. reported that the micropores in coal are the main places where methane occurs, that is, the larger the micropore volume, the greater the methane adsorption capacity.

The adsorption heat of methane in coal is an important thermodynamic parameter for characterizing the adsorption behaviour of methane in coal; it serves as a quantitative indicator to evaluate the adsorption affinity of coal to methane [48]. The isothermal adsorption heat of methane in coal is generally calculated by the Clausius–Clapeyron equation [46,48]. Based on the methane adsorption as an exothermic process, Tang et al. used negative values to characterize the methane adsorption heat [48]. The simulation results show that the adsorption heat is expressed as a positive value. Figure 3 shows the isothermal adsorption heat simulation results of methane in the anthracite and coking coal at 273.15–313.15 K. The methane adsorption heat of anthracite is 28.13–28.57 Kj/mol, and that of coking coal is 26.23–26.48 Kj/mol. As the adsorption temperature increases from 273.15 K to 313.15 K, the methane adsorption heat of both coals shows an increasing trend, which is consistent with the results of Tang et al. and Zhang et al. [48,49].

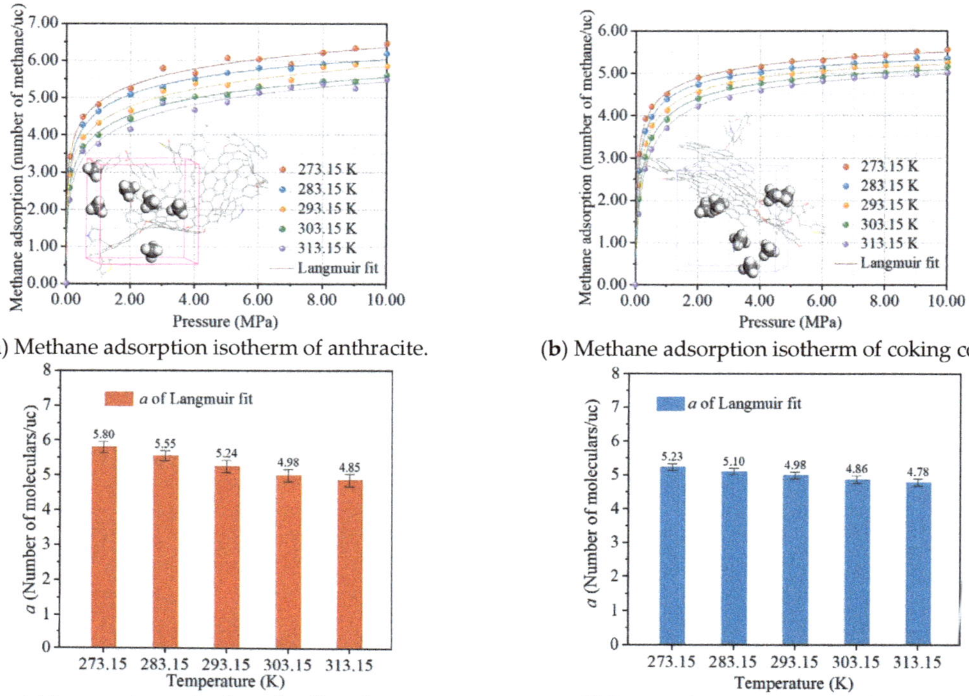

(a) Methane adsorption isotherm of anthracite. (b) Methane adsorption isotherm of coking coal.

(c) Langmuir constant, a, of anthracite. (d) Langmuir constant, a, of coking coal.

Figure 2. Methane adsorption isotherm of anthracite (**a**) and coking coal (**b**) at temperatures in the range of 273.15–313.15 k, and Langmuir fitting constant, a, of anthracite (**c**) and coking coal (**d**).

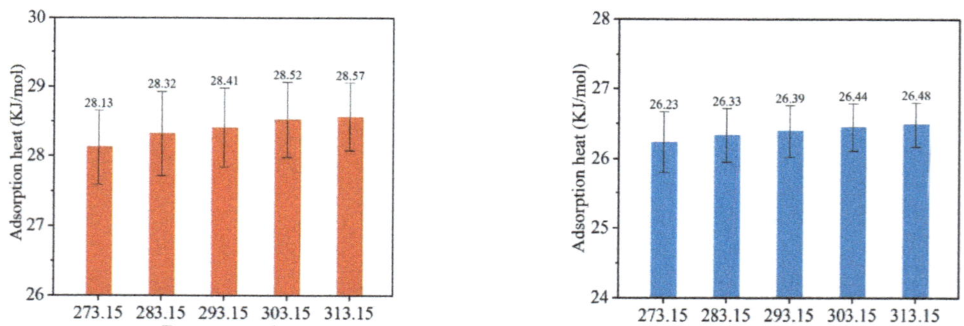

(a) (b)

Figure 3. Adsorption heat of methane at different temperatures in anthracite and coking coal models. (**a**) Adsorption heat of methane in anthracite. (**b**) Adsorption heat of methane in coking coal.

4.2. Effect of Minerals on Methane Adsorption of Coal-Rock Models

The adsorption of methane in the coal–rock compound model can also be fitted by the Langmuir equation [49]. To express the adsorption results clearly, a temperature of 273.15 K was selected to simulate the methane adsorption of compound coal–mineral models. Figure 4 shows the methane isotherm adsorption curves and Langmuir fitting parameter, a, in anthracite–rock and coking coal–rock compound models at 273 K. When the adsorption temperature is constant, the adsorption capacity of methane molecules in the coal–rock compound molecular structure model rapidly increases and then stabilizes with increasing adsorption pressure, which still conforms to the Langmuir adsorption model. When the adsorption pressure is constant, the methane adsorption capacity increases with increasing mineral content. In the anthracite–rock molecular structure model, the saturated adsorption capacity of methane reaches a maximum of 39.82 wt% calcite, which is 24.82 N/uc. In the coking coal–rock molecular structure model, the methane adsorption reaches a maximum of 40.11 wt% calcite, which is 15.18 N/uc. A probe with a diameter equal to the methane molecular diameter (0.38 nm) was selected to detect and count the micropores of the two coal–rock molecular structure models. In the molecular structure model of anthracite–rock, when the mineral content is 39.82 wt% calcite, the micropore volume reaches 0.70 nm^3/uc, which shows the large amount of methane adsorption when minerals are present. The results indicate that the highly developed micropores in clay minerals in coal are the main places for methane adsorption in coal, which is also reported by Feng et al. [50]. As can be seen in our previous paper [37,49] and the simulation results from Wang et al. [29], the surface of clay minerals like kaolinite also has the capacity to adsorb some methane. Besides, Feng et al. reported that, from the Scanning Electron Microscopy (SEM) micrographs of the same coal sample with different densities determined by the X-ray CT scan, the mesostructures of cell cavity pores with non-compact packing of the clay minerals appear to be the primary sites of methane adsorption in coal. Therefore, it is more generally accepted by researchers that the microporosity and surface structure parameters in clay minerals jointly affect the adsorption capacity.

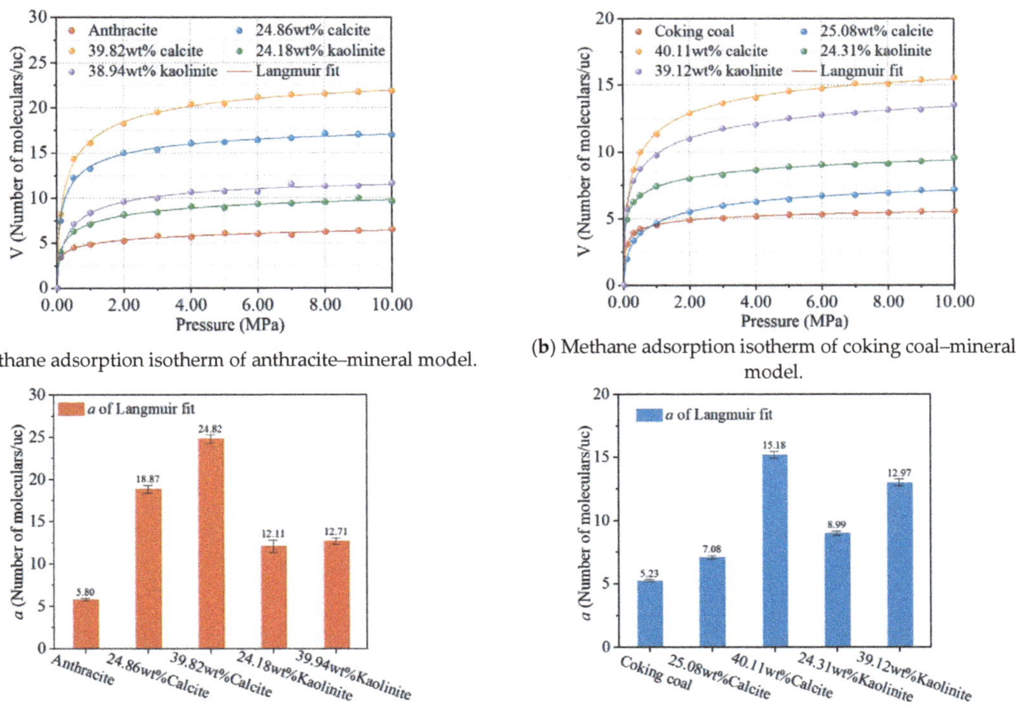

(**a**) Methane adsorption isotherm of anthracite–mineral model.

(**b**) Methane adsorption isotherm of coking coal–mineral model.

(**c**) Langmuir constant, a, of anthracite–mineral model.

(**d**) Langmuir constant, a, of coking coal–mineral model.

Figure 4. Methane adsorption isotherm of the anthracite–mineral model (**a**) and coking coal–mineral model (**b**) at a temperature of 273.15 k, and Langmuir fitting constant, a, of the anthracite–mineral model (**c**) and coking coal–mineral model (**d**).

The isothermal adsorption heat of methane in the coal–rock compound model can also be calculated by the Clausius–Clapeyron equation [49,51]. Zhang et al. used the methane adsorption heat to characterize the energy release information of methane in kaolinite and determined that the reduction in the adsorption heat under the condition of high water content means that the interaction energy between methane molecules and kaolinite molecules was weakened [49]. Figure 5 shows the isothermal adsorption thermal simulation results of methane in the anthracite–rock model and coking coal–rock model at 273.15 K. When calcite and kaolinite minerals are added, the methane adsorption heat of anthracite is 28.13–23.09 Kj/mol, and that of coking coal is 26.23–23.28 Kj/mol. With an increase in the mineral content, the methane adsorption heat of the two coal rock compounds decreases, and the contribution of calcite is greater than that of kaolinite.

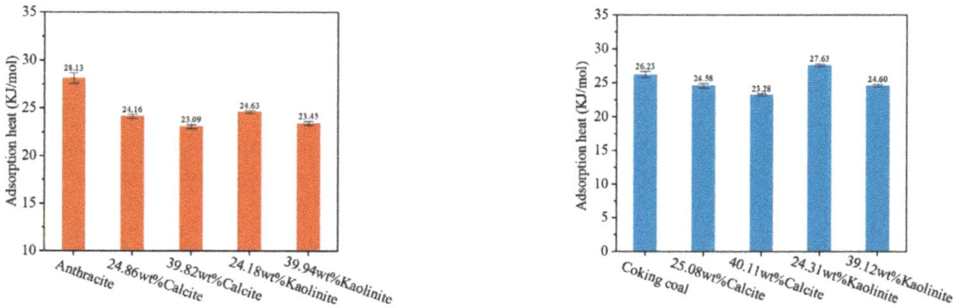

(**a**) Adsorption heat of methane in anthracite–mineral model. (**b**) Adsorption heat of methane in coking coal–mineral model.

Figure 5. Adsorption heat of methane at 273.15 K in anthracite–mineral model (**a**) and coking coal–mineral model (**b**).

4.3. Mechanical Properties of Coal-Rock Models

The bulk modulus and shear modulus in the coal–rock compound models were simulated in the Mechanical Properties of the Force module. Figure 6 shows the simulation results of the bulk modulus and shear modulus of the anthracite–rock model and coking coal–rock model under different calcite and kaolinite mineral contents. The bulk modulus of the anthracite–rock model decreases with the addition of calcite and tends to decrease with increasing calcite mineral content. The bulk modulus increases with the addition of the kaolinite mineral and tends to increase with increasing mineral content. The bulk modulus of the coking coal–rock model increases with the addition of calcite and kaolinite and tends to increase with increasing mineral content. Note that under the same mass percentage, the influence of kaolinite on the change in the bulk modulus and shear modulus of the two coal rock models is greater than that of calcite on the change in the two mechanical properties. The ratio of bulk modulus to shear modulus (E/G) is often used to estimate the brittle or ductile behaviour of materials. A high E/G represents plasticity, and a low E/G represents brittleness [52]. When minerals exist, the brittleness of the molecular model increases, and the brittleness index increases with increasing mineral content [53]. In addition, the atomic representation of coal–rock compound models is presented on a nanometre scale, which cannot illustrate the macroscopic mechanism of rockbursts and coal and gas outbursts. Larger coal–rock compound models will be constructed in our future work.

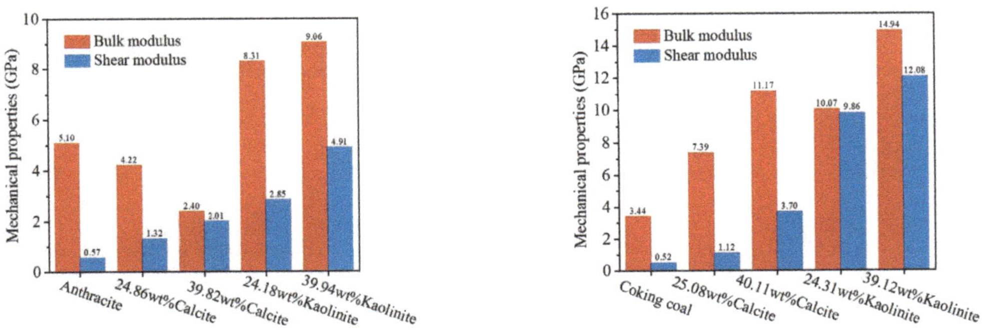

(**a**) Mechanical properties of anthracite–mineral model. (**b**) Mechanical properties of coking coal–mineral model.

Figure 6. Change in mechanical properties after the addition of calcite and kaolinite.

5. Conclusions

(1) The atomistic representations of the coking coal model and anthracite coal model considering the influence of minerals were constructed, which is an approach to characterize the methane adsorption and mechanical characteristics in coal.

(2) The amount of methane adsorption follows the order anthracite coal > coking coal, and the presence of minerals increases the methane adsorption capacity; the increased amount of methane adsorption follows the order calcite > kaolinite.

(3) The presence of calcite and kaolinite greatly increased the shear modulus of compound coal and mineral models. Notably, after the addition of calcite and kaolinite minerals, the brittleness of the model increases, and the brittleness index increases with increasing mineral content.

(4) Only calcite and kaolinite were considered typical co-associated minerals in coal, which is limited to explaining the methane adsorption and mechanical properties in all the coal–rock compound models. In addition, there are limited coal molecular structure models for molecular dynamic simulation work, and the construction of diversified coal molecular structure models is difficult but imperative. Thus, based on Graph Representation Learning and Graph Neural Networks, our team is committed to building diverse molecular structure models for almost all coal and rock, providing a theoretical basis for coal and rock dynamic disasters.

Author Contributions: Methodology, writing—original draft, X.Z.; writing—review and editing, Y.L.; investigation, T.Z.; conceptualization, writing—review and editing, Y.C.; formal analysis, B.Z. All authors have read and agreed to the published version of the manuscript.

Funding: This study was supported by the youth project of National Natural Science Foundation of China (No. 52204220, No. 52304088, No. 42302202), the Science and Technology Innovation and Entrepreneurship project of TDTEC (No. 2022-QN001, No. 2022-2-MS003, No. 2023-TD-MS010), the Natural Science Foundation of Liaoning Province, China (No. 2022-KF-23-09), the Independent Research fund of Joint National-Local Engineering Research Centre for Safe and Precise Coal Mining(Anhui University of Science and Technology) (No. EC2022021).

Institutional Review Board Statement: Not applicable.

Informed Consent Statement: Not applicable.

Data Availability Statement: The data presented in this study are available on request from the corresponding author. The data are not publicly available due to privacy.

Conflicts of Interest: Authors Xiaoyu Zhang, Yingjie Liu, Tianbai Zhou, Yongbo Cai, were employed by the company Chinese Institute of Coal Science. The remaining author Bin Zhang declare that the research was conducted in the absence of any commercial or financial relationships that could be construed as a potential conflict of interest.

References

1. Qi, Q.; Li, Y.; Zhao, S.; Zhang, N.; Zheng, W.; Li, H.; Li, H. Seventy Years Development of Coal Mine Rockburst in Chin Establishment and Consideration of Theory and Technology System. *Coal Sci. Technol.* **2019**, *47*, 1–40.
2. Pan, Y. *Rockburst in Coal Mine*; Science Press: Beijing, China, 2018.
3. Sun, J.; Cheng, J. Study on the Perception and Alarm Method of Coal Mine Rock Burst and Coal and Gas Outburst. *Ind. Mi Autom.* **2022**, *48*, 1–6.
4. Pan, Y. Integrated Study on Compound Dynamic Disaster of Coal-gas Outburst and Rockburst. *J. China Coal Soc.* **2016**, *4* 105–112.
5. Yu, M. Linear and Nonlinear Unified Strength Theory. *Chin. J. Rock Mech. Eng.* **2007**, *26*, 662–669.
6. Petukhov, M.; Linkov, M. The Theory of Post-Failure Deformations and the Problem of Stability in Rock Mechanics. *Int. J. Ro Mech. Min. Sci. Geomech. Abstr.* **1979**, *16*, 57–76. [CrossRef]
7. Calder, P.N.; Madsen, D. High Frequency Precursor Analysis Prior to a Rockburst. *Int. J. Rock Mech. Min. Sci. Geomech. Abs* **1989**, *26*, 3–4.
8. Singh, S.P. Burst Energy Release Index. *Rock Mech. Rock Eng.* **1988**, *21*, 149–155. [CrossRef]
9. Li, Y. Rockburst Mechanism and its Preliminary Application. *J. China Univ. Min. Technol.* **1985**, *14*, 37–43.
10. Zhang, M. Instability Theory and Mathematical Model for Coal/Rock Bursts. *Chin. J. Rock Mech. Eng.* **1987**, *6*, 197–204.
11. Li, H.; Qi, Q.; Zhao, S.; Li, H.; Shu, H.; Chen, L. Discussion on Generalized "Three Factors" Mechanism of Coal Mine Dynam Disaster. *Coal Sci. Technol.* **2021**, *49*, 42–52.
12. Qi, Q.; Pan, Y.; Li, H.; Jiang, D.; Shu, L.; Zhao, S.; Zhang, Y.; Pan, J.; Li, H.; Pan, P. Theoretical Basis and Key Technology Prevention and Control of Coal-Rock Dynamic Disasters in Deep Coal Mining. *J. China Coal Soc.* **2020**, *45*, 1567–1584.
13. Li, X.; Lin, B. Status of Research and Analysis on Coal and Gas Outburst Mechanism. *Coal Geol. Explor.* **2010**, *38*, 7–13.
14. Ma, Y.; Nie, B.; He, X.; Li, X.; Meng, J.; Song, D. Mechanism Investigation on Coal and Gas Outburst: An Overview. *Int. J. Min Metall. Mater.* **2020**, *27*, 872–887. [CrossRef]
15. Jiang, C.; Yu, Q. *Spherical Shell Losing Stability" Hypothesis about Outburst Mechanism and Prevention Technology*; China University Mining and Technology Press: Xuzhou, China, 1998.
16. Liang, B.; Zhang, M.; Pan, Y.; Wang, L. Theory of Instability of Flow Fixation Coupling for Coal and Gas Outburst. *J. China Co Soc.* **1995**, *20*, 492–496.
17. Wang, E.; Zhang, G.; Zhang, C.; Li, Z. Research Progress and Prospect on Theory and Technology for Coal and Gas Outbur Control and Protection in China. *J. China Coal Soc.* **2022**, *47*, 297–322.
18. Hou, Q.; Li, H.; Fan, J.; Ju, Y.; Wang, T.; Li, X.; Wu, Y. Structure and Coalbed Methane Occurrence in Tectonically Deformed Coa *Sci. China Earth Sci.* **2012**, *55*, 1755–1763. [CrossRef]
19. Hou, Q.; Luo, Y.; Song, C.; Han, Y.; Du, J.; Xu, R. Gas Generation during Middle-Rank Coal Deformation and the Preliminar Discussion of the Mechanism. *J. China Coal Soc.* **2014**, *39*, 1675–1682.
20. Wang, J.; Guo, G.J.; Han, Y.; Hou, Q.; Geng, M.; Zhang, Z. Mechanolysis Mechanisms of the Fused Aromatic Rings of Anthraci Coal under Shear Stress. *Fuel* **2019**, *253*, 1247–1255. [CrossRef]
21. Wang, J.; Hou, Q.; Zeng, F.; Guo, G.-J. Stress Sensitivity for the Occurrence of Coalbed Gas Outbursts: A Reactive Force Fiel Molecular Dynamics Study. *Energy Fuels* **2021**, *35*, 5801–5807. [CrossRef]
22. Alder, B.J.; Wainwright, T.E. Phase Transition for a Hard Sphere System. *J. Chem. Phys.* **1957**, *27*, 1208–1209. [CrossRef]
23. Dai, X.; Bai, J.; Yuan, P.; Du, S.; Li, D.; Wen, X.; Li, W. The Application of Molecular Simulation in Ash Chemistry of Coal. *Chin. Chem. Eng.* **2020**, *28*, 2723–2732. [CrossRef]
24. Jiang, J.; Babarao, R.; Hu, Z. Molecular Simulations for Energy, Environmental and Pharmaceutical Applications of Nanoporou Materials: From Zeolites, Metal-Organic Frameworks to Protein Crystals. *Chem. Soc. Rev.* **2011**, *40*, 3599–3612. [CrossRe [PubMed]
25. Yang, Y.; Liu, J.; Yao, J.; Kou, J.; Li, Z.; Wu, T.; Zhang, K.; Zhang, L.; Sun, H. Adsorption Behaviors of Shale Oil in Kerogen Slit b Molecular Simulation. *Chem. Eng. J.* **2020**, *387*, 124054. [CrossRef]
26. Ungerer, P.; Nieto-Draghi, C.; Lachet, V.; Wender, A.; di Lella, A.; Boutin, A.; Rousseau, B.; Fuchs, A.H. Molecular Simulatio Applied to Fluid Properties in the Oil and Gas Industry. *Mol. Simul.* **2007**, *33*, 287–304. [CrossRef]
27. Fuchs, W.; Sandhoff, A.G. Theory of Coal Pyrolysis. *Ind. Eng. Chem.* **1942**, *34*, 567–571. [CrossRef]
28. Xiang, J.; Zeng, F.; Liang, H.; Li, B.; Song, X. Molecular Simulation of the $CH_4/CO_2/H_2O$ Adsorption onto the Molecular Structu of Coal. *Sci. China Earth Sci.* **2014**, *57*, 1749–1759. [CrossRef]
29. Wang, K.; Zhang, B.; Kang, T. Effect of Geological Depths on CH4 Adsorption, Diffusion, and Swelling in Kaolinite by Molecula Simulations. *Energy Fuels* **2020**, *34*, 1620–1626. [CrossRef]
30. Zhang, B.; Zeng, F.; Wang, D.; Kang, G.; Zhang, X.; Kang, T. Macromolecular Structure Model of Anthracite in Southern Qinshu Basin and its Methane Bearing Mechanical Properties. *J. China Coal* **2021**, *46*, 534–543.
31. Taheri, Z.; Pour, A.N. Studying of the Adsorption and Diffusion Behaviors of Methane on Graphene Oxide by Molecula Dynamics Simulation. *J. Mol. Model.* **2021**, *27*, 1–8. [CrossRef]
32. Zhang, X.; Kang, T.; Zhang, B.; Kang, G.; Kang, J.; Zhang, R.; Zhang, L. Combined Molecular Simulations and Experiment Study of Methane Adsorption on Anthracite. *Energy Fuels* **2020**, *34*, 12118–12125. [CrossRef]

Ma, R.; Zhang, S.; Hou, D.; Liu, W.; Yuan, L.; Liu, X. Model Construction and Optimization of Molecule Structure of High-Rank Coal in Feng County, Shaanxi Province. *J. China Coal Soc.* **2019**, *44*, 1827–1835.

Zhang, L.; Kang, T.; Kang, J.; Zhang, X.; Zhang, B.; Guo, J.; Chai, Z. Response of Molecular Structures and Methane Adsorption Behaviors in Coals Subjected to Cyclical Microwave Exposure. *ACS Omega* **2021**, *6*, 31566–31577. [CrossRef] [PubMed]

Zhang, X.; Zhang, R.; Kang, T.; Hu, Y.; Li, C. Experimental and Mechanistic Research on Methane Adsorption in Anthracite Modified by Electrochemical Treatment using Selected Electrode Materials. *Sci. Rep.* **2019**, *9*, 1–12. [CrossRef] [PubMed]

Materials Studio. 2020. Available online: https://cgzb.tyut.edu.cn/info/1078/4523.htm (accessed on 16 December 2020).

Zhang, B.; Kang, J.; Kang, T. Effect of Water on Methane Adsorption on the Kaolinite (0 0 1) Surface based on Molecular Simulations. *Appl. Surf. Sci.* **2018**, *439*, 792–800. [CrossRef]

Skipper, N.T.; Refson, K.; McConnell, J.D.C. Computer Calculation of Water-clay Interactions using Atomic Pair Potentials. *Clay Miner.* **1989**, *24*, 411–425. [CrossRef]

Cygan, R.T.; Liang, J.J.; Kalinichev, A.G. Molecular Models of Hydroxide, Oxyhydroxide, and Clay Phases and the Development of a General Force Field. *J. Phys. Chem. B* **2004**, *108*, 1255–1266. [CrossRef]

Heinz, H.; Koerner, H.; Anderson, K.L.; Vaia, R.A.; Farmer, B.L. Force Field for Mica-Type Silicates and Dynamics of Octadecylammonium Chains Grafted to Montmorillonite. *Chem. Mater.* **2005**, *17*, 5658–5669. [CrossRef]

Xiao, S.; Edwards, S.A.; Gräter, F. A New Transferable Forcefield for Simulating the Mechanics of $CaCO_3$ Crystals. *J. Phys. Chem. C* **2011**, *115*, 20067–20075. [CrossRef]

Abramov, A.; Iglauer, S. Application of the CLAYFF and the DREIDING Force Fields for Modeling of Alkylated Quartz Surfaces. *Langmuir* **2019**, *35*, 5746–5752. [CrossRef]

Lammers, L.N.; Bourg, I.C.; Okumura, M.; Kolluri, K.; Sposito, G.; Machida, M. Molecular Dynamics Simulations of Cesium Adsorption on Illite Nanoparticles. *J. Colloid Interface Sci.* **2017**, *490*, 608–620. [CrossRef]

Pouvreau, M.; Greathouse, J.A.; Cygan, R.T.; Kalinichev, A.G. Structure of Hydrated Kaolinite Edge Surfaces: Dft Results and Further Development of the Clayff Classical Force Field with Metal–O–H Angle Bending Terms. *J. Phys. Chem. C* **2019**, *123*, 11628–11638. [CrossRef]

Meng, J.; Niu, J.; Meng, H.; Xia, J.; Zhong, R. Insight on Adsorption Mechanism of Coal Molecules at Different Ranks. *Fuel* **2020**, *267*, 117234. [CrossRef]

Zhu, H.; Zhang, Y.; Qu, B.; Liao, Q.; Wang, H.; Gao, R. Thermodynamic Characteristics of Methane Adsorption about Coking Coal Molecular with Different Sulfur Components: Considering the Influence of Moisture Contents. *J. Nat. Gas Sci. Eng.* **2021**, *94*, 104053. [CrossRef]

Cheng, Y.; Hu, B. Main Occurrence Form of Methane in Coal: Micropore filling. *J. China Coal Soc.* **2021**, *46*, 2933–2948.

Tang, X.; Wang, Z.; Ripepi, N.; Kang, B.; Yue, G. Adsorption Affinity of Different Types of Coal: Mean Isosteric Heat of Adsorption. *Energy Fuels* **2015**, *29*, 3609–3615. [CrossRef]

Zhang, B.; Kang, J.; Kang, T. Molecular Simulation of Methane Adsorption and its Effect on Kaolinite Swelling as Functions of Pressure and Temperature. *Mol. Simul.* **2018**, *44*, 789–796. [CrossRef]

Feng, Z.; Cai, T.; Zhou, D.; Zhao, Y.-S.; Wang, C. Temperature and Deformation Changes in Anthracite Coal after Methane Adsorption. *Fuel* **2017**, *192*, 27–34. [CrossRef]

Chen, J.; Min, F.; Liu, L.; Liu, C. Mechanism research on surface hydration of kaolinite, insights from DFT and MD simulations. *Appl. Surf. Sci.* **2019**, *476*, 6–15. [CrossRef]

Zhou, X.; Zhang, X.; Xu, S.; Wu, S.; Liu, Q.; Fan, Z. Evaluation of Thermo-Mechanical Properties of Graphene/Carbon-Nanotubes Modified Asphalt with Molecular Simulation. *Mol. Simul.* **2017**, *43*, 312–319. [CrossRef]

Guo, H.; Yu, Y.; Wang, K.; Yang, Z.; Wang, L.; Xu, C. Kinetic characteristics of desorption and diffusion in raw coal and tectonic coal and their influence on coal and gas outburst. *Fuel* **2023**, *343*, 127883. [CrossRef] *applied sciences*

MDPI

Article

Experimental Study of the Multiple Fractalisation of Coal and Rock Failure Subjected to the Coupled Effects of Water, Temperature and Dynamic Loads

Tingxu Jin [1], Xiaoyuan Sun [1,2,*], Kai Liu [1], Shurong Lin [3], Shaoqiang Yang [1,2] and Jianlin Xie [1]

1 College of Safety and Emergency Management Engineering, Taiyuan University of Science and Technology, Taiyuan 030024, China; jintingxu1997@163.com (T.J.)
2 Intelligent Monitoring and Control of Coal Mine Dust Key Laboratory of Shanxi Province, Taiyuan University of Science and Technology, Taiyuan 030024, China
3 School of Emergency Management and Safety Engineering, China University of Mining and Technology (Beijing), Beijing 100083, China
* Correspondence: sunxy253@126.com

Abstract: The mechanical properties of water-rich coal and rock in a subzero environment are very different from those at room temperature, which causes many unexpected hazards for projects. In this study, coal and rock samples subjected to the coupled effects of water, temperature, and dynamic loads were taken as the research object, and the discussion was shaped around their mechanical properties. The crack evolution trend and different gradient impact velocities were determined using a split-Hopkinson pressure bar (SHPB). Multiple fractals were based on high-speed digital image correlation (HS-DIC) technology and the quality-screening method; the evolution trend of the surface cracks in the crushing process and the distribution characteristics of the specimen fragments after crushing were examined from the perspective of fractals. This provided a powerful supplement to the existing research system on the problem of mining via the freezing method, and it accounted for the shortcomings of the existing research to a certain extent. In this research, the results mainly showed four points: (1) The coal samples were determined to have a wave velocity between 1.68 and 2.01 km/s, while the rock samples were between 2.24 and 2.61 km/s. Under the same conditions, the rock's resistance to deformation and damage was greater than that of coal. (2) In the saturated state, the plastic strength of the coal and rock samples was greater than that in the dry state, due to the strengthening of their internal stresses caused by the presence of fissure water. (3) With decreasing temperature, the degree of the dynamic compression factor of coal and rock showed a trend of initially increasing, then decreasing, and then increasing. With the increase in the loading rate, the destruction of the coal and rock was more intense, and the destruction process was accelerated. (4) After the saturated coal and rock samples were frozen, their interiors were affected by the dual factors of contraction under the influence of temperature and expansion under the influence of the freezing expansion force. The internal fissures closed or shrank, and the water in the pores turned into ice, leading to an increase in pore volume.

Keywords: frozen coal and rock mass; impact failure; dynamic mechanical properties; crack propagation; box dimension; crushing form

1. Introduction

In China, the focus of resource extraction is gradually shifting to the west. Underground mining in Xinjiang, Qinghai, and other cold regions has been increasing. The artificial freezing method in mining and digging has received widespread attention from scholars due to its adaptability to various complex engineering conditions and improvement of safety in construction [1], such as the construction of railways on the Tibetan Plateau and the application and development of coal mine freezing methods for well drilling technology [2]. The freezing method of drilling is the application of artificial refrigeration technology to vertical- or inclined-shaft engineering through the temporary freezing of groundwater to consolidate the stability of the underground coal and rock to reduce the occurrence of collapse accidents during the mining process [3]. However, the kinds of mechanical properties that occur on impact have not been clearly explored. In addition, rock structures in cold regions are often affected by dynamic loads. For example, in open-pit coal mines in high-altitude and water-rich areas, the mechanical properties of coal and rock are deteriorated by low temperatures and water, combined with blasting and mining disturbances, which will have an impact on the macroscopic and microscopic structure of coal and rock, leading to slope geological disasters [4,5]. Therefore, a comprehensive understanding of coal and rock subjected to the coupled effects of water, temperature, and dynamic loads has great theoretical and engineering significance for the safe mining of rock structures in cold regions and for the study of the impacts of dynamic loads on coal and rock during freezing [6].

The freezing method has been studied and analysed from many perspectives. For static loading, Fan et al. [7] used computed tomography (CT) to analyse frozen sand and concluded that as the temperature decreased, the specimen changed from tensile–shear to shear damage mode, and the volumetric porosity and crack area of the damaged specimen decreased with decreasing temperature. Kodama et al. [8] investigated the effects of water content, temperature, and loading rate on the strength and damage process of frozen rock and concluded that the rock's strength increased with increasing water content and loading rate, with the effects being more pronounced at lower temperatures. Sun et al. [9] explored the damage characteristics of frozen coal by conducting acoustic emission tests on frozen coal, and they proposed early warning parameters for the characterisation of the evolution of the variance curve, where the compressive strength of frozen coal samples significantly decreased and the saturated frozen coal samples had the lowest compressive strength. Wang et al. [10], by observing the microscopic pore structure of sandstone, also found that low temperatures had this promoting effect on coal and rock samples.

Under dynamic loading, Wang et al. [11] investigated the effects of the dynamic mechanical properties of dry sandstone and saturated siltstone at different temperatures and rates. The results showed that the saturated specimens exhibited shorter compaction stages on the dynamic stress–strain curve due to the presence of pore water or ice. The dry specimens were more sensitive to the strain-rate effects than the uniaxial compressive strength. Wang et al. [2], through the SHPB test, found that the peak stress of the saturated specimen first increased and then decreased as the temperature decreased. Yang et al. [12] studied the deformation and damage characteristics of red sandstone at low temperatures by combining the effects of the rock strength properties, fractal dimension, and dissipated energy with microfracture morphology; they concluded that low temperatures led to a rapid loss of bearing capacity and a sharp decrease in the strength of the red sandstone under high-strain-rate loading, the dissipation energy of the frozen rock specimens was positively correlated with the fractal dimension, and the low-temperature effect produced significant cracks at the material interface of the red sandstone. Xu et al. [13] used a method that employed a combination of low-field nuclear magnetic resonance (LF-NMR) and scanning electron microscopy (SEM) to investigate the dynamic evolution of the microstructure of frozen sandstones as a function of saturation and determine the increase in saturation; the elastic modulus showed a trend of increasing and then decreasing, and the ultimate deformation capacity showed the opposite trend. Yang et al. [14] compared the uniaxial compression of sandstone, marble, and granite at different temperatures for different materials; the peak inflection point of the stress–strain curve was at -5 °C, while the second inflection point temperature increased with the material hardness. Shan et al. [15] conducted SHPB impact tests on frozen red sandstone specimens at -15 °C, compared them with the static uniaxial compression tests, and found that the failure form was closely correlated to the average strain rate.

The presence of water in the interior of coal and rock has also received widespread attention. Zhao et al. [16,17] analysed the fracture of rocks under water-force coupling by means of experiments and simulations, on the basis of which a two-medium model was proposed. Regarding the effects of saturation and temperature on the wave speed of the corresponding ultrasonic wave, previous studies reached different conclusions. Generally, with increasing saturation, the internal gap of the sample is filled with a large number of water molecules, and the propagation of the ultrasonic waves in the water molecules is greater than that in air; thus, the wave speed of the sample significantly increases [18]. Moreover, as water saturation increases, the squeezing of internal moisture leads to a reduction in the internal strength and modulus of elasticity, ultimately leading to a reduction in the rate of wave propagation [19]. Zhao et al. [20] found that the stress in sandstone had a trend of increasing and then decreasing with decreasing temperature. With the decrease in temperature, the stress initially increased, then decreased and finally increased, similar to an "N" shape. Wang et al. [21] concluded that the peak strength of water-saturated granite with decreasing temperature initially remained unchanged, then rapidly decreased and eventually converged to a stable value, similar to a "⌉\" shape.

Based on the above literature, studies of temperature and saturation as variables usually use rocks and rarely involve coal. However, some differences between rock and coal exist; rock is strong and hard with weak water absorption, while coal is a softer material with greater porosity. The mineral composition of coal is also very different from that of rock. Frozen coal and frozen rock in the static-load state have been extensively studied. However, research on dynamic-load impact is limited, and research on the destruction process is even more limited. In underground frozen mining, the difference in mining temperature and water saturation rate often leads to certain deformations. In view of this, in this study, coal and rock specimens from the same location were selected for SHPB uniaxial impact damage tests at five different temperature gradients (20 °C, 10 °C, 0 °C, −10 °C, and −20 °C) and in dry and saturated states; the damage evolution process of the specimens was comprehensively shown in terms of both surface cracking characteristics and fragmentation characteristics. The change trend of the dynamic mechanical parameters of the coal samples from room temperature to colder temperatures was analysed, and the mechanical properties of the coal samples under saturated and dry conditions were compared. Then, based on the HS-DIC technique and mass-screening method, the evolution trend of the surface cracks during crushing and the distribution characteristics of the broken blocks of the crushed specimens were investigated from the perspective of multiple fractalisation, and their correlation was explored to comprehensively determine the impact damage process on coal and rock. This study is expected to further reveal the failure mechanism of coal and rock materials under the coupling of temperature, water, and dynamic loads, providing a certain theoretical basis for low-temperature mining technology and the surrounding rock support of water-rich mines and other geotechnical engineering.

2. Materials and Methods

2.1. Specimen Preparation

As illustrated in Figure 1a, coal and rock samples were collected from the 14030 large-height working face in the first disc area of the Zhaogu No. 2 Mine of Henan Energy and Chemical Group Jiaozuo Coal Industry Group Co., Ltd., Xinxiang City, Henan Province, China. The stratigraphic logging records of the borehole near the sampling point are shown in Figure 1b. The depth of the face was more than 800 m, the main coal was anthracite, the average thickness of the coal seam was 6.5 m, the direct top was interbedded mudstone and sandy mudstone, the thickness was 10.5 m, the basic top was Dazhan sandstone, and the thickness was 8.0 m. There were several aquifers above the face, the alluvial dive flowed to the face along the mining fissures during the advancement of the face, and localised water and sand breakage risks existed [22]. In addition, the dynamic pressure on the top plate of the 14030 working face was evident, and the impact dynamic load effect triggered by the breakage of the top plate occasionally occurred [23]. Coupled with the strongly disturbed work processes present at the working face, the coupled effect of water content and dynamic loads had a significant impact on the mechanical properties of the coal and rock [24].

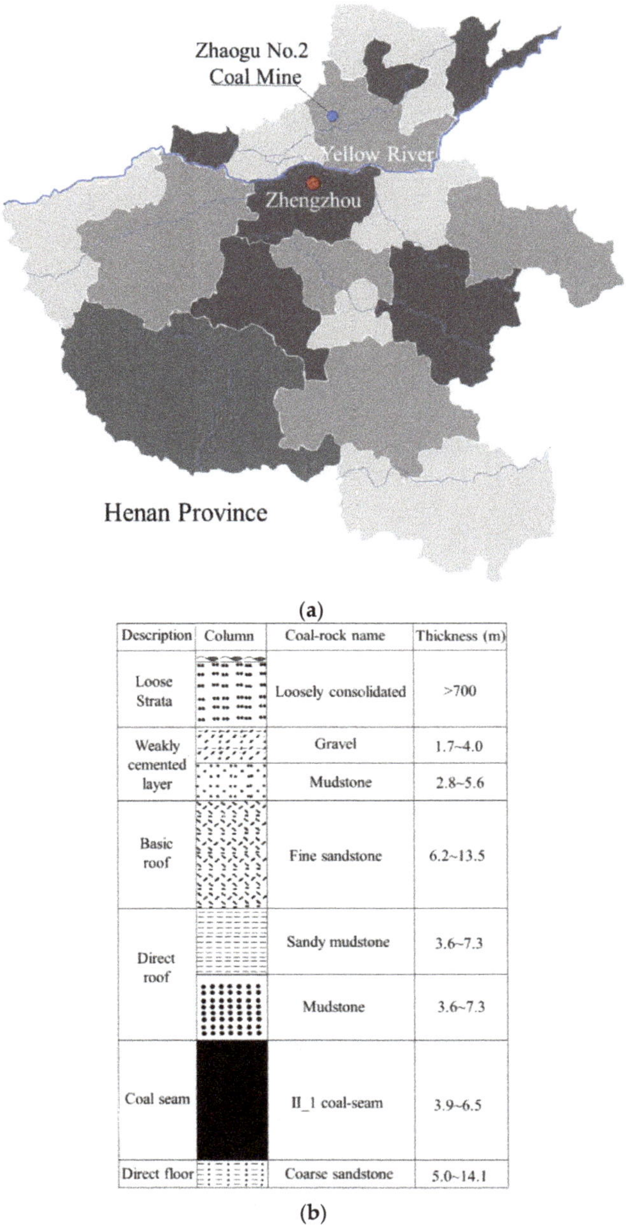

Figure 1. Schematic diagram of coal and rock sampling: (**a**) location of the Zhaogu No. 2 Coal Mine; (**b**) stratigraphic logging records.

The test methods follow the regulations of the Rock Dynamics ISRM Committee [25]. Large samples were selected from the II_1 coal seam and the roof (sandy mudstone) of the 14030 working face and were processed into 40 and 30 original coal and rock cylindrical specimens with dimensions of φ50 mm × 50 mm, respectively. The upper and lower surfaces of the test specimens were precisely polished and sanded to ensure that the non-parallelism of the two ends was less than 0.02 mm and the non-perpendicularity of the circumference and the end face was less than 0.25°. The final prepared coal and rock specimens are shown in Figure 2.

Figure 2. An illustration of coal and rock specimens used in the experiment. (**a**) Coal specimens; (**b**) rock specimens.

2.2. Testing Equipment

Figure 3a shows the φ50 mm SHPB experiment system independently developed by the coal or rock dynamic load damage parameter testing laboratory, which lies in the school of emergency management and safety engineering, China University of Mining and Technology (Beijing, China). The system consisted of a power supply subsystem, a rod subsystem, an information acquisition subsystem, and a data processing subsystem. As shown in Figure 3a, the power supply subsystem, i.e., the stress-wave generator, used a combination of an air compressor, high-pressure air storage chamber, air cavity, and SHPB electronic control system; this controlled the impact velocity of the bullet by setting the magnitude of the air pressure. The rod subsystem consisted of a cylindrical impact rod (bullet), incidence rod, transmission rod, and energy-absorbing rod. Each rod was cylindrical with a diameter of 50 mm and lengths of 400 mm, 3000 mm, 2500 mm, and 1000 mm, respectively. The rod was made of mild steel with a density of 7740 kg/m^3 and a modulus of elasticity of 206 GPa. The data acquisition subsystem consisted of an infrared velocimetry device and a strain signal acquisition system. To ensure the accuracy of the test, double strain gauges were affixed to the incident and transmission rods, dual channels were set up for data acquisition, and the incident, reflected, and transmission wave signals were recorded with an ultra-dynamic strain gauge and researched in a targeted way using the data processing subsystem.

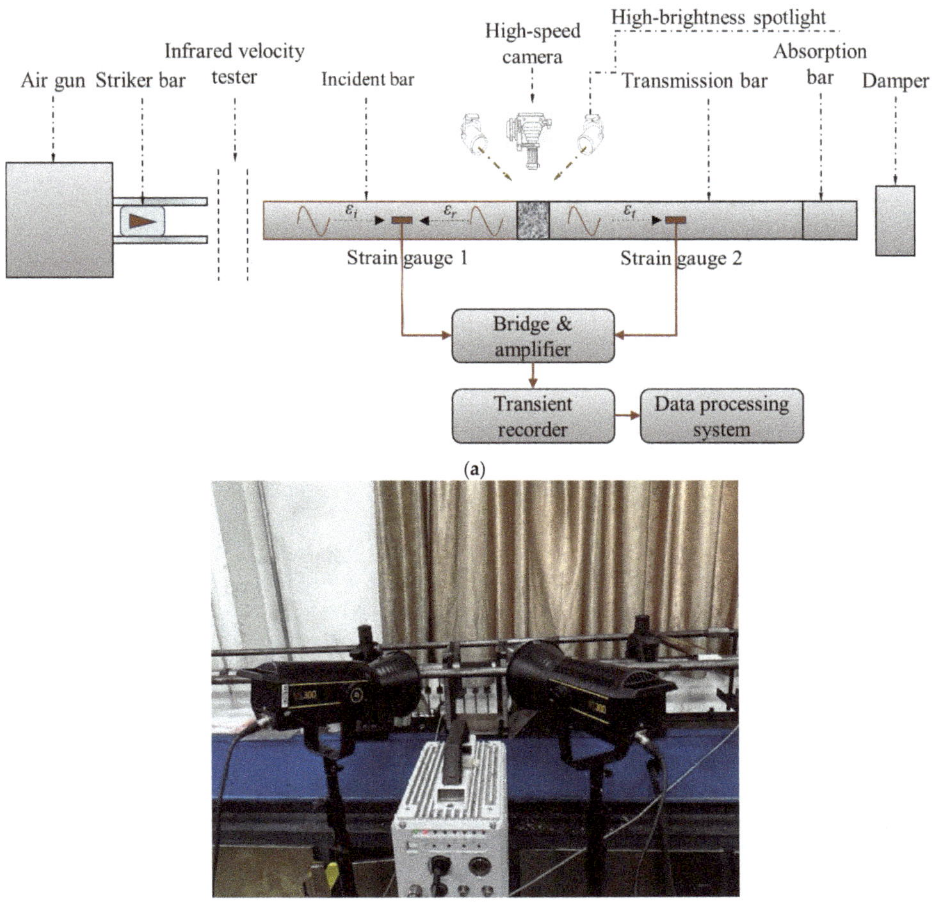

(a)

(b)

Figure 3. Experimental system diagram. (**a**) Schematic diagram; (**b**) physical diagram.

Notably, to be able to accurately reflect the destruction process of the coal and rock specimens during impact dynamic loading, the data acquisition subsystem also included a high-speed photography device for recording the rupture process of the sample. As shown in Figure 3b, the device consisted of a FASTCAM SA5 high-speed camera developed by PHOTRON, China Shenzhen Shenniu Photographic Equipment Co., Ltd. (Shenzhen, China), a produced VL300 fill light and synchronous trigger device, and other components. The camera frame speed was set to 100,000 frames per second with a pixel resolution of 320 × 192, and a shield made of high-transmittance acrylic panels was installed around the coal and rock samples to prevent the camera and the filler lamp from being damaged by the splashing of the samples during the destruction of the coal and rock, as well as to facilitate the collection and sieving of crushed coal and rock fragments [26].

Figure 4 shows the instruments used in the pre-treatment of coal and rock. Figure 4a shows the electric blast dryer, which adopts a hot air circulation system to ensure uniform heating of coal and rock samples. The temperature control range is room temperature (RT) +50–250 °C, and the constant temperature fluctuation is ±1 °C. Figure 4b is a natural saturated container. Figure 4c is the constant-temperature test chamber, which adopts an intelligent and high-precision proportional integral derivative temperature control system with strong temperature stability and a temperature control range from −40 °C to

100 °C. Figure 4d shows the ZBL-U5100 nonmetallic ultrasonic detector produced by Beijing Zhibo Lian Technology Co. (Beijing, China). The acoustic time measurement range is ±1,677,700 μs, the gain accuracy is 0.5 db, and the acceptance sensitivity is less than 10 μv.

(**a**)

(**b**)

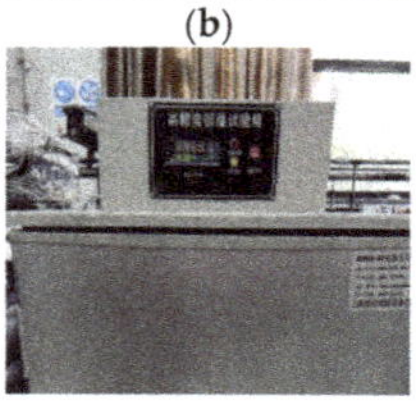

(**c**)

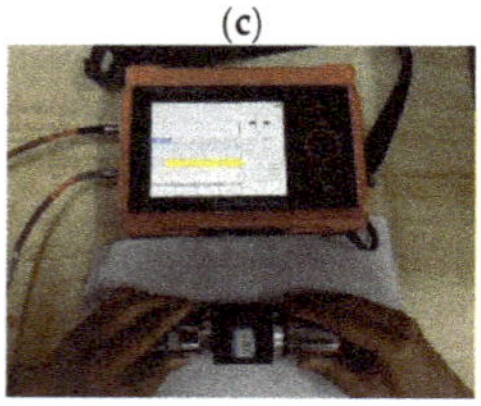

(**d**)

Figure 4. Testing equipment. (**a**) Electric blast drying oven; (**b**) natural saturated container; (**c**) constant-temperature test chamber; and (**d**) wave velocity measuring instrument.

2.3. Test Condition and Procedure

As shown in Figure 5, this experiment was conducted in four stages. In the first stage, all the raw coal and rock samples prepared in the natural water content state were categorised and numbered, and drying was defined as D, saturated as S, and frozen as F. Coal was defined as C, and rock was defined as R. Next, the samples were placed in an electric blast drying oven for drying, the temperature of the drying oven was set at 105 °C, and the drying time was greater than 24 h. The mass of the sample was measured at 1 h intervals after the 24 h interval, and the specimens were considered to have reached a completely dry state when the results of the two weighings before and after were ≤0.01 g. Afterwards, the relevant provisions in the "Methods for the Determination of Physical and Mechanical Properties of Coal and Rock" were followed [27]. Half of the coal and rock samples were naturally saturated with water; specifically, the coal and rock specimens were immersed in a container containing distilled water, keeping the water level 1–2 cm above the top surface of the specimens. The immersion time was greater than 24 h, the samples were weighed at 24 h intervals, and the samples were considered to be fully saturated when the difference in mass between the two times was ≤0.01 g. The saturated and unsaturated coal and rock samples were placed in a high-precision thermostatic chamber with the temperatures set at −20 °C, −10 °C, 0 °C, 10 °C, or 20 °C, and the thermostatic time was 24 h. The longitudinal wave velocity (C_P) of the samples was measured with a nonmetallic ultrasonic detector after each operation to determine the effect of the above steps on the physicomechanical properties of the coal and rock specimens [28]. Finally, they were rearranged in order of wave speed, from highest to lowest.

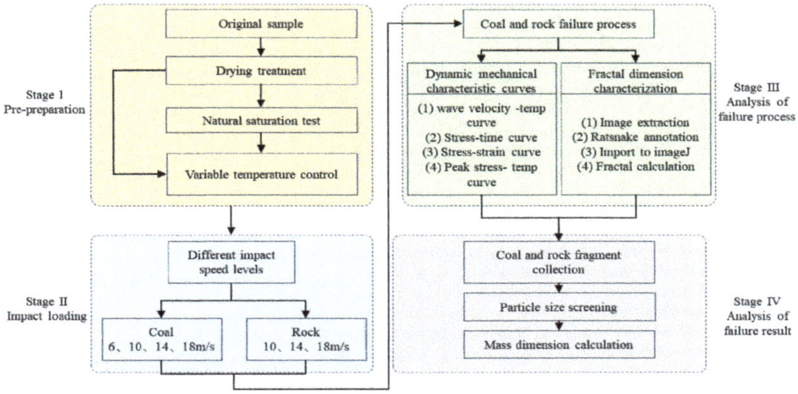

Figure 5. Experimental process flowchart.

In Stage II, the coal and rock specimens were placed in the SHPB device, and the pressure of the air compressor was adjusted such that the impact rod (bullet) struck the incidence rod at different velocities and the energy was finally transferred to the specimen. For the coal specimen, the velocity of the bullet was set to four different levels: 6 m/s, 10 m/s, 14 m/s, and 18 m/s. Due to the high density of the sandstone specimens, the samples did not show significant changes at an impact velocity of 6 m/s; thus, only the impact velocities of the three different grades of 10 m/s, 14 m/s, and 18 m/s were examined. During the experiment, a small-diameter rubber sheet was pasted between the bullet and the incident rod to shape the incident wave [29,30]. Molybdenum disulfide lubrication was also applied to the ends of the specimen in contact with the incidence and transmission rods [31] to reduce friction effects.

During the impact process, the strain signal acquisition system and high-speed photography device shown in Stage III were used to record the stress–strain signals and surface crack evolution of the coal and rock specimens during the impact process to quantitatively characterise the dynamic mechanics and fracture process of the different media using SHPB and then to assess the stability of the coal and rock system and the damage process [32]. After the impact experiments were completed, the coal and rock fragments were collected and sieved using a split-sample sieve, and the mass sieve dimension was used to portray the coal and rock impact damage results, as shown in Stage IV.

2.4. *Introduction to the Theory*

2.4.1. Principles of the SHPB Test

The incident, transmitted, and transmissive wave signals were recorded with the strain gauges in the SHPB rods, based on the one-dimensional stress-wave theory and the assumption of stress homogeneity. The stress, strain, and strain rate of the coal and rock samples can be obtained as a function of time [33]:

$$\sigma(t) = E_0 \varepsilon_t(t) \tag{1}$$

$$\varepsilon(t) = -\frac{2C_0}{L}\int_0^t \varepsilon_r(t)dt \tag{2}$$

$$\dot{\varepsilon}(t) = -\frac{2C_0}{L}\varepsilon_r(t) \tag{3}$$

In the above equation, $\dot{\varepsilon}(t)$, $\varepsilon(t)$, $\sigma(t)$ are the average strain rate, average strain, and stress, respectively; $\varepsilon_r(t)$ and $\varepsilon_t(t)$ are the reflected wave strain and transmitted wave strain, respectively; E_0, C_0 are the modulus of elasticity and wave velocity of the rod, respectively; L is the length of the sample.

2.4.2. Box Dimension Calculation Model

The fractal dimension is a statistic that describes the complexity of a geometric shape. There are many definitions of fractal dimension, among which the counting box dimension is the most widely used and relatively easy to calculate. The box dimension is the number of boxes consisting of N squares (with side length $\delta \times \delta$) to cover the coal and rock cracks and count the number of non-empty boxes containing pixels of the associated cracks N_δ. It is calculated by the formula:

$$N_\delta = a\delta^{-D} \quad (4)$$

The prefactor a comes from the scaling rule for fractal dimensions D. For a fixed crack image, a is a constant. If Equation (4) is expressed in logarithmic form, the expression reads:

$$\ln N_\delta = \ln a - D \ln \delta \quad (5)$$

From Equation (5), the slope with respect to $\ln \delta$ is obtained using linear regression $\ln N_\delta$. The result is the fractal dimension D of the crack image of the coal and rock.

The box dimension was analysed using the tool for box dimension analysis in the FracLac plugin in ImageJ (5.0 version) software. In order to reduce the error, the original image was first extracted with cracks using the Ratsnake (1.4 version) annotation software to improve the accuracy and resolution. Finally, it was imported into the FracLac plug-in for box dimension analysis. The specific process is shown in Figure 6. In combination with the Ratsnake software and the FracLac plug-in, a quantitative description of the surface cracks in the coal and rock samples was achieved. In particular, the ImageJ software evaluated whether the crack sketch was a binary image. If it was a binary image, the FracLac plug-in was opened manually for analysis. Otherwise, the image was binarised by setting a threshold. In Fraclac, the control panel parameters for analysing the cracks were set in the region of interest (ROI). Finally, the box dimension was obtained from the image scanning and the statistics of Equation (5).

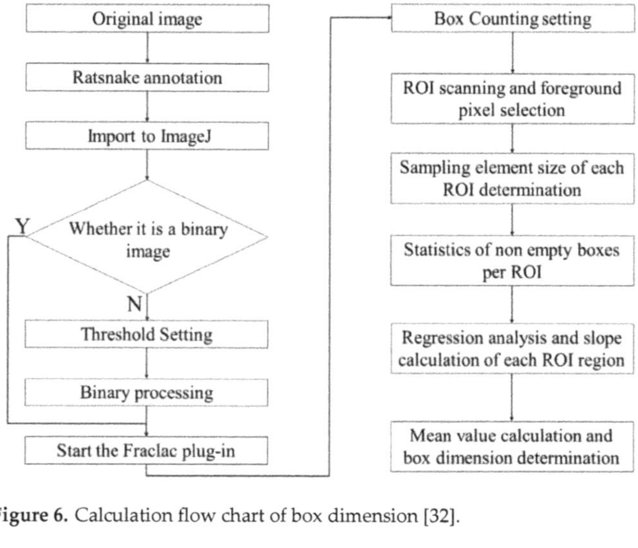

Figure 6. Calculation flow chart of box dimension [32].

2.4.3. Mass–Frequency Relationship Model

In fact, the fractal theory can be used not only to describe the crack extension law during the destruction of coal and rock, but also to analyse the characteristics of the block size distribution after destruction. A quantitative expression for the degree of coal and rock fragmentation can be established with the help of the mass–frequency relationship metric model for samples after impact damage [34]:

$$M(x)/M_T = (x/x_m)^{3-D} \tag{6}$$

where x is the particle size; D is the fractal dimension of the block size distribution; $M(x)$ is the mass of the coal and rock crusher with a diameter smaller than x; x_m is the average scale; M_T is the total mass of the coal and rock crusher. The total mass of coal and rock fragments is the total mass of coal and rock fragments:

$$ln[M(x)/M_T] = (3-D)ln(x/x_m) \tag{7}$$

According to Equation (7), through the double logarithmic scatterplot composed of $ln[M(x)/M_T]$ and $ln(x/x_m)$, we used the least-squares method to fit the data, found the fitted slope and the fitted intercept, and then subtracted the slope of the straight line from 3, which was calculated as the mass fractal dimension D.

3. Analysis of Test Results

As mentioned in the previous section, this experiment analysed the dynamic mechanical properties and crack evolution process of coal and rock materials under SHPB impact conditions at the following four levels: (1) longitudinal wave velocity curve analysis, (2) stress–strain curve relationship, (3) crack evolution trend, and (4) sieving of coal and rock fragments.

3.1. Basic Physical Parameters and Wave Velocity Determination of Coal Stone Samples

3.1.1. Basic Physical Parameters

To better analyse the dynamic mechanical properties of the coal and rock specimens, firstly, the basic physical parameters such as mass, height, and diameter of each specimen were measured. Secondly, the arithmetic mean of the longitudinal wave velocity was calculated after several trials using the ultrasonic method test with reference to ASTM D 2845 [35]. The arithmetic mean of the longitudinal wave speed was calculated after several tests [36,37]. The uniaxial compression damage experiments were measured to determine the compressive coefficient, modulus of elasticity, and Poisson's ratio using the microcomputer-controlled electrohydraulic servo press YAW6106. The final results are summarised in Tables 1 and 2.

Table 1. Test results of coal samples.

Temperature /°C	Sample ID	Size (Diameter × Height)/mm	Density ρ(g/cm³)	Initial Wave Speed	Drying Wave Velocity	Saturation Wave Speed	P Wave Velocity v(km/s)	Average Wave Velocity v(km/s)	Uniaxial Compressive Strength (MPa)	Elastic Modulus Etan (GPa)	Poisson's Ratio μ	Impact Speed v(m/s)	Experimental Result
−20	DFC-20-1	50 × 50	1.53	1.85	1.74	—	2.12	1.96	—	—	—	6.16	impact smash
	DFC-20-2	50 × 50	1.54	1.79	1.64	—	2.12		—	—	—	9.31	impact smash
	DFC-20-3	50 × 50	1.57	1.74	1.46	—	2.04		—	—	—	15.29	impact smash
	DFC-20-4	50 × 50	1.48	1.41	1.36	—	1.56		—	—	—	18.00	impact smash
	DFC-20-5	50 × 100	—	—	—	—	—		14.40	2.51	0.35		
	SFC-20-1	50 × 50	1.45	1.55	1.52	1.56	1.64	2.01	—	—	—	6.17	impact smash
	SFC-20-2	50 × 50	1.45	1.52	1.60	1.88	2.33		—	—	—	9.54	impact smash
	SFC-20-3	50 × 50	1.82	1.51	1.40	1.5	1.81		—	—	—	14.68	impact smash
	SFC-20-4	50 × 50	1.64	2.03	1.64	1.93	2.27		—	—	—	18.07	impact smash
	SFC-20-5	50 × 100	—	—	—	—	—		18.24	2.74	0.33		
−10	DFC-10-1	50 × 50	1.54	1.70	1.56	—	1.8	1.68	—	—	—	6.15	impact smash
	DFC-10-2	50 × 50	1.58	1.67	1.33	—	1.72		—	—	—	9.43	impact smash
	DFC-10-3	50 × 50	1.45	1.64	1.56	—	1.67		—	—	—	14.48	impact smash
	DFC-10-4	50 × 50	1.48	1.37	1.36	—	1.52		—	—	—	17.96	impact smash
	SFC-10-1	50 × 50	1.60	1.51	1.20	1.48	1.9	1.75	—	—	—	6.35	impact smash
	SFC-10-2	50 × 50	1.45	1.49	1.45	1.64	1.67		—	—	—	10.26	impact smash
	SFC-10-3	50 × 50	1.62	1.46	1.42	1.51	1.72		—	—	—	14.08	impact smash
	SFC-10-4	50 × 50	1.59	1.48	1.32	1.46	1.7		—	—	—	18.07	impact smash
0	DC-0-1	50 × 50	1.62	1.63	1.38	—	1.69	1.74	—	—	—	6.38	impact smash
	DC-0-2	50 × 50	1.64	1.62	1.42	—	1.68		—	—	—	9.74	impact smash
	DC-0-3	50 × 50	1.61	1.59	1.49	—	1.79		—	—	—	14.55	impact smash
	DC-0-4	50 × 50	1.45	1.77	1.68	—	1.80		—	—	—	18.24	impact smash
	DC-0-5	50 × 100	—	—	—	—	—		24.78	3.75	0.3		
	SC-0-1	50 × 50	1.46	1.44	1.28	1.44	1.90	1.99	—	—	—	6.03	impact smash
	SC-0-2	50 × 50	1.61	1.90	1.62	1.86	2.19		—	—	—	9.30	impact smash
	SC-0-3	50 × 50	1.45	1.38	1.54	1.86	1.98		—	—	—	14.26	impact smash
	SC-0-4	50 × 50	1.54	1.7	1.28	1.6	1.88		—	—	—	17.92	impact smash
	SC-0-5	50 × 100	—	—	—	—	—		28.20	3.98	0.29		impact smash
10	DC-10-1	50 × 50	1.67	1.59	1.54	—	1.78	1.73	—	—	—	5.55	impact smash
	DC-10-2	50 × 50	1.48	1.58	1.39	—	1.67		—	—	—	9.54	impact smash
	DC-10-3	50 × 50	1.49	1.58	1.44	—	1.77		—	—	—	15.51	impact smash
	DC-10-4	50 × 50	1.43	1.81	1.60	—	1.69		—	—	—	17.96	impact smash
	SC-10-1	50 × 50	1.59	1.38	1.32	1.69	1.71	1.87	—	—	—	5.99	impact smash
	SC-10-2	50 × 50	1.66	1.33	1.49	1.70	1.95		—	—	—	9.47	impact smash
	SC-10-3	50 × 50	1.62	1.33	1.64	1.79	1.88		—	—	—	15.18	impact smash
	SC-10-4	50 × 50	1.42	1.02	1.14	1.51	1.92		—	—	—	17.92	impact smash

Table 1. *Cont.*

Temperature /°C	Sample ID	Size (Diameter × Height)/mm	Density ρ(g/cm³)	Initial Wave Speed	Drying Wave Velocity	Saturation Wave Speed	P Wave Velocity v(km/s)	Average Wave Velocity v(km/s)	Uniaxial Compressive Strength (MPa)	Elastic Modulus Etan (GPa)	Poisson's Ratio μ	Impact Speed v(m/s)	Experimental Result
20	DC-20-1	50 × 50	1.55	1.56	1.48	—	1.59	1.68	—	—	—	6.07	impact smash
	DC-20-2	50 × 50	1.60	1.56	1.54	—	1.71		—	—	—	9.80	impact smash
	DC-20-3	50 × 50	1.46	1.55	1.44	—	1.62		—	—	—	12.79	impact smash
	DC-20-4	50 × 50	1.52	1.81	1.61	—	1.79		—	—	—	17.86	impact smash
	DC-20-5	50 × 100	—	—	—	—	—		16.37	3.21	0.33		
	SC-20-1	50 × 50	1.36	1.33	1.27	1.30	1.42	1.69	—	—	—	6.15	impact smash
	SC-20-2	50 × 50	1.50	1.33	1.35	1.86	1.97		—	—	—	9.12	impact smash
	SC-20-3	50 × 50	1.47	1.31	1.47	1.54	1.68		—	—	—	14.40	impact smash
	SC-20-4	50 × 50		1.87	1.55	1.89	1.67		—	—	—	17.37	impact smash
	SC-20-5	50 × 100	—	—	—	—	—		20.02	3.97	0.31		

Table 2. Test results of rock sample.

Temperature /°C	Sample ID	Size (Diameter × Height)/mm	Density ρ(g/cm3)	Initial Wave Speed	Drying Wave Velocity	Saturation Wave Speed	P Wave Velocity v(km/s)	Average Wave Velocity v(km/s)	Uniaxial Compressive Strength (MPa)	Elastic Modulus Etan (GPa)	Poisson's Ratio μ	Impact Speed v(m/s)	Experimental Result
−20	DFR-20-1	50 × 50	2.76	2.36	2.11	—	2.37	2.38	—	—	—	9.70	impact fracture
	DFR-20-2	50 × 50	2.74	2.36	2.12	—	2.35		—	—	—	14.74	impact smash
	DFR-20-3	50 × 50	2.73	2.24	2.14	—	2.41		—	—	—	18.95	impact smash
	DFR-20-4	50 × 100	—	—	—	—	—		31.68	8.37	0.26		impact smash
	SFR-20-1	50 × 50	2.75	2.16	2.12	2.26	2.48	2.44	—	—	—	10.71	impact fracture
	SFR-20-2	50 × 50	2.77	2.14	2.06	2.29	2.46		—	—	—	14.80	impact smash
	SFR-20-3	50 × 50	2.71	2.14	2.09	2.21	2.38		—	—	—	18.13	impact smash
	SFR-20-4	50 × 100	—	—	—	—	—	—	20.55	8.59	0.24		impact smash
−10	DFR-10-1	50 × 50	2.73	2.22	2.05	—	2.21	2.24	—	—	—	10.21	impact fracture
	DFR-10-2	50 × 50	2.75	2.22	2.11	—	2.25		—	—	—	14.07	impact fracture
	DFR-10-3	50 × 50	2.70	2.22	2.14	—	2.27		—	—	—	17.21	impact smash
	SFR-10-1	50 × 50	2.71	2.12	2.10	2.32	2.3	2.33	—	—	—	10.20	impact smash
	SFR-10-2	50 × 50	2.73	2.12	1.90	2.24	2.35		—	—	—	14.20	impact smash
	SFR-10-3	50 × 50	2.67	2.12	1.82	2.29	2.34		—	—	—	17.80	impact smash

Table 2. *Cont.*

Temperature /°C	Sample ID	Size (Diameter × Height)/mm	Density ρ(g/cm3)	Initial Wave Speed	Drying Wave Velocity	Saturation Wave Speed	P Wave Velocity v(km/s)	Average Wave Velocity v(km/s)	Uniaxial Compressive Strength (MPa)	Elastic Modulus Etan (GPa)	Poisson's Ratio μ	Impact Speed v(m/s)	Experimental Result
0	DR-0-1	50 × 50	2.67	2.21	2.11	—	2.42	2.40	—	—	—	9.38	impact fracture
	DR-0-2	50 × 50	2.71	2.21	2.03	—	2.40		—	—	—	14.02	impact smash
	DR-0-3	50 × 50	2.71	2.19	2.13	—	2.38		—	—	—	17.30	impact smash
	DR-0-4	50 × 100	—	—	—	—	—		39.36	8.64	0.23		impact smash
	SR-0-1	50 × 50	2.70	2.10	1.98	2.30	2.60	2.61	—	—	—	9.12	impact smash
	SR-0-2	50 × 50	2.67	2.10	2.08	2.24	2.64		—	—	—	14.02	impact smash
	SR-0-3	50 × 50	2.67	2.10	1.95	2.17	2.60		—	—	—	17.81	impact smash
	SR-0-4	50 × 100	—	—	—	—	—		20.83	8.82	0.21		impact smash
10	DR-10-1	50 × 50	2.74	2.19	2.04	—	2.50	2.37	—	—	—	10.61	impact fracture
	DR-10-2	50 × 50	2.66	2.19	2.10	—	2.28		—	—	—	13.80	impact smash
	DR-10-3	50 × 50	2.73	2.19	1.90	—	2.32	2.51	—	—	—	17.87	impact smash
	SR-10-1	50 × 50	2.70	2.08	1.85	2.19	2.40		—	—	—	9.69	impact smash
	SR-10-2	50 × 50	2.73	2.08	1.77	2.20	2.60		—	—	—	13.91	impact smash
	SR-10-3	50 × 50	2.75	2.08	1.91	2.31	2.52		—	—	—	10.30	impact smash
20	DR-20-1	50 × 50	2.70	2.19	1.96	—	2.29	2.34	—	—	—	10.39	impact fracture
	DR-20-2	50 × 50	2.75	2.18	2.02	—	2.34		—	—	—	15.80	impact smash
	DR-20-3	50 × 50	2.72	2.17	1.98	—	2.38		—	—	—	17.96	impact smash
	DR-20-4	50 × 100	—	—	—	—	—	—	38.50	8.72	0.25		impact smash
	SR-20-1	50 × 50	2.72	2.08	2.07	2.38	2.43	2.43	—	—	—	9.92	impact fracture
	SR-20-2	50 × 50	2.71	2.07	2.02	2.41	2.44		—	—	—	14.76	impact smash
	SR-20-3	50 × 50	2.69	2.07	1.95	2.34	2.41		—	—	—	18.32	impact smash
	SR-20-4	50 × 100	—	—	—	—	—	—	23.19	8.75	0.23		impact smash

3.1.2. Characteristics of the Wave Velocity Measurements at Different Temperatures and Saturation Levels

In this study, after several measurements of the sample, the average wave speed was used and plotted as a graph of the variation in the wave speed of the longitudinal wave.

Figure 7 shows the ultrasonic wave velocity diagram of the dry and saturated coal and rock samples at different temperatures. As shown in the figure, the degree of wave speed can directly reflect the degree of material densification. The ratio of coal samples (1.68–2.01 km/s) to rock samples (2.24–2.61 km/s) shows that the degree of densification of coal is lower than that of rock.

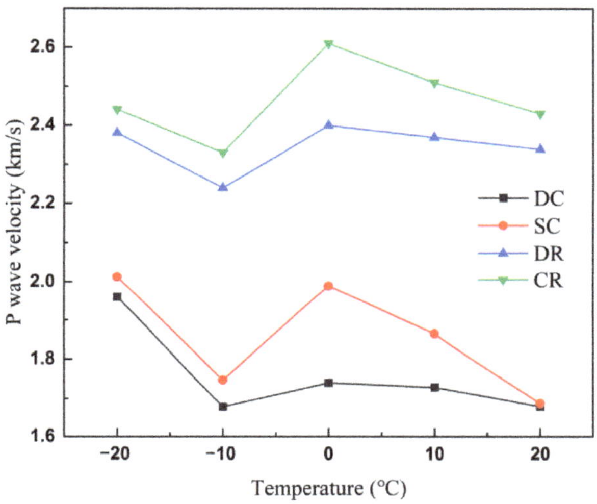

Figure 7. Wave velocity profiles of dry and saturated coal and rock samples at different temperatures.

The wave speeds of all coal and rock samples showed a trend of initially increasing, then decreasing, and then increasing as the temperature decreases. The inflection points were at 0 °C and −10 °C, and an "N"-shaped pattern was observed. Among them, the highest slope of the increase or decrease in wave velocity was observed when the temperature was reduced from 0 °C to −10 °C and then to −20 °C.

Comparing the samples in the dry and saturated states, the wave speeds of the coal and rock samples in the saturated state were all greater than those of the coal and rock samples in the dry state. The degree of rapidity of growth or decrease in the saturated state of the coal and rock was higher than that in the dry state. The wave speed of coal increases more than that of rock from 20 °C to 0 °C and from −10 °C to −20 °C. At 0 °C to −10 °C, rocks decrease more than coal.

The increase in wave speed potentially occurred because the contraction of mineral particles or the phase change of water ice changed the longitudinal wave speed of coal and rock samples. As the pore ice filled the voids inside the coal and rock, the medium was solid–liquid–solid before, and it became solid–solid–solid. The wave impedance of ice was closer to that of the solid medium of coal and rock compared to that of water in the crevices; this caused the propagation of ultrasonic waves inside the coal and rock, and the ice in the pores reduced the wave diffraction and reflection between the original coal and original rock interfaces and the surface of the pores, thus causing a great increase in the longitudinal wave speed. The ice in the pore space reduced the diffraction and reflection of longitudinal waves at the interface between the coal or rock and the pore surface, which led to a significant increase in the longitudinal wave speed. This observation is in high agreement with Li's findings [38].

3.2. *Dynamic Mechanical Characteristics*

3.2.1. Time History Curve of Stress Wave

At the beginning of the experiment, to ensure the accuracy of the SHPB experimental results, stress balance verification was performed before processing the data [32]. Specifically, the two end faces of the specimen needed to reach stress equilibrium under the impact load, and the stress equilibrium curve of the specimens are shown in Figure 8. As shown in the figure, the superposition of the incident wave, reflected wave, and transmitted wave basically coincided with each other, indicating that the coal and rock basically reached the stress equilibrium state.

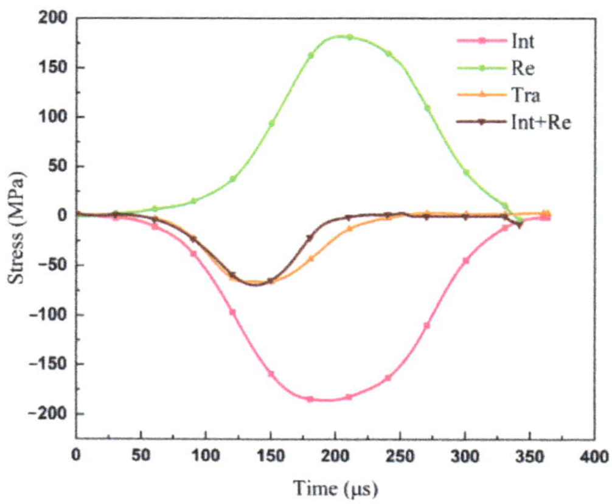

Figure 8. Typical dynamic stress equilibrium checks in dynamic compression tests. (Int: incident wave, Re: reflected wave, and Tra: transmitted wave).

Due to the large noise in the original signal, the data in this study were denoised and analysed using the Hilbert–Huang transformation (HHT) method [39]. Due to space limitations, only some data with different variables were analysed as an example.

Figure 9a,b show the stress time course curves of raw coal at the same freezing temperature of −10 °C with different saturations and speeds. As shown in the figure, (1) all curves of the incident waves have sinusoidal or half-sinusoidal curves. With the increase in the impact velocity, the peak points of the incident, reflection, and transmission curves gradually increase, the peak arrival time is advanced, and the overall duration of the impact is also shortened. At lower velocities, the reflected stress tail reaches a negative value. The reason for this phenomenon is that the specimen is not destroyed completely under incident wave loading conditions, and a large amount of strain energy is stored in the medium instead. These strain energies are released in the unloading phase of the incident wave and rebound to the bar as compression waves. (2) In the dry state and natural water-saturated state, the peak value of the transmission wave of the raw coal in the water-saturated state is higher than that in the dry state. Compared with Table 1, a denser sample in the water-saturated state occurs because a large amount of liquid water or solid water fills the internal cracks. The large amount of liquid or solid water in the internal clefts leads to a reduction in energy dissipation within the sample due to the presence of the Stefan effect.

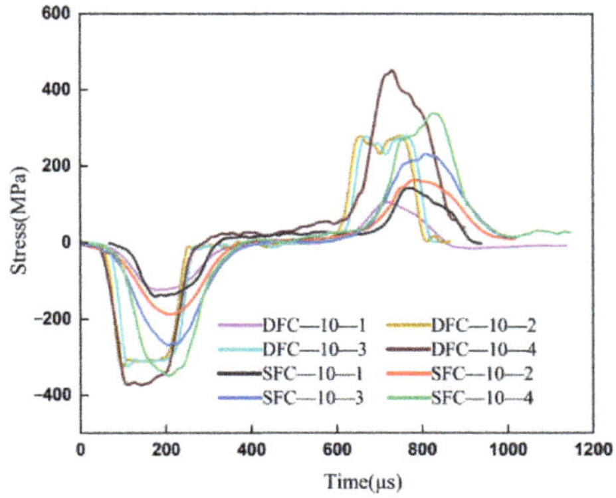

(**a**)

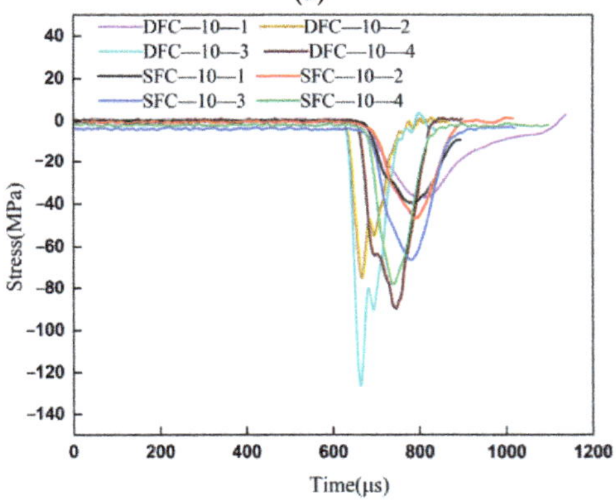

(**b**)

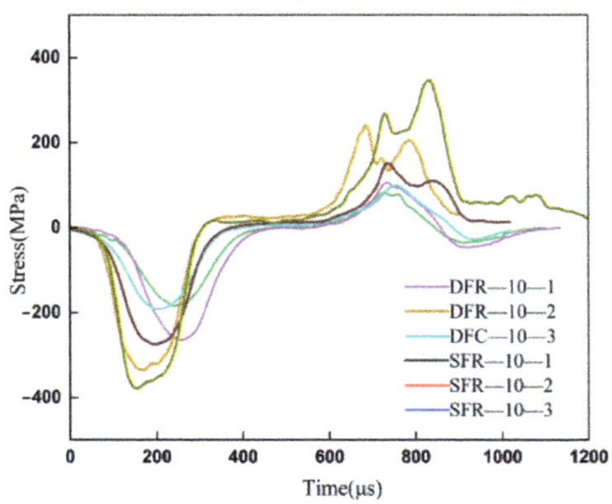

(**c**)

Figure 9. *Cont.*

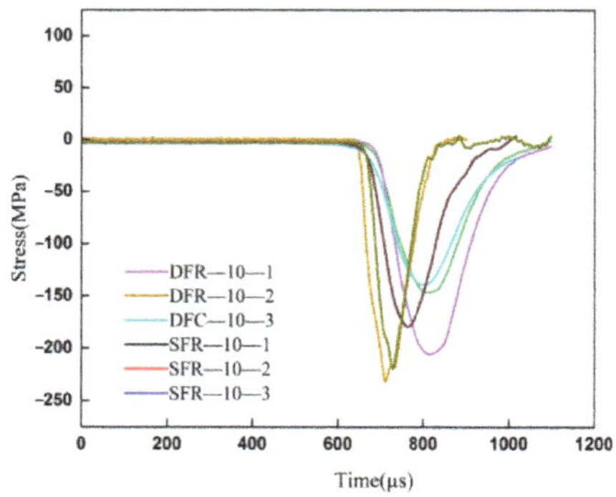

(d)

Figure 9. Denoised stress-wave signals at different impact velocities. (**a**) Incident-reflected waves of coal at −10 °C; (**b**) transmitted waves of coal at −10 °C; (**c**) incident-reflected waves of rock at −10 °C; and (**d**) transmitted waves of rock at −10 °C.

Using the same means, the same noise reduction was applied to the experimental data of the raw rock. As shown in Figure 9c,d, the graphs show the stress time-course curves of the raw rock at the same freezing temperature of −10 °C for different saturations and velocities. The overall regularity of the coal and rock bodies has a high degree of overlap. The trend shows that with increasing speed, the peak value of the curve increases; the sample reaches the peak time earlier. Raw rock in the dry and saturated contrasting cases also had a similar pattern to raw coal, both of which had higher curves in the dry case than in the saturated state.

Notably, the peak of the transmittance wave of the coal samples was significantly lower than that of the rock samples. Similarly, the process of reaching the peak of the stress wave in coal samples tended to be relatively slow. This was caused by the different densities of the coal and rock samples, which were determined through the wave speeds in Tables 1 and 2. A denser sample causes a faster propagation of the ultrasonic wave. This curve and wave speed could be used to validate each other.

3.2.2. Dynamic Mechanical Characteristics of Coal and Rock during Impact

To better analyse the effect of temperature on the coal and rock samples, the stress–strain diagrams and peak stress diagrams were plotted for analysis. Using the three-wave method, the stress–strain curve was obtained after the calculation of the stress time-course curve. As shown in Figure 10a,b, the stress–strain curves in the water-saturated state had velocities of 18 m/s for coal and rock. As shown in the figures, the trend process of the coal and rock stress–strain curves at different temperatures is very similar; however, some differences in the characteristics of each stage are present. First, the coal samples at 10 °C, 20 °C, and −20 °C show a kind of "saddle shape", which was evidently a manifestation of the strain intensification, and the brittleness of the samples was also more significant in this stage; rocks did not have this shape. Second, at a freezing temperature of −10 °C, the slope of the curve rises slowly in the initial phase, and the strain duration was shorter. The slow decrease in the unloading stage also showed that at −10 °C, the sample had the worst load-carrying capacity and the most severe deformation. In the last part of the curve, a certain "rebound" phenomenon occurred at the end stage because after the peak value, the coal and rock samples still had the ability to resist external loads.

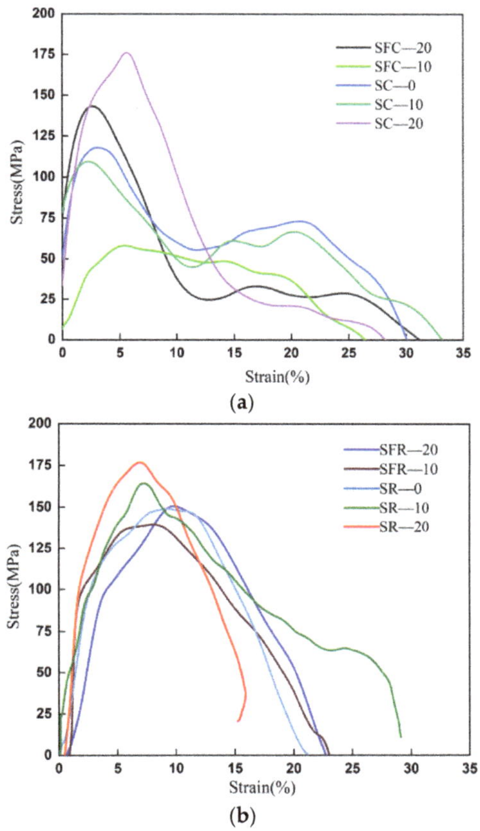

Figure 10. Dynamic mechanical characteristic curve of coal and rock impact. (**a**) 18 m/s for coal; (**b**) 18 m/s or rock.

In the analysis of the dynamic load mechanics, a fixed trend for the change in stress at different temperatures among studies by different scholars has not been shown. In this study, the variation curve of the peak stress with the temperature gradient in water-saturated coal and rock was plotted based on the stress–strain curve in Figure 10a,b, as shown in Figure 11a,b. In Figure 11a, for the coal samples at temperatures from −20 °C to −10 °C, the peak strength rapidly decreased. From −10 °C to 10 °C, the peak intensity slowly increased. A temperature of −10 °C was an inflection point of change.

In Figure 11b, the rock samples also maintained a similar pattern. The overall stress peaks all increased with increasing impact velocity. The inflection point decreased from −10 °C to 0 °C. The same pattern was maintained with the peak stress curve of coal at 18 m/s. Overall, the peak strength of water-saturated rock maintained the trend of rising, falling, and then rising with the decrease in temperature, which is similar to an "N" shape. However, the coal and rock samples are also slightly different. The 10 m/s impact velocity for the rock shows a U-shaped pattern.

Based on the above results, two aspects of the joint role are as follows: (1) With a decrease in the temperature, the cold coal or rock body lead to a gradual reduction in its own strength and volume shrinkage, resulting in the closure of the internal fissure. (2) In the freezing process, liquid water transforms into solid ice; it expands in volume and generates freezing expansion force. Under the action of frost heave force, micro-cracks sprout or expand the existing cracks at the cracks. The two factors are mutually restraining and mutually exist. At different temperatures, the change in its own strength dominates the strength of the material. At 0 °C, the freezing expansion forces within the material dominate, increasing the degree of internal damage to the rock and leading to an instantaneous decrease in its strength. For coal, the value of this "critical" temperature is −10 °C.

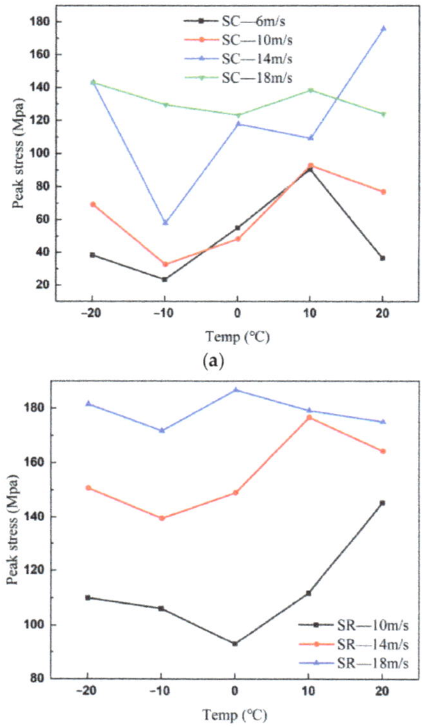

(**b**)

Figure 11. Variation of peak intensity of water-saturated coal and rock with temperature. (**a**) Coal; (**b**) rock.

3.3. Process of Crack Propagation and Specimen Failure

3.3.1. Recognition and Feature Extraction of Surface Cracks

Currently, several methods can be used to study the damage process in impact dynamics, and the most popular method is high-speed (HS) photography to record the entire damage process, including fractal analysis [40,41] and deep learning [42,43]. However, in the analysis of the images, more scholars tend to analyse them qualitatively but fail to quantify them. Because the generation of the cracks in the image is very complex, the extraction of all cracks is difficult. Therefore, in this study, we used the Ratsnake annotation software [44] to extract the cracks from each original image during the fracture process. Rastnake can quickly segment the image with annotations and customise the annotation. RGB images captured using the HS camera were converted to black-and-white binarised images; this improved the resolution and accuracy [45]. Due to space limitations, in this paper, raw rock samples at different temperatures and saturations were selected as examples, and deformation was generated during the destruction process until fragments were ejected during the camera recording. The shooting frequency of the high-speed camera was 100,000 frames/s, and each set of data was intercepted every 1 µs. The final damage image and the crack extension process are shown in Figure 12a,b. Figure 12c is last frame of the results extracted from the original crack images and Ratsnake software. The dry rock at 20 °C (DR-20) is taken as an example; its impact velocity is 17.96 m/s. It can be seen that the sample has developed large cracks with debris ejection.

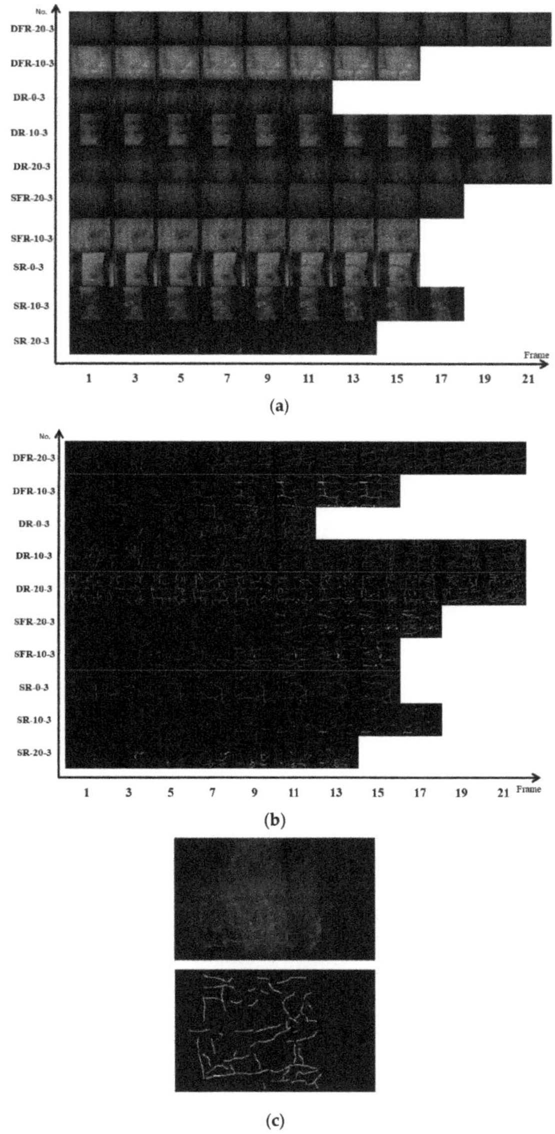

Figure 12. The whole process of image destruction. (**a**) Original crack images; (**b**) Ratsnake software extraction results; and (**c**) last frame with comparison and Ratsnake results with image.

3.3.2. Crack Quantification and Box Dimension Calculation

Referring to the process shown in Figure 6, based on the binarised images identified from the cracks in the Ratsnake surface, the processed images were quantified using FracLac embedded in ImageJ to analyse the evolutionary pattern of each specimen from crack generation to rupture injection. The average values were found using linear regression and multiple calculations. Due to the low degree of failure of some rocks at a speed of 10 m/s, the screening results are not representative; a total of 40 groups of coal samples and 24 groups of rock samples were obtained after rounding off part of the data to obtain the dimensionality-change values. Due to space limitations, the partial curves of the dimensionality change of the coal and rock boxes were plotted, as shown in Figure 13.

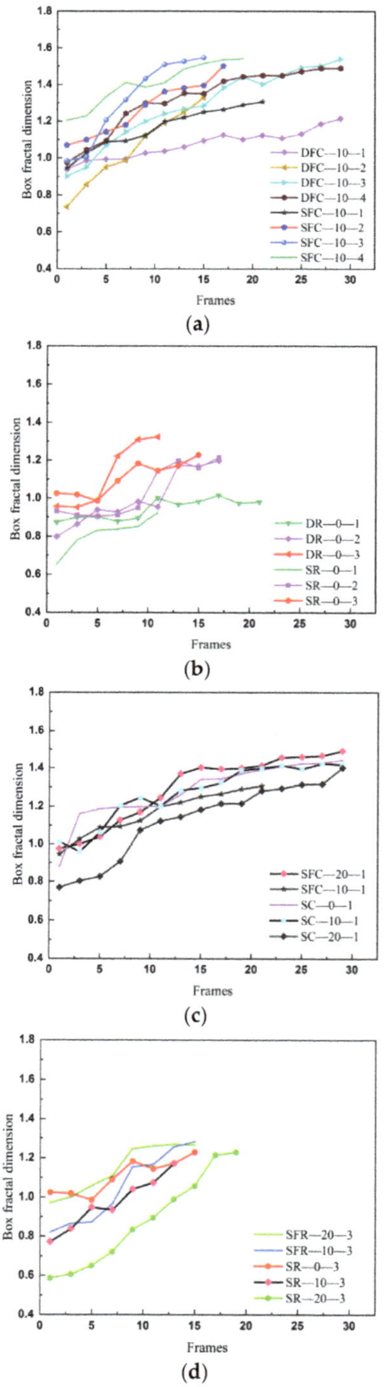

Figure 13. Box dimension evolutions of cracks in coal and rock specimens. (**a**) Different saturation levels and impact speeds for coal samples; (**b**) different saturation levels and impact speeds for rock samples; (**c**) different temperatures for coal samples; and (**d**) different temperatures for rock samples.

Figure 13a shows the distribution of box dimensions for different saturation and impact velocities, all at a temperature of −10 °C. The following can be observed from this figure: (1) For the individual curves, an overall trend from a rapid increase in the initial phase to a slow increase and then to localised oscillations in the later phase is shown. The increase is due to the entire process being characterised by an increase in stresses on the coal or rock body over time and a continuous increase in damage. The oscillations can be caused by the transient closure of the internal fissures due to the compression of the internal fissures after the impact on the coal or rock body. (2) For the coal simples at an impact velocity speed of 6 m/s, the dimension of the dried sample is higher than that in the saturated state, with a faster end time. This indicates that the number of cracks and the damage degree of the sample in the dry state are greater than those in the saturated state. (3) Under the same saturation conditions, comparing the crack evolution characteristics under different impact velocities, with an increase in the impact velocity, the value of the box dimension increases, but the duration is shortened. This shows that the change in velocity can cause a large change in surface cracks and presents a positive correlation. A faster impact velocity correlates to a faster damage process, a faster evolution of the cracks, and an earlier ejection of the number of coal and rock breakers.

Figure 13b shows the evolution pattern of the box dimension of the surface cracks of the rock samples under different saturation and velocity conditions at 0 °C. Figure 13b exhibits a similar pattern to Figure 13a: it is shown that, under different speed and saturation conditions, the box dimension shows the typical characteristics of fast in the early stage, slow in the late stage, and an oscillating increase. In addition, the box dimension as a whole increases with the decrease in saturation and increases with the increase in speed.

Notably, on the one hand, the overall holding time of the rock is shorter than that of the coal sample; on the other hand, the final value of the box dimension is smaller than that of the original coal. This reflects several factors: (1) By the nature of the material, the rock is somewhat more brittle, more resistant to deformation, and less destructive at the time of destruction. (2) Rocky materials have a high rebound energy. For raw coal, the vast majority of the energy from the incident rod is absorbed by the sample and used to increase its surface energy. In the case of rock, the energy passes through and reaches the transmissive rod quickly, and a large portion of the energy is converted into the ejection energy of the fragments.

Correspondingly, Figure 13c reflects the change trend of the surface crack box dimension of coal samples under the same velocity gradient (6 m/s) and different temperature conditions; here, the curves still show the change characteristics of overall increase and local oscillation. In addition, Figure 13c also reflects the effect of temperature on the crack evolution. Specifically, the amplitude follows this trend: 20 °C > −10 °C > 10 °C > 0 °C > −20 °C, and the slope shows the opposite trend. A comparison with the longitudinal wave speed in Figure 7 in the previous section shows that the magnitude of the box dimension number shows a significant negative correlation with the average wave speed value. This occurs because a higher wave speed correlates to a higher degree of densification of the sample, indicating that the sample is more resistant to damage.

Figure 13d shows the effect of temperature on the box dimension number of cracks on the surface of the rock samples under the condition of the same impact velocity (18 m/s) and the same saturation (100%). As shown in the figure, the amplitude and the growth rate follow these trends: −10 °C > 20 °C > −20 °C > 10 °C > 0 °C, and 20 °C > −20 °C > −10 °C > 10 °C > 0 °C, respectively. Compared with Figure 13c, the amplitude of the box dimension still shows the characteristic of a negative correlation with the value of the mean wave velocity. The difference is that the rock holding time is short because the shock velocity in Figure 13d is much larger than that in Figure 13c. However, the overall box dimension of the rock samples is still lower than that of the coal samples, indicating that for the box dimension, the effect of velocity (external loading) is much greater than that of temperature (physical influence of the medium).

Notably, the box dimension characterisation method used in this study has a somewhat defective nature. (1) The original cracks are manually marked using Ratsnake. The filling of cracks and the ejection criteria of fragments of samples have certain subjective factors. (2) The sample is a three-dimensional body, and the photo taken is a plane plan. The cracks in the curved part of the sample may have an angle with the analysis plane, resulting in errors. (3) In the shooting process, HS technology can only detect the evolution process of surface cracks and cannot fully reflect the change in internal cracks. Based on this, we selected sample sieves with different apertures to fully screen the impacted coal and rock crushed blocks and discusses the impact damage from another perspective.

3.4. Fractal Characteristics of Crushing Products

To better analyse the result after sample destruction, and referring to the relevant standards [32], the sample sieves with 9.50 mm, 4.75 mm, 2.36 mm, 1.18 mm, 0.60 mm, 0.30 mm, 0.15 mm, 0.075 mm, and eight grades were combined in the order of largest to smallest. The coal and rock after SHPB impact damage were sieved on site. A high-precision balance was used to weigh the broken pieces of coal and rock for each rank. The statistical results are summarised in Tables 3 and 4.

Table 3. Calculation results of the mass dimension of coal.

Sample No.	Impact Speed (m/s)	Cumulative Mass Percent of Each Particle Size Under the Sieve/%								Fitting Results			Fractal Dimension
		0.075	0.15	0.3	0.6	1.18	2.36	4.75	9.5	Fitting Slope	Intercept Distance	Decision Factor (R^2 Value)	
DFC-20-1	6.16	0.21	0.67	2.15	5.23	10.49	29.17	65.95	100.00	1.28	0.19	0.99	1.72
DFC-20-2	9.31	0.51	1.52	4.44	9.97	19.03	42.49	75.10	100.00	1.10	0.22	0.99	1.90
DFC-20-3	15.29	0.70	2.31	7.49	14.54	28.62	58.06	92.60	100.00	1.03	0.29	0.95	1.97
DFC-20-4	18.00	3.10	6.54	14.89	26.42	45.02	75.91	95.11	100.00	0.74	0.22	0.95	2.26
SFC-20-1	6.17	0.00	0.04	0.25	0.80	2.18	5.91	18.88	100.00	1.78	−0.01	0.99	1.22
SFC-20-2	9.54	0.41	1.32	4.50	10.48	19.60	46.08	84.13	100.00	1.15	0.08	0.97	1.85
SFC-20-3	14.68	0.37	1.00	3.08	6.65	12.41	27.17	58.75	100.00	1.15	0.07	0.99	1.85
SFC-20-4	18.07	1.72	3.71	9.83	19.84	35.06	69.46	97.71	100.00	0.88	0.21	0.96	2.12
DFC-10-1	6.15	0.05	0.23	0.82	1.94	3.96	9.63	28.76	100.00	1.47	0.18	0.99	1.53
DFC-10-2	9.43	0.28	0.73	2.05	4.47	8.48	20.71	52.05	100.00	1.21	0.20	1.00	1.79
DFC-10-3	14.48	0.84	1.81	4.49	8.97	15.93	29.82	50.91	100.00	0.97	0.11	0.99	2.03
DFC-10-4	17.96	3.04	6.46	14.82	25.14	43.04	71.13	91.55	100.00	0.74	0.26	0.95	2.26
SFC-10-1	6.35	0.42	0.94	2.33	4.43	8.93	37.02	53.82	100.00	1.17	−0.01	0.99	1.83
SFC-10-2	10.26	0.39	0.93	2.34	4.71	9.29	23.54	60.56	100.00	1.16	0.17	1.00	1.84
SFC-10-3	14.08	0.90	2.41	6.38	11.72	22.28	50.18	84.28	100.00	0.99	0.17	0.98	2.01
SFC-10-4	17.69	1.88	4.30	11.05	19.55	33.77	55.64	82.94	100.00	0.83	0.23	0.96	2.17
DC-0-1	6.38	0.48	1.08	3.02	6.55	13.33	36.50	74.99	100.00	1.15	−0.05	0.99	1.85
DC-0-2	9.74	0.55	1.35	4.11	9.22	17.81	38.89	73.21	100.00	1.10	0.12	0.98	1.90
DC-0-3	14.55	0.81	1.84	4.56	8.75	15.33	33.39	64.75	100.00	1.00	0.14	0.99	2.00
DC-0-4	18.24	2.28	4.93	12.65	25.34	42.90	75.82	96.59	100.00	0.81	0.17	0.94	2.19
SC-0-1	6.03	0.07	0.21	0.93	2.35	5.32	13.77	27.94	100.00	1.46	−0.03	0.99	1.54
SC-0-2	9.30	0.53	1.31	3.56	6.84	13.79	29.97	64.21	100.00	1.09	0.27	0.99	1.91
SC-0-3	14.26	1.98	4.16	10.28	20.03	33.90	64.08	95.43	100.00	0.85	0.13	0.96	2.15
SC-0-4	17.92	1.48	3.15	7.79	15.37	27.23	54.01	89.19	100.00	0.91	0.26	0.98	2.09
DC-10-1	5.55	0.18	0.47	1.38	2.77	5.69	15.02	38.49	100.00	1.28	0.11	1.00	1.72
DC-10-2	9.54	0.69	1.63	4.54	8.95	18.16	39.68	70.26	100.00	1.05	0.09	0.99	1.95
DC-10-3	15.51	1.12	2.72	6.50	11.61	22.10	45.45	80.37	100.00	0.94	0.21	0.98	2.06

Table 3. *Cont.*

Sample No.	Impact Speed (m/s)	Cumulative Mass Percent of Each Particle Size Under the Sieve/%								Fitting Results			Fractal Dimension
		0.075	0.15	0.3	0.6	1.18	2.36	4.75	9.5	Fitting Slope	Intercept Distance	Decision Factor (R^2 Value)	
DC-10-4	18.32	1.71	3.81	9.77	18.75	37.00	65.68	84.80	100.00	0.87	0.19	0.96	2.13
SC-10-1	5.99	0.32	0.78	2.17	4.23	9.05	23.44	55.65	100.00	1.20	0.04	1.00	1.80
SC-10-2	9.47	0.81	1.95	5.35	11.53	20.77	43.81	75.21	100.00	1.01	0.13	0.98	1.99
SC-10-3	15.18	1.22	2.77	6.83	11.64	20.86	40.67	76.00	100.00	0.92	0.23	0.99	2.08
SC-10-4	17.93	1.33	2.83	6.53	12.17	20.45	42.24	70.68	100.00	0.99	0.21	0.90	2.01
DC-20-1	6.15	0.99	2.57	7.59	15.95	28.22	52.72	80.24	100.00	1.25	0.09	0.98	1.75
DC-20-2	9.80	1.50	3.67	8.79	15.71	23.85	40.37	68.05	100.00	0.84	0.19	0.98	2.16
DC-20-3	12.79	0.88	2.08	5.26	9.96	17.87	37.37	70.80	100.00	0.99	0.14	0.99	2.01
DC-20-4	17.86	2.24	4.49	10.34	17.84	31.53	56.07	83.96	100.00	0.81	0.13	0.98	2.19
SC-20-1	6.15	0.99	2.57	7.59	15.95	28.22	52.72	80.24	100.00	0.96	0.23	0.96	2.04
SC-20-2	9.12	0.83	1.94	5.27	10.97	20.42	45.56	79.36	100.00	1.02	0.20	0.98	1.98
SC-20-3	14.40	0.75	2.15	5.25	9.29	17.33	37.46	70.73	100.00	1.00	0.15	0.99	2.00
SC-20-4	17.37	1.25	2.61	6.32	11.78	19.37	38.71	74.77	100.00	0.99	0.13	0.92	2.01

Table 4. Calculation results of the mass dimension of rock.

Sample No.	Impact Speed (m/s)	Cumulative Mass Percent of Each Particle Size Under the Sieve/%								Fitting Results			Fractal Dimension
		0.075	0.15	0.3	0.6	1.18	2.36	4.75	9.5	Fitting Slope	Intercept Distance	Decision Factor (R^2 Value)	
DFR-20-1	9.70	0.00	0.03	0.07	0.14	0.23	0.41	0.61	100.00	1.52	−0.98	0.75	1.48
DFR-20-2	14.74	0.01	1.35	2.33	3.36	5.02	8.31	22.73	100.00	1.47	0.06	0.79	1.53
DFR-20-3	18.95	0.01	2.38	4.22	6.28	8.78	14.68	37.59	100.00	1.39	0.19	0.76	1.61
SFR-20-1	10.71	–	–	–	–	–	–	–	–	–		–	–
SFR-20-2	14.80	0.01	1.59	2.71	3.72	5.23	9.68	24.77	100.00	1.44	−1.53	0.78	1.56
SFR-20-3	18.13	0.01	2.61	4.69	6.76	9.97	16.18	36.63	100.00	1.38	−1.51	0.75	1.62
DFR-10-1	10.21	0.00	0.01	0.02	0.02	0.03	0.03	0.03	100.00	1.80	0.05	0.64	1.20
DFR-10-2	14.07	0.00	0.02	0.03	0.03	0.04	0.04	0.04	100.00	1.74	−0.49	0.62	1.26
DFR-10-3	17.21	0.01	1.95	3.30	4.75	6.47	10.46	19.87	100.00	1.37	−0.17	0.75	1.63
SFR-10-1	10.20	0.00	0.08	0.11	0.13	0.22	0.48	0.78	100.00	1.73	0.15	0.79	1.27
SFR-10-2	14.20	0.01	1.25	2.05	2.76	3.85	6.37	16.33	100.00	1.64	−0.41	0.79	1.36
SFR-10-3	17.80	0.01	2.23	3.76	5.24	7.58	13.31	33.56	100.00	1.40	0.17	0.77	1.60
DR-0-1	9.38	–	–	–	–	–	–	–	–	–		–	–
DR-0-2	14.02	0.01	1.36	2.16	2.97	4.07	6.31	19.06	100.00	1.71	−0.03	0.83	1.29
DR-0-3	17.30	0.01	2.38	4.06	5.92	8.38	15.24	30.00	100.00	1.46	0.15	0.78	1.54
SR-0-1	9.12	0.00	0.04	0.08	0.14	0.22	0.32	0.88	100.00	1.95	0.08	0.80	1.05
SR-0-2	14.02	0.01	1.20	1.99	2.78	3.63	6.21	18.15	100.00	1.46	0.20	0.78	1.54
SR-0-3	17.81	0.01	1.52	2.47	3.57	5.14	8.83	21.52	100.00	1.43	−0.80	0.78	1.57
DR-10-1	10.61	–	–	–	–	–	–	–	–	–		–	–
DR-10-2	13.80	0.00	0.23	0.39	0.58	0.82	1.33	3.41	100.00	1.68	0.16	0.80	1.32
DR-10-3	17.87	0.01	2.55	4.36	6.26	8.49	14.20	35.17	100.00	1.37	−0.70	0.76	1.63
SR-10-1	9.69	0.00	0.04	0.09	0.13	0.20	0.28	0.43	100.00	1.88	−0.02	0.76	1.12
SR-10-2	13.91	0.00	0.57	0.99	1.45	2.01	3.43	8.99	100.00	1.56	0.05	0.80	1.44
SR-10-3	10.30	0.01	1.87	3.24	4.72	6.65	11.43	26.42	100.00	1.41		0.77	1.59
DR-20-1	10.39	–	–	–	–	–	–	–	–	–	−0.83	–	–
DR-20-2	15.80	0.00	0.32	0.59	0.88	1.38	2.50	9.73	100.00	1.42	−0.18	0.78	1.58
DR-20-3	17.96	0.01	1.65	3.07	4.68	6.97	12.68	28.54	100.00	1.37	0.11	0.76	1.63
SR-20-1	9.92	0.00	0.04	0.07	0.11	0.15	0.35	1.18	100.00	2.00	−0.67	0.83	1.00
SR-20-2	14.76	0.01	1.60	2.73	3.96	5.77	9.91	25.98	100.00	1.45	0.10	0.78	1.55
SR-20-3	18.32	0.01	2.59	4.60	6.89	9.81	17.59	41.69	100.00	1.38	0.23	0.76	1.62

Notably, some of these values were discarded because most of the rock samples did not undergo significant damage at an impact of 10 m/s and were not of screening significance. Referring to Equation (7), a double logarithmic expression for the impacted samples was established and regression analyses were performed. The results are shown in Figure 14. Combining Tables 3 and 4 shows that there is a good logarithmic correlation between the cumulative mass percentage of the sub-screen particle size and the sample sieve diameter. The correlation coefficient was very high, which indicated that the post-crushing lump size distribution of this coal and rock samples were self-similar to a fractal distribution. The fractal dimension fluctuates between 1.29 and 2.26.

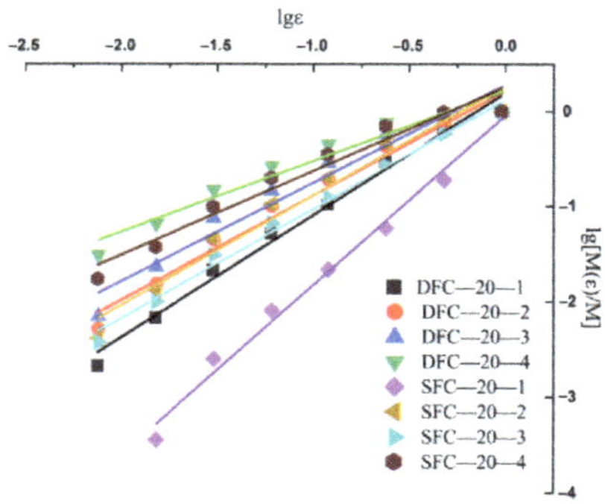

(**a**)

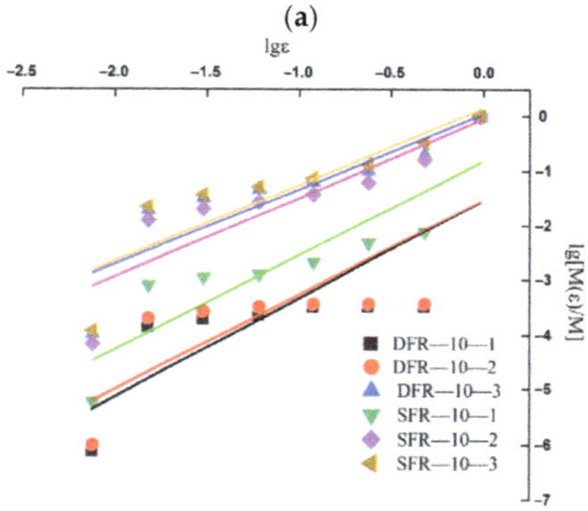

(**b**)

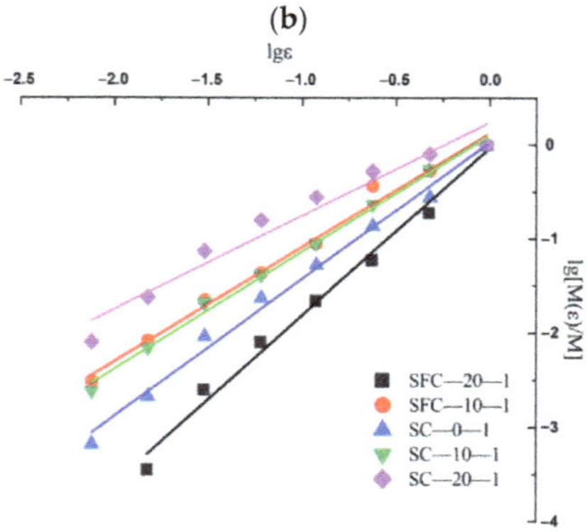

(**c**)

Figure 14. *Cont.*

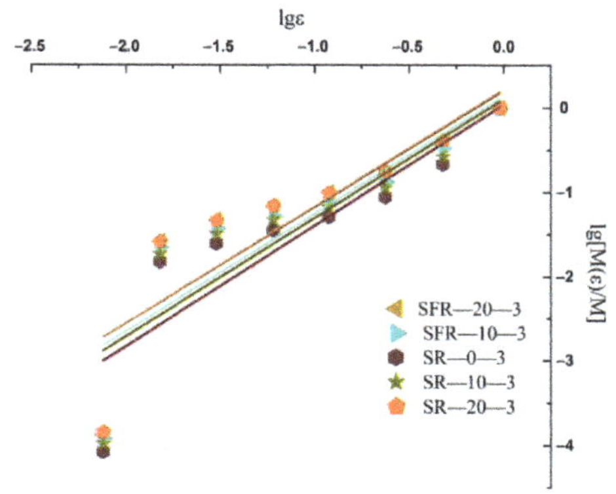

(d)

Figure 14. Statistical graphics of cumulative particle size distribution. (**a**) Different saturation levels and impact speeds for coal samples; (**b**) different saturation levels and impact speeds for rock samples; (**c**) different temperatures for coal samples; and (**d**) different temperatures for rock samples.

Figure 14a shows the mass fractal dimension of coal at the same temperature (−20 °C) and different saturation and velocity conditions. From the figure, the following can be observed: (1) Under the same saturation conditions, with increasing impact velocity, the slope of the fitted curve of the raw coal sample gradually decreases and the value of the mass dimension gradually increases. This result indicates that with increasing velocity, the degree of sample devastation increases and the degree of fragmentation rises. (2) At an impact velocity of 6 m/s, the fitted slope of the dry state is lower than that of the saturated state, which indicates that the mass dimension number in the saturated state is lower than that in the dry state. The crushing volume is larger in the saturated state.

Figure 14b shows the mass fractal dimension of the rocks at the same temperature (−10 °C) under different saturation and velocity conditions. Figure 14b exhibits a similar pattern to Figure 14a, where both velocity and saturation have a similar effect on the coal and rock samples. However, the slopes of the raw rock samples are higher than those of the raw coal samples, indicating that at the inflection point of −10 °C, the rock is still more brittle than the raw coal.

Figure 14c shows the mass fractal dimension of coal under the same saturation and velocity (6 m/s) and different temperature conditions; from this, the slopes follow the following trend: −20 °C > 0 °C > 10 °C > −10 °C > 20 °C. Comparison with the longitudinal wave speed shows that the magnitude of the wave speed is inversely proportional to the slope. The wave speed reflects the denseness of the material, and the slope is negatively correlated with the mass dimension number; thus, the lower the wave speed, the smaller the slope of the mass dimension regression curve and the larger the mass size number.

Figure 14d shows the mass fractal dimension of the rock for the same saturation and velocity (18 m/s) and different temperature conditions. The pattern is similar to that of the raw coal, and both show that the temperature can cause some changes in the samples. The slopes follow this trend: 0 °C < 10 °C < 20 °C < −20 °C < −10 °C. The difference is that (1) At the same saturation and temperature, the slope of the maximum velocity of the original rock is higher than the slope of the minimum velocity of the original coal. This result shows that the degree of dynamic compressive resistance of the rock is evidently much higher than that of the raw coal. The energy absorption of the rock is greater than that of raw coal. The rock sample has stronger resistance to deformation. (2) At different temperatures, the change in the original rock is not as sensitive as that of the original coal. The slope change is minimal. This result shows that the rock is less affected by temperature than the original coal. Under the influence of the temperature and velocity, the mass dimension of rock is dominated by the influence of velocity.

Overall, the slope of the fitted curves decreases with increasing impact velocity and decreasing saturation. Raw coal and raw rock were also affected to varying degrees at different temperatures. The two inflection points at −10 and 0 °C are the most pronounced. The decrease in the slope of the fit and, correspondingly, the increase in the fractal dimension indicate an increase in the degree of fragmentation of the coal or rock mass and a smaller volume. By observing the collected sample fragmentation and quantifying the fractal dimension D, it can be concluded and mutually corroborated by the waveform plots in Figure 14 that coal bodies are more brittle and rocks are somewhat more plastic.

4. Discussion

To better analyse the effects of temperature and saturation on the dynamic mechanics of coal and rock, the changes that occur within the coal and rock samples were analysed through micro-mechanisms.

4.1. Mechanism of Water Action on Cracking Microcosm

In the dry state, the cracks inside the coal or rock body are all in the form of air. When the coal and rock samples reach the water-saturated state, the cracks and crevices of the coal and rock samples are filled with free water. The cracks in the coal or rock body are mostly open (type I) cracks. The presence of free water in coal and rock has a decelerating effect on the high-speed crack expansion [25]. Figure 15 shows the different mechanical analyses of the internal cracks suffered by the coal and rock bodies under different velocities. There are three main factors for the presence of water in the dynamic impact.

(1) σ_1 is the curved-moon-face effect, which can be shown as a curve created by the surface of a liquid close to the surface of the coal or rock and produced by surface tension.

During rapid loading, the free water in the crack does not easily reach the tip of the crack due to the faster crack extension; additionally, the presence of surface forces on the curved-moon face of the free water is equivalent to acting as a beneficial tensile force on the crack face, whose presence hinders the crack extension. This effect has been extensively studied in rock mechanics and plays an important role in the understanding of rock stress and deformation in practical engineering and geological settings.

(2) σ_2 is the Stefan effect, which is the presence of free water in the fractures of the coal or rock body. A certain amount of viscous flow leads to hydrostatic pressure. When a change in the volume of free water occurs, it causes a change in the hydrostatic pressure of the water between the cracks, subsequently causing the flow of viscous fluid and the creation of resistance to the movement of the cracks. The manifestation of the Stenfan effect depends on the rate of change in the water content, the rate of change in temperature, and the physical properties of the material.

(3) σ_3 is the Newtonian internal friction effect, the interaction caused by the parallel relative motion of the water molecules when the internal particles of coal or rock undergo displacement changes due to impact.

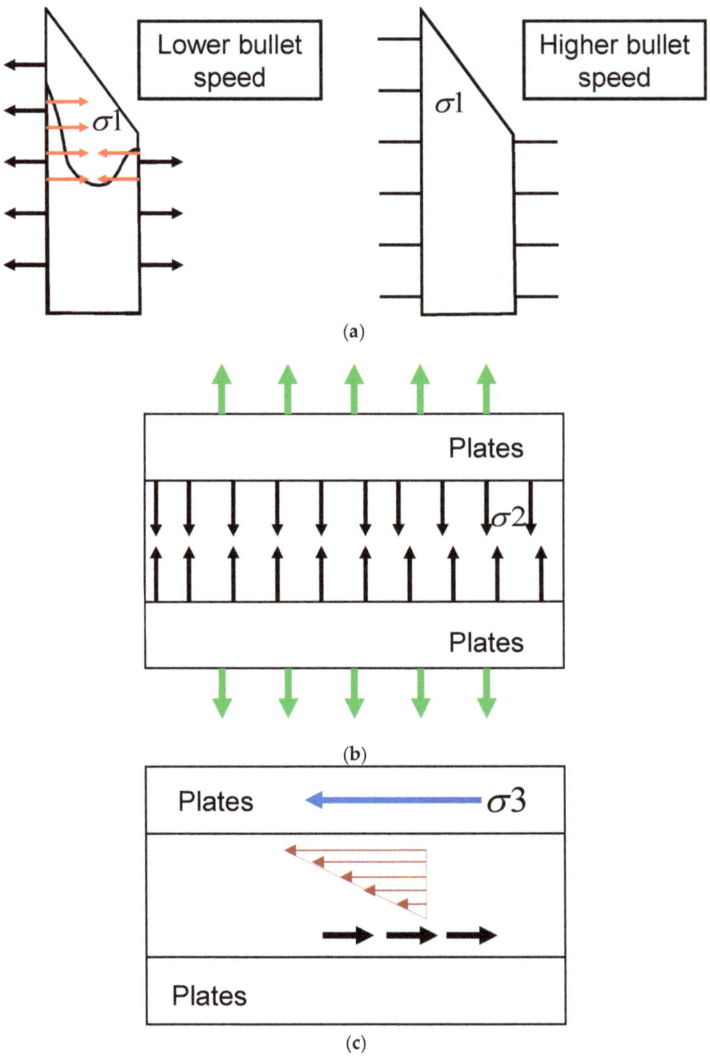

Figure 15. Mechanistic study of strength effect in sandstone at different impact velocities. (**a**) Curved-moon-face effect; (**b**) Stefan effect; and (**c**) Newtonian internal friction effect.

4.2. Mechanism of Temperature Effect on the Crack Microcosm

The influence of temperature on coal and rock bodies is mainly in two aspects: (1) As the temperature decreases, the volume of the coal or rock body gradually decreases, the internal cracks shrink or close, and the internal matrix of the coal or rock shrinks; this process leads to its densification. (2) At temperatures above 0 °C, the internal particles of the coal or rock and the water combine with each other, thus promoting the agglomeration of the coal or rock particles and increasing the overall bonding force of the coal or rock. Below 0 °C, the water molecules inside the saturated coal or rock are converted from liquid water to solid ice, the water–ice phase transition leads to micro-cracks in the coal and rock, and the primary cracks develop further; this is considered the freezing and expansion force. Solid ice occupies the liquid space, which in turn increases its strength.

The two are positively and negatively correlated but coexist to provide the strength of the coal or rock body. The influence degree of the two on the coal or rock is also inconsistent at different temperatures. They validate each other in the wave velocity curve in Figure 7. Therefore, the freezing contraction that enhances the strength of the coal or rock body should satisfy two conditions: a certain degree of saturation and high freezing temperature. As long as the coal or rock body reaches a certain degree of saturation, the coal or rock body below the freezing point will be frozen with the reduction in the freezing temperature, and the coal samples that have been frozen will be further frozen with a continuous reduction in the temperature; this is called the "freezing and expansion of the saturated coal samples critical temperature". If the relationship between the two is balanced, the safety factor is greatly improved when excavating.

5. Conclusions

In this study, coal and rock are taken as research objects; water, temperature, and dynamic load are taken as variables; and multifractal technology is used to analyse the failure trends of coal and rock. The following conclusions are drawn:

(1) The dynamic mechanical properties of coal and rock samples under the SHPB impact process are affected by their own physical properties, temperature, saturation degree, and impact velocity under joint action. When measuring the longitudinal wave velocity, the rock samples (2.24–2.61 km/s) have lower wave velocities than the coal samples (1.68–2.01 km/s). With decreasing temperature, both the coal and rock samples show a trend of initially increasing and then decreasing, with an "N"-shaped pattern. Comparing the samples in the dry state and saturated state, the wave velocities of the coal and rock samples in the saturated state are higher than those of the coal and rock samples in the dry state.

(2) The peak stress–temperature curve and stress–strain curve are plotted after noise reduction of the stress curve. The higher density of the specimen at the same impact velocity correlates to a faster propagation of the stress wave. With the increase in impact velocity, the strain intensification gradually becomes evident. Under the influence of speed and temperature, the influence factor of speed is greater.

(3) In the dynamic impact damage process of the coal and rock, the box dimension has evident self-similarity. The quantitative description of the cracks in the box dimension of the coal and rock specimens shows an overall increase and local oscillation; additionally, the dimension number of the dry samples is higher than that of the saturated condition under the same impact velocity with a fast end time. In the same saturated condition, the value of the box dimension increases with increasing impact velocity without shortening the duration. Comparing coal and rock, the holding time of the whole rock is shorter than that of the coal sample, and the final value of the box dimension number is smaller than that of the original coal.

(4) After sieving the impacted coal and rock samples and calculating the mass dimension number, a linear positive logarithmic correlation is observed between the cumulative mass percentage of the sub-screen grain size and the sample sieve diameter. The correlation coefficient is large and indicates that the post-crushing mass distribution of the coal and rock samples is self-similar to a fractal distribution. The fractal dimension fluctuates between 1.29 and 2.26. The slope of the fitted curve decreases with increasing impact velocity and decreasing saturation. Raw coal and raw rock are also affected to different degrees by different temperatures.

(5) Through microscopic analyses, it can be concluded that when the inside of the coal or rock is filled with a large number of water molecules, the decrease in temperature will cause the internal coal or rock matrix to contract, thus increasing its densification, and when the water freezes into ice, due to its ice uplift force, it results in expansion and deformation of the inside of the coal or rock. The two influence each other and the dominant factors are different at different temperatures.

It should be noted that this paper has initially explored the damage characteristics of coal and rock bodies with temperature, water, and dynamic loading and its influencing factors using SHPB; however, further exploration, such as triaxial impact experiments under variable temperatures, freeze–thaw cyclic impact experiments, the main influencing factors subjected to the freezing method when drilling wells and tunnelling, etc., still need to be studied more deeply.

Author Contributions: Conceptualisation, X.S.; Methodology, K.L.; Software, S.L.; Validation, S.Y.; Writing—review and editing, J.X.; Writing—original draft, T.J. All authors have read and agreed to the published version of the manuscript.

Funding: This research is supported by the Fundamental Research Program of Shanxi Province (202103021224277), by the Scientific and Technological Innovation Programs of Higher Education Institutions in Shanxi (2021L334), and by the Taiyuan University of Science and Technology Scientific Research Initial Funding (20222028).

Institutional Review Board Statement: Not applicable.

Informed Consent Statement: Not applicable.

Data Availability Statement: The data used to support the findings of this study are available from the corresponding author upon request. The data are not publicly available due to privacy.

Conflicts of Interest: The authors declare that they have no conflict of interest regarding the publication of this study.

eferences

Zhao, Y.; Liu, Q.; Lin, H.; Wang, Y.; Tang, W.; Liao, J.; Wang, X. A review of hydromechanical coupling tests, theoretical and numerical analyses in rock materials. *Water* **2023**, *15*, 2309. [CrossRef]

Wang, Y.; Yang, Y.; Zhang, Y.; Wang, J. Dynamic mechanical properties of coals subject to the low temperature-impact load coupling effect. *Sci. Rep.* **2019**, *9*, 20218. [CrossRef] [PubMed]

Zhang, C.; Yu, Y.; Li, W.; Lu, Q. Study on temperature field evolution and frozen wall closure judgment criteria of inclined shaft under inclined holes freezing condition. *Geofluids* **2022**, *2022*, 8002667. [CrossRef]

Paul, J.; Sengupta, M. Computer aided blasting-round design in an arctic coal mine. *Int. J. Min. Sci. Technol.* **1987**, *5*, 299–308. [CrossRef]

Chang, Z.; Zhang, W.; Zhao, G.; Dong, F.; Geng, X. Aging Stability Analysis of Slope Considering Cumulative Effect of Freeze–Thaw Damage—A Case Study. *Minerals* **2022**, *12*, 598. [CrossRef]

Li, H. Study on construction technology for tunnelling in the Plateau permafrost region. In *Cold Regions Engineering. Cold Regions Impacts on Research, Design, and Construction*; American Society Civil Engineers: Reston, VA, USA, 2009; pp. 16–21.

Fan, L.; Fan, Y.; Xi, Y.; Gao, J. Spatial failure mode analysis of frozen sandstone under uniaxial compression based on CT technology. *Rock Mech. Rock Eng.* **2022**, *55*, 4123–4138. [CrossRef]

Kodama, J.; Goto, T.; Fujii, Y.; Hagan, P. The effects of water content, temperature and loading rate on strength and failure process of frozen rocks. *Int. J. Rock Mech. Min.* **2013**, *62*, 1–13. [CrossRef]

Sun, X.; Liu, B.; Dong, Q.; Shi, F. Experimental investigation on damage characteristics and early warning of frozen coal under uniaxial compression. *Adv. Mater. Sci. Eng.* **2023**, *2023*, 1607529. [CrossRef]

Wang, T.; Sun, Q.; Jia, H.; Ren, J.; Luo, T. Linking the mechanical properties of frozen sandstone to phase composition of pore water measured by LF-NMR at subzero temperatures. *Bull. Eng. Geol. Environ.* **2021**, *80*, 4501–4513. [CrossRef]

Wang, L.; Wu, Z.; Liu, Q. Dynamic mechanical properties of dry and water-saturated siltstones under sub-zero temperatures. *Rock Mech. Rock Eng.* **2020**, *53*, 4381–4401. [CrossRef]

Yang, Y.; Zhang, N.; Wang, J. Fracture morphology analysis of frozen red sandstone under impact. *Shock Vib.* **2021**, *2021*, 4388132. [CrossRef]

Xu, J.; Pu, H.; Sha, Z. Dynamic mechanical behavior of the frozen red sandstone under coupling of saturation and impact loading. *Appl. Sci.* **2022**, *12*, 7767. [CrossRef]

Yang, R.; Fang, S.; Li, W.; Yang, Y.; Yue, Z. Experimental study on the dynamic properties of three types of rock at negative temperature. *Geotech. Geol. Eng.* **2019**, *37*, 455–464. [CrossRef]

Shan, R.; Song, Y.; Song, L.; Bai, Y. Dynamic property tests of frozen red sandstone using a split hopkinson pressure bar. *Earthq. Eng. Eng. Vib.* **2019**, *18*, 511–519.

Zhao, Y.; Liu, Q.; Zhang, C.; Liao, J.; Lin, H.; Wang, Y. Coupled seepage-damage effect in fractured rock masses: Model development and a case study. *Int. J. Rock Mech. Min.* **2021**, *144*, 104822. [CrossRef]

Zhao, Y.; Wang, Y.; Wang, W.; Tang, L.; Liu, Q.; Cheng, G. Modeling of rheological fracture behavior of rock cracks subjected to hydraulic pressure and far field stresses. *Theor. Appl. Fract. Mech.* **2019**, *101*, 59–66. [CrossRef]

Yu, G.; Vozoff, K.; Durney, D. Effects of confining pressure and water saturation on ultrasonic compressional wave velocities in coals. *Int. J. Rock Mech. Min.* **1991**, *28*, 28–36. [CrossRef]

19. Liu, J.; Liu, D.; Cai, Y.; Gan, Q.; Yao, Y. Effects of water saturation on P-wave propagation in fractured coals: An experimen perspective. *J. Appl. Geophys.* **2017**, *144*, 94–103. [CrossRef]
20. Zhao, E. *Dynamic Characteristics and Damage Mechanism of Coal and Rock under Dynamic-Static Load and Temperature Coupling Effe* China University of Mining and Technology: Beijing, China, 2023.
21. Wang, J.; Yang, Y.; Guo, Y. Low temperature effect of saturated granite on dynamic characteristics at high strain rates. *Rock S Mech.* **2017**, *38* (Suppl. S2), 163–169.
22. Wang, Z.; Tang, Y.; Li, M. Development and application of overburden structure composed of caving arch and towering rc beam in deep lonawall panel with thin bedrock. *Chin. Coal J.* **2023**, *48*, 563–575.
23. Yang, S.; Wu, S.; Wang, Z. Surface subsidence and its prediction method of mining deep-buried seam with thick alluvial lay and thin bedrock. *Chin. Coal J.* **2023**, *48*, 523–537.
24. Chen, Y.; Kang, T.; Wu, C. Study on mechanical behavior and mechanism of sandstone under the coupling effect of water conte and dynamic load. *Processes* **2023**, *11*, 2318. [CrossRef]
25. Kuruppu, M.D.; Obara, Y.; Ayatollahi, M.R. ISRM-suggested method for determining the mode I static fracture toughness usi semi-circular bend specimen. *Rock Mech. Rock Eng.* **2014**, *47*, 267–274. [CrossRef]
26. Zhou, Y.; Xia, K.; Li, X.; Li, H.; Ma, G.; Zhao, J. Suggested methods for determining the dynamic strength parameters and mod fracture toughness of rock materials. *Int. J. Rock Mech. Min. Sci.* **2012**, *49*, 105–112. [CrossRef]
27. *GB/T 23561.5-2009*; National Standard Compilation Groups of People's Republic of China. Methods for Determining the Physi and Mechanical Properties of Coal and Rock. Standards Press of China: Beijing, China, 2009.
28. Sun, X.; Chen, G.; Li, J.; Xu, X.; Fu, S.; Xie, J.; Liang, L. Propagation characteristics of ultrasonic P-wave velocity in artific jointed coal briquettes. *J. Geophys. Eng.* **2020**, *5*, 827–837. [CrossRef]
29. Xie, B.; Chen, M.; Chen, S.; Luan, Z.; Li, X.; Liang, T. Experimental study on dynamic impact failure mechanical properties of samples. *Chin. J. Disaster Prev. Mitig. Eng.* **2023**, 1–8. [CrossRef]
30. Deshpande, V.; Chakraborty, P.; Chakraborty, T.; Tiwari, V. Application of copper as a pulse shaper in SHPB tests on brittle materia experimental study, constitutive parameters identification, and numerical simulations. *Mech. Mater.* **2022**, *171*, 104336. [CrossRef]
31. Feng, D.; Peng, Y.; Liu, Y. Study on SHPB technique. *Chin. Prog. Geophys.* **2006**, 273–278.
32. Sun, X.; Jin, T.; Li, J.; Xie, J.; Li, C.; Li, X. Dynamic characteristics and crack evolution laws of coal and rock under split Hopkins pressure bar impact loading. *Meas. Sci. Technol. E.* **2023**, *34*, 075601. [CrossRef]
33. Chen, W.; Bo, S. *Split Hopkinson (Kolsky) Bar: Design, Testing and Applications*; Springer Science & Business Media: Berlin/Heidel Germany, 2010.
34. Mandelbrot, B.B. *The Fractal Geometry of Nature*; WH Freeman: New York, NY, USA, 1982; Volume 1.
35. *ASTM D 2845*; ASTM Committee D-18 on Soil and Rock. Standard Test Method for Laboratory Determination of Pulse Velocit and Ultrasonic Elastic Constants of Rock. ASTM International: West Conshohocken, PA, USA, 2008.
36. Zhao, Y.; Zhang, C.; Wang, Y.; Lin, H. Shear-related roughness classification and strength model of natural rock joint based fuzzy comprehensive evaluation. *Int. J. Rock Mech. Min. Sci.* **2021**, *137*, 104550. [CrossRef]
37. Zhao, Y.; Liu, J.; Zhang, C.; Zhang, H.; Liao, J.; Zhu, S.; Zhang, L. Mechanical behavior of sandstone during post-peak cyc loading and unloading under hydromechanical coupling. *Int. J. Rock Mech. Min. Sci.* **2023**, *33*, 927–947. [CrossRef]
38. Li, Y.; Chang, X.; Dou, K. Research of ice mechanical properties parameters based on ultrasonic wave. *China Meas. Test* **20** *45*, 36–40.
39. Sun, X.; He, Y.; Jin, T.; Xie, J.; Li, C.; Pang, J. Microseismic signal characteristics of the coal failure process under weak-ener and low-frequency disturbance. *Sustainability* **2023**, *15*, 14387. [CrossRef]
40. Iakovidis, D.K. Ratsnake: A versatile image annotation tool with application to computer-aided diagnosis. *Sci. World J.* **20** *2014*, 286856. [CrossRef]
41. Li, Y.; Zhao, Y.; Jiang, Y.; Zhang, B.; Song, H.; Liu, B. Characteristics of pore and fracture of coal with bursting proneness bas on DIC and fractal theory. *Energies* **2020**, *13*, 5404. [CrossRef]
42. Yang, R.; Li, Y.; Zeng, D.; Guo, P. Deep DIC: Deep learning-based digital image correlation for end-to-end displacement a strain measurement. *J. Mater. Process Tech.* **2022**, *302*, 117474. [CrossRef]
43. Gerdes, L.; Berger, S.; Saelzer, J.; Franck, P.; Helwing, R.; Zabel, A.; Walther, F. Application-Oriented Digital Image Correlation the High-Speed Deformation and Fracture Analysis of AISI 1045 and Ti6Al4V Materials. *Appl. Mech.* **2022**, *3*, 1190–1205. [CrossRe
44. Ai, D.; Zhao, Y.; Wang, Q.; Li, C. Experimental and numerical investigation of crack propagation and dynamic properties of ro in shpb indirect tension test. *Int. Impact Eng.* **2019**, *126*, 135–146. [CrossRef]
45. Wu, N.; Fu, J.; Xiong, C. Studying the characteristics of chaos and fractals of construction rocks under different loading velociti *Materials* **2022**, *15*, 7890. [CrossRef]

applied sciences

MDPI

rticle

Study on Gas Extraction Technology for Goaf Using L-Shaped Borehole on the Ground

ongyong Shu

China Coal Research Institute, Beijing 100013, China; slyccri@163.com; Tel.: +86-13717519923

Abstract: This study aimed to examine gas extraction technology in the goaf of an L-shaped borehole in the mining fissure zone of a short-distance coal seam group. The numerical simulation method was used to analyze the failure law of overlying rock during mining, and a mathematical model was established for gas migration in the mining overburden. Finally, gas extraction tests were performed for the L-shaped borehole in the mining fissure zone. The results showed that as the coal mining project progressed, the damage area of the overlying strata in the goaf became larger, and the plastic damage area of the overlying rock along the strike had a saddle shape, being concave in the middle and convex at both ends. The closer the L-shaped borehole in the mining fissure zone was to the coal seam roof, the greater the amount of air leaking from the working face into the goaf, and the lower the overall gas concentration in the goaf. When the vertical distance of the L-shaped borehole was too high, the ability of the L-shaped borehole to control the gas concentration in the lower goaf was weakened. Moreover, the mining fracture zone was a good space for gas migration and storage. Thus, arranging the L-shaped borehole in this zone can greatly improve the efficiency of borehole gas extraction. According to the overlying rock conditions and mining conditions of Tunlan Mine, the L-shaped borehole was positioned 43 m away from the roof of the coal seam. The extraction rate of the L-shaped borehole reached 9.30 $m^3 \cdot min^{-1}$, and the gas concentration in the corners of the working face was kept below 0.4%, yielding an excellent extraction effect.

Keywords: gas extraction; L-shaped borehole; short-distance coal seam group; mining fissure zone; numerical simulation

1. Introduction

Coal is China's main energy source, accounting for 54.5% of the country's total primary energy consumption in 2022 [1]. With its economic development, China's demand for coal is also growing stronger, and the intensity and depth of coal mining are also increasing [2,3]. Mine gas is one of the important factors restricting efficient coal production. Especially with the increase in mining intensity and gas content in coal seams, a large amount of gas emission accompanies the coal mining process. If it is not controlled, it will cause gas explosions in the working face [4,5]. At this stage, most mines can no longer effectively control mine gas only through ventilation, so the method of goaf drainage has been widely used [6–8].

At present, there are various methods for gas drainage in goafs, among which underground borehole and gas drainage by buried pipes in goafs are often considered as the simplest methods [9]. Xu et al. achieved efficient gas drainage in the goaf of Pingshu Coal Mine by optimizing the location of the low-level roadway [10]. Chai et al. took Tingnan Coal Mine as their research background and explored the rational layout of the high-level drainage roadway on the roof of the working face and its drainage effect through theoretical analysis, fluent numerical simulation, on-site verification, and other methods [11]. Lei used field experiments to verify the effectiveness of large-diameter boreholes in reducing gas concentration in a goaf and the upper corner [12]. Wang et al. studied the gas production data from vertical boreholes in Australia and conducted a detailed analysis of the correlation between suction pressure, total flow rate, gas flow rate, and air leakage rate as the spacing between boreholes changed [13]. Qu et al. conducted research on gas drainage data from directional boreholes in a goaf, analyzing the technical advantages of this method compared to vertical boreholes drilled from the surface [14].

To improve the efficiency of gas drainage a goaf, it is necessary to further clarify the laws of fractures and damage in the overlying strata of the goaf, as well as the migration patterns of gas in the goaf [15,16]. After mining a coal seam, the layer-separated cracks and break-through cracks in the overlying strata become connected to each other, forming a mining crack zone that changes dynamically [17,18]. This zone provides a channel and space for pressure-relief gas flow and storage in the coal seam or in adjacent coal strata. Extracting the gas in a fracture zone during mining can greatly improve the gas drainage efficiency [19]. Dou et al. analyzed the transfer of elastic energy in a coal seam roof under varying conditions and revealed the development characteristics of fractures [20]. Liu et al. introduced the Sigmoid function to establish a more accurate porosity model based on the theory of overlying strata movement, reflecting the distribution characteristics of the "three horizontal zones" in a goaf, and developed a model for measuring porosity in the goaf [19]. Wang et al. used PFC software to establish a three-dimensional single working-face model and calculated the distribution of porosity values in the overlying strata [21]. Tian et al. employed theoretical analysis, numerical simulation, and formula-fitting methods to obtain the distribution law of porosity in the overlying strata of fully mechanized caving mining under different types of coal pillars in various sections [22].

The study of gas distribution patterns in goafs has significant implications for gas control measures in such regions. Qin et al. employed CFD simulations to elucidate the impact of gas release characteristics from gas sources and the positioning of drainage boreholes on the distribution and migration patterns of gas in a goaf [23]. Rao et al. utilized a combination of numerical simulations and field tests to investigate the gas migration characteristics in the goaf of longwall mining faces [24]. Zhang et al. introduced gas state equations, continuity equations, momentum equations, porosity evolution equations, and permeability evolution equations to simulate gas concentration fields under natural conditions and during gas drainage in a goaf [25]. Li et al. examined the accumulation patterns of gas in a goaf, considering factors such as desorption from residual coal and gas emission from upper adjacent layers [26]. Zhou et al. established a three-dimensional gas distribution model for a goaf, analyzing the influence of pipe spacing and drainage negative pressure on gas distribution in the mined-out area [27]. Li et al. used numerical simulations and air fluid mechanics to analyze the gas behavior in a goaf; the authors calculated the minimum induced air volume and distributed air volume required for the safe production of a tailings roadway [28]. Zhang et al. developed a simplified three-dimensional engineering model of a fracture field based on simulations, and they also simulated the distribution law of the pressure relief gas in a goaf using a high-extraction roadway [15]. Yu et al. conducted "U" and "U + I" ventilation tests on a three-dimensional, fully mechanized caving face simulation platform. The distribution law of the gob pressure field and gas concentration field was obtained [16]. The literature does not currently contain any systematic research on gas drainage technology for the goaf of an L-shaped borehole in the mining fissure zone of a short-distance coal seam. Thus, in this study, I used numerical simulations to investigate the failure characteristics of the overlying strata of a short-distance coal seam group, analyzed the gas extraction effect of the L-shaped borehole goaf in different layers, and performed practical engineering tests to verify the technical advantages of an L-shaped borehole in the mining of a fissure zone.

2. Overview of Mining Area

The Xishan mining area is located in the central part of Shanxi Province and is currently the largest coking coal production base in China. The coal seam has a widely varying buried depth and contains an uneven gas volume. The distance between minable coal seams in the same coal group is close, generally within 10 m. When production activities are ongoing in the mining layer, a significant amount of gas is emitted from the upper and lower adjacent layers, and a string of mutual interference occurs. To study the gas extraction technology in the goaf of an L-shaped borehole in the mining fracture zone of a short-distance coal seam group, the gas extraction in the goaf of the L-shaped borehole in the mining fracture zone of the 12507 working face in Tunlan Mine, Xishan mining area, was used as our research background. The 12507 working face of Tunlan Mine is located in the southern five-pan area of Tunlan Mine (Figure 1). The main coal seams are the 2# and 3# coal seams. The average mining thickness is 4.57 m, the strike length is 806 m, and the inclination length is 220 m. The mine uses a full-height mining method, and the roof is managed using the full caving method.

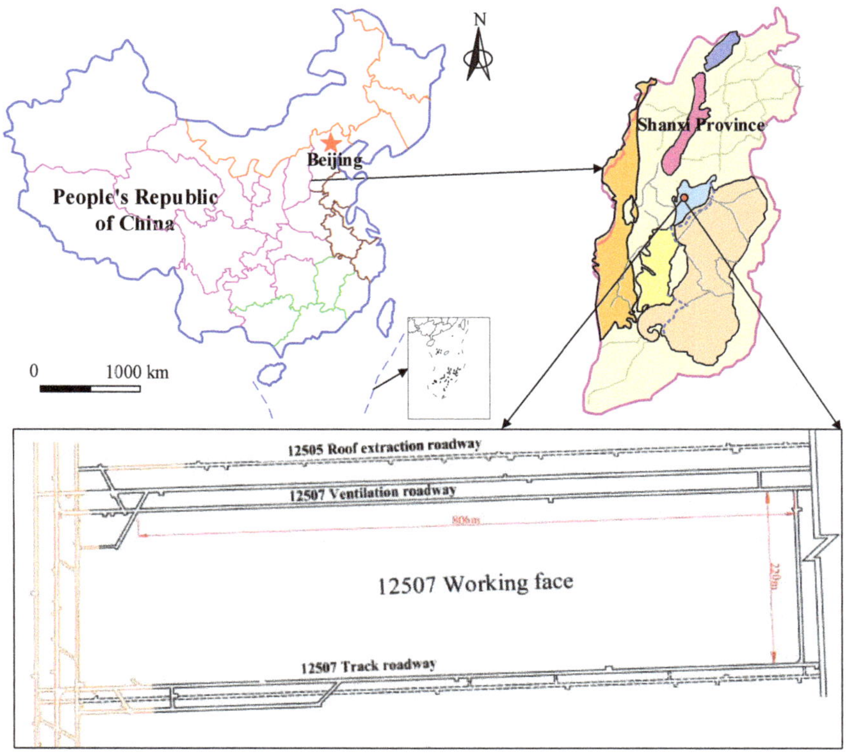

Figure 1. Working face overview.

3. Study on the Failure Law of Mining Overburden

To analyze the evolution law of the mining overburden stress field in the 12507 working face of Tunlan Mine, COMSOL Multiphysics 3.5 software was used to establish a three-dimensional digital simulation model. The size of the model was 400 m × 400 m × 150 m (Figure 2). The action of part of the overburden was equivalent to a uniformly distributed load of 10 MPa. The physical and mechanical parameters of each rock formation in the model are shown in Table 1.

As illustrated in Figure 3, the roof failure of the working face begins as shear failure, whereby cracks are developed in the roof. Then, the failure develops into tensile failure, and fracture or caving finally occurs. The top to the bottom of the coal seam roof contains a tensile failure area, a shear failure area, and an undamaged area. As the working face advances, the tensile failure area gradually increases, and the upper shear failure area also expands; this phenomenon is more obvious when the key layer is broken. The rock stratum in the mining fracture zone is in a plastic failure state, and the mining fracture is developed in this state. Above the mining fracture zone and up to the bedrock surface, the rock stratum is basically undamaged. At the edge of the goaf, because of the boundary coal pillars, the rock mass is in the tensile and compressive stress zones, the mining faults are fully developed, and the plastic zone develops the most. These phenomena form a saddle-shaped distribution pattern with high convexity at both ends and low concavity in the middle. When the working face is mined up to 180 m, the failure height of the plastic zone reaches 55 m on the roof of the coal seam.

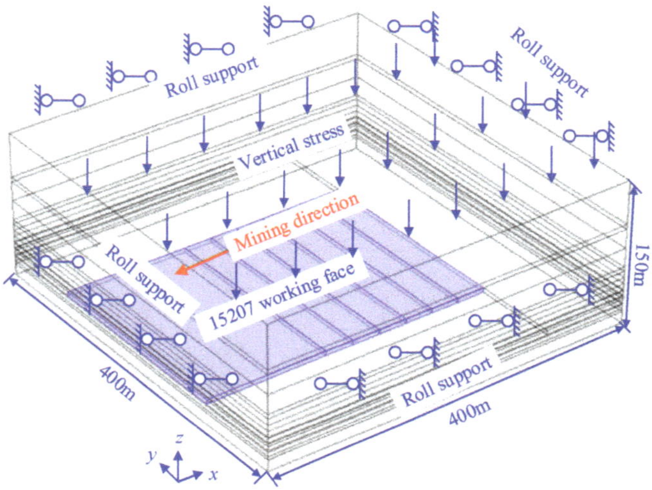

Figure 2. Numerical simulation model.

Table 1. Physical and mechanical parameters of each rock formation in the model.

No.	Lithology	Thickness (m)	Bulk Modulus (MPa)	Poisson's Ratio	Elastic Modulus (MPa)	Density (kg/m^3)	Uniaxial Tensile Strength	Internal Friction Angle	Cohesion (MPa)
1	Limestone	12.39	4523	0.19	8412.78	2700	4.03	42	4.8
2	Medium-grained sandstone	3.34	3240	0.17	6415.2	2510	2.35	38	4.07
3	No. 7 coal seam	4.57	1580	0.28	2085.6	1380	0.52	30	1.42
3	Fine-grained sandstone	4.57	4176	0.18	8017.92	2680	2.62	40	4.53
4	Sandy mudstone	1.35	2800	0.24	4368	2450	1.55	35	3.61
5	No. 2 + 3 coal seam	4.57	1580	0.28	2085.6	1380	0.52	30	1.42
6	Fine-grained sandstone	2.75	4176	0.18	8017.92	2680	2.62	40	4.53
7	Sandy mudstone	0.78	2800	0.24	4368	2450	1.55	35	3.61
8	Siltstone	5.13	2530	0.25	3795	2550	1.66	36	3.3
9	No. 4 coal seam	1.12	1580	0.28	2085.6	1380	0.52	30	1.42
10	Medium-grained sandstone	6.46	3240	0.17	6415.2	2510	2.35	38	4.07
11	Sandy mudstone	6.27	2800	0.24	4368	2450	1.55	35	3.61
12	Fine-grained sandstone	6.75	4176	0.18	8017.92	2680	2.62	40	4.53
13	Medium-grained sandstone	4.57	3240	0.17	6415.2	2510	2.35	38	4.07
14	Fine-grained sandstone	19.92	4176	0.18	8017.92	2680	2.62	40	4.53
15	Mudstone	20	2230	0.26	3211.2	2230	0.89	32	2.98
16	Fine sandstone	4.97	4176	0.18	8017.92	2680	2.62	40	4.53
17	Medium-grained sandstone	45.06	3240	0.17	6415.2	2510	2.35	38	4.07

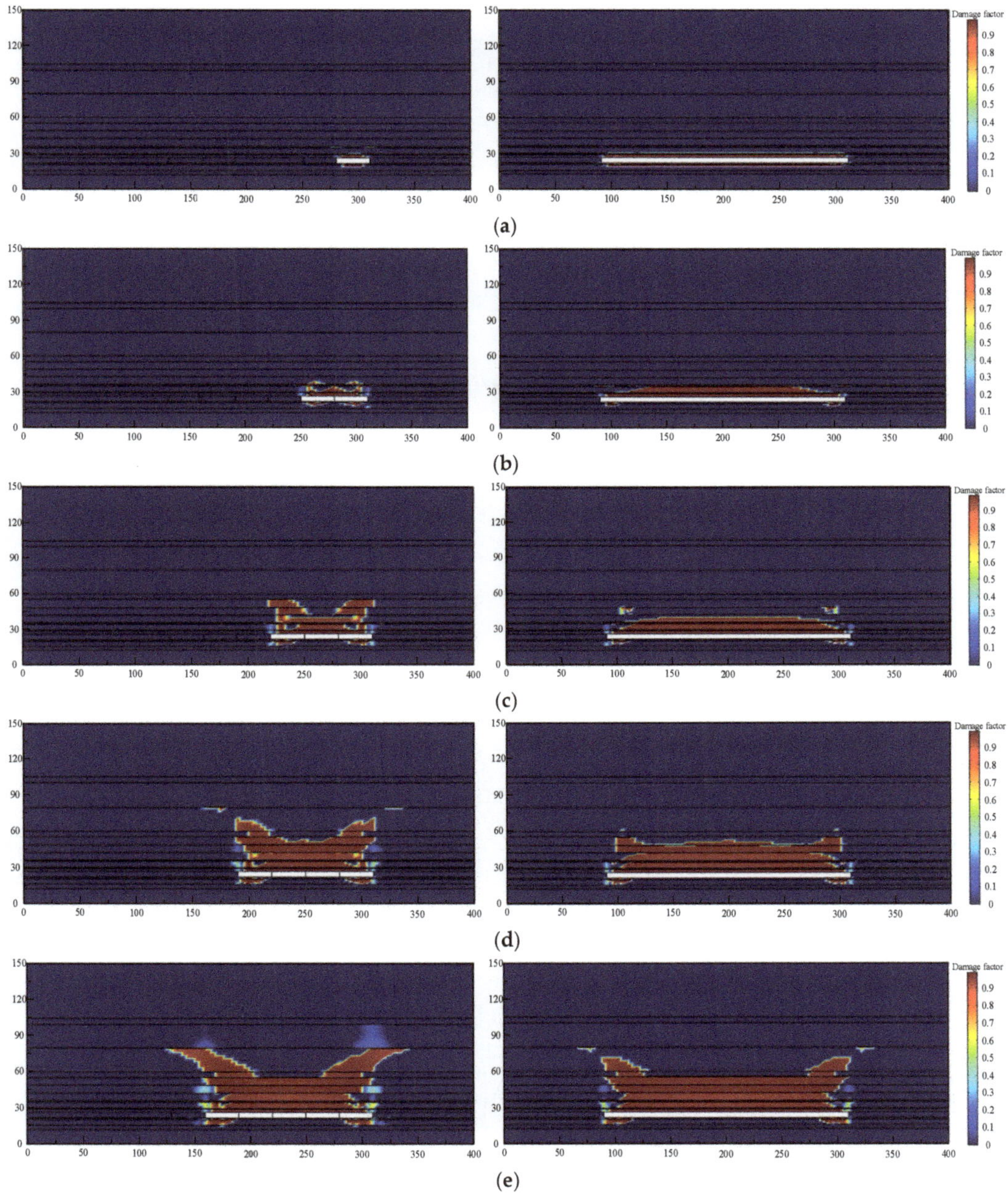

Figure 3. Distribution of transverse–longitudinal plastic zone at different mining distances: (**a**) mining distance of 30 m; (**b**) mining distance of 60 m; (**c**) mining distance of 90 m; (**d**) mining distance of 120 m; (**e**) mining distance of 150 m.

4. Numerical Simulation Study on Gas Extraction of L-Shaped Borehole in Mining Fissure Zone

After the coal seam is mined, the separation fractures in the overburden layer are connected to fractures in the through layer to form a dynamic mining fissure zone. This area provides channels and spaces for the flow and storage of the pressure-relieved gas in the coal seam or adjacent coal seams, and the gas migration is very complex. In this chapter, a gas migration model for the mining overburden is established, the gas migration law in the mining fracture field is examined, the gas drainage effect of the L-shaped borehole in different layers is analyzed, and a reasonable location for the L-shaped borehole is determined.

4.1. Mathematical Model of Gas Migration in Mining Overburden

4.1.1. Gas Transport Equation

The gas flow in the roadway and working face can be equivalent to the fluid flow in the pipeline. The Navier–Stokes equation is used to describe the gas flow state in the working face and roadway.

$$-\nabla \cdot \eta\left(\nabla \mu + (\nabla \mu)^T\right) + \rho\mu \cdot \nabla \mu + \nabla p = 0 \tag{1}$$

where η is the gas flow viscosity coefficient, kg/(m·s); μ is the fluid flow velocity, m/s; ρ is the fluid density, kg/m^3; and p is the fluid pressure, MPa.

4.1.2. Gas Flow Equation in the Goaf

After the coal seam is mined, the overlying roof is broken, filling the entire goaf. Additionally, several voids are formed within the gangue, resulting in a porous-medium structure. Therefore, the Brinkman equation is used to describe the gas flow in the goaf:

$$\frac{\rho}{\varepsilon}\left(\frac{\partial \mu}{\partial t}\right) + \left(\frac{\mu}{k}\right)u = -\nabla p + \nabla\left(\frac{\mu}{\varepsilon}\left(\nabla u + (\nabla u)^T\right)\right) + F \tag{2}$$

where ε is porosity, and k is permeability, m^2.

4.1.3. Gas Diffusion Equation

The gas in the goaf obeys the dispersion law and the diffusion law. However, none of the above equations reflect the gas diffusion and migration; hence, it is necessary to use the flow–diffusion equation for a supplementary description:

$$\theta_s \frac{\partial c}{\partial t} + \nabla \cdot (-\theta_s D_L \nabla c + uc) = S_C \tag{3}$$

where θ_s is the gas volume fraction; c is the gas concentration, mol/m^3; D_L is the gas pressure diffusion tensor, m^2/d; and S_C is the gas source.

4.2. Physical Model Building

According to the actual situation of the working face, a physical model of the L-shaped well gas extraction in the mining fissure zone was established (Figure 4). The size of the goaf model was 300 m × 220 m × 60 m, the size of the working face was 5 m × 220 m × 3.5 m, and the dimensions of the two tunnels were 20 m × 4.5 m × 3.5 m. The diameter of the L-shaped borehole was 0.1714 m. To analyze the extraction effect of the L-shaped borehole at different positions, the drilling positions were set at 23 m, 33 m, 43 m, and 53 m from the roof of the coal seam, respectively.

According to the above research on the stress and damage law of the mining overburden, the overburden strata in the goaf caved and an O-type rupture occurred while the working face advanced. The mining fissures and free caving rocks in the middle of the goaf were basically compacted, and the coal-pillar-side separation fissures and caving fissures around the goaf remained. The distribution of the caving dilatation coefficient in the goaf is described by the following expression (the initial caving dilatation coefficient of the roof is taken to be 1.6, and the compacted dilatation coefficient is taken to be 1.1):

$$K_{(x,y)} = K_{\min} + (K_{\max} - K_{\min}) \times \exp(-m_0 d_0 \times (1 - \exp(-\xi m_1 (d_1 + \phi)))) \tag{4}$$

where $K_{(x,y)}$ is the caving and fragmentation coefficient of the goaf; K_{max} is the crushing expansion coefficient of the initial caving; K_{min} is the crushing expansion coefficient after compaction; m_0 and m_1 are the attenuation rates from the solid wall and the working face, respectively; d_0 and d_1 are the distances from the point (x, y) to the working face boundary and solid wall, m.

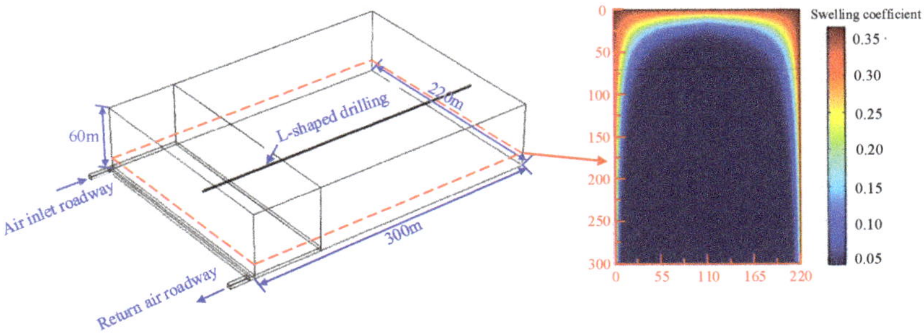

Figure 4. The physical model of the L-shaped well gas extraction.

The specific values can be obtained through a trial-and-error calculation. Assuming m_1 = 0.0368, m_0 = 0.268, and ξ = 0.233, the $K_{(x,y)}$ distribution results are illustrated in Figure 4.

The position of the air inlet was set as the inlet boundary, and the inlet boundary contained the wind speed and gas concentration values. According to the actual situation of the working face, the airflow speed of the air inlet was set to 2.2 m/s, and the gas concentration was 0, that is, the incoming fresh airflow was considered to not contain any gases. According to the calculation process of the field test data of Tunlan Mine, the gas emission from the coal wall of the working face was set at 7.48×10^{-5} kg/(m^3·s), and the gas emission from the residual coal in the goaf was set at 7.48×10^{-5} kg/(m^3·s); the suction pressure of L-shaped borehole drainage was 50 kPa.

As shown in Figure 5, as the vertical distance increases, the gas concentration in the upper corners first decreases, and then, increases, indicating that when the vertical distance parameter of the ground L-shaped borehole is within a certain range, the ground L-shaped borehole is extracted from the goaf. A gas with a higher concentration was generated, and the gas concentration at the points investigated in the goaf was reduced to varying degrees. When the vertical distance of the local L-shaped borehole is too high, the ground L-shaped borehole has a weakened ability to control the gas concentration in the lower goaf. Although the ground L-shaped borehole can extract the gas with a higher concentration in the goaf, it has little impact on the change in the gas concentration near the upper corner. When the vertical distance reaches 53 m, the gas concentration in the upper corner reaches 1.2%. In summary, the L-shaped borehole in the mining fissure zone of the 12507 working face of Tunlan Mine should be located at about 43 m from the roof of the coal seam.

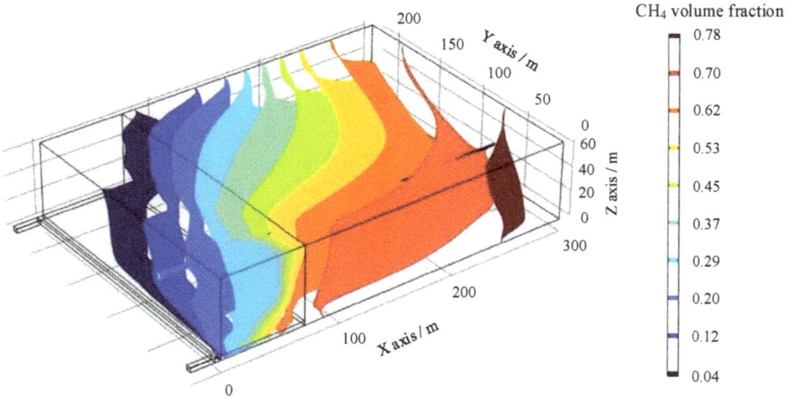

(**a**)

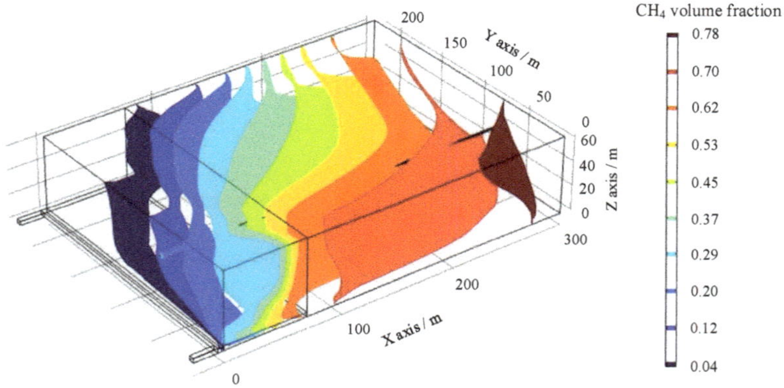

(**b**)

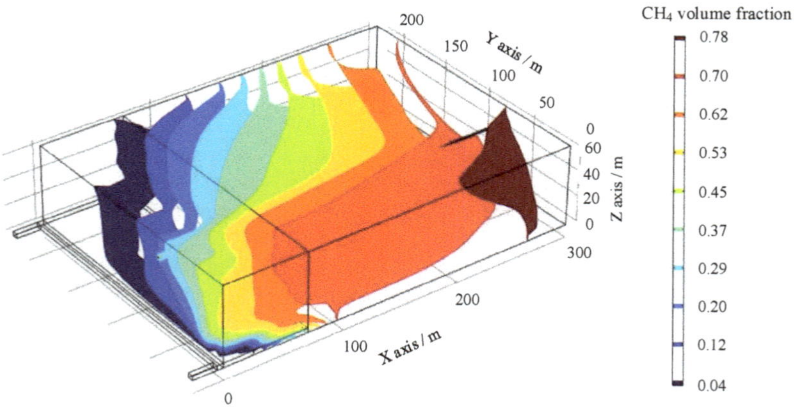

(**c**)

Figure 5. *Cont.*

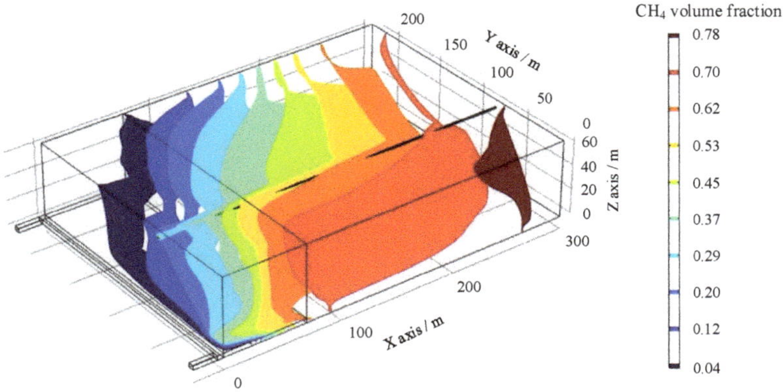

(d)

Figure 5. Distribution of transverse–longitudinal plastic zone at different mining distances: (**a**) L-shaped borehole 23 m away from coal seam roof; (**b**) L-shaped borehole 33 m away from coal seam roof; (**c**) L-shaped borehole 43 m away from coal seam roof; (**d**) L-shaped borehole 53 m away from coal seam roof.

5. Analysis of Gas Extraction Effect in Goaf of L-Shaped Borehole in Mining Fissure Zone

Considering the development law of the 12507 working face mining overburden fissure, the gas migration law in the goaf under the action of the L-shaped borehole in the mining fissure zone, the actual gas control situation in the Tunlan Mine goaf, and the repeated on-site field explorations, we recommend that the location of the L-shaped borehole in the mining fissure zone that should be selected is that west of Dupo Village, Gujiao City. This location corresponds to the surface of the 12507 working face (X: 4,193,380.212, Y: 37,596,000.438, H: 1212.894). The terrain is relatively flat, and the traffic is relatively moderate, which is convenient for construction, transportation, and subsequent gas extraction. The vertical depth of the borehole is 502.00 m, the well depth is 1117.16 m, the plane projection distance from the orifice to the target point is 724.62 m, the landing point is 48.30 m on the 2# coal roof, and the target point is 38.19 m on the 2# coal roof (Figures 6 and 7).

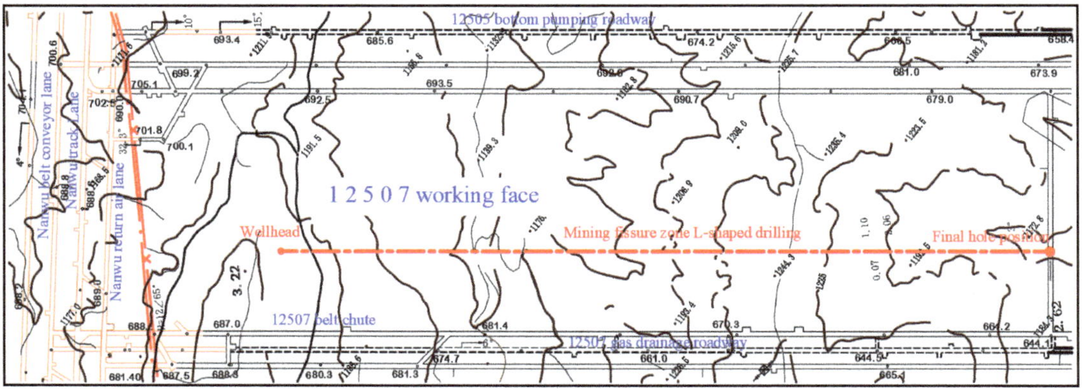

Figure 6. Upper and lower comparisons of L-shaped drilled wells in mining fracture zone of 12507 working face in Tunlan Mine.

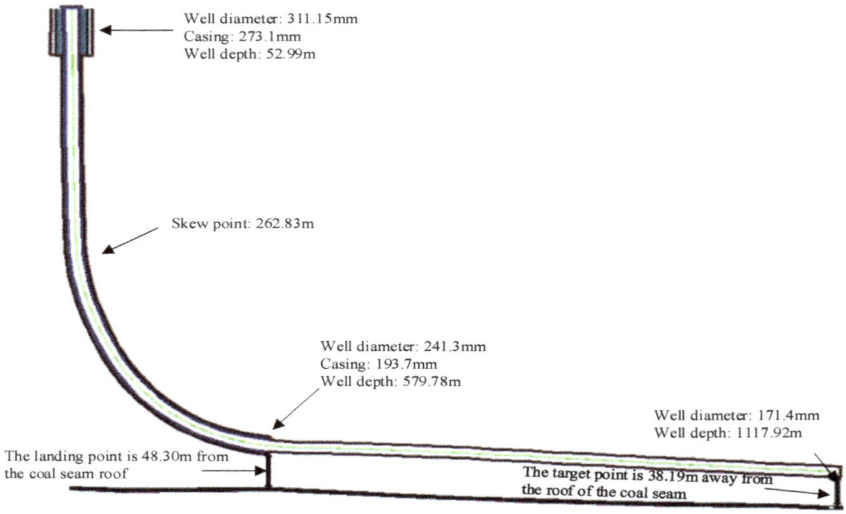

Figure 7. Sectional view of L-shaped borehole in mining fissure zone of 12507 working face in Tunlan Mine.

Figures 8 and 9 show the variation law of the gas production rate, cumulative production, and extraction concentration of the L-shaped borehole in the mining fissure zone. According to downhole monitoring, on 17 February, the roof collapsed across a large area, the overlying stratum fissures in the goaf were connected, and the L-shaped borehole on the ground in the mining fissure zone began to emit gas. After 149 days of production, the cumulative production reached 2,009,318 m^3. The gas production rate and concentration of the boreholes were unstable, exhibiting an increase–decrease–increase cycle. This is because the lithology of the coal seam roof is complex, and the degree of continuity of the overlying fissures varies with position. The gas production rate and the extraction concentration of the L-shaped borehole in the fissure zone during the mining stage were high, with average values of 9.30 $m^3 \cdot min^{-1}$ and 53.20%, respectively. These values are considered to indicate an excellent extraction effect.

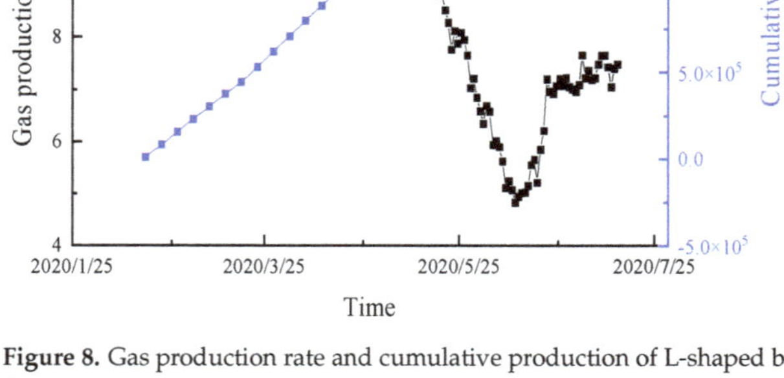

Figure 8. Gas production rate and cumulative production of L-shaped borehole on ground of mining fissure zone.

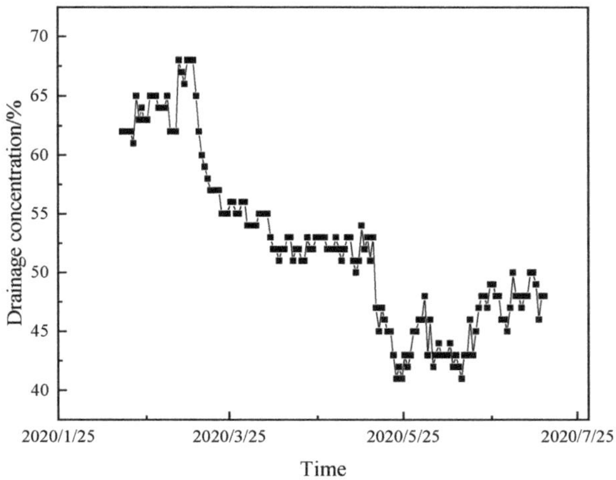

Figure 9. CBM extraction concentration of surface L-shaped boreholes in mining fissure zone.

6. Conclusions

(1) As coal mining progresses, the damage area of the overlying strata in the goaf enlarges, and the plastic damage area of the overlying rock along the strike takes the shape of a saddle that is concave in the middle and convex at both ends

(2) The closer the L-shaped borehole in the mining fissure zone is to the coal seam roof, the greater the amount of air leaking from the working face into the goaf, and the overall gas concentration in the goaf becomes relatively lower. When the vertical distance of the L-shaped borehole is too high, the L-shaped borehole has a weakened ability to control the gas concentration in the lower goaf.

(3) The mining fracture zone is a good space for gas migration and storage. Setting the L-shaped borehole in this position can greatly improve the efficiency of the borehole gas extraction. According to the overlying rock and mining conditions of Tunlan Mine, the L-shaped borehole is positioned 43 m away from the roof of the coal seam. The extraction rate of the L-shaped borehole can reach 9.30 $m^3 \cdot min^{-1}$, and the gas concentration in the corners of the working face is kept below 0.4%, which indicates an excellent extraction effect.

Funding: This study was supported by the China Coal Research Institute Technology Innovation Fund (2022CX-I-05).

Institutional Review Board Statement: Not applicable.

Informed Consent Statement: Not applicable.

Data Availability Statement: The data presented in this study are available on request from the corresponding author. The data are not publicly available due to containing information that could compromise the privacy of research participants.

Conflicts of Interest: The author declare no conflict of interest. The funders had no role in the design of the study; in the collection, analyses, or interpretation of the data; in the writing of the manuscript; or in the decision to publish the results.

eferences

Kao, X.; Wang, W.; Zhu, X.; Zhang, J. Spatial and Temporal Characteristics of Coal Consumption and Carbon Emissions in China. *Environ. Sci. Pollut. Res.* **2023**, *30*, 105770–105780. [CrossRef]

Fan, Y.; Deng, C.; Zhang, X.; Li, F.; Wang, X.; Qiao, L. Numerical Study of CO_2-Enhanced Coalbed Methane Recovery. *Int. J. Greenh. Gas Control* **2018**, *76*, 12–23. [CrossRef]

Shu, L.; Wang, K.; Liu, Z.; Zhao, W.; Zhu, N.; Lei, Y. A Novel Physical Model of Coal and Gas Outbursts Mechanism: Insights into the Process and Initiation Criterion of Outbursts. *Fuel* **2022**, *323*, 124305. [CrossRef]

4. Gao, A.; Qin, B.; Zhang, L.; Ma, D.; Li, L. Experimental Study on Gas Migration Laws at Return Air Side of Goaf unc High-Temperature Conditions. *Combust. Sci. Technol.* **2023**, *195*, 1930–1944. [CrossRef]
5. Ren, T.X.; Balusu, R. Proactive Goaf Inertisation for Controlling Longwall Goaf Heatings. *Procedia Earth Planet. Sci.* **2009**, 309–315. [CrossRef]
6. Wang, Y.; Si, G.; Oh, J.; Belle, B. CFD Modelling of Longwall Goaf Atmosphere under Vertical Boreholes Gas Drainage. *Int. J. C Geol.* **2023**, *280*, 104400. [CrossRef]
7. Si, G.; Belle, B. Performance Analysis of Vertical Goaf Gas Drainage Holes Using Gas Indicators in Australian Coal Mines. *Int Coal Geol.* **2019**, *216*, 103301. [CrossRef]
8. Ma, Q.; Xue, J.; Shi, Y.; Zeng, X. Characteristics of Porosity Distribution and Gas Migration in Different Layers of Comprehensi Working Face Goaf. *Energies* **2023**, *16*, 2325. [CrossRef]
9. Zhou, F.; Xia, T.; Wang, X.; Zhang, Y.; Sun, Y.; Liu, J. Recent Developments in Coal Mine Methane Extraction and Utilization China: A Review. *J. Nat. Gas Sci. Eng.* **2016**, *31*, 437–458. [CrossRef]
10. Xu, C.; Wang, K.; Li, X.; Yuan, L.; Zhao, C.; Guo, H. Collaborative Gas Drainage Technology of High and Low Level Roadways Highly-Gassy Coal Seam Mining. *Fuel* **2022**, *323*, 124325. [CrossRef]
11. Chai, J. Study on Gas Control in Goaf of High Gas Coal Seam: A Case Study of Tingnan Coal Mine, China. *Geofluids* **20** *2022*, 5262173. [CrossRef]
12. Lei, Y. Movement Law of Methane Drained by Large-Diameter Borehole Drilling Machine in the Goaf. *Processes* **2022**, *10*, 16 [CrossRef]
13. Wang, Y.; Si, G.; Xiang, Z.; Oh, J.; Belle, B.; Webb, D. A Theoretical Goaf Resistance Model Based on Gas Production Analysis Goaf Gas Drainage. *Int. J. Coal Geol.* **2022**, *264*, 104140. [CrossRef]
14. Qu, Q.; Guo, H.; Loney, M. Analysis of Longwall Goaf Gas Drainage Trials with Surface Directional Boreholes. *Int. J. Coal Ge* **2016**, *156*, 59–73. [CrossRef]
15. Zhang, M.; Wu, S.; Wang, Y. Research and Application of Drainage Parameters for Gas Accumulation Zone in Overlying Strata Goaf Area. *Saf. Sci.* **2012**, *50*, 778–782. [CrossRef]
16. Yu, Z.; Yang, S.; Qin, Y.; Hu, X.; Cheng, J. Experimental Study on the Goaf Flow Field of the "U + I" Type Ventilation System fo Comprehensive Mechanized Mining Face. *Int. J. Min. Sci. Technol.* **2015**, *25*, 1003–1010. [CrossRef]
17. Xie, J.; Zhao, Y. A Mathematical Model to Study the Coupling Effect of Deformation-Seepage-Heat Transfer on Coalbed Metha Transport and Its Simulative Application. *Math. Probl. Eng.* **2020**, *2020*, 1247240. [CrossRef]
18. Yang, H.; Liu, Z.; Zhu, D.; Yang, W.; Zhao, D.; Wang, W. Study on the Fractal Characteristics of Coal Body Fissure Developme and the Law of Coalbed Methane Migration of around the Stope. *Geofluids* **2020**, *2020*, 9856904. [CrossRef]
19. Liu, Q.; Lin, B.; Zhou, Y.; Li, Y. Porosity Model of the Goaf Based on Overlying Strata Movement and Deformation. *Environ. Ea Sci.* **2022**, *81*, 214. [CrossRef]
20. Dou, L.; Yang, K.; Liu, W.; Chi, X. Mining-Induced Stress-Fissure Field Evolution and the Disaster-Causing Mechanism in t High Gas Working Face of the Deep Hard Strata. *Geofluids* **2020**, *2020*, 8849666. [CrossRef]
21. Wang, G.; Xu, H.; Wu, M.; Wang, Y.; Wang, R.; Zhang, X. Porosity Model and Air Leakage Flow Field Simulation of Goaf Bas on DEM-CFD. *Arab. J. Geosci.* **2018**, *11*, 148. [CrossRef]
22. Tian, S.; Mao, J.; Li, H. Porosity Distribution Law of Overlying Strata in the Goaf of the Adjacent Working Face: From t Perspective of Section Coal Pillar Types. *Minerals* **2022**, *12*, 782. [CrossRef]
23. Qin, Z.; Yuan, L.; Guo, H.; Qu, Q. Investigation of Longwall Goaf Gas Flows and Borehole Drainage Performance by CF Simulation. *Int. J. Coal Geol.* **2015**, *150–151*, 51–63. [CrossRef]
24. Balusu, R.; Belle, B.; Tanguturi, K. Development of Goaf Gas Drainage and Inertisation Strategies in 1.0-Km- and 3.0-Km-Lo Panels. *Min. Metall. Explor.* **2019**, *36*, 1127–1136. [CrossRef]
25. Zhang, F.; Xu, H.; Qin, Y.; Guo, M.; He, S.; Wang, K.; Shi, Y.; Xiang, Z. Numerical Simulation and Investigation of Methane G Distribution and Extraction in Goaf with U-Type Ventilation of Working Face. *Environ. Sci. Pollut. Res.* **2023**, *30*, 59510–595 [CrossRef] [PubMed]
26. Li, F.; Shang, Y.; Kong, D.; Zhang, G.; Wang, Y.; Wang, Y.; Zhang, Z. Gas Migration Law and Precision Extraction in Close Distan Coal Seam Goaf: A Case Study. *Geotech. Geol. Eng.* **2023**, *41*, 3781–3801. [CrossRef]
27. Zhou, A.; Xu, Z.; Wang, K.; Wang, Y.; An, J.; Shi, Z. Coal Mine Gas Migration Model Establishment and Gas Extraction Technolo Field Application Research. *Fuel* **2023**, *349*, 128650. [CrossRef]
28. Li, Z.-X.; Wang, J.-R.; Zhou, X. Numerical Simulation of Gas Drainage during Open Region Movement in Goaf. *J. China Un Min. Technol.* **2004**, *33*, 74–78.

applied sciences

MDPI

rticle

Double Unloading Gas Control Technology for Fracturing Soft Coal Seams in Overlying Key Strata

ın Xie [1], Feng Li [2,*], Zhengxu Yan [2] and Jingjing Huo [1]

[1] Lu'an Chemical Sijiazhuang Co., Ltd., Jinzhong 045300, China; xiejun147852@163.com (J.X.); huo-jingjing@163.com (J.H.)
[2] School of Emergency Management and Safety Engineering, China University of Mining & Technology (Beijing), Beijing 100083, China; zqt2110102063@student.cumtb.edu.cn
* Correspondence: lifengcumtb@126.com

Abstract: Based on the 'three highs and one low' geological conditions of high gas pressure, high gas content, high ground stress, and low permeability in deep coal seams, this study proposes a dual method of hydraulic fracturing of key layers of overlying rock layers combined with pre-extraction of gas via large-diameter caving boreholes. The aim is to unload and dissipate the coal seam by fracturing the overlying key strata, allowing the stress and energy from the excavation working face to be transmitted and transferred to the deep coal seam. Additionally, large-diameter drilling effectively increases the effective drainage radius of the coal seam, resulting in a shorter extraction time. To validate this approach, a fracturing model and a gas extraction model were established for the key layers of the overlying rock layer using the engineering background of the 15,111 excavation working face of a mine in Shanxi. FLAC3D software v.6.0 was utilized to simulate the stress and energy changes of the coal seam before and after fracturing of the key layers, while COMSOL software v.6.0 was used to analyze the gas migration conditions, permeability, and effective drainage radius changes before and after drilling and caving drilling. The findings, combined with the engineering test results, conclude that key strata fracturing combined with large-diameter caving can effectively increase the permeability of coal seams and improve gas extraction. This study serves as a theoretical basis for guiding the design of gas drainage technology under the effects of coal seam pressure relief and permeability enhancement.

Keywords: key strata fracturing; gas extraction; large-diameter borehole; permeability; effective extraction radius

1. Introduction

About 50% of mines in China are classified as high gas or coal and gas outburst mines, with soft and low-permeability coal seams accounting for as much as 82% of these mines [1,2]. As mining operations go deeper, the in situ stress, gas content, and gas pressure gradually increase, while the permeability further decreases. The permeability of most outburst mines is only 1×10^{-4} to 1×10^{-3} md [3–5]. The poor gas permeability of coal seams makes it challenging to pre-extract gas, resulting in a high amount of residual gas. During the mining process, frequent abnormal gas outflow leads to the gas concentration in the working surface exceeding the limit, posing an increasingly serious threat of gas disasters [6–8]. Conventional drilling has limited effectiveness in dealing with this type of coal seam, as it has a small extraction radius, low drainage efficiency, and poor pressure relief effect on the original coal seam, making gas drainage more difficult [9,10]. Therefore, the main approach for gas control is to utilize external force measures to release surrounding rock stress, modify coal seam gas migration characteristics, and enhance drainage efficiency [11–13].

At present, various technologies are used to address high gas and low permeability coal seams, including mining protective layers [14,15], hydraulic slotting [16,17], hydraulic fracturing [18–21], hydraulic punching [22,23], and intensive gas drainage drilling [24]. Protective layer mining is currently the most effective method to prevent coal and gas outbursts, which has been successful in field practice [15,25]. Ni et al. [26] conducted industrial experiments on pulse hydraulic fracturing in both structural and non-structural zones, analyzing the extension direction of coal seam cracks during the pulse hydraulic fracturing process. Li et al. [27] conducted experiments using large-diameter boreholes in soft coal seams in geological structures, reducing the risks of collapse and coal and gas outbursts. However, conventional pressure relief, permeability enhancement, and drainage methods face new challenges in deep coal seam gas control. Mining protective layers requires favorable geological conditions and may not be feasible when the adjacent protective layer is too thin or the distance between coal seams is inappropriate. Additionally, cracks formed via hydraulic measures have limitations, as they have a short maintenance time and can be re-compacted under high stress conditions [28–30]. The 'soft coal' undergoes plastic deformation internally when subjected to high-pressure water, resulting in blocked cracks and reduced gas permeability [31]. In the soft coal seam, ordinary extraction drill holes are prone to problems such as jamming and plugging during the construction process, making gas extraction more difficult. This is mainly due to the large ground stress in the deep soft coal seam [32], and the current technology and process methods cannot realize effective pressure relief on the soft coal seam continuously, and it is difficult to fundamentally prevent the occurrence of coal and gas protrusion accidents. In addition, the above-mentioned technologies also have environmental impacts. During hydraulic fracturing, a certain amount of fracturing fluid is used and, when it enters underground drinking water sources through rock fractures, it may pose a threat to human health [33].

This paper aims to address the above issues by adopting a method that combines hydraulic fracturing in the overlying key layer with large-diameter borehole in the coal seam for comprehensive penetration enhancement management. The study involves hydraulic fracturing and decompression in the overlying key strata while simultaneously conducting large-diameter borehole drilling within the coal seam. An analysis is conducted on stress transfer and energy dissipation pertaining to pre- and post-fracturing unloading of the coal seam, and we study the change rules in permeability, gas pressure, and the effective extraction radius of the coal seam under double decompression conditions. The findings of this study will provide a theoretical basis for improving gas extraction and preventing coal and gas outbursts in soft coal seams.

2. Method

2.1. Theory

The research indicates that the overlying rock layer of the coal seam is a composite rock mass consisting of multiple layers. The transmission of stress in this layer primarily originates from the main/inferior key strata of the overlying rock. The key strata is a hard rock layer that can control the collapse activity of the overlying rock strata on the mining face partially or completely [34], and it is the most important path for the downward transmission of geostress to the coal seam [35]. The "weak layer" area in the coal seam excavation will form a high concentration of vertical stress and, at the same time, it is characterized by high gas pressure, low permeability of the coal seam, and a large gas pressure gradient. Based on these characteristics, a dual pressure relief gas control technology is proposed, involving hydraulic fracturing of the key strata overlying the coal seam (primary pressure relief of the coal seam) and large-diameter hole drilling (secondary pressure relief of the coal seam), as shown in Figure 1. The main key strata and inferior kay strata are determined based on the properties of the overlying rock and theoretical calculations. Hydraulic fracturing technology is used to fracture and relieve pressure on the rock mass overlying the key layers of the coal seam, thereby alleviating the pressure on the coal seam. Simultaneously, large-diameter hole drilling is employed in the coal seam to increase the diameter of the borehole, which provides the secondary decompression for the coal seam, causing the coal body to expand and enhancing its permeability. This approach significantly improves the gas drainage effect in the coal seam and provides theoretical support for the study of permeability in soft coal seams under pressure-relief conditions.

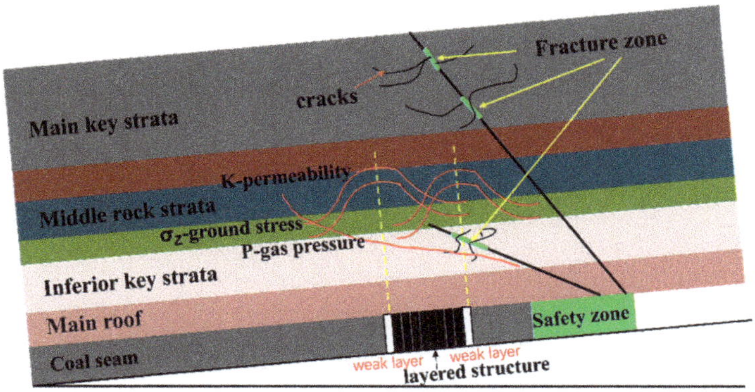

Figure 1. Structural model of laminated combined coal rock body.

2.2. Coal Seam General Situation and Model Parameters

The 15,111 working face of a mine in Shanxi is located in the central disc area and its comprehensive geological histogram is shown in Figure 2. The working face has an elevation range of 557–613 m and is designed to be mined with a strike length of 3529 m and an inclined length of 161 m–322 m. It has a recoverable reserve of 6,466,700 t. The 15,111 working face is a coal and gas outburst working face and, in order to carry out production, it must meet the requirement of eliminating the danger of gas proliferation as stated in the relevant laws and regulations. The original coal seam has a gas content of 11.22 m^3/t, which needs to be reduced to less than 8 m^3/t to eliminate the danger of gas proliferation. Therefore, the target pre-pumping rate of the 15,111 working face is set at 30%. However, due to the lack of systematic research, the previous gas management work relied on field experience to formulate the extraction program, resulting in unstable extraction results with an extraction rate of less than 30% and difficulty in meeting the standard. The physical and mechanical parameters of the rock layer measured through tests are presented in Table 1, while the gas-related parameters of the coal seam are shown in Table 2.

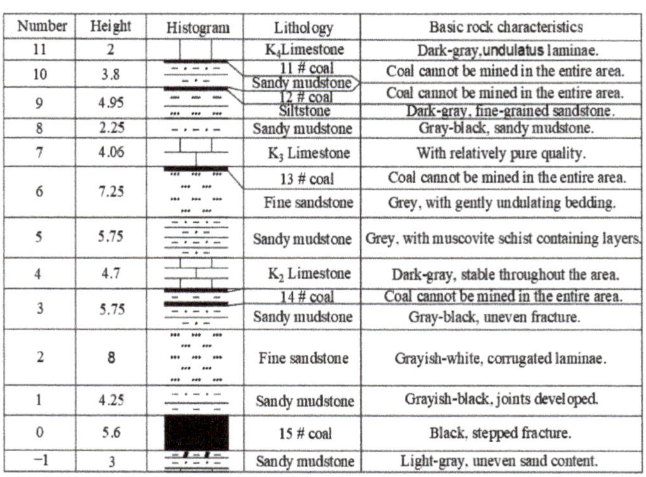

Number	Height	Histogram	Lithology	Basic rock characteristics
11	2		K_4Limestone	Dark-gray, undulatus laminae.
10	3.8		11 # coal Sandy mudstone	Coal cannot be mined in the entire area.
9	4.95		12 # coal Siltstone	Coal cannot be mined in the entire area. Dark-gray, fine-grained sandstone.
8	2.25		Sandy mudstone	Gray-black, sandy mudstone.
7	4.06		K_3 Limestone	With relatively pure quality.
6	7.25		13 # coal Fine sandstone	Coal cannot be mined in the entire area. Grey, with gently undulating bedding.
5	5.75		Sandy mudstone	Grey, with muscovite schist containing layers.
4	4.7		K_2 Limestone	Dark-gray, stable throughout the area.
3	5.75		14 # coal Sandy mudstone	Coal cannot be mined in the entire area. Gray-black, uneven fracture.
2	8		Fine sandstone	Grayish-white, corrugated laminae.
1	4.25		Sandy mudstone	Grayish-black, joints developed.
0	5.6		15 # coal	Black, stepped fracture.
−1	3		Sandy mudstone	Light-gray, uneven sand content.

Figure 2. Comprehensive geological histogram of 15,111 working face.

Table 1. Physical and mechanical parameters of rock layers.

Lithology	Unit Weight kN/m^3	Modulus of Elasticity/GPa	Poisson's Ratio/μ	Compressive Strength/MPa	Tensile Strength MPa	Internal Friction Angle °
Fine sandstone	26	18	0.22	100	10	42
Limestone	26	40	0.20	120	18	45
Sandy mudstone	25	15	0.28	40	3	38
Coal	14	0.3	0.34	10	0.6	28

Table 2. Coal seam gas related parameters.

Parameter Name	Numerical Value	Parameter Name	Numerical Value
Initial crack rate	0.012	Langmuir pressure constant (MPa)	2
Initial porosity	0.049	Maximum value of adsorbed gas (m^3/t)	28.8
Initial permeability (mD)	0.004	Molar mass of coal gas (L/mol)	22.4
Initial gas pressure (MPa)	1.6	Density of coal (kg/m^3)	1380
Dynamic viscosity of gas (Pa·s)	1.08×10^{-5}	Negative pressure (kPa)	21
Elastic modulus of coal matrix (MPa)	8139	Poisson ratio (υ)	0.34
Elastic modulus of coal (MPa)	2713	Single pore diffusion coefficient	5.599×10^{-12}
Klinkenberg factor (Pa)	1.4 × 105	Constant amount of adsorbed gas b (MPa^{-1})	0.494

2.3. Hydraulic Fracturing Radius Determination

2.3.1. Determination of Key Strata of Overlying Rock Strata

1. Determine hard rock stratum according to load

According to masonry beam theory, the nth layer of the overlying rock layer bears the load (q_m) of the nth layer to the mth layer $(q_m)_n$:

$$(q_m)_n = \frac{E_n h_n^3 \sum_{i=n}^{m} \gamma_i h_i}{\sum_{i=n}^{m} E_i h_i^3}. \quad (1)$$

In Formula (1), E_i represents the elastic modulus of the ith layer of the rock layer, GPa; γ_i is the unit weight of the ith layer, kg/m^3; and h_i is the thickness of layer i, m.

If layer $m + 1$ is a hard rock layer, it should have the following supporting characteristics:

$$(q_m)_n > (q_{m+1})_n. \quad (2)$$

According to Equation (2), the load (q_1) of the first layer itself is 106.3 kPa, while the load $(q_2)_1$ from the first layer to the second layer is 31.7 kPa. Based on Equation (2), it can be inferred that the second layer of overburden rock does not contribute to the loading of the first layer and can be considered as a hard rock stratum.

The load (q_2) applied to the second layer itself is 208 kPa, the load $(q_3)_2$ applied to the second layer up to the third layer is 275.1 kPa, the load $(q_4)_2$ applied to the second layer up to the fourth layer is 281.4 kPa, the load $(q_5)_2$ applied to the second layer up to the fifth layer is 314.7 kPa, and the load (q_6) applied to the second layer up to the sixth layer is 297.8 kPa. It can be seen that the sixth layer is a hard rock layer, and the load on the second layer is 314.7 kPa, which controls the upper three layers.

Similarly, (q_6) = 188.5 kPa, $(q_7)_6$ = 217.6 kPa, $(q_8)_6$ = 255 kPa, $(q_9)_6$ = 261.4 kPa, $(q_{10})_6$ = 296 kPa, $(q_{11})_6$ = 316.1 kPa, and $(q_{12})_6$ = 290.1 kPa; therefore, the twelfth layer is a hard rock layer, and the load that had been applied to the sixth layer is 316.1 kPa, controlling the upper five layers.

Therefore, according to the theory related to the key strata [36], the second, sixth, and twelfth layers of the upper part of the coal seam were determined to be hard rock layers.

2. Calculate the breaking distance of the hard rock stratum

Assuming that the initial breaking distance of layer i is $(L_a)_i$ and the periodic breaking distance of layer i is $(L_b)_i$, the calculation formula is as follow:

$$(L_a)_i = h_i \times \sqrt{\frac{2R_i}{q_i}}, \tag{3}$$

$$(L_b)_i = h_i \times \sqrt{\frac{R_i}{3q_i}} \tag{4}$$

where h_i is the height of the ith layer, m; R_i is the tensile strength of the ith layer, MPa; and q_i is the load of the ith layer of hard rock bearing its control rock layer, MPa.

Based on the formula, it can be inferred that $(L_b)_i = (L_a)_i/2.45$. The initial breakage distances of the second and sixth layers are determined as 78.5 m and 74.7 m, respectively, while the periodic breakage distances are measured at 32.1 m and 30.5 m, correspondingly.

3. Determine the key strata according to the breaking distance

The initial and periodic breakage distances of the second and sixth layers in the overlying hard rock strata were calculated using Equations (3) and (4), respectively. Based on the calculation results, it is evident that the magnitude of the initial breakage distance for the hard rock layer follows this order: $(L_a)_2 > (L_a)_6$; similarly, periodic breakage distance: $(L_b)_2 > (L_b)_6$. Consequently, we can conclude that the second layer of hard rock acts as a primary key strata with synchronous failure occurring in both layers. Therefore, considering the overburdened rock strata, it is essential to recognize the second layer as the main key strata while regarding the sixth layer as the inferior one.

2.3.2. Determination of Crack Initiation Pressure

Based on the theory of linear elastic tensile damage, the H-W model was employed to calculate the rupture pressure. The model assumes the following: (1) the rock is impermeable; (2) the rupture takes place in the borehole wall; and (3) the rock experiences tensile damage. According to this criterion, the rupture equation is derived without considering the pore pressure of the top sandstone:

$$p_b = 3\sigma_h - \sigma_H + \sigma_t \tag{5}$$

where P_b is the rock initiation pressure, σ_h is the minimum horizontal stress, σ_H is the maximum horizontal stress, and σ_t is the tensile strength of rock.

According to the field measurement and test results, the minimum horizontal stress σ_h = 8.08 MPa, the maximum horizontal stress σ_H = 14.98 MPa, and the tensile strength σ_t = 10 MPa of the rock mass in the key strata can be calculated to obtain the crack initiation pressure P_b of the rock mass in the key strata—19.26 MPa.

2.3.3. Determination of Fracturing Radius

As shown in Figure 3, different colors are used to indicate different rock layers, the roadway section is 4.5 m wide and 3.5 m high. The fracturing test borehole, labeled as ①, is located in the roadway gang and has a height of 1.5 m. It has an azimuth angle of 0° (angle with the center line of the roadway), an inclination angle of 12°, and a depth of 52 m. The observation borehole, labeled as ②, is positioned 3 m horizontally from the fracturing borehole. It has a height of 1.5 m, an azimuth angle of 20°, an inclination angle of 12°, and a depth of 55 m. The water seepage zone of the observation borehole after fracturing is labeled as ③. ④ shows a diagram of a hydraulic fracturing unit. The maximum pumping pressure of the emulsion pump is 30 MPa, which fulfills the requirement for the fracturing pressure of the key layer of the top plate. To achieve automatic water pressure sealing and maintain the water injection pressure, the MA ZF-A64-enhanced ME rubber water injection sealing device shown in Figure ⑤ was used, with a length of 2 m and a maximum expansion diameter of 120 mm. ⑥ shows a graph of the change in pressure in the pressure gauge with time during the fracturing process. The on-site water injection process is divided into four stages, as shown in Figure 3: (a) the water pressure naturally rises and reaches equilibrium at approximately 3.5 MPa; (b) the water injection pressure of the pump is manually increased to 20.15 MPa; (c) the water injection pressure is maintained at around 18 MPa until the pressure plummets to ensure continuous water injection and sufficient extension of rock cracks; (d) the water level in the hole is observed, and the pressure is manually released to switch off the injection pump. The comprehensive analysis of water injection pressure, water injection time, location of fracturing section, and water seepage section indicates that the sandstone fracturing pressure is 20.15 MPa, the sandstone holding pressure stage requires continuous water injection for 17.5 min, and the hydraulic fracturing radius is 15 m.

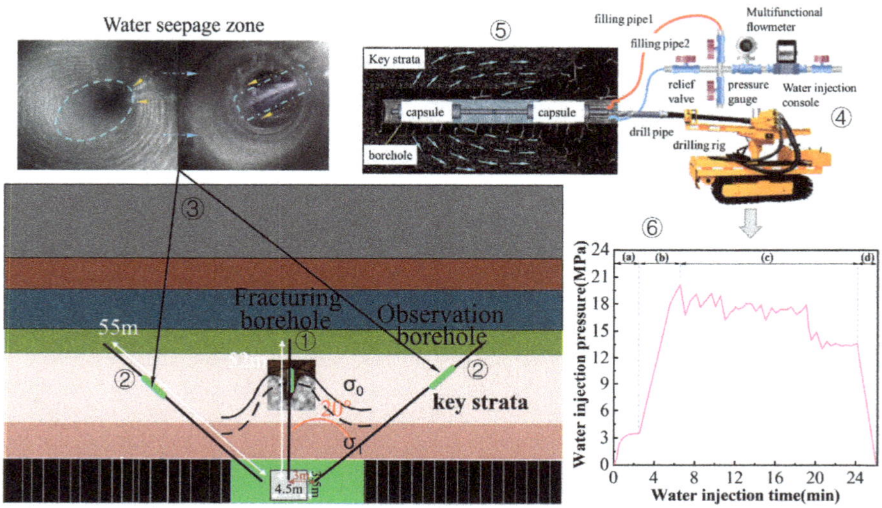

Figure 3. Design and test of fracturing scheme for overlying strata.

2.4. Establishing Numerical Models

2.4.1. Mathematical Modeling

1. Seepage equations of motion

The gas in the coal is mainly composed of two parts: free gas Q_b and adsorbed gas Q_a. Therefore, the total gas storage capacity is

$$Q = Q_\alpha + Q_b = \frac{\varphi p M_c}{RT} + \frac{abp}{1+bp}\frac{100 - A - M}{100 + 31M}\rho_s\rho_a \tag{6}$$

where φ is the porosity of coal,%; p is the coal seam gas pressure, MPa; R is the gas molar constant, J/(mol·K); T is the coal seam temperature, K; M_c is the molar mass of methane, kg/mol; a is the maximum adsorbed gas amount of coal, m^3/t; b is the adsorption constant, MPa^{-1}; M is the proportion of coal moisture, %; A is the proportion of coal ash content,%; ρ_s is the coal density, kg/m^3; and ρ_a is the gas density under standard conditions, kg/m^3.

The gas flow in coal seam conforms to Darcy's law of seepage, and the calculation equation is

$$V = -\frac{k_0}{\mu}\nabla p \tag{7}$$

where V is the gas seepage velocity, m/s; k_0 is the initial permeability of the coal seam, and m^2; μ is the gas dynamic viscosity coefficient, Pa s.

The flow of gas in coal seams conforms to the law of conservation of mass, and the calculation equation [37]:

$$\frac{\partial Q}{\partial t} + \nabla \bullet (\rho V) = 0. \tag{8}$$

Combine Equations (6)–(8) to obtain the following formula:

$$\frac{\partial(\frac{\varphi p M_c}{RT} + \frac{abp}{1+bp}\frac{100-A-M}{100+31M}\rho_s \rho_a)}{\partial t} + \nabla(\frac{pM_c}{RT})(-\frac{k_0}{\mu}\nabla p) = 0. \tag{9}$$

2. Control equation for coal deformation

Assuming that the coal seam is isotropic and considering the adsorption expansion effect, combined with Terzaghi's effective stress formula and constitutive equation, the deformation control equation of the coal body can be obtained [38]:

$$\frac{G}{(1-2v)}u_{j,ij} + Gu_{i,jj} + (a + \frac{2G}{3K_s})p_{,i} + F_i - \{[\frac{4G}{9V_m K_s} - \frac{2(1-2v)}{3V_m}]\frac{ab\rho_s RT}{1+bp}\}p_{,i} = 0 \tag{10}$$

where G is the shear modulus, MPa; u_i is the displacement of the coal body in the i direction, m; $u_{i,jj}$, $u_{j,ij}$ is the displacement component; F_i is the body stress, MPa/m^3; $p_{,i}$ is the gas pressure in all directions, MPa; v is Poisson's ratio of coal body; V_m is the molar volume of methane under standard conditions, L/mol; and K_s is the volume modulus of the coal skeleton.

3. Other equations

The dynamic model of coal seam porosity can be expressed as [39]

$$\varphi = \frac{\varphi_0 + \varepsilon_v}{1+\varepsilon_v} \tag{11}$$

where φ_0 is the initial porosity of the coal seam, %; ε_v is the volumetric strain of the coal seam.

$$k = \frac{k_0}{1+\varepsilon_v}(1+\frac{\varepsilon_v}{\varphi_0})^3 \tag{12}$$

where k_0 is the initial permeability of the coal seam.

2.4.2. Establishing Physical Models

Based on the FLAC3D software (http://www.itascacg.com/software/FLAC3D (accessed on 23 December 2023)), the study utilized a three-dimensional overburden rock fracturing model with 10 layers of different lithologies, as shown in Figure 4a. The model had dimensions of 80 × 65 × 50.4 (m) and the cutting face had dimensions of 3.5 × 4.5 (m). The specific physico-mechanical parameters of the rock layers can be found in Table 1. The initial ground stress field was established by applying a vertical stress of 13 MPa and a horizontal stress of 14.5 MPa based on on-site field measurements. A grid model was used to simulate the advancement of the working face, with the roadway being excavated for a distance of 10~70 m. Instead of hydraulic fracture cracks, crack lines were used, with a long crack of 30 m located between 40~70 m. The crack had a width of 15 m and a crack gap width of 0.2 m. This simulated the stress and energy distribution of the coal bed after the destruction of the key strata.

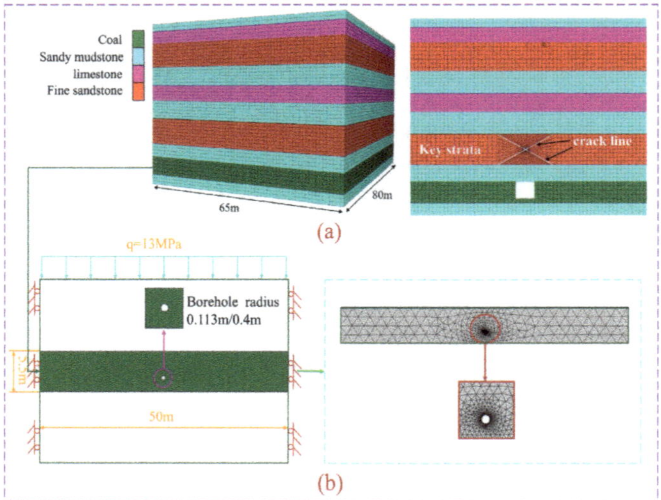

Figure 4. Model building and meshing. (**a**) Overburden fracturing model. (**b**) Geometric model of gas extraction.

Based on the COMSOL software (http://cn.comsol.com/ (accessed on 2 January 2024)), the geometric model of gas extraction was established, as shown in Figure 4b. The length of the coal wall section is 50 m, and the thickness of the coal body is 5.5 m. The opening height of the extraction hole is 1.8 m, with a diameter of 113 mm for the ordinary drilling and 0.4 m for the hole making drilling. The bottom of the numerical model is a fixed boundary, while the boundaries of the left and right flanks are sliding boundaries. The top of the model represents the boundary of the constant load, with an overlying load of 13 MPa. Once the geometric model is generated, the model is then meshed.

3. Analysis of Simulation Results

3.1. Coal Seam Stress–Energy Change Rule before and after Key Strata Stress Removal

Under the influence of ground stress and mining disturbance, the stress field and energy field of the coal seam are disrupted. This disruption results in the formation of three distinct areas: the broken pressure relief area, the stress concentration area, and the original stress area. In the broken pressure relief area, the vertical stress on the coal body is lower than the original stress, leading to the creation of more fissure channels and a reduction in stress concentration caused by the mining disturbance. The stress concentration area experiences an initial increase and then decrease in vertical stress on the coal body, causing compaction of pores and fissures and a decrease in permeability. This area is also prone to impact dynamics phenomena under the influence of the mining disturbance. The original stress area, on the other hand, is less affected by mining activities, resulting in a lower stress field for both the key layer and the coal seam. Here, the coal seam retains its complete layer structure, and the stress load is evenly transmitted downwards.

3.1.1. Coal Seam Stress Change Rule

When the excavation face reaches a depth of 40 m, hydraulic fracturing is carried out to decompress the overlying key strata. Figure 5a,b illustrates the stress cloud map before and after the coal seam is excavated to a depth of 70 m. After fracturing the overlying key strata, a distinct pressure relief area forms within the coal seam, spanning a range of 40×30 m. In Figure 5a, the depressurization of the key layer results in the formation of a stress relief zone extending up to 28 m in front of the excavation face, with the maximum reduction in stress being 6.32 MPa. In Figure 5b, when mining continues in the coal seam after relieving the key strata, the phenomenon of stress concentration on both sides of the roadway disappears, and the range of coal seam unloading is between 6–22 m.

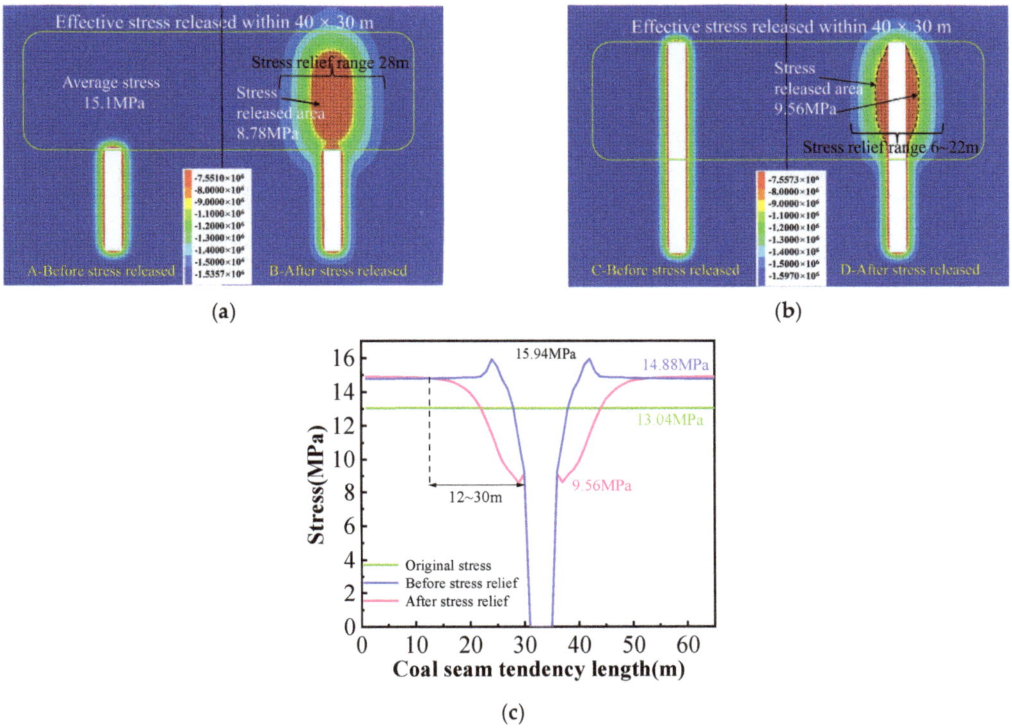

Figure 5. Stress change rule of coal seam before and after unloading of key strata. (**a**) Advanced stress nephogram of coal seam. (**b**) Stress nephogram after coal seam excavation. (**c**) Stress curve after coal seam excavation.

The variation in stress in the coal seam at y = 56 m in Figure 5b is studied and presented in Figure 5c. From the figure, it is evident that the initial stress value of the coal seam is 13.04 MPa when unaffected by mining disturbance and pressure unloading. However, when mining is conducted without pressure unloading, there is a concentration of stress in the two gangs of the roadway, with the maximum value reaching 15.94 MPa. After pressure unloading of the key layer, the minimum stress value in the roadway gang within the coal seam is 9.56 MPa, with the maximum unloading stress being 7.4 MPa. This unloading accounts for 46% and has an average value of 3 MPa. The unloading range is observed to be between 12–30 m, which is consistent with the radius of hydraulic fracturing.

3.1.2. Coal Seam Energy Change Rule

The change rule of coal seam energy is similar to the change rule of coal seam stress. Key strata fracturing causes the redistribution of coal seam energy. The energy distribution cloud diagram of the coal seam is shown in Figure 6a,b. From Figure 6a, it can be observed that there is an energy peak area at 3~4 m in front of the roadway after the coal seam is excavated along the tendency to 40 m. After fracturing the key strata, the energy is redistributed, resulting in an energy dissipation zone of about 40 × 30 m in front of the excavation working face. Figure 6b shows the coal seam after key layer decompression. The energy concentration phenomenon disappeared in the two gangs of the roadway, and an energy dissipation zone formed within 3 m of the two gangs of the roadway. The energy dissipation range of the coal seam was 6~24 m.

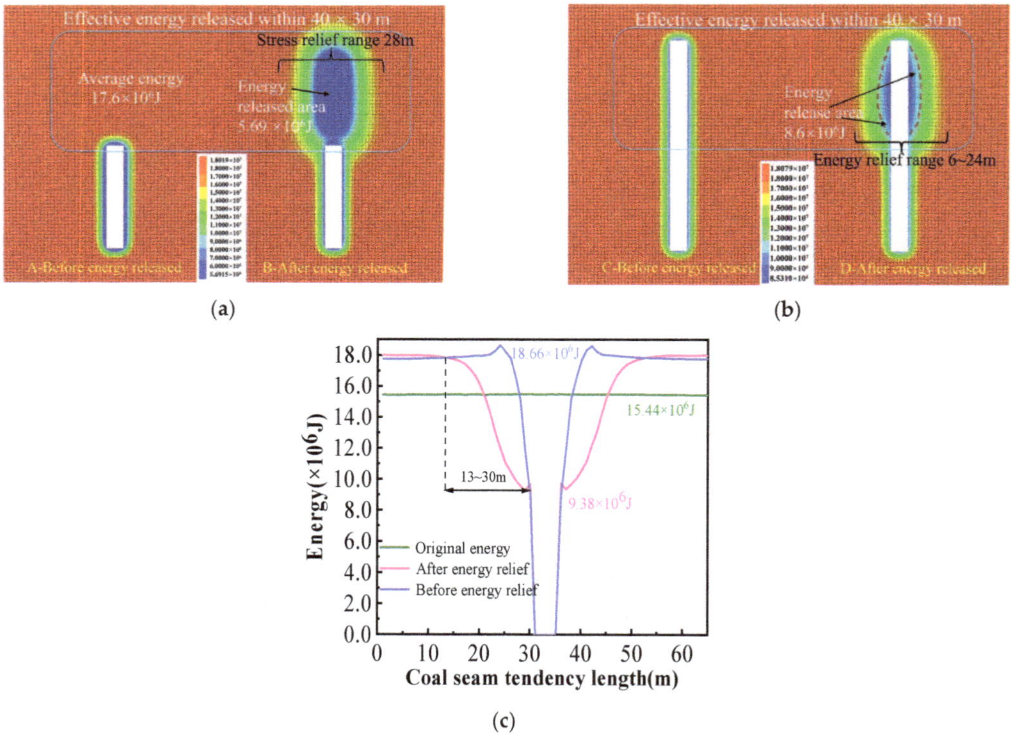

Figure 6. Energy change rule of coal seam before and after unloading of key strata. (**a**) Advanced energy nephogram of coal seam. (**b**) Energy nephogram after coal seam excavation. (**c**) Energy curve after coal seam excavation.

The variation in energy in the coal seam at y = 56 m in Figure 6b is studied and presented in Figure 6c. Initially, the coal seam has an energy value of 15.44 × 10^6 J when it is not affected by mining disturbance and pressure unloading. However, when mining without pressure unloading occurs, the peak energy value of the two gangs of the roadway increases to 18.66 × 10^6 J. This indicates that the energy is transferred to the two gangs of the roadway after mining the coal seam, resulting in an energy concentration area and potential power phenomena. Furthermore, the peak energy at 3 m of the two gangs of the roadway decreases by more than 40% after the key strata is unpressurized, with a maximum energy dissipation of 9.28 × 10^6 J. The unilateral energy dissipation range extends up to 17 m, which is consistent with the radius of hydraulic fracturing.

3.2. Coal Seam Permeability Change Rule

3.2.1. Change Rule of Key Strata Unloading Permeability

After hydraulic fracturing of the key layer, the permeability of the coal seam undergoes significant changes, as depicted in Figure 7a,b. Analyzing the changes in coal seam permeability during the stress unloading process in the key layer reveals that hydraulic fracturing disrupts the stress transfer path of the overlying rock layer, thereby enhancing the permeability of the coal seam. The extent of stress unloading directly impacts the degree of permeability enhancement. For overlying rock stresses of 13 MPa, 10 MPa, 8 MPa, and 5 MPa, the corresponding permeabilities were 3.95 × 10^{-18} m^2, 1.56 × 10^{-17} m^2, 3.14 × 10^{-17} m^2, and 4.32 × 10^{-17} m^2, respectively. Notably, permeability exhibits a linear relationship with stress, represented by the equation y = −0.5087x + 6.9376. This implies that higher overlying rock stresses result in lower permeability and, consequently, reduced gas extraction efficiency. Therefore, fracturing the key layer to relieve stress in the coal seam significantly enhances gas extraction effectiveness.

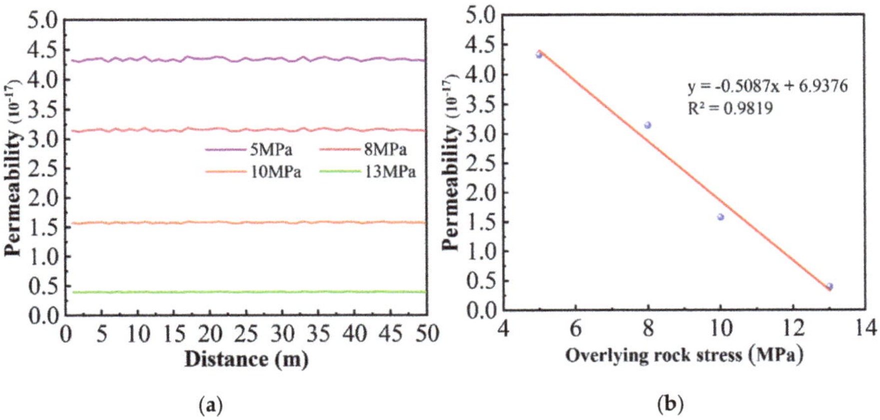

Figure 7. Variation in permeability of coal seams. (**a**) Stress permeability of different overburden rocks. (**b**) Stress and permeability fitting curves.

3.2.2. Change Rule in Permeability of Large-Diameter Borehole

In the process of gas drainage, the pore pressure of the coal seam continuously decreases, leading to desorption of the coal matrix and changes in effective stress, which, in turn, results in continuous changes in coal permeability. Figure 8 illustrates the simulated change in coal seam permeability during the extraction process using ordinary drilling and hole drilling. The results show that hole drilling has a greater impact on improving coal seam permeability compared to ordinary drilling, and the enhanced permeability range is also larger. The permeability of coal seams around ordinary boreholes reached 5.1×10^{-17} m^2, with an influence range of approximately 5 m. However, when hole drilling was used, the permeability increased by about 1×10^{-17} m^2 compared to ordinary drilling, and the influence range of permeability expanded to 10 m, showing a significant improvement.

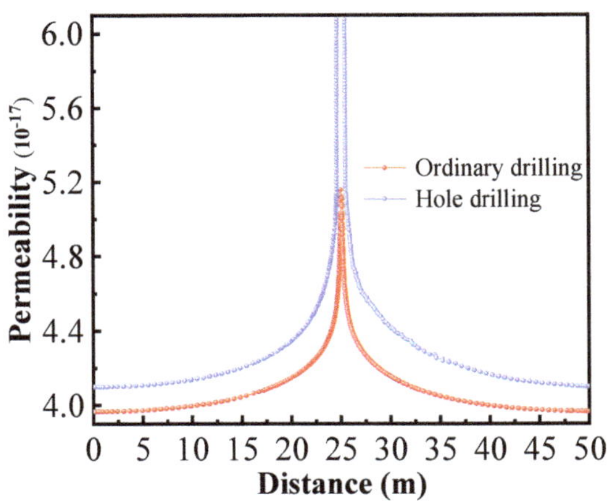

Figure 8. Variation in permeability around boreholes.

3.2.3. Change Rule of Double Unloading Permeability

The method of hydraulic fracturing was employed to reduce the overburden rock stress from 13 MPa to 9.65 MPa. The unloading pressure was approximately 3.35 MPa. The permeability of the coal seams was monitored for different borehole spacing (4 m, 5 m, 6 m, and 7 m), and the comprehensive permeability of the coal seams is illustrated in Figure 9. Initially, the permeability of the coal seam was 3.95×10^{-18} m^2 when it was not fractured or subjected to large-diameter burrowing. However, by combining the method of fracturing the key strata with large-diameter burrowing, the permeability increased to a range of 5.59×10^{-17} m^2 to 7.46×10^{-17} m^2. When the spacing of drilling holes was 4 m, the permeability of coal seam was the largest, reaching 7.46×10^{-17} m^2, but its anti-reflection range was also the smallest. As the spacing of drilling holes increases to 5 m and 6 m, the permeability of coal seam gradually decreases. When the spacing of drilling holes was 7 m, the average permeability of coal seam was 6.45×10^{-17} m^2 and the anti-reflection range was up to 30 m. For the soft coal seam boring process, the use of key layer hydraulic fracturing and the coal seam large-diameter hole-making method for field tests can overall improve the level of gas management in the working face.

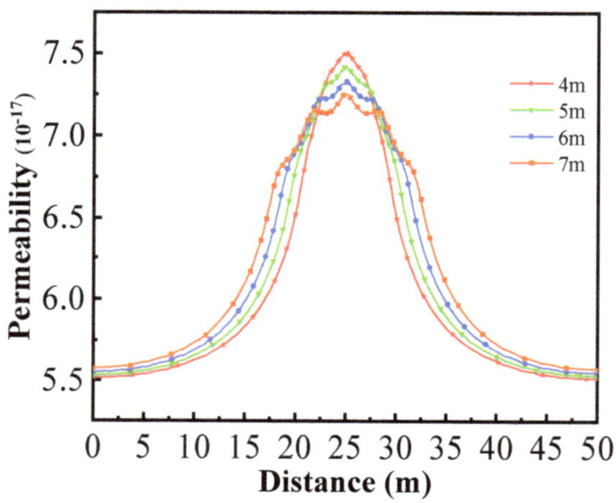

Figure 9. Variation in coal seam permeability.

3.3. Drilling and Extraction Rules before and after Stress Relief of Key Strata

3.3.1. Gas Pressure Distribution in Extraction Boreholes

Gas extraction is a long-term process; usually coal mines use 180 d as an extraction cycle, so the number of days of extraction for the coal seam was monitored for 30 d, 90 d, and 180 d, and the gas pressure distribution cloud map was plotted when the stress of the overlying rock layer on the coal seam was gradually depressurized from 13 MPa to 10 MPa and 8 MPa as shown in Figure 10. Subfigures a–c illustrate the cloud maps of the extraction pressure for ordinary boreholes over time, while subfigures d–f depict the cloud maps of the extraction pressure for hole drilling over time. The results demonstrate that the pressure relief effect of the overlying strata of the coal seam can enhance the gas extraction range of the coal seam drilling with increasing extraction time. Furthermore, within the same extraction time, the gas extraction pressure reduction range of the hole drilling was significantly greater than that of the ordinary extraction drilling. As shown in Figure 10d, the rate of decrease in gas pressure during the extraction time from 30 d to 90 d is significantly larger than that from 90 d to 180 d, and it can be predicted that the rate of decrease of gas pressure after 180 d is even slower.

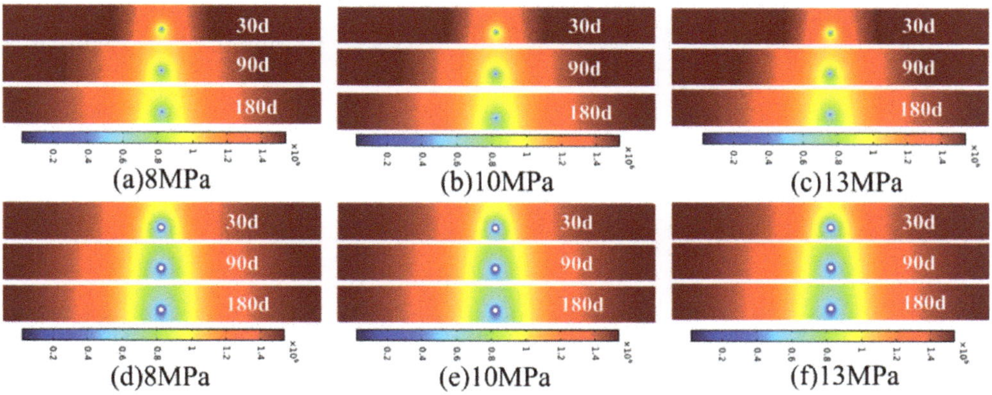

Figure 10. Gas pressure distribution under different overburden stresses.

According to the difference in stress of the overlying rock layer, the gas pressure extraction curves of hole drilling and ordinary drilling at 180 d of extraction were plotted, as shown in Figure 11. There is a difference in the gas pressure drop rate under different stress levels of the overlying rock layer, and the gas pressure drop rate of coal seam with overburden stress of 8 MPa was faster than that of coal seam with overburden stress of 13 MPa; this indicates that the greater the degree of decompression of overburden, the more obvious the effect of gas pressure drop of the coal seam is. As shown in Figure 11a, with the overburden stress of 8 MPa, the gas pressure at 25 m away from the extraction borehole is 1.29 MPa, which is lower than that of the ordinary borehole of 1.35 MPa. Therefore, under the same stress of the overlying rock layer, the range and speed of gas pressure decrease of the hole drilling is larger than that of the ordinary drilling. As shown in Figure 12, when the stress of the overlying rock layer is 13 MPa, with the increase in the number of days of extraction, the decrease in gas pressure in the hole drilling is larger than that in the ordinary drilling; the gas pressure in the hole making drilling decreases from 1.6 MPa to 1.33 MPa, while the gas pressure in the ordinary drilling decreases from 1.6 MPa to 1.4 MPa. In the same period of time of extraction, the decrease in the gas pressure of the two sides of the hole drilling is greater than that of the ordinary drilling.

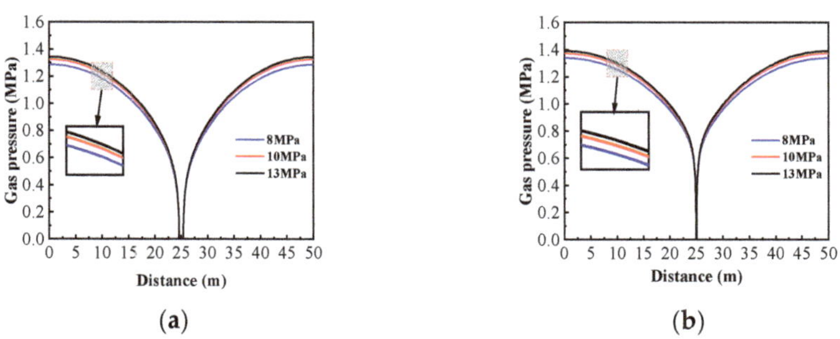

Figure 11. Variation curve of overlying rock stress and gas pressure overlying rock layers. (**a**) Hole drilling. (**b**) Ordinary drilling.

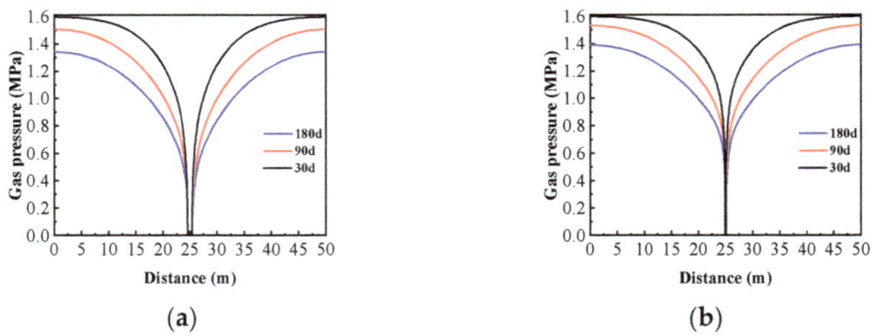

Figure 12. Variation curve of extraction time and gas pressure. (**a**) Hole drilling. (**b**) Ordinary drilling.

3.3.2. Effective Extraction Radius

According to the findings in Figure 13, it is evident that the effective extraction radius increases as the stress load of the overlying rock formation decreases and as the borehole diameter increases. For stress loads of 13 MPa, 10 MPa, 8 MPa, and 5 MPa in the overlying rock layer, the effective extraction radius of the hole drilling was measured to be 3 m, 3.2 m, 3.4 m, and 3.6 m, respectively. On the other hand, the extraction radius of the ordinary borehole was found to be 1.15 m, 1.23 m, 1.3 m, and 1.4 m. These results can be attributed to the expansion and deformation of the coal matrix and the opening up of pore fissures in the coal layer. This suggests that an increase in overburden stress within the coal seam significantly hinders gas extraction, while increasing the diameter of boreholes can help expand the effective extraction radius. Consequently, unloading coal seam stress through hydraulic fracturing and large-diameter borehole technology in overlying strata can effectively improve the extraction effect.

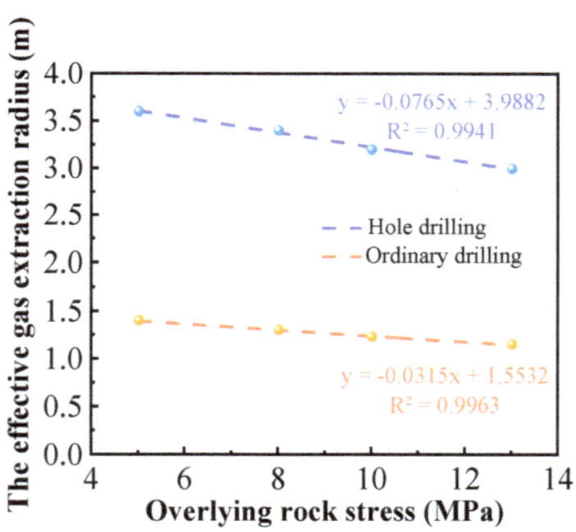

Figure 13. Fitting curve between effective gas extraction radius and overburden stress.

3.3.3. Optimization of Hole Drilling Arrangement Parameters

Through a numerical simulation study of single-hole extraction, it has been concluded that the effective extraction radius and gas pressure drop rate of coal seam hole drilling are significantly larger compared to ordinary drilling. Furthermore, after fracturing the key strata of the overlying rock layer, the effective extraction radius of hole drilling becomes even larger. Based on this, it is determined that hole drilling should be adopted and the process of drilling holes should be optimally arranged. The simulation in Figure 14 shows the use of single-row and double-row drill holes, with varying spacing of 4 m, 5 m, 6 m, and 7 m. The double-row drill holes are arranged in a 'triple-eyelet' configuration, with a spacing of 2.5 m between the upper and lower rows. According to Figure 14a,b, it can be seen that the effective range enhancement of borehole extraction is greater than 2 times the effective extraction radius, due to the superposition effect of the effective extraction radius of the borehole. Additionally, the use of the triangular double-row layout expands the extraction range of the holes, effectively addressing the gas control blank area at the top and bottom plates of the coal seam, which is not adequately addressed by a single row of holes.

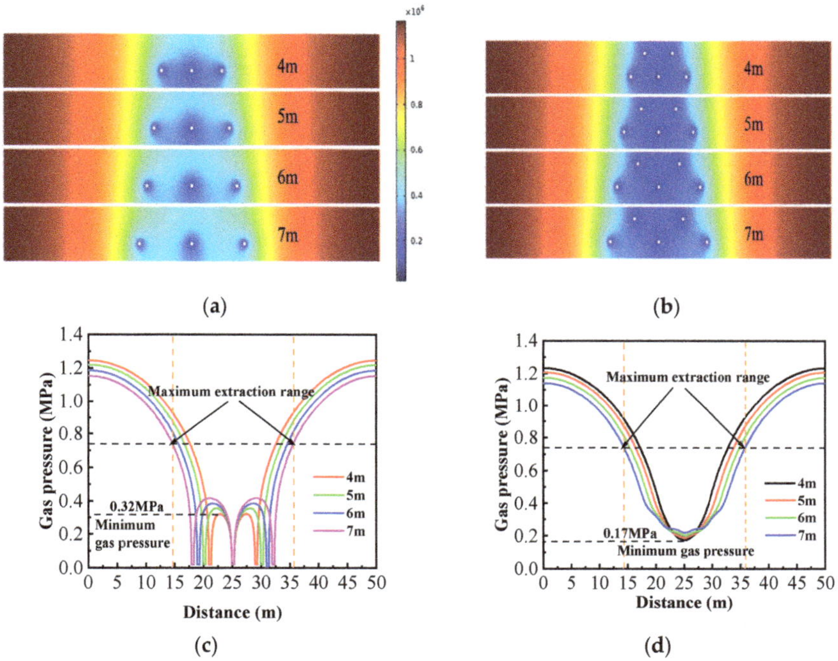

Figure 14. Gas pressure distribution and change curve of different drill hole spacing and multi-row drill hole arrangements. (**a**) Gas pressure distribution in single row of holes. (**b**) Gas pressure distribution in double row of holes. (**c**) Variation curve of gas pressure in a single row of holes. (**d**) Variation curve of gas pressure in double row of holes.

The results shown in Figure 14c indicate that the gas pressure at the center of the adjacent drill holes is 0.32 MPa, 0.34 MPa, 0.38 MPa, and 0.41 MPa for drill hole spacings of 4 m, 5 m, 6 m, and 7 m, respectively. The maximum pressure relief range is found to be 21 m. Similarly, in the case of Figure 14d with double rows of holes, the gas pressures at the center of the adjacent holes are 0.17 MPa, 0.19 MPa, 0.21 MPa, and 0.23 MPa for drill hole spacings of 4 m, 5 m, 6 m, and 7 m and the maximum pressure relief range is also 21 m. It can be concluded that the pressure relief range of double-row holes is greater in the longitudinal direction, while remaining consistent with that of the single-row holes in the transverse direction. Additionally, the gas pressure decreases faster in double-row drill holes within the same extraction time, with the lowest gas pressure reaching 0.17 MPa. This suggests that double-row drill holes can more effectively reduce the gas level in the coal seam below the safety threshold.

4. Field Practice Effect Analysis

4.1. Working Face Conditions and Drilling Arrangement

The test site is located in the boring face of a mine in Shanxi, as depicted in Figure 15. The designed fracturing drill hole has a height of 1.5 m, an azimuth angle (angle with the centerline of the roadway) of 0°, an inclination angle of 12°, and a hole depth of 52 m. On the other hand, the observation drill hole has a height of 1.5 m, a horizontal distance of 3 m from the fracturing hole, an azimuth angle of 20°, an inclination angle of 12°, and a hole depth of 55 m. Following the hydraulic fracturing of the key layer, the extraction is carried out using the 'triangular shape' drilling method with double rows of holes. Six large-diameter coal seam drill holes are arranged parallel to the vertical coal wall face in the boring face. The design parameters of the extraction drill holes are as follows: negative pressure of 20 kPa, hole diameter of 400 mm, hole spacing of 5.0 m, and hole depth of 70 m.

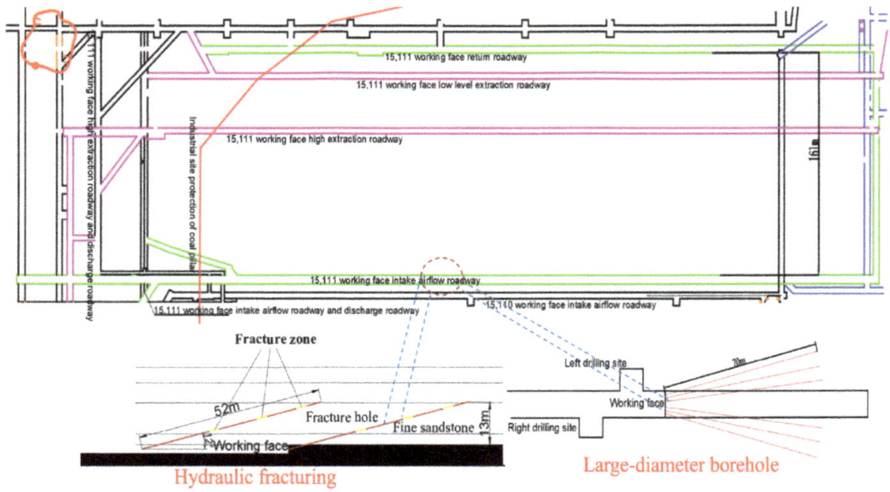

Figure 15. Design layout of drill holes on-site at the working face.

4.2. On-Site Implementation

By monitoring the extraction data before and after fracturing of ordinary drilling and hole drilling, the pure amount of gas extracted from ordinary drilling and hole drilling before and after fracturing of the key strata, the cumulative number of coal blasts before and after unloading pressure, and the comparison of the amount of gas extracted before and after unloading and the concentration of the gas extracted are plotted, as shown in Figure 16. The analysis reveals that before fracturing in the key strata the cumulative 30-day pure gas extraction volume from ordinary drilling was 1713 m^3, whereas hole drilling yielded a 30-day pure gas extraction volume of 8412 m^3 after fracturing. Furthermore, the cumulative 30-day pure gas extraction volume at the drilling site before unloading pressure was 10,891 m^3, which significantly increased to 31,183 m^3 after unloading pressure was applied. Comparing the cumulative number of coal cannons before and after fracturing, the key strata indicates that decompressing the key strata can greatly reduce the occurrence of coal cannons, resulting in a reduction of more than 90% in their frequency.

The results of the field application indicate that adopting hydraulic fracturing of key layers and implementing large-diameter borehole technology in coal seams can significantly enhance gas extraction. The cumulative 30-day pure amount of gas extracted from using large-diameter gas extraction boreholes was approximately five times higher compared to ordinary drilling without fracturing. Furthermore, the gas extraction concentration at the hole drilling field after fracturing was 1.6 times higher than that of the ordinary drilling field without depressurization. Additionally, the cumulative 30-day pure amount of gas extraction at the fracturing drilling site was 2.9 times higher than the non-fractured drilling field. These findings demonstrate that hydraulic fracturing of key strata and the use of large-diameter borehole technology in soft coal seams effectively depressurizes and increases the permeability of the coal body, resulting in a significant improvement in gas extraction efficiency

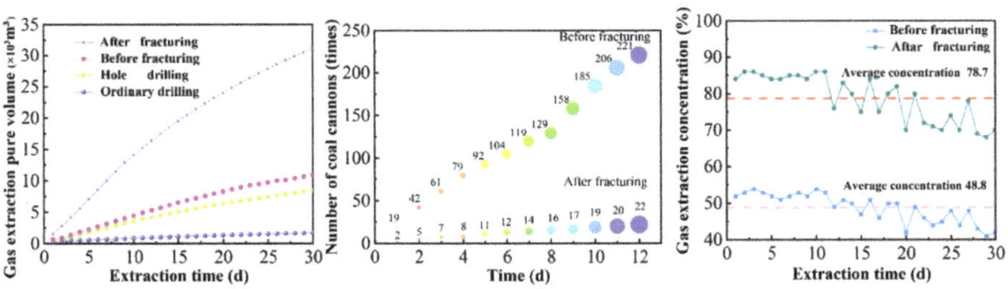

(**a**) (**b**) (**c**)

Figure 16. Changes in pure volume of gas extraction and extraction concentration before and after fracturing of key strata. (**a**) Gas extraction. (**b**) Cumulative number of coal cannons before and after fracturing. (c) Gas extraction concentration before and after fracturing.

5. Conclusions

1. Following hydraulic fracturing of the overlying key strata, a new stress field known as the 'three zones' was created. The average peak stress on both sides of the coal seam roadway decreased by 3.35 MPa, with the energy peak decreasing by over 40% and the stress concentration area shrinking by approximately 2 m. These findings suggest that hydraulic fracturing of the key strata has a substantial impact on the unloading and dissipation of stress and energy within the coal seam.
2. The study of stress effects on coal seam permeability revealed a significant increase in permeability through hydraulic fracturing and large-diameter cavity drilling technology. The maximum enhancement observed was up to 190 times, suggesting that reducing overlying rock stress and increasing drill hole diameter can enhance gas extraction from coal seams.
3. A gas extraction model was developed to address the issue of multi-row drilling in thick coal beds, considering the impact of overburden stress. Through simulations of various hole spacing configurations and on-site validation, 'triangular' hole spacing of 5 m was determined as the most effective solution. This spacing successfully addressed the limitation of single-row hole spacing in influencing the coal bed longitudinally.

Author Contributions: Methodology, F.L.; software, Z.Y.; investigation, J.X.; resources, J.H.; Data curation, F.L.; writing—original draft preparation, Z.Y.; writing—review and editing, F.L.; project administration, J.X.; funding acquisition, F.L. All authors have read and agreed to the published version of the manuscript.

Funding: This study was financially supported by National Natural Science Foundation of China (52064046, 51804311), The Fundamental Research Funds for the Central Universities (2023ZKPYAQ03) and the Special Project for Regional Collaborative Innovation of Xinjiang Autonomous Region [Science and technology assistance plan for Xinjiang] (2020E0258).

Institutional Review Board Statement: Not applicable.

Informed Consent Statement: Not applicable.

Data Availability Statement: The data used to support the findings of this study are available from the corresponding authors upon request.

Conflicts of Interest: Authors Xie, J. and Huo, J. were employed by the company "Lu'an Chemical Sijiazhuang Co., Ltd.". The remaining authors declare that the research was conducted in the absence of any commercial or financial relationships that could be construed as a potential conflict of interest.

References

1. Wang, L.; Cheng, Y.; Wang, L.; Guo, P.; Li, W. Safety line method for the prediction of deep coal-seam gas pressure and i application in coal mines. *Saf. Sci.* **2012**, *50*, 523–529. [CrossRef]
2. Wang, L.; Liu, S.; Cheng, Y.; Yin, G.; Guo, P.; Mou, J. The effects of magma intrusion on localized stress distribution and i implications for coal mine outburst hazards. *Eng. Geol.* **2017**, *218*, 12–21. [CrossRef]
3. Cheng, Y.; Pan, Z. Reservoir properties of Chinese tectonic coal: A review. *Fuel* **2020**, *260*, 116350. [CrossRef]
4. Kursunoglu, N.; Onder, M. Application of structural equation modeling to evaluate coal and gas outbursts. *Tunn. Undergr. Spa Technol.* **2019**, *88*, 63–72. [CrossRef]
5. Zhao, B.; Cao, J.; Sun, H.; Wen, G.; Wang, B. Experimental investigations of stress-gas pressure evolution rules of coal and g outburst: A case study in Ding ji coal mine, China. *Energy Sci. Eng.* **2020**, *8*, 61–73. [CrossRef]
6. Zhou, A.; Xu, Z.; Wang, K.; Wang, Y.; An, J.; Shi, Z. Coal mine gas migration model establishment and gas extraction technolo field application research. *Fuel* **2023**, *349*, 128650. [CrossRef]
7. Zhao, P.; Zhuo, R.; Li, S.; Shu, C.; Laiwang, B.; Jia, Y.; Suo, L. Analysis of advancing speed effect in gas safety extraction channe and pressure-relief gas extraction. *Fuel* **2020**, *265*, 116825. [CrossRef]
8. Liu, Y.; Zhang, C.; Song, Z. Numerical simulation of surface gas venthole extraction and the effect of ventilation mode pressure-relief mining. *Processes* **2022**, *10*, 750. [CrossRef]
9. Zhou, B.; Xu, J.; Yan, F.; Peng, S.; Gao, Y.; Li, Q.; Cheng, L. Effects of gas pressure on dynamic response of two-phase flow f coal-gas outburst. *Powder Technol.* **2021**, *377*, 55–69. [CrossRef]
10. Zou, Q.; Liu, H.; Zhang, Y.; Li, Q.; Fu, J.; Hu, Q. Rationality evaluation of production deployment of outburst-prone coal mines: case study of nantong coal mine in Chongqing, China. *Saf. Sci.* **2020**, *122*, 104515. [CrossRef]
11. Chen, H.; Cheng, Y.; Ren, T.; Zhou, H.; Liu, Q. Permeability distribution characteristics of protected coal seams during unloadir of the coal body. *Int. J. Rock Mech. Min.* **2014**, *71*, 105–116. [CrossRef]
12. Lou, Z.; Wang, K.; Zang, J.; Zhao, W.; Qin, B.; Kan, T. Effects of permeability anisotropy on coal mine methane drainag performance. *J. Nat. Gas Sci. Eng.* **2021**, *86*, 103733. [CrossRef]
13. Chen, H.; Chen, Y.; Zhou, H.; Li, W. Damage and permeability development in coal during unloading. *Rock Mech. Rock Eng.* **201** *46*, 1377–1390. [CrossRef]
14. Wang, L.; Lu, Z.; Chen, D.; Liu, Q.; Chu, P.; Shu, L.; Wen, Z. Safe strategy for coal and gas outburst prevention in deep-and-thic coal seams using a soft rock protective layer mining. *Saf. Sci.* **2020**, *129*, 104800. [CrossRef]
15. Cheng, Y.; Wang, L.; Liu, H.; Kong, S.; Yang, Q.; Zhu, J.; Tu, Q. Definition, theory, methods, and applications of the safe an efficient simultaneous extraction of coal and gas. *IJCST* **2015**, *2*, 52–65. [CrossRef]
16. Zou, Q.; Liu, H.; Jiang, Z.; Wu, X. Gas flow laws in coal subjected to hydraulic slotting and a prediction model for its permeabilit enhancing effect. *Energy Sources Part A* **2021**, *7*, 1–15. [CrossRef]
17. Lin, B.; Yan, F.; Zhu, C.; Zhou, Y.; Zou, Q.; Guo, C.; Liu, T. Cross-borehole hydraulic slotting technique for preventing an controlling coal and gas outbursts during coal roadway excavation. *J. Nat. Gas Sci. Eng.* **2015**, *26*, 518–525. [CrossRef]
18. Fan, C.; Li, S.; Luo, M.; Yang, Z.; Lan, T. Numerical simulation of hydraulic fracturing in coal seam for enhancing undergroun gas drainage. *Energy Explor. Exploit.* **2019**, *37*, 166–193. [CrossRef]
19. Wang, T.; Zhou, W.; Chen, J.; Xiao, X.; Li, Y.; Zhao, X. Simulation of hydraulic fracturing using particle flow method an application in a coal mine. *Int. J. Coal Geol.* **2014**, *121*, 1–13. [CrossRef]
20. Wang, W.; Li, X.; Lin, B.; Zhai, C. Pulsating hydraulic fracturing technology in low permeability coal seams. *IJMST* **2015**, 2 681–685. [CrossRef]
21. Yew, C.H. *Mechanics of Hydraulic Fracturing*, 1st ed.; Gulf Publishing: Houston, TX, USA, 1997; pp. 39–43.
22. Wang, L.; Xu, Y. Study of the law of gradual change of the influence of hydraulic punching under a rational coal output. *Arab. Geosci.* **2019**, *12*, 427. [CrossRef]
23. Gu, B.; Wu, Y. Research and application of hydraulic punching pressure relief antireflection mechanism in deep "Three-Sof outburst coal seam. *Shock Vib.* **2021**, *2021*, 7241538. [CrossRef]
24. Li, Y.; Jiang, T.; Guo, X. Research on gas extraction technology of directional long borehole in ultrathick coal seam. *Geofluids* **202** *2022*, 6957896. [CrossRef]
25. Wang, H.; Cheng, Y.; Yuan, L. Gas outburst disasters and the mining technology of key protective seam in coal seam group in th Huainan coalfield. *Nat. Hazards* **2013**, *67*, 763–782. [CrossRef]
26. Ni, G.; Lin, B.; Zhai, C. Impact of the geological structure on pulsating hydraulic fracturing. *Arab. J. Geosci.* **2015**, *8*, 10381–1038
27. Liu, L.; Liu, M.; Wei, J. Regional outburst-prevention technique by gas pre-drainage based on large diameter boreholes along co seams under deep mining. *Procedia Eng.* **2011**, *26*, 623–629.
28. Wang, T.; Hu, W.; Elsworth, D.; Zhou, W.; Zhou, W.; Zhao, X.; Zhao, L. The effect of natural fractures on hydraulic fracturin propagation in coal seams. *J. Petrol. Sci. Eng.* **2017**, *150*, 180–190. [CrossRef]
29. Ma, Y.; Mao, X.; Yang, K.; Liu, J.; Zhao, A. Improvement on gas drainage of soft gassy coal seam with underground hydrauli flushing and fracturing: A case study in Huainan. *Arab. J. Geosci.* **2020**, *13*, 178. [CrossRef]
30. Zhang, R.; Hao, C. Research on the development of hydraulic flushing caverning technology and equipment for gas extraction i soft and low permeability tectonic coal seams in China. *ACS Omega* **2022**, *7*, 21615–21623. [CrossRef]

. Lu, W.; Huang, B.; Zhao, X. A review of recent research and development of the effect of hydraulic fracturing on gas adsorption and desorption in coal seams. *Adsorpt. Sci. Technol.* **2019**, *37*, 509–529. [CrossRef]
. Zhou, B.; Xu, J.; Peng, S.; Yan, F.; Yang, W.; Cheng, L.; Ni, G. Influence of geo-stress on dynamic response characteristics of coal and gas outburst. *Rock Mech. Rock Eng.* **2020**, *53*, 4819–4837. [CrossRef]
. Zhang, D.; Yang, T. Environmental impacts of hydraulic fracturing in shale gas development in the United States. *Petrol. Explor. Dev.* **2015**, *42*, 876–883. [CrossRef]
. Xie, J.; Xu, J. Effect of key stratum on the mining abutment pressure of a coal seam. *Geosci. J.* **2017**, *21*, 267–276. [CrossRef]
. Li, M.; Zhang, J.X.; Huang, Y.; Gao, R. Measurement and numerical analysis of influence of key stratum breakage on mine pressure in top-coal caving face with super great mining height. *J. Cent. South Univ.* **2017**, *24*, 1881–1888. [CrossRef]
. Li, F.; Liu, H.; Wang, C.; Sun, R.; Xiang, G.; Ren, B.; Wang, G. Stress Relief and Permeability Enhancement with Hydraulic Fracturing in Overlying Key Strata of Deep and Soft Coal Seams. *ACS Omega* **2023**, *8*, 12183–12193. [CrossRef]
. Wei, P.; Huang, C.; Li, X.; Peng, S.; Lu, Y. Numerical simulation of boreholes for gas extraction and effective range of gas extraction in soft coal seams. *Energy Sci. Eng.* **2019**, *7*, 1632–1648. [CrossRef]
. Xia, T.; Gao, F.; Kang, J.; Wang, X. A fully coupling coal–gas model associated with inertia and slip effects for CBM migration. *Environ. Earth Sci.* **2016**, *75*, 582. [CrossRef]
. Hu, G.; Huang, X.; Xu, J.; Qin, W.; Wang, H. A co-extraction technology of coal and CBM based on the law of gas advanced relieving pressure of in-seam coalface. *Disaster Adv.* **2012**, *5*, 1351–1356.

applied sciences

MDPI

Article

Numerical Study on the Dynamic Response of Gas Explosion in Uneven Coal Mine Tunnels Using CESE Reaction Dynamics Model

Jiaqi Du [1], Jian Chen [1,*], Lingqi Zhu [1], Liwen Guo [1], Fusheng Wang [1] and Xiangming Hu [1,2]

[1] School of Emergency Management and Safety Engineering, North China University of Science and Technology, Tangshan 063210, China; dujiaqi@stu.ncst.edu.cn (J.D.); tszhulq1972@126.com (L.Z.); guoliwen64@163.com (L.G.); fswang3418@163.com (F.W.); xiangming0727@163.com (X.H.)

[2] College of Mining and Safety Engineering, Shandong University of Science and Technology, Qingdao 266590, China

* Correspondence: chenjian@ncst.edu.cn

Featured Application: (1) A gas explosion simulation combining the chemical reaction mechanism and tunnel structure is described; (2) the shock wave flow fields in tunnels with uneven walls and smooth walls are compared; and (3) the dynamic responses in tunnels with uneven walls and smooth walls are analyzed.

Abstract: A numerical simulation method combining the detailed chemical reaction mechanism of methane deflagration with an approximate real tunnel structure was proposed to confirm whether the unevenness of the tunnel wall during a coal mine gas explosion can be ignored. The approximate real tunnel model and smooth wall tunnel model were developed using 3D modeling methods. The propagation and attenuation processes of shock waves in the two tunnel models, as well as the different dynamic responses of the two tunnel walls, were compared and analyzed. Research results show that the non-uniformity of the tunnel wall decreases the shock wave overpressure and propagation velocity. The peak overpressure reduction value of the shock wave reaches 81.91 kPa, and the shock wave overpressure reaches its peak at an extended maximum time of 7.4 ms. The stress distribution on the approximate real tunnel wall is discontinuous, the propagation speed of stress waves in the bend tunnel is slower, and the duration of high load is relatively low. The displacement of the approximate real tunnel after gas explosion is lower than that of tunnels with smooth walls, and the displacement of most measuring points on the tunnel on the right is only 1/3–1/2 that of the smooth tunnel.

Keywords: CESE solver; methane deflagration; shock wave overpressure; tunnel with uneven walls; structural response

1. Introduction

Gas dynamic accidents in deep mines seriously threaten the safety of underground workers and are a key concern of China's coal industry [1–3], and among them, gas explosion is the most serious dynamic disaster in coal mines. Gas explosion is the most serious dynamic disaster in coal mines. Gas explosion accidents often lead to a large number of casualties. Among the 556 serious coal mine accidents from 2000 to 2022, 285 gas explosion accidents occurred, resulting in 6378 deaths [4]. At the same time, the high-intensity shock wave generated by gas explosion damages the tunnel walls and internal facilities, posing a serious threat to the safety production of coal mines in China.

In recent years, extensive research has been conducted on gas explosions in different scenarios [5–7]. Compared with gas explosion experiments, numerical simulation tools not only support the rapid modeling of gas explosions in various enclosed or semi-enclosed spaces, but, at the same time, can also accurately capture the shock wave flow field at every moment. Among them, computational fluid dynamics (CFD) is widely used to accurately simulate the evolution of a shock wave flow field in a confined space following a methane/air premixed gas deflagration. Zhu et al. [8] used a flame accelerator simulator (FLACS) to establish a large-scale tunnel model. The effects of methane volume concentration, blockage rate, tunnel length, and cross-sectional area on gas explosion overpressure were studied. Zhang et al. [9] studied the variation in shock wave overpressure in the main tunnel under bifurcated tunnels with different lengths and angles using the CFD program. Gao et al. [10] established a numerical model of gas explosion using OpenFOAM open-source CFD code based on the spatial structure characteristics of goaf. The flame and shock wave propagation characteristics induced by the gas explosion were analyzed under different arrangements for gradient diameter rocks. Liu et al. [11] explored the influence of cavity structure on explosion wave propagation through theoretical analysis, experimental research, and numerical simulation. A numerical model of shock wave propagation in a cavity was established to study the propagation process of explosion shock waves in cavities of different sizes. These studies have revealed the propagation and attenuation laws of shock waves in different gas explosion scenarios. However, the structural displacement and deformation under the impact load of gas explosion cause energy loss, and changes in the structural plane also alter the direction of shock wave reflection. For situations with structural deformation, analyzing only the explosion flow field lacks accuracy [12,13].

The Arbitrary Lagrangian–Euler (ALE) algorithm can accurately and efficiently solve the problem of the large deformation of structures caused by explosion [14–16]. Among its types, empirical methods, such as the TNT equivalent method [17] and equivalent gas explosion pressure method [18,19], are often used as the explosion sources of gas explosion. Sochet et al. [20] simplified the gas explosion load to an equivalent TNT explosive to simulate the scene of accidental explosions in industrial sites. The propagation and interaction of shock waves in complex environments, such as in gas storage, were studied. Jiang et al. [21] used improved material constitutive and state equations to characterize gas explosion and reproduced the June 13 gas explosion accident in Shiyan, Hubei Province. The propagation process of shock wave inside buildings and the evolution process of building damage related to the accident were studied. However, the gas detonation model based on the C-J detonation theory ignores the complex intermediate reaction of gas deflagration [22]. In the case of a large combustible gas cloud volume, the accuracy cannot be guaranteed [23]. Currently, studies on the complete coupling process between chemical reaction of methane deflagration and structure are limited.

In different gas explosion scenarios, due to the complex spatial structure of an underground coal tunnel and its harsh working environment, combustible gas is easy to accumulate, and the risk of coal mine gas explosion is significant. The shock wave generated by the rapid expansion of combustible gas in the semi-closed space of a coal mining face not only harms human health and damages valuable equipment such as shearers, but also cause the severe vibration of the coal tunnel walls in a short period of time, seriously affecting the emergency rescue and production recovery after the accident. Zhu et al. [24] used a CFD program modeled to simulate gas explosions in tunnel working faces, coal mining working faces, and transverse tunnels to study the characteristics of methane–air explosions in large tunnels with different structures. Pang et al. [25] conducted a numerical study on the methane/air explosion process in straight tunnels with different support spacings. Gao et al. [26] simulated the gas explosion process in tunnels containing pipeline systems or lines and explored the influence of flexible obstacles in tunnels on the propagation law of gas explosions. Zhang et al. [27] established a model of a coal mine refuge chamber to analyze its structural safety after a gas explosion. These researchers simulated the underground tunnel gas explosion process from the perspective of the tunnel structure and obstacles. However, the tunnel walls are not smooth in the actual gas explosion scene, and the entire coal tunnel wall has evident bulges and depressions. The uneven tunnel walls lead to a relatively complex propagation of gas explosion shock waves. The unevenness of the real coal mine tunnel walls leads to a significant increase in the analysis difficulty and calculation cost. Therefore, reports on this aspect are limited.

The coupling analysis of the methane-deflagration chemical reaction and structural deformation is realized, in this paper, to address these issues. Meanwhile, whether the dents and protrusions on the tunnel wall can be ignored is confirmed. The immersion boundary layer algorithm (IBM) is used to combine the tunnel structure with a finite-rate model of the CESE solver to simulate gas explosions in the approximate real tunnel model and smooth wall tunnel model. The formation and dissipation process of reactants and reaction products in the tunnel after explosion is analyzed. Subsequently, the characteristics of shock wave propagation and attenuation, as well as the dynamic response of the tunnel wall, are discussed in detail for approximate real tunnels and tunnels with smooth walls.

2. Numerical Simulation Algorithm and Constitutive Model

2.1. CESE-IBM Coupling Model

In this study, a high-precision, multidimensional compressible-flow CESE solver based on the spatiotemporal conservation element/solution element method was used [28]. Compared with traditional methods, this solver can handle space and time uniformly, achieving second-order accuracy in space and time. The flow field and its spatial derivatives are simultaneously considered variables for the solution, and high-speed shock waves generated by gas explosions are accurately captured.

During the coupling solution of chemical models, CESE flow fields, and FEM structure, information exchange is carried out between the chemical solver and CESE solver, which transfer fluid density, energy, temperature, and pressure information to each other, as shown in Figure 1. The structural mesh and fluid should be solved independently. The CESE solver uses a Eulerian mesh, and the structural finite-element solver uses a Lagrangian mesh. The interface between the fluid and the structure is determined by the Lagrange grid element. In each time step, the interface serves as the carrier of information transmission between the CESE solver and the structure solver. The Eulerian grid applies fluid pressure to the structure interface whereas the structure interface feeds back the displacement and velocity information of the structure to the fluid [29–31]. The moving mesh method (MMM) and immersed boundary method (IBM) are introduced by the CESE solver for fluid structure coupling. The MMM has high calculation accuracy and is suitable for small deformation. However, at each time step, grid information should be updated; thus, the approach is time-consuming. In the IBM, the fluid grid is fixed, and the structural FEM grid moves in the fluid region. This method is suitable for large deformation, and the calculation is relatively stable [32]. The IBM method was adopted considering the stability and calculation cost.

Figure 1. Coupling mechanism between the chemical solver of CESE and structure.

2.2. Finite-Rate Chemical Reaction Model

A finite-rate chemical model is used, and the detailed chemical reaction mechanism of methane deflagration is considered to accurately simulate the shock wave of methane explosion.

The following equation is used for the system involving n_s species and n_r basic reactions [33]:

$$\sum_{k=1}^{n_s} v'_{kl} X_k \Leftrightarrow \sum_{k=1}^{n_s} v''_{kl} X_k \quad l = 1,2,3\ldots,n_r \tag{1}$$

Here, v'_{kl} and v''_{kl} are the forward and reverse stoichiometric coefficients of species k in the reaction l, respectively, and X_k is the chemical symbol of this species k. The net molar formation rate w_k of species k can be obtained from the following Equation (2):

$$\dot{w}_k = \sum_{k=1}^{n_s} (v''_{kl} - v'_{kl}) \left(k_{fl} \prod_{k=1}^{n_s} [X_k] - k_{bl} \prod_{k=1}^{n_s} [X_k] \right) \tag{2}$$

Here, $[X_k]$ is the molar concentration of species k, and k_{fl} and k_{bl} are the forward and reverse reaction rate coefficients, respectively; they are calculated from Equations (3) and (4), as follows:

$$k_{fl} = A_l T^{\beta l} \exp\left(-\frac{E_{al}}{RT} \right) \tag{3}$$

$$k_{bl} = \frac{k_{fl}}{k_{cl}} \tag{4}$$

In the above equation, A_t stands for the pre-index factor, β_l is the temperature index, E_{al} is the activation energy per unit mole, R is the general gas constant, and T is the temperature. K_{cl} is the equilibrium constant, which is obtained from the following Equation (5):

$$K_{cl} = \left(\frac{P_{atm}}{RT} \right)^{\sum_{k=1}^{n_s} v_{kl}} \exp\left(\sum_{k=1}^{n_s} v_{kl} \left(\frac{S^0{}_k}{R} - \frac{H^0{}_k}{RT} \right) \right) \tag{5}$$

$S^0{}_k$ and $H^0{}_k$ in Equation (5) represent the entropy and enthalpy in the standard state, respectively, and the EOS of gas adopts the ideal gas state equation:

$$P = \sum_{k=1}^{n_s} [X_k] RT \tag{6}$$

The thermal conductivity, viscosity, and diffusion coefficients of species k required by the Navier–Stokes solver can be obtained through the transfer data files in the reference literature [34]. The calculation cost is expensive, considering that the current detailed kinetic mechanism of methane combustion, GRI-Mech 3.0 [35], contains 53 species and 325 elementary reactions. Therefore, the simplified 18-step and 53-step methane combustion models based on GRI-Mech 3.0 were adopted [36], and these models will be verified in subsequent studies to select the best methane combustion reaction mechanism suitable for simulation analysis. Among them, the 18 step basic reaction mechanism extracted the skeleton reaction of complex methane combustion mechanism, which has high prediction accuracy for methane combustion, as follows [36–38]:

$$H + O_2 <=> O + OH \tag{7}$$

$$O + H_2 <=> H + OH \tag{8}$$

$$H + OH + M <=> H_2O + M \tag{9}$$

$$OH + H_2 <=> H_2O + H \tag{10}$$

$$2OH <=> O + H_2O \tag{11}$$

$$CHO + M <=> CO + H + M \tag{12}$$

$$CH_2O + H <=> CHO + H_2 \tag{13}$$

$$CH_2O + OH <=> CHO + H_2O \tag{14}$$

$$CH_3 + O <=> CH_2O + H \tag{15}$$

$$CH_3 + O_2 <=> CH_3O + O \tag{16}$$

$$CHO + H <=> CO + H_2 \tag{17}$$

$$CHO + OH <=> CO + H_2O \tag{18}$$

$$H + CH_3(+M) <=> CH_4(+M) \quad (19)$$

$$CH_4 + H <=> CH_3 + H_2 \quad (20)$$

$$CH_4 + OH <=> CH_3 + H_2O \quad (21)$$

$$CH_4 + O <=> CH_3 + OH \quad (22)$$

$$H + CH_2O(+M) <=> CH_3O(+M) \quad (23)$$

$$CO + OH <=> CO_2 + H \quad (24)$$

In the above equations, M represents catalyst, derived from mixed gas participating in the reaction or N_2.

2.3. Material Failure Criterion

When studying the full coupling between a gas explosion and structure, the MAT_PLASTIC_KINEMATIC material model can effectively describe the strain rate effect and element failure behavior when the explosion wave acts on the structure. The materials used for the surface of the coal tunnel and the steel pipe are coal materials and Q345 steel, respectively [39,40]. The material parameters include static mechanical parameters and dynamic mechanical parameters, which need to be obtained through uniaxial load experiments and dynamic impact experiments, respectively. The relationship between yield stress and plastic strain and strain rate in the material model is shown in Formula (25), as follows:

$$\sigma_y = \left(\frac{\dot{\varepsilon}}{C}\right)^{\frac{1}{P}} \left(\sigma_0 + \beta E_p \varepsilon^{eff}{}_p\right) + \left(\sigma_0 + \beta E_p \varepsilon^{eff}{}_p\right) \quad (25)$$

Here, σ_y is the yield stress; σ_0 is the initial yield stress; ε is the strain rate; ε^{eff} is the effective plastic strain; β and E_p are hardening parameters and the hardening modulus, respectively; and C and P are the strain rate parameters. The material parameters of the tunnel wall and steel pipe are shown in Table 1.

Table 1. Material parameters of coal and steel.

Material	ρ/kg·m^{-3}	E/GPa	μ	σ_y/MPa	E_{tan}/GPa	C/s^{-1}	P	B	F_s
Tunnel surface	1860	2.61	0.30	12.3	0.25	-	-	0.5	0.8
Steel pipe	7850	200	0.33	473	75.19	6844	3.91	1	0.25

2.4. Chemical Model Validation

Qu et al. [41] conducted a 9.5% concentration 100 m^3 volume gas explosion experiment in a large square tunnel with a cross-sectional area of 7.2 m^2 and a length of 900 m, and the ignition source was located at the closed end of the tunnel. According to the experiment, a 3D numerical model of the same size tunnel was established to verify the accuracy of the 18-element reaction, 53 element reaction kinetic model, and simplified detonation model of methane deflagration, as shown in Figure 2.

The simulation results of the simplified explosion model that simplifies the complex methane combustion reaction mechanism into a one-step reaction differ greatly from the experimental data. The simulation results of the 18- and 53-element reaction kinetic models are very close, and after 120 m, they are close to the experimental data. After mesh refinement (with an average of 5.8 elements/m), the numerical simulation results using the 18-step reaction kinetic models in a gas explosion within 120 m of the front of the tunnel are closer to the experimental data. Therefore, when the density of grid cells is sufficient, using the 18-step methane reaction mechanism can reduce computational costs while ensuring the accuracy of simulation results.

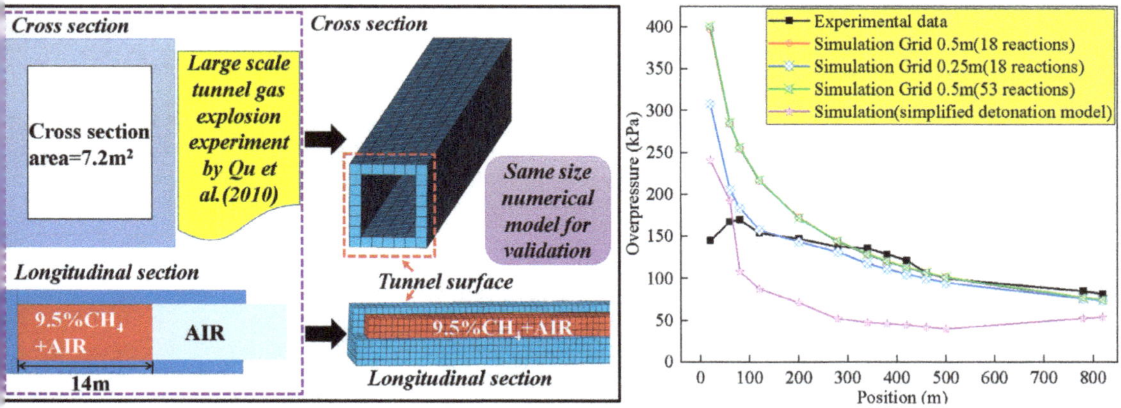

Figure 2. Verification of different gas explosion models based on experimental data.

3. Numerical Model of Gas Deflagration in Tunnel

The approximate real tunnel model obtained through 3D modeling is shown in Figure 3. The model mainly includes four parts: a T-shaped tunnel, two-section straight tunnel, and a bend.

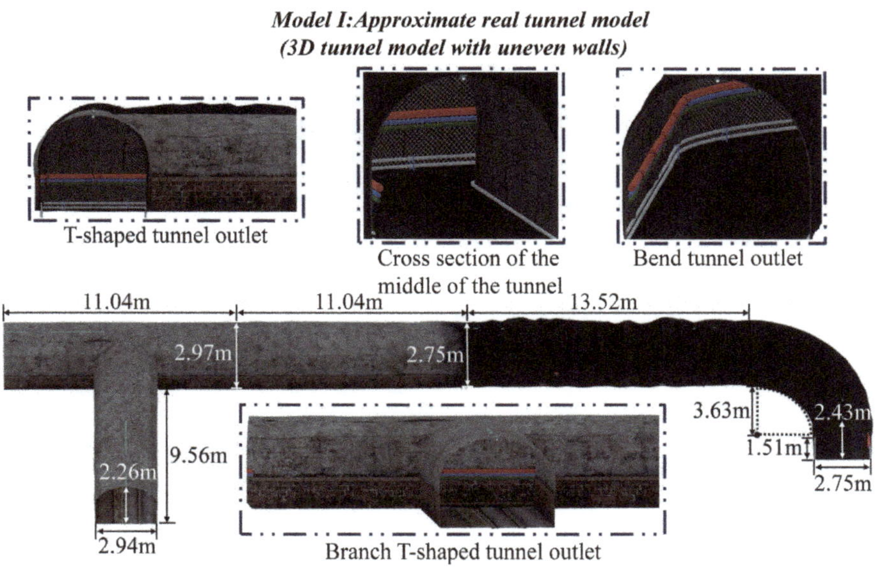

Figure 3. Model 1: approximate real tunnel model with uneven tunnel walls.

The walls of the left T-shaped tunnel and straight tunnel have little concave–convex fluctuation. By contrast, the walls of the right straight tunnel and bend are evidently concave–convex and uneven. The straight tunnel and branch tunnel of the T-type tunnel are 11.04 and 9.56 m long, respectively, the widths of the exit end are 2.97 and 2.94 m, and the height is 2.26 m. The straight tunnel connected with the T-shaped tunnel is also 11.04 m long; this tunnel is a transition section, and the rear end width is 2.75 m, which is different from the width of the front end. The length of the second straight tunnel is 13.52 m, and the connected bend includes a 1/4 arc with an inner diameter of 3.63 m and a tunnel section with a length of 1.51 m, with a total length of 9.37 m. The width and height of the tunnel are not fixed due to the evident depressions and protrusions on the right tunnel wall, and the cross-sectional area of the tunnel is between 5.8 m^2 and 5.9 m^2.

A tunnel model with smooth walls of the same size is established as a control to clarify the impact of the concave–convex fluctuation of tunnel walls on gas explosion simulation, as shown in Figure 4. The lengths of the straight tunnel, T-shaped tunnel, and bend are consistent with the approximate real tunnel model, but the tunnel walls of the model do not have concave–convex fluctuation. The height and width remain unchanged, and the cross-sectional areas of the tunnels are always 5.88 m^2.

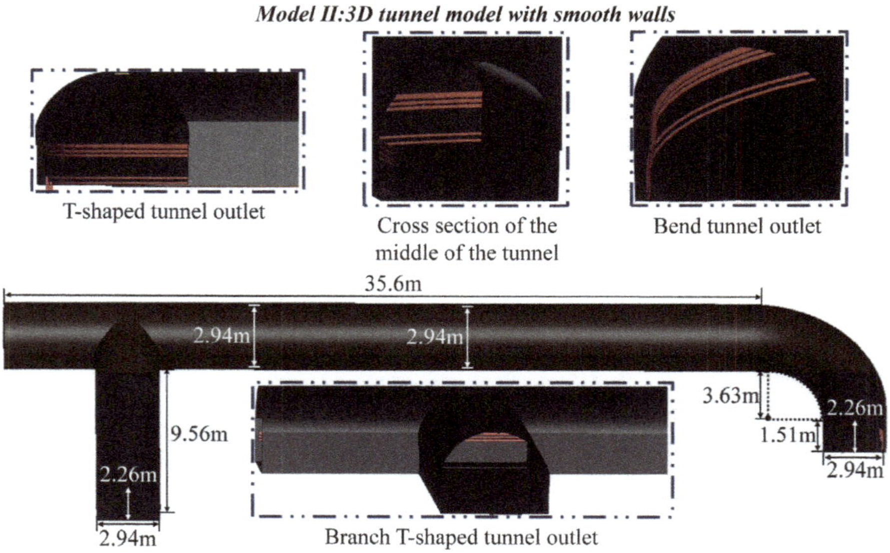

Figure 4. Model 2: tunnel model with smooth walls.

The characteristic lengths of small parts, such as the tunnel pipe support, cable, and miner's lamp, are extremely small, greatly increasing the analysis difficulty and calculation time of the numerical model. These small obstacles have a low blockage rate of gas and a limited impact on the shock wave of gas explosion [42]. Therefore, such parts are ignored, and only tunnel and large pipes are reserved. The tunnel walls and pipes are divided by a 2D quadrilateral grid, as shown in Figure 5. The size of the tunnel wall grid transits from 0.05 m in the middle of the straight tunnel to 0.075 m on both sides (19.2 elements/m in the transverse direction of the tunnel; 20.7 elements/m in the vertical direction). The CESE fluid part uses high-quality structured grids, and the computational domain covers the entire tunnel. The fluid grid is divided into two parts: a methane–air mixture and air. The grid size from the combustible-gas part in the middle to the air part transits from 0.045 m to 0.090 m (12.5–18.7 elements/m in the transverse direction of the tunnel; 15.2 elements/m in the vertical direction). The total number of grid elements is 1,451,496.

In the CESE fluid element, the position of the ignition point and the distribution of the 9.5% methane–air mixture and air are shown in Figure 6. The ignition point is located in the middle of the whole numerical model. The initial pressure and temperature of the ignition point are 5 bar and 3500 K, respectively. The ignition point is surrounded by the methane–air mixture at normal temperature and pressure, with a length of 6.17 m. The remaining component of the fluid grid is air under normal temperature and pressure conditions and the outflow boundary is set at the exit end of the tunnel.

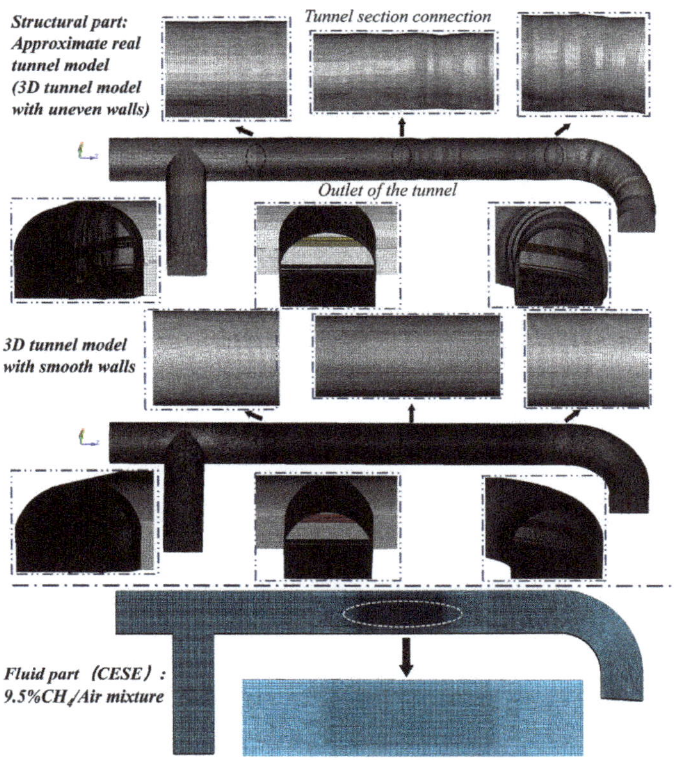

Figure 5. Mesh division of the finite-element numerical model.

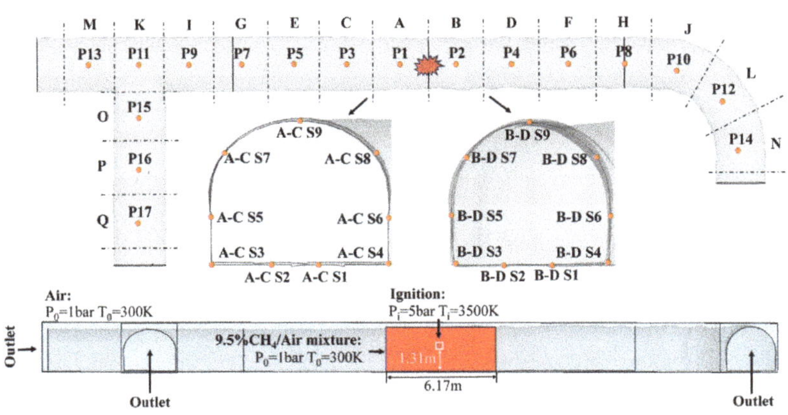

Figure 6. Numerical model boundary conditions and monitoring locations.

For the convenience of analysis, the straight tunnel, T-shaped tunnel, and bend of the approximate real tunnel model and the tunnel model with smooth walls are divided into a–q tunnel sections. The center position of each section of the tunnel is set as the overpressure and gas mass-fraction monitoring point, numbered P1–P17. In the cross-section between each tunnel section, nine locations are selected as monitoring points for stress and displacement, and these are located at the tunnel bottom (S1, S2), tunnel corner (S3, S4), tunnel wall (S5–S8), and tunnel top (S9).

4. Results and Discussion

4.1. Formation and Dissipation of Reactants and Reaction Products

The multistep elementary reaction of methane deflagration is a very short process. Figure 7I shows the mass fraction changes in the intermediate products CH_3 and CH_2O and final product CO_2 when methane reactants are consumed whereas Figure 7II shows the changes in overpressure and temperature during methane combustion.

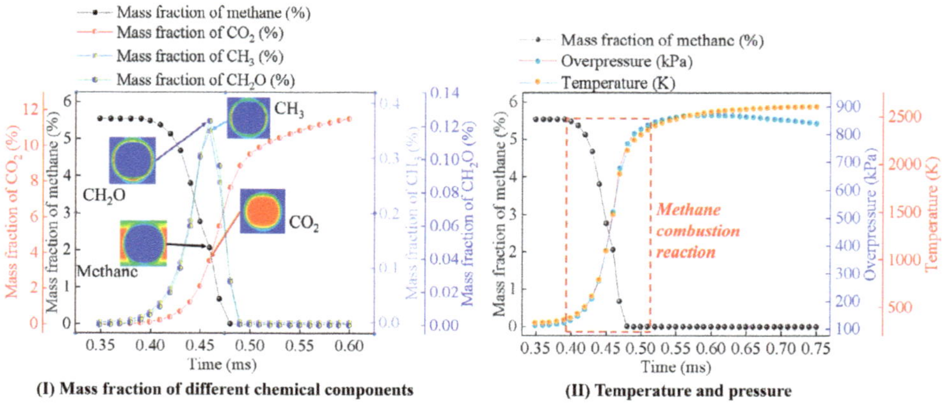

Figure 7. Methane deflagration reaction.

The initial reaction rate of methane is slow, but the chain reaction starts after generating CH_3, the product of the initial elementary reaction, and CH_2O, the oxidation product of CH_3. As a result, the methane combustion reaction rate sharply increases, and the final product CO_2 starts to be generated. At the same time, the pressure and temperature of the mixed gas sharply increase. The mass fraction of intermediate products reaches its peak during the intense consumption of methane. The components rapidly participate in the next reaction after generation because of the unstable chemical properties of intermediate products. Moreover, the peak mass fraction is significantly lower than the initial reactants and final products. At this time, the distribution of intermediate products CH_3 and CH_2O is a thin circle, representing the region of intense methane reaction. The final product CO_2 is inside the circle whereas methane that has not participated in the reaction is outside the circle. When methane is completely consumed, the intermediate products are almost simultaneously consumed. However, the mass fraction of the final product CO_2 still slowly increases, indicating the remaining CO that still continues the reaction at this time. Therefore, the pressure and temperature of the mixed gas slowly increase for a short period of time after the complete consumption of methane.

Subsequently, the formation and diffusion of harmful gases CO and CO_2 after explosion were analyzed, as shown in Figure 8I,II. At 0.5 ms, the CO mass fraction at points P1 and P2 at the centers of tunnel sections A and B reached 4.77% and 3.07%, respectively, and the CO mass fraction decreased sharply after reaching the peak. The trend of CO mass-fraction time curves at the P3 and P4 points, which are 3 m apart from the P1 and P2 points, is similar to that of the P1 and P2 points, but the peak value is less than 0.09%, and the CO mass fraction at points P5 and P6 (with an interval of 7.5–10.5 m from the ignition source) is less than 0.02%. This finding shows that the CO produced via methane deflagration is consumed in a large amount before it diffuses to the distance. The CO_2 mass fraction at the P1 and P2 points reached 11.98% and 12.11%, respectively. After reaching the peak value, the CO_2 gas began to diffuse slowly to the distance. At 35 ms, the CO_2 gas mass fraction at P1 and P2 decreased to 8.49% and 5.89%, respectively. At the same time, the CO_2 mass fraction at the P3–P8 points in the further distance (4.5–10.5 m from ignition point) increased to 1.29–5.54%.

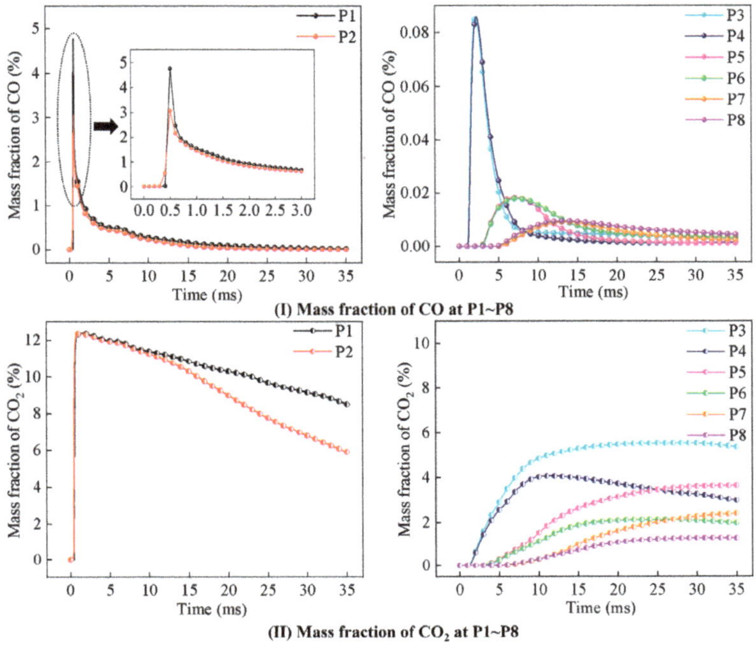

Figure 8. CO and CO_2 concentration time-variation curves at different positions after methane explosion.

4.2. *Propagation Characteristics of Shock Wave in Approximate Real Tunnel Model and Tunnel Model with Smooth Walls*

In real coal mine gas explosion scenarios, the unevenness of the tunnel walls and floor caused by coal mining has a significant im-pact on the propagation of shock waves in the tunnel. Therefore, we compared and analyzed the propagation and attenuation processes of gas explosion shock waves in ap-proximately real tunnel models with uneven walls and floors, as well as in smooth tunnel models with smooth walls and floors, as shown in Figure 9.

During the time period from 1.5 ms to 10 ms, the explosion shock wave propagates in the straight tunnels on both sides simultaneously. At 1.5 ms, the initial shock wave is superimposed with the reflection wave from the tunnel wall, and the overpressure reaches the maximum value of 884.5 kPa. The uneven wall and floor changes the incident angle of the shock wave and the propagation direction of the reflected wave. Subsequently, the superposition of the shock wave changes. Therefore, the overpressure distribution in an approximate real tunnel is asymmetric. At 5–10 ms, the complete consumption of methane reactants causes the shock wave to lose its energy replenishment, thereby reducing the shock wave overpressure. The orange–red high-pressure area in the approximate real tunnel inclines to the right straight tunnel with severe concave and convex fluctuations. The overpressure distribution of the fluid in the tunnel with smooth walls maintains symmetry. The propagation distance of the shock wave in two different tunnel models during this time period is the same.

After 15 ms, the shock wave propagating to the left has reached the intersection of the straight tunnel and the branch tunnel whereas the shock wave on the right has entered the bend tunnel. Within 15 ms to 20 ms, the propagation range and overpressure distribution of the shock wave in the left of the approximate real tunnel with only slight fluctuations are almost the same as those in the tunnel with smooth walls. The propagation distance of shock waves in the bend tunnel of the approximately real tunnel is lower than that in a smooth wall tunnel. At the same time, the area of the orange high-pressure zone in the bend tunnel of the approximately real tunnel gradually declines. In 35 ms, the shock waves have reached the tunnel exit and the maximum overpressure is 220.5 kPa.

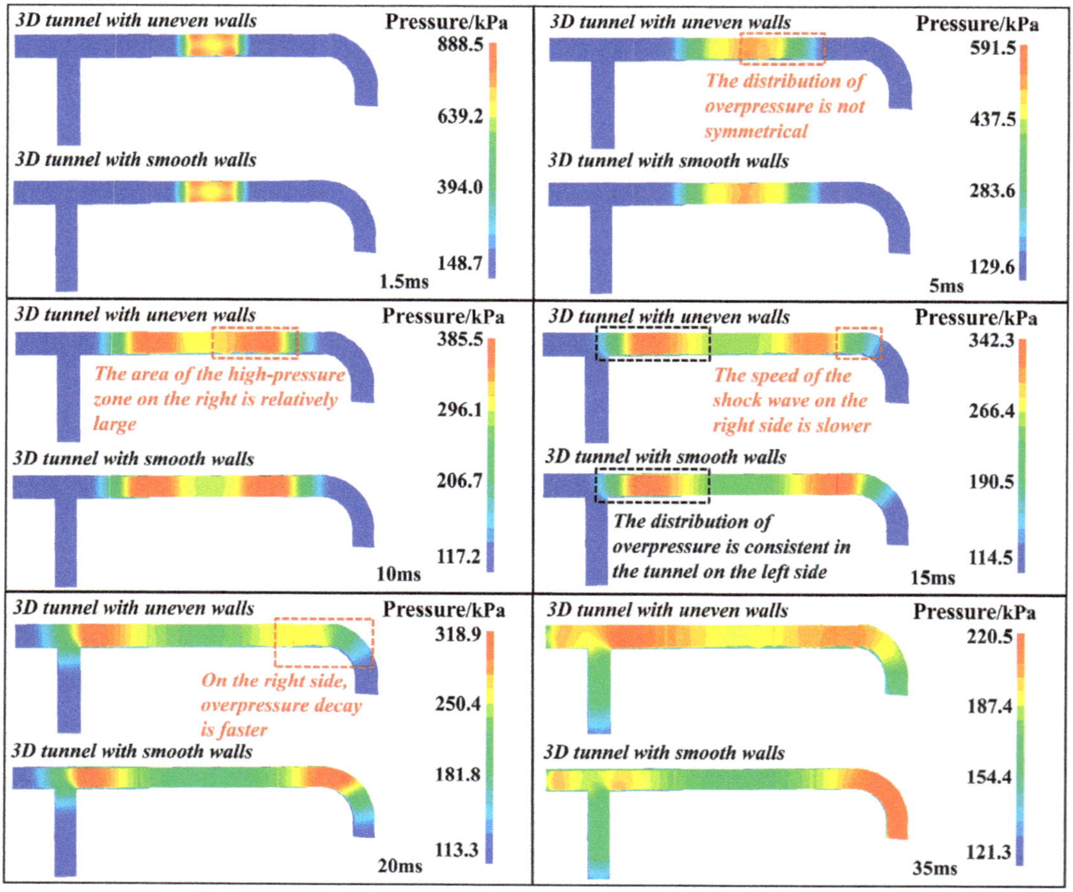

Figure 9. Propagation process of shock waves in approximate real tunnel and smooth wall tunnel.

The changes in overpressure over time in the A-Q section of the tunnel were monitored to further explore the reasons for the different distributions of overpressure in the two different tunnel models. At the same time, the changes in shock wave velocity field and vorticity in straight tunnels, bend tunnels, and T-shaped tunnels in the two models were analyzed separately, as shown in Figure 10.

When the shock wave propagates in the straight tunnel, the velocity direction of the fluid near the right tunnel in the approximate real tunnel changes significantly, increasing the local resistance of the fluid and energy dissipation. Ultimately, it leads to a concentric distribution of shock wave velocity in the right straight channel. After 11 ms, compared with the left straight lane, the shock wave velocity and overpressure attenuation amplitude increase. As the distance between the overpressure monitoring points and the center of the tunnel increases, the difference between the overpressure time curves of P2, P4, P6, and P8 on the right side of the approximate real tunnel and those of P1, P3, P5, and P7 on the left side gradually increases. The peak overpressure at points P2, P4, P6, and P8 is low, and considerable time is required to reach the peak overpressure. For the smooth wall tunnel model, the change in the velocity direction of the fluid near the wall is unremarkable, the local resistance inside the fluid is small, and the shock wave velocity is distributed in a layered manner. At 11 ms, the velocity and overpressure attenuation amplitude of the shock wave in the straight tunnels on both sides are relatively small. The overpressure time curves of monitoring points P1, P3, P5, P7 and P2, P4, and P6 and P8 are close to overlapping.

When the shock wave enters the bend, the direction of fluid velocity deviates significantly. The approximate real tunnel has higher vorticity near the wall and at the turning of the tunnel. In the smooth wall tunnel, only the fluid at the turning point of the tunnel has high vorticity. The propagation speed of shock waves in the approximate real tunnel is only half of that in the smooth wall tunnel. The differences in the peak overpressure and peak overpressure arrival time of the overpressure time curves of monitoring points P10, P12, and P14 in the two models increase. The peak overpressure differences were 73.78, 81.91, and 81.38 kPa, respectively and the peak overpressure arrival time differences were 3.4, 4.1, and 7.4 ms, respectively.

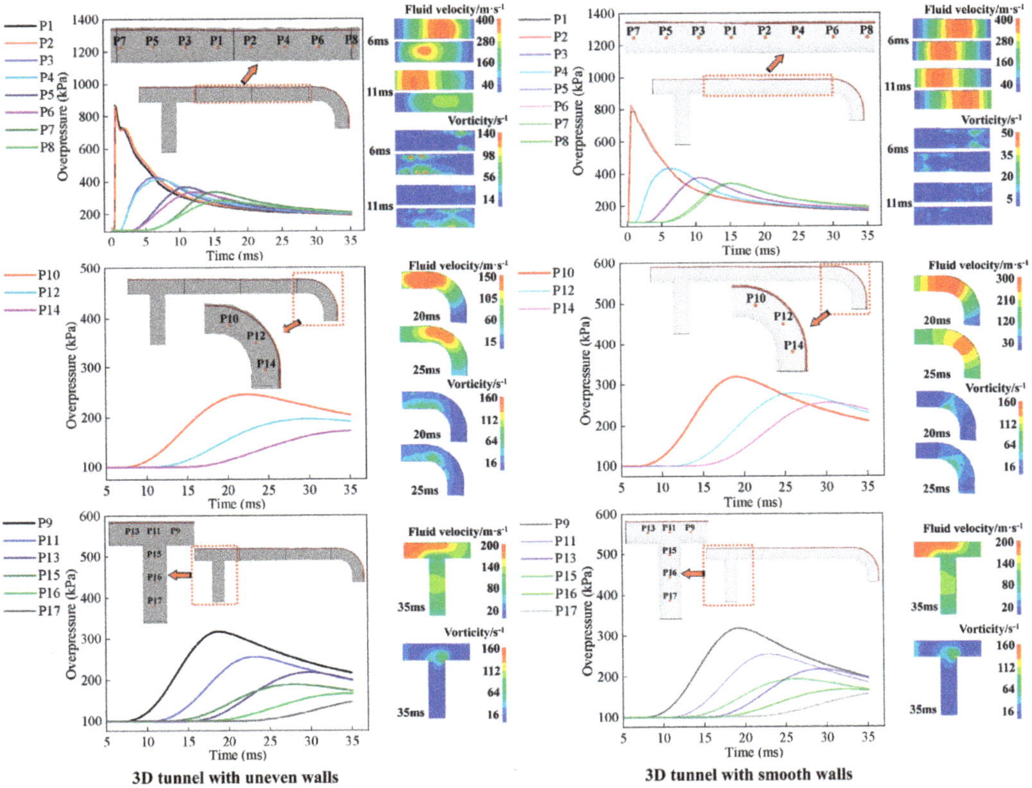

Figure 10. Overpressure time curve of monitoring points P1–P17 in approximate real tunnel and smooth wall tunnel.

When the shock wave enters the T-shaped tunnel on the left, the presence of a branch causes the velocity direction of the shock wave to deviate, and at the intersection of the tunnel, the vorticity of the fluid is relatively high. Among them, the shock wave velocity in the main tunnel is significantly higher than that in the branch tunnel. The flow velocity and vorticity cloud maps within the T-shaped tunnel in both models are similar. The overpressure peaks at the monitoring points located in the T-shaped tunnel in both models are similar. These findings indicate that when the unevenness of the tunnel walls and floors is low, the tunnel can be regarded as smooth.

4.3. Stress Response of Approximate Real Tunnel Model and Tunnel Model with Smooth Walls

The results of the previous analysis show that the uneven walls of the tunnel have a certain impact on the propagation and overpressure distribution of the shock wave, leading to the difference in load between the approximate real tunnel and the tunnel with smooth walls.

Figure 11 shows the effective stress distribution of the outer surface and inner walls of the tunnel in different time periods for the two tunnel models. At 1.5 ms, the high-intensity explosion shock wave initially acts on the tunnel walls near the explosion gas. The stress distribution of the approximate real tunnel is the same as in the tunnel with smooth walls; the former bears a high stress. The stress in most areas ranges from 1.5 MPa to 3 MPa. The stress concentration areas are located at the left and right corners of the tunnel, with stresses ranging from 4 MPa to 5 MPa. After 5 ms, the shock wave is reflected many times between the tunnel walls, the stress on the tunnel walls increases significantly, and the maximum stress reaches more than 18 MPa. And this is already higher than the yield limit of 12.3MPa of the coal tunnel material, causing plastic deformation in the coal tunnel (when the plastic deformation reaches 0.8, the material will fail). At 10 ms, the red stress concentration area appears in the center of the tunnel bottom. For the approximate real tunnel model, the phenomenon of discontinuous stress distribution becomes prominent, especially in the walls with more evident concave–convex fluctuation on the right side. The stress distribution on the tunnel surface with smooth walls is continuous and symmetrical. At 15 ms, the stress wave has diffused to the T-shaped structure and bend part of the tunnel. The diffusion of the stress wave in the bend of the approximate real tunnel is delayed compared with that of the tunnel with smooth walls. After 20 ms, the range of the red stress concentration zone and maximum stress on the approximate real tunnel surface are lower than those in smooth-walled tunnels. At 35 ms, the stress wave has spread to the whole tunnel, the stress on the upper wall of the T-shaped tunnel and bend is low, and the stress load on the bottom of the tunnel is high. The green stress zone on the wall of the approximate real tunnel is split by the blue unloading zone and distributed in a serrated shape whereas the transition between the green stress zone on the wall of the tunnel with smooth walls and the blue unloading zone is relatively good. At this time, the maximum stress values of the two tunnels still increase, reaching 23.42 and 24.10 MPa.

Figure 12 shows the variation curve of stress with time at the S1–S9 measuring points at the sections A–C, B–D, G–I, H–J, and N in the approximate real tunnel model and the tunnel model with smooth walls. In each section in the two tunnel models, the point with the highest stress and the first to reach the peak always includes the measuring points S3 and S4 at the corner of the tunnel, followed by the measuring points S1 and S2 located at the bottom of the tunnel. However, the stress of measuring points S5 and S6 of and S7 and S8 on the tunnel wall is relatively low. For sections A–C and B–D, which are nearest to the explosion center, the stress time curve of measuring point S9 on top of the tunnel reaches the highest point the slowest, but its peak value is close to those of S1 and S2. The trend and stress range of the stress time curves at each measuring point in sections A–C, B–D, and G–I of the approximate real tunnel and the smooth wall tunnel are consistent. The curves of the symmetric measuring points S1 and S2, S5 and S6, and S7 and S8 in the smooth wall tunnel almost overlap. However, a certain gap is found in the stress time curves of these mutually symmetric measuring points in the approximate real tunnel. At the same time, the high stress-application time of each stress-monitoring point of the approximate real tunnel is lower than that of the tunnel model with smooth walls, and its curve falls faster, and this phenomenon is more evident for sections B–D and H–J of the right tunnel. In addition, the stress loading amount of sections H–J and N located in the bend in the approximate real tunnel is lower than that of sections H–J and N of the tunnel with smooth walls. These results show that the propagation ability of stress waves in the approximate real tunnel is worse than that in the smooth tunnel.

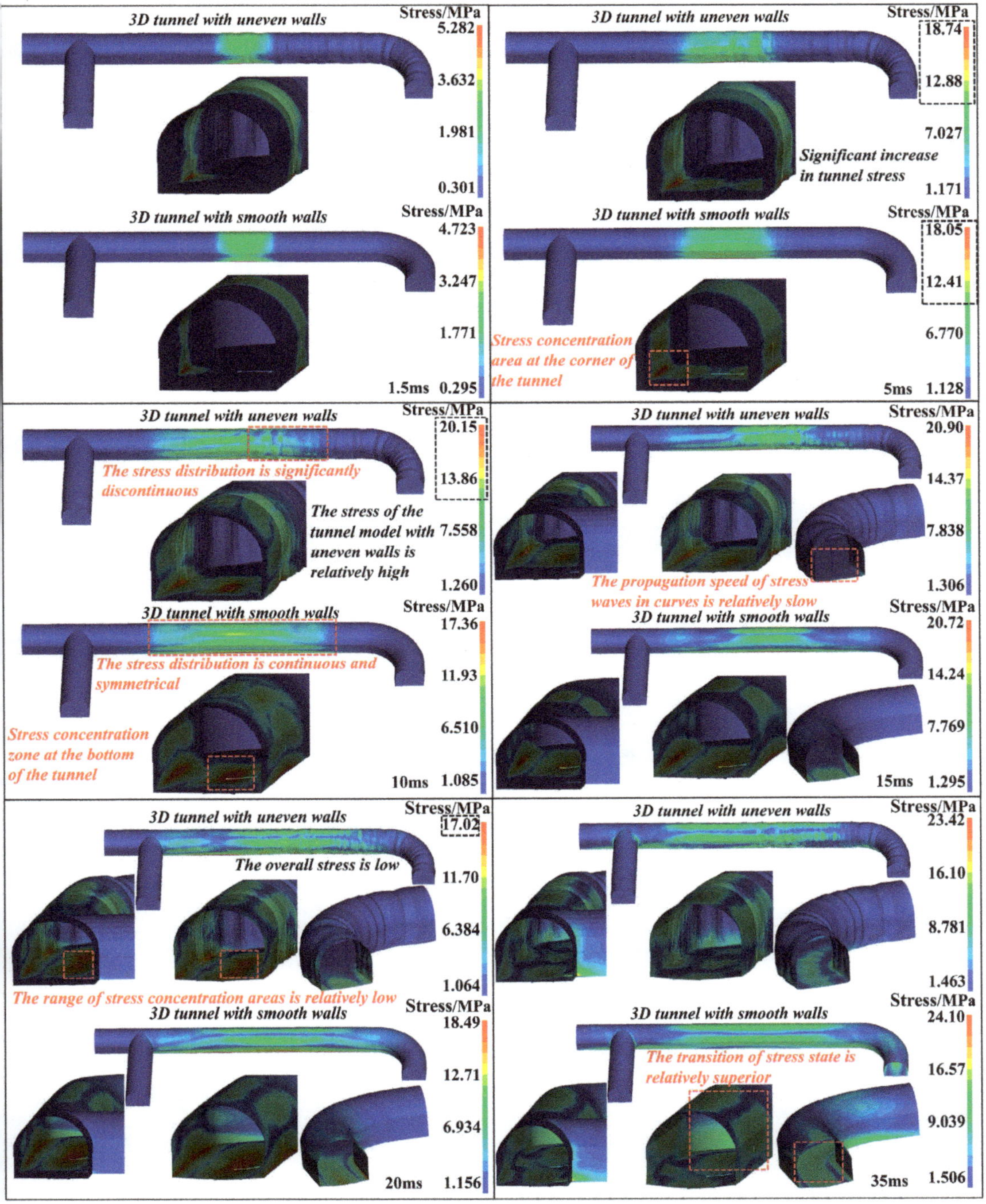

Figure 11. Stress distribution characteristics in the approximate real tunnel and smooth wall tunnel.

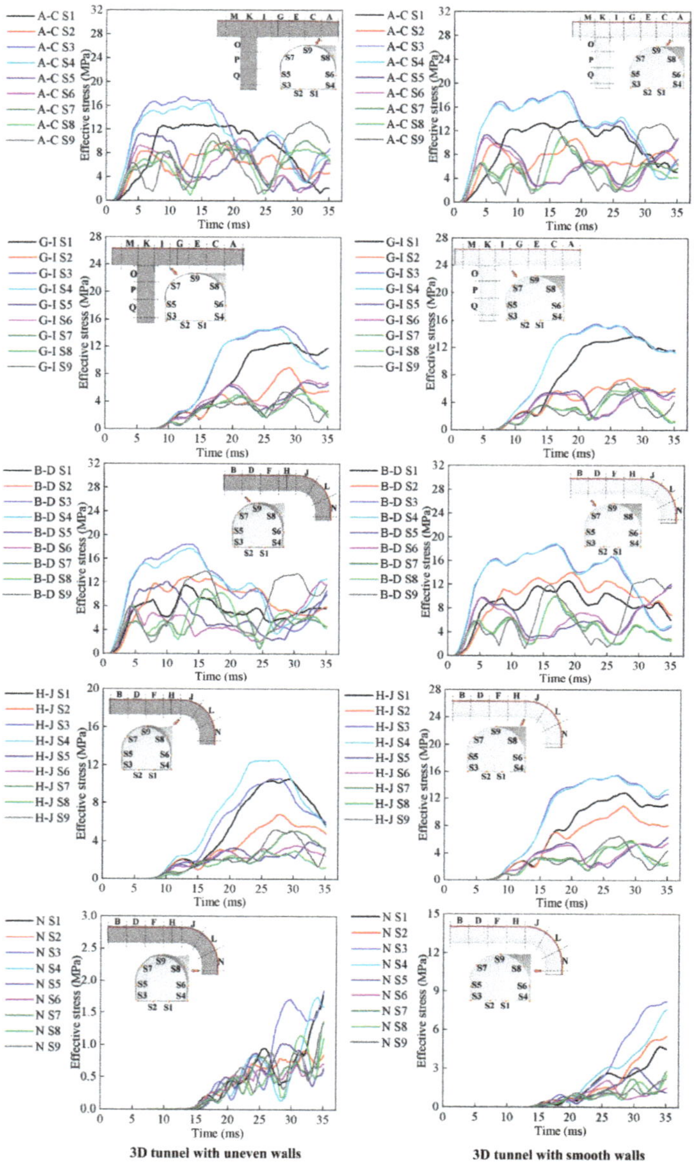

Figure 12. Stress time curves of S1–S9 monitoring points in approximate real tunnel and smooth wall tunnel.

Overall, the development and diffusion processes of stress waves are closely synchronized with the propagation of shock waves. Except for bends, the positions and ranges of the green stress zone and the red stress concentration zone in the two types of tunnels are the same, and the stress distribution of the tunnel section are similar, as shown in Figure 13. The stress concentration area is located at the corner of the tunnel, the high stress area is located on the top surface of the tunnel near the explosive gas, and the general stress area is located on the wall of the tunnel. However, compared with the tunnel with smooth walls, the stress of the approximate real tunnel wall presents the characteristics of discontinuous stress distribution and low maximum stress. Moreover, the stress unloading on the wall of the approximate real tunnel occurs earlier than that of the smooth wall tunnel, and the propagation speed of stress waves in the bend of the former is relatively slow, leading to varying degrees of deformation in the two tunnel types.

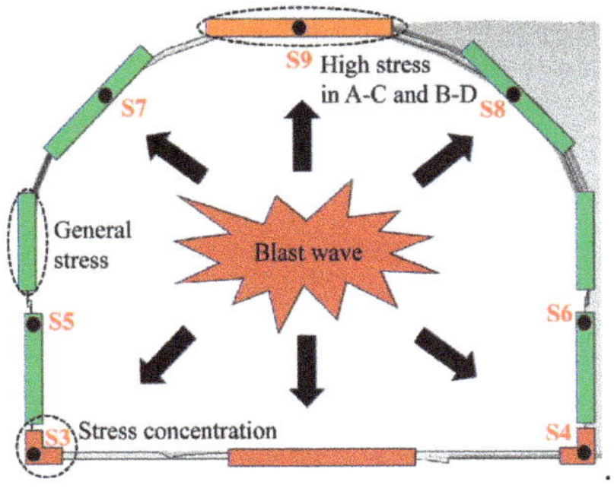

Figure 13. Stress distribution in tunnel cross-section.

4.4. Displacement Response of Approximate Real Tunnel Model and Tunnel Model with Smooth Walls

The final deformation of the approximate real tunnel model and the smooth wall tunnel model is shown in Figure 14.

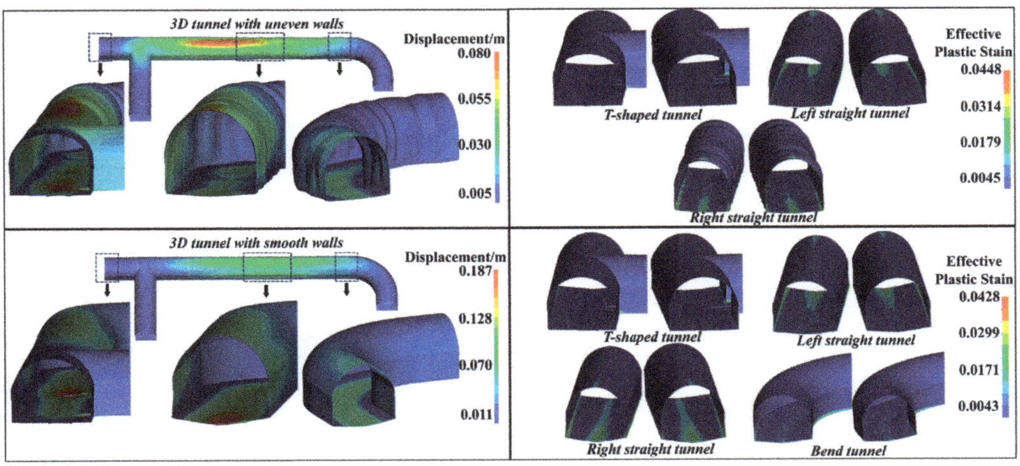

Figure 14. Final deformation situation in approximate real tunnel and smooth wall tunnel.

For the approximate real tunnel, the deformation degree of the tunnel on the right is lower than that of the tunnel on the left. When the left straight tunnel is considered the center, the deformation degree of the tunnel gradually decreases toward both sides. The top and bottom of the left straight tunnel have the most severe deformation, with a maximum deflection of 0.08 m. The bend tunnel and the branch of the T-shaped structure have the smallest deformation. The displacement distribution of the smooth wall tunnel is uniform, and the deformation degrees of the left and right tunnels are similar. The part with the highest degree of deformation is located at the bottom of the center of the tunnel, with a maximum deflection of 0.187 m, which is more than twice that of an approximate real tunnel. The minimum total displacement on the surface of the light-green deformation zone exceeds 0.07 m. The plastic deformation areas in the two types of tunnel models are mostly distributed in the middle and corner of the bottom of the straight tunnel, and small plastic damage areas are found in the top of the middle of the straight tunnel and the corner of the T-shaped tunnel. The plastic deformation area in the approximate real tunnel is smaller than that in the smooth wall tunnel, and no plastic deformation is found in its bend tunnel.

Figure 15 shows the displacement curves of the S1–S9 measuring points at the sections A–C, B–D, G–I, H–J, and N in the approximate real tunnel model (left-hand side of Figure 13) and the smooth wall tunnel model (right-hand side of Figure 13). In each section of the two tunnel models, except for the N section of the bend, the points with higher displacement are always S1 and S2 at the bottom of the tunnel and S9 at the top of the tunnel whereas S3 and S4 at the corner of the tunnel have the lowest displacement.

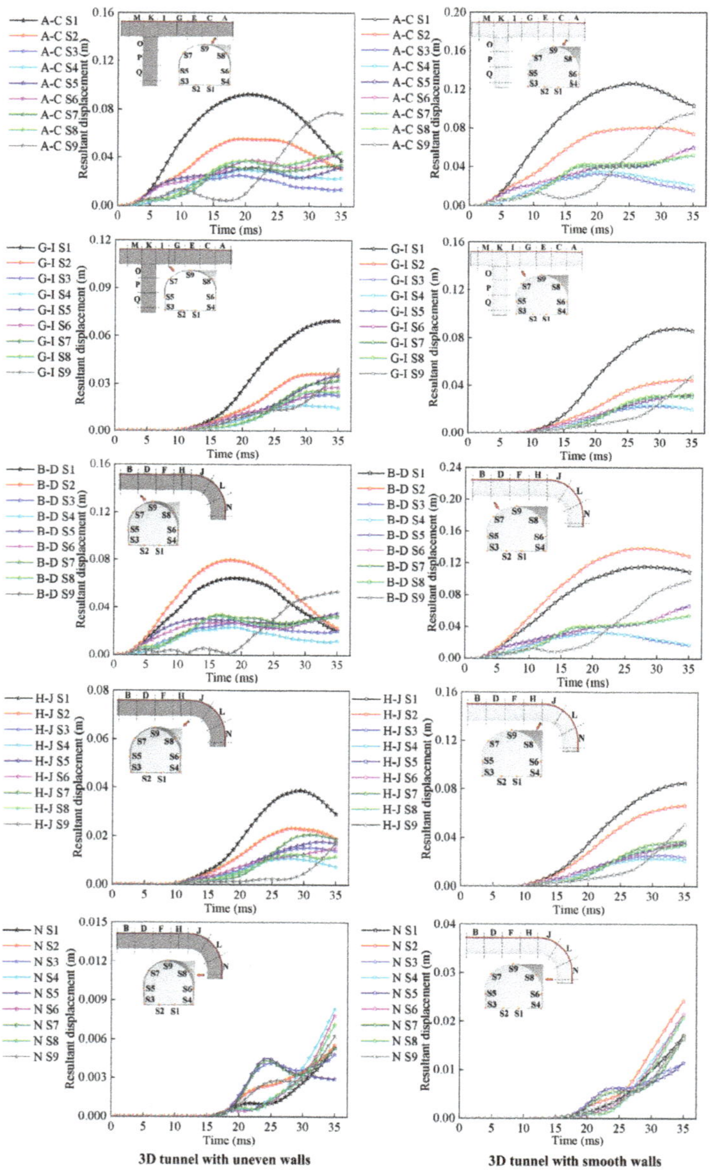

Figure 15. Displacement time curve of S1–S9 monitoring points in approximate real tunnel and smooth wall tunnel.

The results of the previous analysis show that the duration of high load on the wall of the approximate real tunnel is lower than that of the tunnel with smooth walls, resulting in the lower displacement of each monitoring point on each section of the approximate real tunnel than in the tunnel with smooth walls in general. The maximum displacement values of monitoring points A–C S1 and G–I S1 are 26.9% and 20.7% lower than in the tunnel with smooth walls, respectively whereas the displacement values of most monitoring points on the right tunnel section are less than half of those in the tunnel with smooth walls. Overall, the maximum deformation displacement of the approximate real tunnel does not exceed 0.1m, and as the shock wave energy decays, most of the deformation of the tunnel gradually recovers. Therefore, the approximate real tunnel will not lose stability as a result of gas explosion.

5. Conclusions

In this study, a high-precision CESE solver and immersion boundary layer algorithm (IBM) were used for the first time to couple tunnel structures and methane explosions. The gas deflagration process and wall dynamic response in the approximate real tunnel model and smooth wall tunnel model were studied, and the following conclusions were drawn:

(1) The highest mass fraction of intermediates CH_3 and CH_2O produced by methane deflagration is much lower than the final product CO_2, and they are almost simultaneously depleted after the complete consumption of methane. At a distance of 1.5 m from the ignition point, the maximum mass fraction of intermediate product CO reaches 4.77%, but CO gas is consumed in large quantities prior to diffusion. At a distance of 1.5–10.5 m from the ignition point, the CO_2 mass fraction reaches 1.29–12.11%.

(2) The uneven walls of a tunnel that are approximately real cause changes in the distribution of shock wave overpressure. On straight tunnels, the distribution of shock wave overpressure is asymmetric, with the red high-pressure zone leaning to the right. Meanwhile, the severely uneven tunnel walls decrease the propagation speed and overpressure of the shock wave. Compared with the tunnels with smooth walls, the peak overpressure reduction value of the shock wave reaches 81.91 kPa, and the time when the shock wave overpressure reaches its peak is extended by a maximum of 7.4 ms.

(3) The development and diffusion processes of stress loads are close to synchronization with the propagation of shock waves. The propagation speed of stress waves in the bend of the approximate real tunnel is slower than that in the smooth wall tunnel model. Except in the bend tunnel, the positions and ranges of the green stress zone and the red stress concentration zone in the two tunnel models are the same. Compared with smooth-walled tunnels, the stress distribution in the approximate real tunnel is discontinuous, and the transition zone between the green stress zone and the blue unloading zone is serrated, with a shorter duration of high loads.

(4) The approximate real tunnel will not lose stability as a result of gas explosion. The displacement of the approximate real tunnel after gas explosion is lower than that in tunnels with smooth walls, and the area of plastic deformation zone is small. The wall displacement of the left tunnel is, at most, 26.9% lower than that of the smooth tunnel. The displacement values of most monitoring points on the right tunnel section are less than half of those in the tunnel with smooth walls.

These findings prove that the depressions and protrusions on the wall surfaces of underground coal mine tunnels cannot be ignored.

Author Contributions: Conceptualization, J.D.; methodology, J.D.; software, J.D. and X.H.; validation, X.H.; investigation, F.W.; data curation, L.Z.; writing—original draft preparation, L.Z.; writing—review and editing, J.C.; visualization, F.W.; supervision, L.G. All authors have read and agreed to the published version of the manuscript.

Funding: This work was financially supported by the National Natural Science Foundation of China—General Project (52274202), Central Guidance for Local Development Funding Projects (226Z4601G), Natural Science Foundation of Hebei (E2022209051), 2023 Scholarship Program for

Introducing Overseas Students (C20230359), Higher Education Youth Top Talent Project of Hebei (BJK2023086), Youth Talent Promotion Program of North China University of Science and Technology (28438101), and Key Scientific Research Project of North China University of Science and Technology (ZD-ST-202313-23).

Institutional Review Board Statement: Not applicable.

Informed Consent Statement: Not applicable.

Data Availability Statement: Data are contained within the article.

Conflicts of Interest: The authors declare no conflicts of interest.

References

1. Wang, G.; Wang, R.; Wu, M.M.; Fan, C.; Song, X. Strength criterion effect of the translator and destabilization model of gas-bearir coal seam. *Int. J. Min. Sci. Technol.* **2019**, *29*, 327–333. [CrossRef]
2. Chen, J.; Cheng, W.; Wang, G.; Li, H.; Li, Y. New method of monitoring the transmission range of coal seam water injection ar correcting the monitoring results. *Measurement* **2021**, *177*, 109334. [CrossRef]
3. Wang, G.; Qin, X.; Zhou, J.; Han, F.; Cai, J. Simulation of coal microstructure characteristics under temperature-pressure couplir based on micro-computer tomography. *J. Nat. Gas. Sci. Eng.* **2021**, *91*, 103906. [CrossRef]
4. Wang, Y.; Fu, G.; Lyu, Q.; Li, X.; Chen, Y.; Wu, Y.; Xie, X. Modelling and analysis of unsafe acts in coal mine gas explosic accidents based on network theory. *Process Saf. Environ. Prot.* **2023**, *170*, 28–44. [CrossRef]
5. Xiao, G.Q.; Wang, S.; Mi, H.F.; Khan, F.; Yu, M.G. Analysis of obstacle shape on gas explosion characteristics. *Process Saf. Enviro Prot.* **2022**, *161*, 78–87. [CrossRef]
6. Hou, Z.H.; Wang, D.M.; Zhang, W.; Luo, S.Y.; Lu, Y.S.; Tian, S.Y.; Zhou, Q.; Xu, Z.M. Study on the influence of ignition positic on the explosion characteristics of methane-air premix in a semi-closed pipeline. *Process Saf. Environ. Prot.* **2023**, *172*, 642–65 [CrossRef]
7. Meng, Q.; Wu, C.; Li, J.; Wu, P.; Xu, S.; Wang, Z. A study of pressure characteristics of methane explosion in a 20 m buried tunn and influence on structural behaviour of concrete elements. *Eng. Fail. Anal.* **2021**, *122*, 105273. [CrossRef]
8. Zhu, Y.; Wang, D.; Shao, Z.; Zhu, X.; Xu, C.; Zhang, Y. Investigation on the overpressure of methane-air mixture gas explosions i straight large-scale tunnels. *Process Saf. Environ. Prot.* **2020**, *135*, 101–112. [CrossRef]
9. Zhang, X.B.; Shen, S.S.; Yang, M.; Wang, H.; Ren, J.Z.; Lu, F.C. Influence of length and angle of bifurcated tunnel on shock wav propagation. *J. Loss Prev. Process Ind.* **2022**, *78*, 104802. [CrossRef]
10. Gao, K.; Li, S.N.; Han, R.; Li, R.Z.; Liu, Z.M.; Qi, Z.P.; Liu, Z.Y. Study on the propagation law of gas explosion in the space base on the goaf characteristic of coal mine. *Saf. Sci.* **2020**, *127*, 104693. [CrossRef]
11. Liu, W.; Mu, C.; Li, Z. Influence of cavity structure on gas explosion characteristics in coal mine. *Powder Technol.* **2022**, *398*, 11708 [CrossRef]
12. Ma, Q.J.; Guo, Y.H.; Zhong, M.Y.; Ya, H.; You, J.F.; Chen, J.H.; Zhang, Z.K. Numerical simulation of hydrogen explosio characteristics and disaster effects of hydrogen fueling station. *Int. J. Hydrogen Energy* **2023**, *51*, 861–879. [CrossRef]
13. Du, Y.; Ma, L.; Zheng, J.; Zhang, F.; Zhang, A. Coupled simulation of explosion-driven fracture of cylindrical shell using SPH-FE method. *Int. J. Press. Vessel. Pip.* **2016**, *139–140*, 28–35. [CrossRef]
14. Valsamos, G.; Casadei, F.; Solomos, G.; Larcher, M. Risk assessment of blast events in a transport infrastructure by fluid-structu interaction analysis. *Saf. Sci.* **2019**, *118*, 887–897. [CrossRef]
15. Gu, Y.; Wang, J.; Li, H.; Tang, K.; Liu, L. Formation characteristics and penetration performance of an underwater shaped charg jet. *Ocean Eng.* **2022**, *258*, 111695. [CrossRef]
16. Jiang, X.; Xue, Y.; Kong, F.; Gong, H.; Fu, Y.; Zhang, W. Dynamic responses and damage mechanism of rock with discontinui subjected to confining stresses and blasting loads. *Int. J. Impact Eng.* **2023**, *172*, 104404. [CrossRef]
17. Guo, Y.; He, L.; Wang, D.; Liu, S. Numerical investigation of surface conduit parallel gas pipeline explosive based on the TN equivalent weight method. *J. Loss Prev. Process Ind.* **2016**, *44*, 360–368. [CrossRef]
18. Xue, Y.; Chen, G.; Zhang, Q.; Xie, M.; Ma, J. Simulation of the dynamic response of an urban utility tunnel under a natural ga explosion. *Tunn. Undergr. Space Technol.* **2020**, *108*, 103713. [CrossRef]
19. Zhu, L.Q.; Du, J.Q.; Chen, J.; Guo, L.W.; Wang, F.S. Dust raising law of gas explosion in a 3D reconstruction real tunnel: Based o ALE-DEM model. *Powder Technol.* **2024**, *434*, 119224. [CrossRef]
20. Sochet, I.; Sauvan, P.E.; Boulanger, R.; Nozeres, F. External explosion in an industrial site. *J. Loss Prev. Process Ind.* **2014**, *29*, 56–6 [CrossRef]
21. Jiang, H.; Chi, M.; Hou, D.; Ding, H.; Xie, Z.; Zeng, X. Numerical investigation and analysis of indoor gas explosion: A case stud of "6·13" major gas explosion accident in Hubei Province, China. *J. Loss Prev. Process Ind.* **2023**, *83*, 105045. [CrossRef]
22. Du, Y.; Zhou, F.; Ma, L.; Zheng, J.; Xu, C.; Chen, G. Consequence analysis of premixed flammable gas explosion occurring in pip using a coupled fluid-structure-fracture approach. *J. Loss Prev. Process Ind.* **2019**, *57*, 81–93. [CrossRef]
23. Yang, S.; Sun, W.; Fang, Q.; Yang, Y.; Xia, C.; Bao, Q. Investigation of practical load model for a natural gas explosion in a unconfined space. *J. Saf. Sci. Resil.* **2022**, *3*, 209–221. [CrossRef]

Zhu, Y.; Wang, D.; Shao, Z.; Xu, C.; Li, M.; Zhang, Y. Characteristics of methane-air explosions in large-scale tunnels with different structures. *Tunn. Undergr. Space Technol.* **2021**, *109*, 103767. [CrossRef]

Pang, L.; Zhang, Q.; Wang, T.; Lin, D.C.; Cheng, L. Influence of laneway support spacing on methane/air explosion shock wave. *Saf. Sci.* **2012**, *50*, 83–89. [CrossRef]

Gao, K.; Liu, Z.; Wu, C.; Li, J.; Liu, K.; Liu, Y.; Li, S. Effect of low gas concentration in underground return tunnels on characteristics of gas explosions. *Process Saf. Environ. Prot.* **2021**, *152*, 679–691. [CrossRef]

Zhang, B.; Zhao, W.; Wan, W.; Zhang, X. Pressure characteristics and dynamic response of coal mine refuge chamber with underground gas explosion. *J. Loss Prev. Process Ind.* **2014**, *30*, 37–46. [CrossRef]

Chang, S.-C. The method of space-time conservation element and solution element—A new approach for solving the Navier-Stokes and Euler equations. *J. Comput. Phys.* **1995**, *119*, 295–324. [CrossRef]

Rokhy, H.; Soury, H. Fluid structure interaction with a finite rate chemistry model for simulation of gaseous detonation metal-forming. *Int. J. Hydrogen Energy* **2019**, *44*, 23289–23302. [CrossRef]

Haghgoo, M.; Babaei, H.; Mostofi, T.M. 3D numerical investigation of the detonation wave propagation influence on the triangular plate deformation using finite rate chemistry model of LS-DYNA CESE method. *Int. J. Impact Eng.* **2022**, *161*, 104108. [CrossRef]

Haghgoo, M.; Babaei, H.; Mostofi, T.M. Numerical simulation of the influence of confined multi-point ignited H_2–O_2 mixture on the propagation of shock waves towards a deformable plate. *Int. J. Hydrogen Energy* **2022**, *47*, 27080–27095. [CrossRef]

Rokhy, H.; Mostofi, T.M. Tracking the explosion characteristics of the hydrogen-air mixture near a concrete barrier wall using CESE IBM FSI solver in LS-DYNA incorporating the reduced chemical kinetic model. *Int. J. Impact Eng.* **2023**, *172*, 104401. [CrossRef]

Im, K.-S.; Joh, S.-T.; Kim, C.-K.; Chang, S.-C.; Jorgenson, C. Application of the CESE method to detonation with realistic finite-rate chemistry. In Proceedings of the 40th AIAA Aerospace Sciences Meeting & Exhibit, Reno, NV, USA, 14–17 January 2002; p. 1020. [CrossRef]

Kee, R.J.; Rupley, F.M.; Miller, J.A. *Chemkin-II: A Fortran Chemical Kinetics Package for the Analysis of Gas-Phase Chemical Kinetics;* Sandia National Laboratories: Livermore, CA, USA, 1989. [CrossRef]

Miller, J.A.; Kee, R.J.; Westbrook, C.K. Chemical Kinetics and Combustion Modeling.*Annu. Rev. Phys. Chem.* **1993**, *41*, 345–387. [CrossRef]

Liu, H.; Chen, F.; Liu, H.; Zheng, Z.H.; Yang, S.H. 18-Step Reduced Mechanism for Methane/Air Premixed Supersonic Combustion. *J. Combust. Sci. Technol.* **2012**, *18*, 467–472.

Wei, Y.; Li, X.; Che, X. Study on the effect of combustion efficiency of variable thrust rocket engines. *J. Phys. Conf. Ser.* **2023**, *2551*, 012025. [CrossRef]

Lu, H.; Liu, F.; Wang, Y.; Fan, X.; Yang, J.; Liu, C.; Xu, G. Mechanism reduction and bunsen burner flame verification of methane. *Energies* **2018**, *12*, 97. [CrossRef]

Gan, L.; Zong, Z.; Lin, J.; Chen, Y.; Xia, M.; Chen, L. Influence of U-shaped stiffeners on the blast-resistance performance of steel plates. *J. Constr. Steel Res.* **2022**, *188*, 107046. [CrossRef]

Liu, F.; Silva, J.; Yang, S.; Lv, H.; Zhang, J. Influence of explosives distribution on coal fragmentation in top-coal caving mining. *Geomech. Eng.* **2019**, *18*, 111. [CrossRef]

Qu, Z. Numerical study on shock wave propagation with obstacles during methane explosion. *Appl. Mech. Mater.* **2010**, *33*, 114–118. [CrossRef]

Wang, S.; Li, Z.; Fang, Q.; Yan, H.; Liu, Y. Numerical simulation of overpressure loads generated by gas explosions in utility tunnels. *Process Saf. Environ. Prot.* **2022**, *161*, 100–117. [CrossRef]

applied sciences

MDPI

Article

Study on the Macro-Fine Mechanical Behavior of Ore Flow Based on the Discrete Element Method

Zhiguo Xia [1,2], Zhe Deng [1], Zengxiang Lu [1,2,*] and Chenglong Ma [1]

[1] School of Mining Engineering, University of Science and Technology Liaoning, Anshan 114051, China; xzgyy88@163.com (Z.X.); 18070516733@163.com (Z.D.); machenglong1231@163.com (C.M.)
[2] Green Mining Engineering Research Center of Metal Mineral Resources of Liaoning Province, Anshan 114051, China
* Correspondence: zengxiang_lu@sohu.com

Abstract: The mechanical behavior associated with the flow of ore-rock bulk materials is an important factor leading to the instability and failure of the shaft wall of the ore storage section in ore passes. It is of great significance for accurately understanding the stability failure characteristics of the shaft wall in the ore storage section in the ore-drawing process to understand the flow characteristics and internal mechanical transfer mechanism of ore-rock bulk. The flow characteristics, contact compactness, stress distribution characteristics, and contact force probability distribution of the ore-rock bulk are analyzed by the discrete element method, which realizes the quantitative characterization of the damage degree of ore-rock flow and reveals the damage mechanism of the shaft wall in the storage section of the ore pass. The results show that (1) in the process of ore-rock particle flow in the ore pass storage section, the macroscopic flow pattern of ore-rock particles changes from a "—" shape to a "V" shape, and the friction between ore-rock particles, particles, and the ore-pass wall is an important reason for the change of the macroscopic flow pattern; (2) the probability distribution of contact force strength between the particles decreases exponentially in the whole ore-drawing process, in which the strong force chains play a major role in the stability of the bulk system; and (3) the overpressure frequency and overpressure coefficient could be used to quantitatively characterize the wall damage degree under the action of ore-rock flow. The dynamic lateral pressure fluctuates periodically in exponential form and decreases, and the dynamic load formed by the ore-rock flow mainly acts on the lower part of the ore storage section.

Keywords: ore storage section of ore passes; discrete element method; ore-rock flow characteristics; lateral pressure of shaft wall; wall damage degree

1. Introduction

The wear and damage of the ore-pass wall is a common problem in mine production, which not only reduces the service life of the ore pass but also greatly increases safety hazards in mine production. The flow characteristics of ore-rock bulk materials (which consist of different-sized ore-rock blocks) and their accompanying mechanical behavior are the main causes of the deformation, instability, and collapse of the ore pass [1–3]. Therefore, it is of great significance to study the distribution law of the main force sources and the distribution characteristics of the damaged areas, to reveal the wear mechanism and to ensure the stability of the ore-pass wall.

In order to improve the service life of the ore pass in mines, many scholars have conducted in-depth research based on the theories of fluid mechanics, contact mechanics, and ore drawing. For example, Liu et al. [4] established a theoretical equation under the non-point drawing-point according to the stochastic medium draw theory, taking the draw-point as the moving boundary based on the fact of the ore-drawing process. Huang et al. [5] proposed an experimental technique for the impact damage of rock blocks based on an impact test, which is helpful in analyzing the fragmentation for impact damage of the ore-pass wall under impact load. Chi et al. [6] qualitatively analyzed the migration law of ore-rocks in the ore storage section via the two-dimensional flow network. Zhao et al. [7], with the help of Hertz contact theory and kinematics theory, conducted the quantitative characterization of the volume damage of the shaft wall. Jin et al. [8] studied the characteristics of particle percolation from the perspective of ore drawing. Classical mechanics has made great achievements in the study of ore-rock flow, but in the actual production process, the complex geological environment and harsh operating conditions of the ore pass make it difficult to directly analyze the flow characteristics and mechanical behavior of the stored materials in the ore pass, which affect the normal production of the mine. For this reason, it is urgent to seek a new method to analyze the ore-rock flow in the storage section of an ore pass.

With the rise of simulation in engineering mechanics, the discrete element method (DEM) has become the mainstream method for studying the mechanical behavior of bulk particles [9]. Many scholars have carried considerable research on discontinuous media based on the discrete element method. For example, Yang et al. [10] created storage materials in the ore pass using randomly distributed spheres and irregular polyhedral and studied the change rule of the impact force on the ore-pass wall and the materials in the ore pass storage section, as well as the change of the voidage of ore-rock. Sato and Tang [11] simulated the characteristics of the ore-rock flow in rectangular and vertical ore passes by using the 3D discrete element method and found that the square-section ore pass is preferable. Based on the summary of a series of numerical experiments using the discrete element method, Hadjigeorgiou and Lessard [12] studied the effects of the ore pass geometry, shape, and size distribution of ore-rock block on the ore-rock flow in the ore pass. According to the established SRM model, Esmaieli et al. [13] analyzed the effects of stress–structure interaction on the stability of the ore pass and the ore-rock impact on the ore-pass wall. Yuan et al. [14], based on the discrete element analysis method, found that the change in wheat-grain flow mode was the main reason for the arch formation inside the bulk and the local overpressure on the silo wall. Montellano et al. [15] analyzed the flow characteristics of glass beads and corn particles in the central unloading process from the perspectives of flow pattern, packing density, and stress distribution. By studying the changes in the force chain network, Pengkai et al. [16] analyzed the mechanism of the influence of the friction coefficient of the silo wall on the lateral pressure of the silo wall during the discharge of the rectangular hopper. Cheng et al. [17] analyzed the variation characteristics of the ore-rock block flow field and the mesoscopic-stress field in the discharge process of a shallow circular silo and explained the mechanical behavior of the ore-rock fragment on the silo wall from the perspective of the ore-rock flow. Feng et al. [18] studied the influence of ore-rock flow on the dynamic lateral pressure of the silo wall and analyzed the stress characteristics of the silo wall under the arch effect during ore drawing. Pacheco-Martinez et al. [19] studied the flow characteristics of the ore-rock particles inside the silos under different vibration frequencies and believed that the tangential motion of the silo wall effectively contained the formation of the Janssen effect. Using physical experiments and DEM numerical simulations, David Žurovec et al. [20] experimentally validated the mathematical derivation of Janssen's theory, providing important information and insights for the design of storage facilities. Considering the problem of the static packing assumption of the Janssen model, Windows-Yule et al. [21] explored the stress redistribution of particles in a container with a laterally moving sidewall, and extended and improved the model for characterizing the constrained dynamic system. Haiyang Zhao et al. [22], using the DEM numerical simulation method, studied the static stress distribution in confined granular columns, and found that the lateral pressure coefficient is not a constant value that decreases gradually to a constant with the packing depth.

Presently, research on granular flow and other related aspects mainly focuses on grain silos, storage bins, and the hanging of the ore pass, and few involve the mechanical damage of the ore-pass wall. Moreover, the existing research still lacks an understanding of the macro-flow and micromechanical behavior of ore-rock bulk particles. Therefore, this paper analyzes the flow characteristics of ore-rock bulk via the discrete element method. On this basis, the contact density, internal stress distribution, contact force distribution, and so on are discussed, and the mechanical evolution mechanism of the bulk system and the stress characteristics of the ore pass of the storage section are revealed in the process of ore-rock flow, which lays a foundation for future study on the stability of the ore-pass wall.

2. Physical Test

2.1. Determination of Ore-Rock Particle Parameters

The mechanical parameters (such as density, particle size distribution, grading, internal friction angle, and cohesion) of the ore-rock bulk materials play an important role in the overall fluidity of the storage materials and the force condition of the ore-pass wall. In order to ensure the accuracy of the experimental results, waste rock from a certain mine was crushed and screened into different sizes in the laboratory to keep its mechanical properties close to the ore-rock bulk in situ. According to the previous studies on the physical parameters of bulk materials, the bulk density of the broken stones, which was obtained by the water injection method [23], was 2050 kg/m^3. On this basis, the quartering method was used to scale the building stone samples, the particle size distribution, and the mass proportion of the scaled samples were obtained by using different grid screens, as shown in Table 1.

Table 1. Particle size distribution and mass proportion of ore-rock bulks.

Particle size/mm	5~10	10~15	15~20	20~25	25~30
Quality Percentage/%	15	25	30	20	10

Using the slump test device, as shown in Figure 1, the natural repose angle of the fine-size particle bulk is 38.6°, and the internal friction angle and cohesion of the ore-rock bulk are obtained indirectly by measuring the natural repose angle [24].

Figure 1. Slump test device for measuring the natural repose angle of the stone bulk. The green arrow in the figure is the pointer of the protractor.

2.2. Construction of the Central Unloading Model

Taking the ore pass of an iron mine in Northeast China as an example, the parameters of the ore pass storage section are 32 m high and 6 m in diameter, and the parameters of the ore-drawing funnel below the storage section are as follows: height—6.4 m, width of ore-drawing port—3 m, and sidewall angle—72°. Combined with the actual geometric size and production conditions of the ore pass, the geometric similarity ratio of 1:20 was selected to construct the physical similarity experiment platform. In the physical model, the ore-pass storage section and the ore-drawing funnel have the same center line; the height of the storage section H is 1.6 m, the diameter D is 0.3 m, the height of the funnel h is 0.32 m, the width of the ore-drawing port A is 0.15 m, and the angle of the funnel side wall is 72°, as shown in Figure 2. In addition, before building the model, the inside of the acrylic silo was polished with sandpaper to simulate the friction effect of the ore pass on the ore-rock bulk in the process of ore drawing.

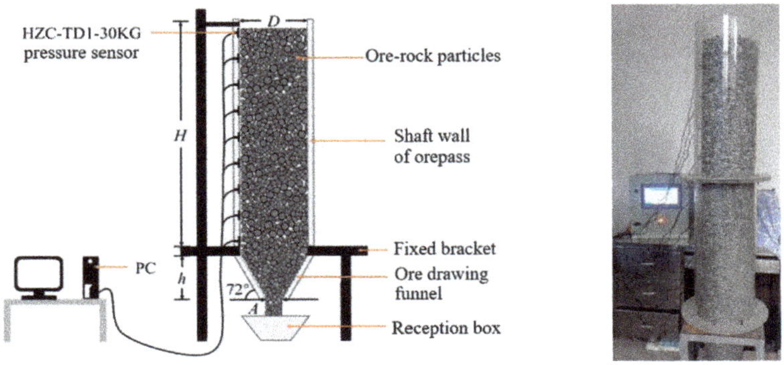

(**a**) Sketch diagram (**b**) Physical diagram

Figure 2. Physical similarity model.

According to the existing research results, for the central drawing model, the lateral pressure on both sides of the ore-pass wall was not different [25]. Therefore, in this paper, nine pressure sensors of the type HZC-TD1-30 KG, which is produced by Chengying Sensor Ltd. in Bengbu City of China, were uniformly arranged on the left side wall along the geometric center line of the ore pass to record the change in the lateral pressure on the ore-pass wall in real time.

3. Construction of the Numerical Model

3.1. Determination of the Contact Model and Detailed Parameters

In the two-dimensional simulation of particle flow, particles are usually simplified to disc-shaped objects, which is the basis of model iterative calculation and the precondition of mechanical equilibrium. Because the ore-rock particles in nature are mostly irregular blocks with sharp edges, this will lead to large errors in mechanical simulation calculation between ore-rock particles. The calculation time can be shortened, and the results can be achieved close to the irregular block motion, by adjusting the rolling friction coefficient between the particles and restricting the rotational ability of the particles [26,27]. Based on this, the anti-rotation linear contact model and the linear contact model built-in PFC were selected, respectively, for the ore-rock particles and the wall in this paper.

The anti-rotation contact model, which is based on the linear contact model, mainly considers the effective modulus, the normal and tangential stiffness ratio K^*, and the anti-rotational friction coefficient of the particles. Specifically, the effective modulus of the ore material in this study is determined to be E^* at 350 MPa according to the elastic modulus, and the normal and tangential stiffness ratio K^* is 1.0. The collapse measurement model established by PFC was used for determining the anti-rotational friction coefficient, which is based on the same value of the numerical model's natural repose angle obtained according to the physical test. After adjusting the anti-rolling friction coefficient and different mesoscopic parameters many times, the anti-rotational friction coefficient was finally determined to be 0.7. The established collapse measurement model is shown in Figure 3, and the mesoscopic parameters are shown in Table 2.

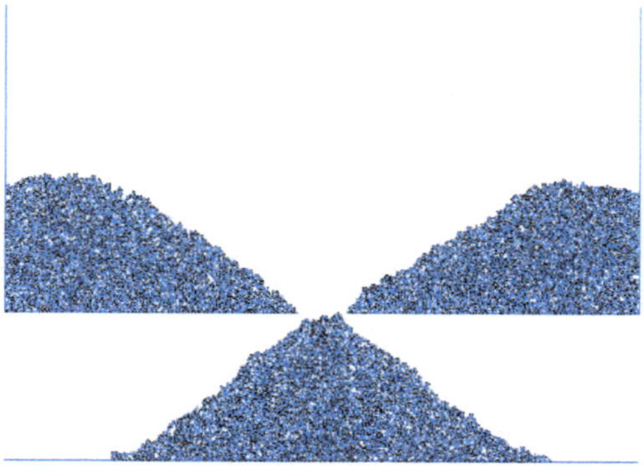

Figure 3. Discrete element collapsed measurement model with the particle size in a range of 0.1 m~0.6 m.

Table 2. Mesoscopic parameters of the numerical model.

Types	Normal Stiffness/(N/m)	Tangential Stiffness/(N/m)	Ore-Rock Bulk Density/(kg/m^3)	Friction Coefficient	Anti-Rotation Friction Coefficient	Particle Size/m	Number of Particles/N
Ore particles	3.33×10^9	3.33×10^9	2050	0.7	0.7	0.1~0.6	13,468
Wall	3.33×10^9	3.33×10^9	—	0.65	—	—	—

It should be pointed out that the ore-rock bulk density given in Table 2 is the same as the physical experimental value. The main reason for this is that the gravity density of the waste-rock block measured by the laboratory is 3050 kg/m^3. According to the loosening coefficient of 1.50 of the ore-rock bulk provided by the mine, the ore-rock bulk density is calculated to be 2033 kg/m^3 simply by the gravity density of the waste rock being divided by the loosening coefficient. Considering that there may be some errors in this treatment method, the ore-rock bulk density is simply adjusted to 2050 kg/m^3, as shown in Table 2, and this is just a coincidence.

3.2. Establishment of the Numerical Model

According to the geometric parameters of the ore pass in situ, as shown in Section 2.2 in this paper, when the numerical model is established, the center of the drawing port is considered as the coordinate origin, and the total height of the ore storage section and the ore-drawing funnel is 38.4 m, as shown in Figure 4a. In order to monitor the lateral pressure of the ore-pass wall, 9 monitoring points, numbered 1–9 from bottom to top, are arranged along the left shaft wall at 4 m intervals. For observing the flow state of the ore-rock bulk, the ore-rock particles with different colors are placed in layers every 4 m upward from the lower to upper in the ore storage section, there are seven layers.

For analyzing the micromechanical changes during the ore-rock bulk flowing in the ore pass, the ore-rock stress is monitored by the method of measuring the circle [28,29]. A total of 18 columns and 53 measurement circles are arranged in the storage section of the ore-pass from bottom to top. Columns 1 to 6 are called Group 1; 7 to 12 are named Group 2; and 13 to 18 are named Group 3. The radius of each measurement circle is set to 1 m. By using surfer software, the coordination number and internal stress recorded via the measurement circle were used to draw the stress distribution cloud map in the Y-direction, so as to quantitatively characterize the micromechanical change characteristics of the stored ore-rock bulk. By using surfer software, the coordination number and internal stress recorded by the measurement circle were used to draw the stress distribution cloud map in the Y-direction so as to quantitatively characterize the micromechanical change characteristics of the stored ore-rock bulk, as shown in Figure 4b.

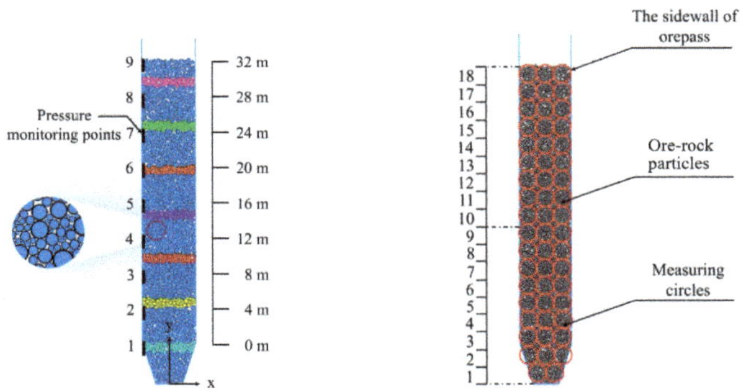

(a) Ore-drawing model of ore pass (b) Measurement circle layout

Figure 4. Ore-drawing model of ore pass and measurement circle layout.

It should be noted that during the physical experiment, the geometric size of the ore pass and the size of the ore-rock particles in the case mine were scaled according to the ratio of 1:20. However, in numerical analysis, these were modeled according to the actual size of the mine, which is very different from the method proposed by Di Renzo, Bierwisch and Radl et al. [30–32]. The authors' intention is to verify the consistency of the results of different methods and to conform to the actual situation of the ore pass in the case mine.

3.3. Reliability Analysis of the Numerical Models

Figure 5 shows the comparison of the static lateral pressures obtained, respectively, by numerical simulation, the physical similarity experiment, and the theoretical calculation. The theoretical lateral pressure value of the shaft wall is calculated according to Janssen's theory [33].

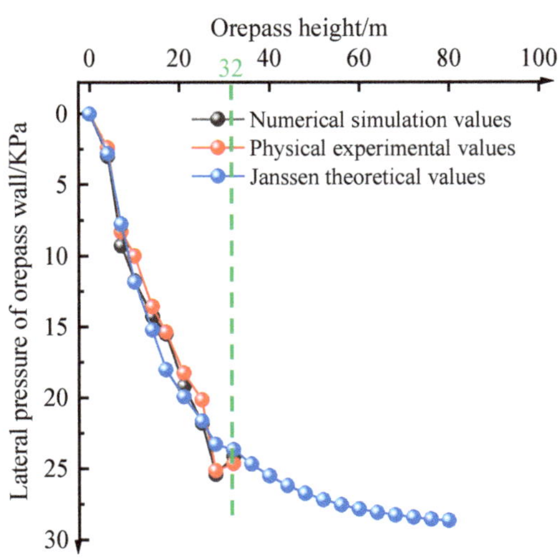

Figure 5. Comparison of lateral pressures on ore-pass wall obtained by numerical simulation, physical test and Janssen theoretical calculation. The green dashed line shows the maximum height of the ore-rock stored in the ore pass and that in 2D and 3D experiments.

Although this study did not carry out more relevant research based on Janssen theory, and only compared the lateral pressure calculated results of the Janssen equation with those of physical experiments and numerical simulations, it can be seen from Figure 5 that the static lateral pressure values obtained by the three methods have a good consistency.

Based on the physical similarity experiment, it is found that the maximum errors between the numerical simulation and the theoretical calculation via the Janssen formula are, respectively, 11.8% and 18.1%; this indicates that the numerical model has a certain reliability in the process of analyzing the characteristics of the pressure distribution on the side wall of the ore pass.

Unfortunately, in the physical experiment and numerical simulation, the ore-rock storage height of the studied ore storage section in the ore pass was limited to 32 m, i.e., the maximum ore storage height of the ore pass in the case mine site, and the pressure curve obtained by these two methods did not show the pressure saturation phenomenon.

The pressure fluctuation phenomenon in the experimental curve may be related to the rotation and mobility of the ore-rock blocks in bulk systems caused by the gravitational compaction of the stored materials [10,34]. The rotation and mobility of the large and irregular blocks in the bulk system will change their contact mode (such as point, line, and surface contact [24]), which results in a decrease in the voidage of the bulk and the changing of the force transfer effect of granular particles.

4. Flow Characteristics and Stress Distribution of Ore-Rock Bulk

4.1. Flow Characteristics of Ore-Rock Bulk

According to the actual process of ore drawing from the ore pass in the mine, a numerical simulation of a one-time drawing was carried out. During the ore drawing, the stored particles leave the ore pass under the action of gravity. When the volume of drawn ore-rock particles is greater than or equal to 5 m^3, the ore drawing is suspended. When the internal force in the storage bulk reaches a new equilibrium, the one-time ore drawing is over. Repeating the above process, it is found that the ore-rock particles were completely drawn out from the ore pass 22 times, and the ore-drawing test is over. When ore drawing for the 21st time, the ore-rock particles near the pressure measuring point of No. 1 have been drawn out; therefore, Figure 6 shows the macroscopic flow characteristics of ore-rock for ore drawing 0~20 times.

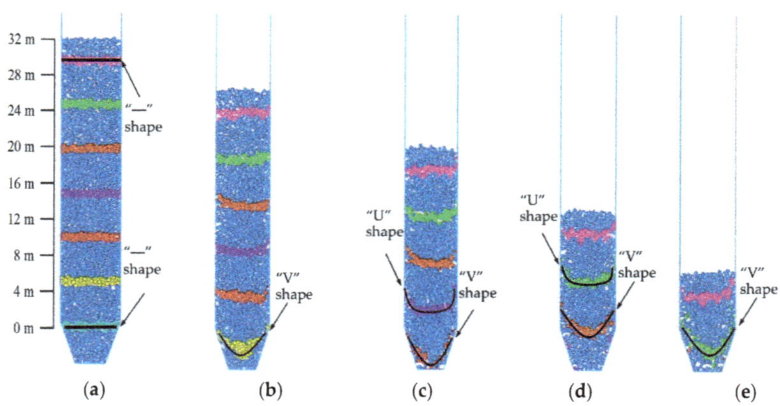

Figure 6. Macroscopic ore-rock flow characteristics at different ore-drawing times. (**a**) Before ore drawing begin; (**b**) Ore drawing for 5 times; (**c**) Ore drawing for 10 times; (**d**) Ore drawing for 15 times and (**e**) Ore drawing for 20 times.

Before the ore drawing begins, because the stress between the ore-rock particles is in a state of equilibrium, there is no movement trend of rotation, slide, slip, etc., and the ore-rock particles keep the original spatial shape and arrangement mode and distribute uniformly in the ore pass storage section. Each layer of labeling particles is distributed as a "—" shape.

At the beginning of the ore drawing, the ore-rock at the ore-drawing port is drawn out first and releases a certain space. Due to the influence of the interaction force between the particles and the friction resistance of the shaft wall, the velocity of the ore-rock decreases with the increase in the distance from the ore-drawing port. As a result, the ore-rock flowing inside the ore pass and the funnel presents the characteristics of high velocity of the ore-rock near the geometric center of the storage section; the low velocity of them near the shaft wall; and leads the labeling-particle layers distributed as a "U" shape. The distribution of labeling-particle layers within the height range of 0 m~10 m in the upper storage section of the drawing funnel is in the transition from a "—" shape to a "U" shape, but within the height range of 10 m~24 m, they are still distributed in the character similar to a "—" shape. With the increase in the ore-drawing times, the "U"-shaped distribution characteristics of each labeling layer become more and more obvious. Due to the influence of the interaction force between the ore-pass wall and the ore-rock particles inside the drawing funnel, and the extra boundary constraints by the funnel, the shape of the labeling layer leads to a change from a "U" to a "V" distribution. Finally, the labeling particles layer is drawn out from the funnel as a "V"-shape flow characteristic.

4.2. Stress Distribution of the Ore-Rock Bulk

4.2.1. Contact Density of Ore-Rock Bulk

The coordination number (Z_i) is the average contact number between the particles of the ore-rock bulk, which reflects the quality and density of the internal contact of ore-rock particles to a certain extent. Therefore, the evolution law of internal micromechanics in ore-rock particles' flow process can be directly studied by analyzing the coordination number [35]. The formula for calculating coordination number is as follows [36]:

$$Z_i = \frac{2N_C}{N_P} \tag{1}$$

where N_C is the actual contact number of bulk particles (normal contact force is greater than 0) and N_P is the total number of particles in the bulk.

It should be noted that when the number of contacts of a single particle is only 1, or there is no other particle in contact with it, it is considered that this particle has no contribution to the micromechanics of the entire particle system and should be ignored. In order to facilitate the statistics of the coordination number changes in each labeling layer, the average value of the coordination number Z_a is introduced, and its calculation formula is as follows:

$$Z_a = \frac{\sum\limits_{i=1}^{3} Z_i}{3} \tag{2}$$

According to the coordination Z_a obtained by the monitoring during the numerical simulation, the change curve of the average value of the labeling-particle layer at different ore-drawing times was drawn, as shown in Figure 7. The coordination numbers monitored by 1~6 measurement circles were divided into Group 1, those measured by 7~12 were divided into Group 2, and those measured by 13~18 were divided into Group 3.

It can be seen from Figure 7 that the average coordination numbers at different positions in the ore-rock bulk are different to some extent during the whole ore-drawing process—the closer to the bottom of the ore pass, the greater the average coordination number, which also indicates that the ore-rock bulk is more compact. This phenomenon reflects the influence of the action of ore-rock particles' gravity.

It can also be seen from Figure 7, that before the ore drawing begins, the average coordination number of Groups 1 to 3 is about 3.84~4.15, 3.72~3.2~4.05, and 3.21~3.68, respectively. The change relationship of the coordination numbers of different groups is Group 3 > Group 2 > Group 1, while the corresponding coordination number of each group has the opposite fluctuation rule. When the ore-rock was drawn five times, the coordination numbers of Group 1 gradually increased from 3.41 to 3.87, that of Group 2 fluctuated to decrease and was similar to the change rule of the coordination numbers without ore drawing, and that of Group 3 decreased rapidly from 3.54 to 0. When ore drawing 10 times, the coordination numbers of Group 1 increased from 3.44 to 3.85, and those of Group 2 and Group 3 decreased exponentially with the increase in the number of measurement circles and finally stabilized at about 0. With the increase in the ore-drawing times, the change rule of the coordination numbers of each group is almost consistent, decreasing exponentially to 0.

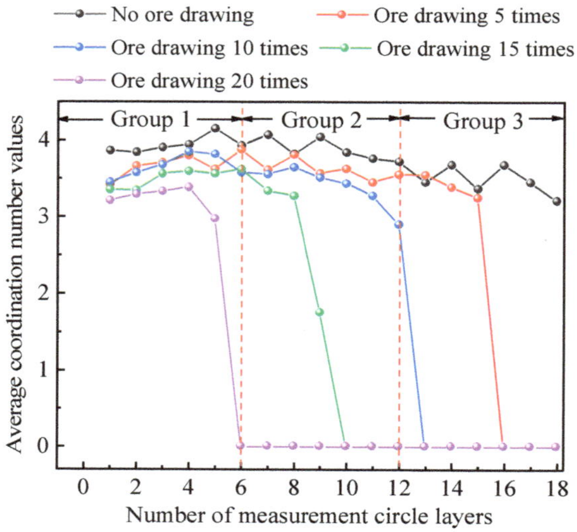

Figure 7. Distribution pattern of coordination numbers inside the ore-rock bulk under different ore-drawing times.

According to the above analysis, it can be seen that due to the effect of the inertia force generated by the particles flowing and the bulk gravity of ore-rock bulk in the range of Groups 2 and 3 during the entire ore-drawing process, the ore-rock particles' coordination numbers in the range of the measurement circles of Group 1 fluctuate and then decrease. This indicates that the ore-rock particles in this region tend to have a state of compact contact with high contact strength compared to that of loose contact with small contact strength. Affected by the ore-rock drawing out, the coordination number of Group 2 is lower than that of Group 1, the ore-rock particles move toward having a dense contact strength with high strength from having loose contact with low strength, but the degree is not as high as that of Group 1. The coordination number of Group 3 was affected by the ore drawing (the height of the stored ore-rock declined and there were no ore-rock particles above Group 3), and the coordination numbers gradually decreased after the ore drawing began.

4.2.2. Stress Distribution Characteristics of Ore-Rock Bulk

A stress distribution nephogram is a common method used to intuitively analyze the characteristics of stress changes within the bulk [28]. During stress monitoring, stress measurement circles are arranged on the surface of ore-rock particles for stress monitoring and surfer software is used to draw the monitored stress data into a stress cloud map. According to the data gained by the stress measurement circle, the stress distribution nephogram is shown in Figure 8.

It can be seen from Figure 8 that, before the ore drawing begins, the ore-rock bulk has a large stress concentration phenomenon in the range of 0–16 m, and the stress increases with the increase in the shaft depth. When the ore-rock was drawn 5 times, the stress value of the ore-rock in the ore pass decreased to a different degree—that is, the high-stress (blue part) area decreased. this indicates that the decrease in the amount of ore-rock in the storage section is the main reason for the stress change and the area decrease in the high stress. With the increase in the ore-drawing times, the total amount of ore-rock in the ore pass is constantly decreasing, and the high-stress concentration area is gradually reduced to the drawing funnel from the range of 0–16 m. When the ore-rock in the storage section is completely moved into the drawing funnel, the high-stress area in the storage section completely disappears.

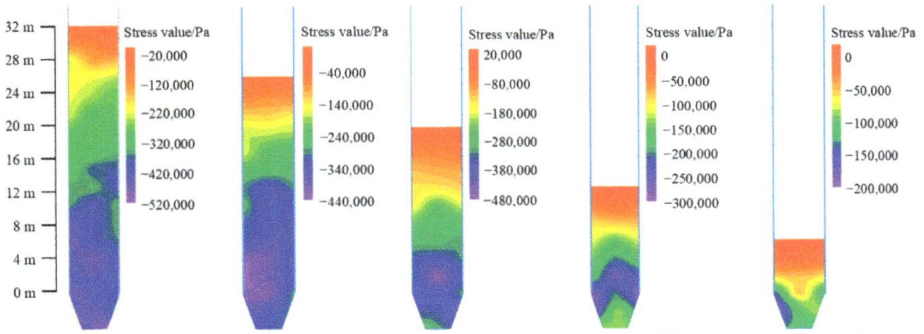

Figure 8. Stress distribution characteristics of ore-rock bulk at different ore-drawing times. The compressive stress obtained in the PFC numerical simulation is negative, and the tensile stress is positive [28]. The greater the compressive stress, the brighter the blue and purple color, indicating that the ore-pass wall is more obviously squeezed by the ore-rock particles. Tensile stress in this figure may be caused by the "arch forming-arch collapsing" phenomenon inside the stored materials when the particles move downward during ore drawing. (**a**) Before ore drawing begin; (**b**) Ore drawing for 5 times; (**c**) Ore drawing for 10 times; (**d**) Ore drawing for 15 times and (**e**) Ore drawing for 20 times.

It is worth noting that in the entire process of ore drawing, although the range of the high-stress concentration area in the ore storage section continues to decrease with the ore-rock drawing out, the storage section in the range of 0~16 m and the drawing funnel are the main areas of high-stress concentration. As a result, the ore-pass wall in this area would bear a larger stress, making it easy to produce a large number of fine cracks and causing damage to the ore-pass wall.

4.2.3. Contact Force Evolution Characteristics of Ore-Rock Bulk

As a macroscopic expression of microscopic contact force, the force chain can reflect the microscopic mechanical behavior of ore-rock particles [10,37]. However, due to the inhomogeneity of the spatial distribution of contact force between the ore-rock particles, it is very challenging to directly characterize the evolution of the contact force network quantitatively, and the application of a contact force probability distribution curve can reflect the distribution of the contact force between particles from the perspective of contact force [38]. Taking the total contact force F between particles as an example, the normal and tangential contact force vector and average contact force $\langle f_n \rangle$ of any particle i in the ore-rock particle system are normalized. The calculation formula is as follows [39]:

$$f = \frac{F}{\langle f_n \rangle} \tag{3}$$

where f is the normalized processing result of the total contact force, $\langle f_n \rangle$ is the average contact force, $\langle f_n \rangle = \sum_{i=1}^{N_i} F/N_i$, i is the random particle number, and N_i is the total number of particles.

In the process of studying the macro- and micromechanical variation characteristics of ore-rock bulk, the normalization treatment of the resultant force is used to quantitatively analyze the contact force distribution characteristics inside the bulk [23]. Based on Formula (3), Figure 9 shows the probability distribution curve of the contact force between ore-rock particles. The horizontal coordinate, which is divided into several intervals at 0.5 intervals, is the normalized result of the total contact force *f*, and the numbers of contact forces in each interval are calculated respectively. The ordinate reflects the normalized result of the contact force strength, which represents the percentage of the number of contact forces *P*(*f*) in the *f* interval of the total number of contact forces.

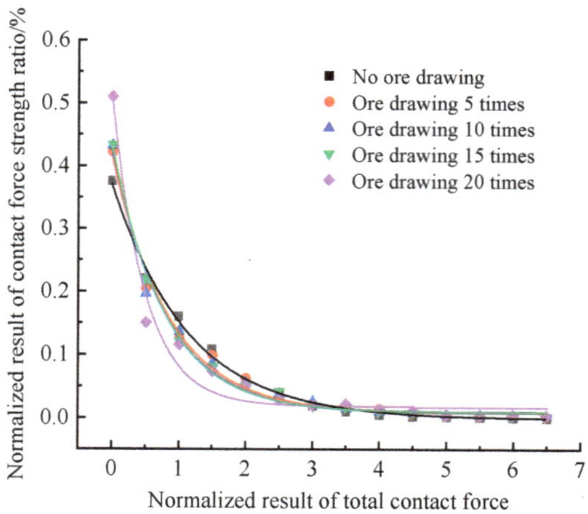

Figure 9. Probability distribution of contact force for the bulk system at different ore-drawing times.

There are five curves in Figure 9, which represent, respectively, the change characteristics of contact force before the ore drawing begins and when the ore draws 5, 10, 15, and 20 times. It can be seen from the probability distribution of contact force strength in Figure 9 that the contact force strength accounts for 85% in the range of 0.0~2.0, while it accounts for less than 15% in the range of 2.0~7.0. This shows that the weak force chains are the main component, while the strong force chain is the important component of the force chain network, and the strong and weak force chains are interwoven to form the force chain network, which maintains the stability of the whole bulk system. At the same time, with the increase in ore-drawing times, the force chains in the whole bulk system continue to break and recombine under the action of the tangential forces, and the proportion of weak force chains decreases exponentially, while that of strong chains tends to be stable. This indicates that the ore-rock particles contained in the weak force chain contribute little to the stability of the whole system, while those in the strong force chain play an important role in the stability of the whole system.

It is found that after normalization, the distribution of contact force strength between ore-rock particles decreased exponentially, which indicates that the ore-rock flow in the ore-drawing process has a significant effect on the Probability Distribution Function (PDF) between particles. In addition, according to the variation characteristics of the PDF, the exponential function $y = A_1 \times e^{-\frac{x}{m}} + A_2$ is used to fit it, and the fitted PDF of the contact force is shown in Table 3. It can be seen from Table 3 that all fitting coefficient R^2 is greater than 0.97, indicating that the PDF of the contact force has a good fitting effect.

Figure 10 shows the characteristics of the contact force chains in ore-rock particles at different ore-drawing times. In this figure, the deeper the black line, the greater the contact force of ore-rock blocks. During the entire ore-drawing process, due to the influence of gravity, the color of the contact force chain gradually deepens from top to down. That is, the contact force gradually increases. According to the statistics of the strong and weak force chains respectively, the ratio of the strong and the weak force chains is about 1:17, which indicates that the weak force chain is the main component of the force chain network in the contact force chain of ore-rock particles, while the proportion of the strong force chain is relatively small. The force chain network composed of the strong and weak force chains interwoven maintains the stability of the whole ore-rock bulk. With the increase in ore-drawing times, the strong and weak force chains inside the particle system continuously break and transfer under the action of ore-rock flow, and the proportion of the weak force chains decreases exponentially, which indicates that the number of ore-rock particles contained in the weak force chains increases gradually, but contributes less to the stability of the whole bulk system. Meanwhile, the number of ore-rock particles contained in the strong force chains gradually decreases but contributes more to the stability of the whole bulk system.

Table 3. Fitting function for the probability distribution of the internal contact force of the bulk at the partial different ore-drawing times.

Ore-Drawing Times/Time	Fitting Functional Equation	Fitting Coefficient R^2
0	$y = 0.37387 \times e^{-\frac{x}{1.13352}} - 0.00329$	0.996
5	$y = 0.40365 \times e^{-\frac{x}{0.85799}} - 0.00615$	0.988
10	$y = 0.41238 \times e^{-\frac{x}{0.80218}} - 0.00789$	0.987
15	$y = 0.42004 \times e^{-\frac{x}{0.79039}} - 0.00720$	0.995
20	$y = 0.48422 \times e^{-\frac{x}{0.49188}} - 0.01676$	0.971

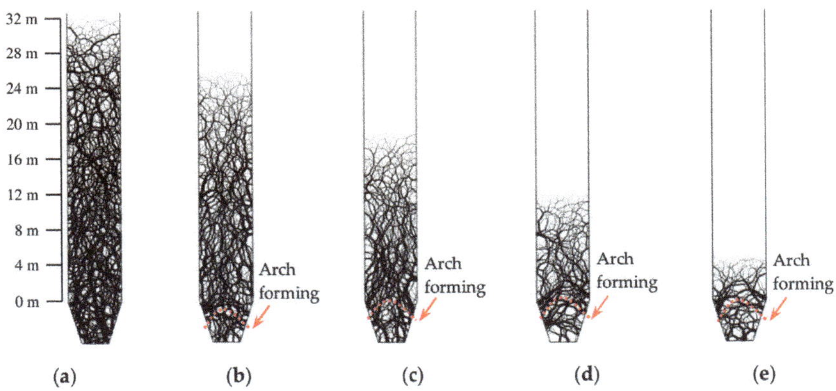

Figure 10. Distribution of force chain contact morphology in ore-rock particles at different ore-drawing times. The red dotted lines in (**b**–**e**) show where the arch might be formed under the action of the strong force chains. (**a**) Before ore drawing begin; (**b**) Ore drawing for 5 times; (**c**) Ore drawing for 10 times; (**d**) Ore drawing for 15 times and (**e**) Ore drawing for 20 times.

It can also be seen from Figure 10 that the distribution characteristics of the strong and weak force chains show that the contact strength between ore-rock particles increases with the increase in the depth of stored ore-rock, and the contact between ore-rock particles becomes closer; especially the characteristic before ore drawing is obvious. This phenomenon reflects that the contact strength between ore-rock particles is obviously affected by the ore-rock's gravity compaction. The effect of gravity compaction, based on the study results of Tempone, P., et al. [40], is mainly manifested in that it changes the contact mode between irregular ore-rock particles and reduces the voidage of the particle system.

It seems that the material stored in the ore pass has a strong compression effect before the ore drawing, which is only a manifestation of the gravity compaction of the stored materials. Before ore drawing begins, as shown in Figure 10a, the stored ore-rock material in the ore pass seems to be in a static state, but in fact, due to the irregularity of the ore-rock blocks, the geometric center of each block does not coincide with its gravity center, resulting in the rotation and movement of the ore-rock block in the particle system and constantly changing its spatial arrangement form and the contact mode between particles until it can no longer overcome the resistance from the other particles. As a result, the contact strength between the ore-rock blocks reaches the maximum value, and the voidage of the bulk system reaches the minimum value. As shown in Figure 10b,c, with the ore-rock near the funnel port drawing out, the upper ore-rocks continue to move toward the funnel, the void ratio of the stored materials in ore pass increases, and the contact degree of ore-rock blocks decreases. So, the distribution state of strong and weak force chains also changes before and during ore drawing.

4.3. Lateral Pressure Distribution of the Shaft Wall in the Storage Section

In the storage section, the important reason for the change of ore-rock stress on both sides of the ore-pass wall is that the particles change their internal spatial distribution and flow characteristics during the flow process [41]. However, because of the effect of the discontinuity of contact between the ore-rock bulk and the ore-pass wall, it is difficult to accurately characterize the change mechanism of the wall stress by the change of the dynamic lateral pressure of the shaft wall. The overpressure coefficient and overpressure times can be used to reveal the mechanical mechanism of the stress change of the ore-pass wall. According to the elevation position of each measuring point, measuring points 1–4 and 5–9 are divided into the lower and the upper part of the storage section, respectively, and the change characteristics of the lateral pressure at each measuring point are gained in real time, as shown in Figure 11.

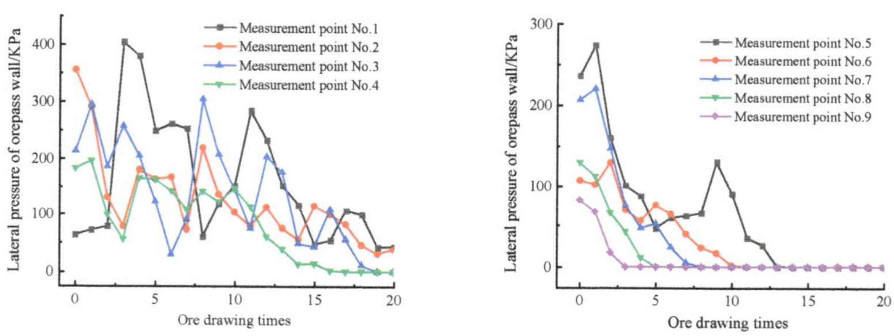

(**a**) Measurement points 1 to 4 (**b**) Measurement points 5 to 9

Figure 11. Characteristics of the dynamic lateral pressure change at each measuring point of the shaft wall.

It can be seen from Figure 11 that, in the process of ore drawing, the dynamic lateral pressure on the shaft wall of each measuring point basically fluctuates exponentially and decreases, and the pressure value increases gradually from top to bottom (point 9 to 1), while that of some measuring points (point 1 and 3~7) increases significantly compared with the static lateral pressure before ore drawing. The maximum pressure value appears after a period of ore-drawing time, and not at the moment of initial drawing. At the beginning of ore drawing, the loose ore-rock body will produce the phenomenon of the particles' rotation, and they will slip under the action of their own gravity, which breaks the initial mechanical equilibrium and causes the mechanical behavior of "arch forming-arch collapsing" continuously inside the stored materials. The formation of the arch, which may be instantaneous and unstable, causes the increase in the lateral pressure value of the measuring point near the arch foot—that is, the "over-pressure phenomenon". The collapse of the arch leads to the instantaneous decrease in the lateral pressure value near the measuring point. The phenomenon of "arch forming- arch collapsing" occurs alternately in the process of ore-rock flow, which results in the exponential fluctuation decreasing of dynamic lateral pressure at each measuring point and the peak lateral pressure appearing after a period of time. This is consistent with the results of previous research on bulk and spontaneous arch formation during ore-rock flow [42,43].

Also, the overpressure data at each measuring point of the ore-pass wall under different ore-drawing times are counted in real time, which is shown in Table 4.

Table 4. Overpressure data of each measurement point under different ore-drawing times.

Region	Measurement Point Number	Number of Overpressure/Time	Maximum Overpressure Factor
Lower part of the storage section	1	15	6.4
	2	0	0
	3	3	1.42
	4	1	1.07
Upper part of the storage section	5	1	1.16
	6	1	1.21
	7	1	1.06
	8	0	0
	9	0	0

It can be seen from Table 4 that the degree of the overpressure phenomenon at each measuring point is different under different ore-drawing times. In the 1~9 measuring points, the accumulated overpressure of point 1 is 15 times, that of point 3 is 3 times, and that of points 4~7 is only 1 time. This indicates that the frequency of the overpressure phenomenon in the lower part of the storage section is much greater than that in the upper part, and the closer the measuring point at the bottom of the storage section, the greater the frequency of the overpressure phenomenon. This is due to the speed and acceleration gradually increasing and the movement intensity of ore-rock particles gradually rising when the ore drawing began. At this time, a strong dynamic load is generated on the wall of the ore storage section, causing the phenomenon of overpressure and the increase in dynamic lateral pressure on the wall. With the continuous increase in ore-drawing times (the total amount of the ore-rock in the shaft is constantly reduced), the dynamic lateral pressure of the wall will inevitably decline even if the movement of the ore-rock particles is intense. When the lower area of the storage section bears the secondary impact load caused by the broken arch in the ore-rock bulk, and the additional gravity compaction of the upper covered ore-rock particles, the dynamic lateral pressure, overpressure coefficient, and overpressure frequency of the shaft wall are led to increase. As a result, the overpressure coefficient of measuring point 1 is the largest, followed by the point 3. The peak overpressure coefficient of measuring point 1 is about five times that of point 6; this further indicates that the overpressure coefficient is positively correlated with the storage depth in the process of ore-rock flow. In a word, with the largest overpressure frequency and the largest overpressure coefficient, the damage degree of the ore-pass wall will be greater. The deeper the storage depth, the greater the overpressure coefficient. The phenomenon of overpressure occurs the most frequently, and in the measuring point area where the overpressure coefficient value is the largest, the damage degree of the well wall will be greater. Therefore, from the perspective of overpressure frequency and overpressure coefficient, the damage degree of the lower storage section is much greater than that of the upper storage section.

5. Conclusions

Based on the discrete element analysis method, the flow characteristics, the mechanical evolution mechanism of the ore-rock particles in the ore-drawing process, and the stress characteristics of the wall of the ore-pass storage section are studied. A method for characterizing the damage degree of the ore-pass wall using the overpressure coefficient and overpressure frequency is proposed. The main conclusions of this study are as follows:

The important factors affecting the ore-rock's macroscopic flow pattern are the friction between ore-rock particles, particles, and the shaft wall, and the constraint of funnel boundary. In the ore-drawing process, the macro-shape of rock particles with different flow rates gradually changes from a "—" shape to a "U" shape, and finally, a "V" shape is drawn out.

The study of coordination number variation characteristics between ore-rock particles shows that under the different ore-drawing times, the average coordination numbers between ore-rock particles near the bottom of the ore pass are larger than that in the upper part of the ore pass. When the ore is drawn 5 times, the average coordination number of measuring circles No. 1–6 (near the bottom of the ore pass) increased from 3.41 times to 3.87; after ore drawing 10 times, it increased to 3.85; and the average coordination numbers measured by measuring circles No. 7–18 (located in the upper part of measuring circles No. 1–6), depending on the position of the measuring circle, decreased continuously from the top to the bottom, until they were 0. This causes the density and stress concentration to become more and more obvious, and the range of the high-stress concentration area decreases with the decrease in the total amount of stored ore-rock. The shaft wall and funnel at the bottom of the storage section are under the action of high stress, which makes it easy to produce a large number of fine cracks, causing damage to the ore-pass wall.

With the increase in the ore-drawing times, the probability distribution of the contact force strength decreases exponentially. The strong and weak force chains in the ore-rock bulk constantly break and transfer under the action of ore-rock flow, and the number of ore-rock particles contained in the weak force chains gradually increased, while that in the strong force chains gradually reduced. However, the strong force chains play a leading role in ensuring the stability of the whole bulk structure system.

Author Contributions: Methodology, software, formal analysis, data curation, writing—original draft, funding acquisition, Z.X.; Software, investigation, resources, data curation, Z.D.; Conceptualization, methodology, formal analysis, writing—review and editing, project administration, funding acquisition, Z.L.; Resources, data curation, visualization, supervision, C.M. All authors have read and agreed to the published version of the manuscript.

Funding: The work was supported by the National Natural Science Foundation of China (Grant No. 52204137, Grant No. 51774176), the Natural Science Foundation of Liaoning Province (Grant No. 2022-BS-281), and the Outstanding Young Scientific and Technological Talents Project of Liaoning University of Science and Technology (Grant No. 2023YQ10).

Data Availability Statement: The detailed data are available upon request. The data are not publicly available due to privacy.

Conflicts of Interest: The authors declare no conflicts of interest.

References

1. Jiang, W. The study of arching mechanism of ores in chute based on discontinuous deformation analysis. *Int. J. Digit. Conte Technol. Its Appl.* **2013**, *7*, 190.
2. Alex, M.R.; Verne, M.; Sanjay, N.; Ting, R. Experimental and numerical investigation of high-yield grout ore pass plugs to resi impact loads. *Int. J. Rock Mech. Min. Sci.* **2014**, *70*, 1–15.
3. Hadjigeorgiou, J.; Stacey, T.R. The absence of strategy in orepass planning, design, and management. *J. South. Afr. Inst. Mi Metall.* **2013**, *113*, 795–801.
4. Liu, H.; He, R.; Li, G.; Sun, D.; Li, P. Development of gravity flow draw theory and determination of its parameters. *Int. J. Ro Mech. Min. Sci.* **2023**, *171*, 105582. [CrossRef]
5. Huang, Y.; Cao, P.; Wang, Y. Motion situation and kinetic energy analysis for ore pass of underground mine. *Appl. Mech. Mat* **2011**, *90–93*, 383–386. [CrossRef]
6. Ma, C.; Lu, Z.X.; Yin, Y.; Cao, P. Prediction model for the migration trajectory and velocity of ore-rock bulks in an orepass storag section. *Chin. J. Eng.* **2021**, *43*, 627–635.
7. Zhao, Y.; Ye, H.W.; Lei, T.; Wang, C.; Wang, Q.Z.; Long, M. Theoretical study of damage characteristics on ore pass wall based c the erosion-wearing theory. *Chin. J. Rock Mech. Eng.* **2017**, *36*, 4002–4007.
8. Jin, A.; Chen, S.; Sun, H.; Zhao, Y.; Qin, X.; Ju, Y.; Gao, Y. Characteristics of particle percolation based on inhomogeneous partic size distribution. *J. Cent. South Univ. (Sci. Technol.)* **2020**, *51*, 1673–1681.
9. Cundall, P.A.; Strack, O.D.L. A discrete numerical model for granular assemblies. *Geotechnique* **1979**, *29*, 47–65. [CrossRef]
10. Yang, Y.; Deng, Z.; Lu, Z. Effects of ore-rock falling velocity on the stored materials and the force on the shaft wall in a vertic orepass. *Mech. Adv. Mater. Struct.* **2023**, *30*, 3455–3462. [CrossRef]
11. Sato, A.; Tang, H. Analysis of Ore Pass Hang-Ups in Long Vertical Ore Passes by 3-D DEM. *Int. J. Min. Eng. Miner. Process.* **202** *9*, 1–11.
12. Hadjigeorgiou, J.; Lessard, J.F. Numerical investigations of ore pass hang-up phenomena. *Int. J. Rock Mech. Min. Sci.* **200** *44*, 820–834. [CrossRef]

- Esmaieli, K.; Hadjigeorgiou, J.; Grenon, M. Stability analysis of the 19A ore pass at Brunswick Mine using a two-stage numerical modeling approach. *Rock Mech. Rock Eng.* **2013**, *46*, 1323–1338. [CrossRef]
- Yuan, F.; Pang, K.; Dong, C.; Xu, Z. The PFC3D numerical simulation on dynamic pressures and flow of sidedraw silos. *Eng. Mech.* **2016**, *33*, 301–305.
- González-Montellano, C.; Gallego, E.; Ramírez-Gómez, Á.; Ayuga, F. Three dimensional discrete element models for simulating the filling and emptying of silos: Analysis of numerical results. *Comput. Chem. Eng.* **2012**, *40*, 22–32. [CrossRef]
- Xu, P.; Duan, X.; Qian, G.; Zhou, X.G. Dependence of wall stress ratio on wall friction coefficient during the discharging of a 3D rectangular hopper. *Powder Technol.* **2015**, *284*, 326–335. [CrossRef]
- Cheng, Q.; Sun, W.; Lu, S. Discrete element analysis of squat silo under eccentric drawing by PFC3D. *J. Civ. Eng. Manag.* **2016**, *33*, 43–47.
- Feng, Y.; Liu, J. Research on the dynamic evolution of overpressure coefficient of grain unloading into arch in silos. *Chin. J. Appl. Mech.* **2020**, *37*, 1036–1042+1388–1389.
- Pacheco-Martinez, H.; Van Gerner, H.J.; Ruiz-Suárez, J.C. Storage and drawing of a granular fluid. *Phys. Rev. E Stat. Nonlinear Soft Matter Phys.* **2008**, *77 Pt 1*, 021303. [CrossRef]
- Žurovec, D.; Hlosta, J.; Nečas, J.; Zegzulka, J. Monitoring bulk material pressure on bottom of storage using DEM. *Open Eng.* **2019**, *9*, 623–630. [CrossRef]
- Windows-Yule CR, K.; Mühlbauer, S.; Cisneros, L.T.; Nair, P.; Marzulli, V.; Pöschel, T. Janssen effect in dynamic particulate systems. *Phys. Rev. E* **2019**, *100*, 022902. [CrossRef] [PubMed]
- Zhao, H.; An, X.; Wu, Y.; Qian, Q. DEM modeling on stress profile and behavior in granular matter. *Powder Technol.* **2018**, *323*, 149–154. [CrossRef]
- Chen, Q.; Liu, E.; Wang, S. Characteristics of ore contact force in ellipsoid ore drawing law. *J. Min. Saf. Eng.* **2021**, *38*, 1210–1219.
- Zhou, J. Study on the natural resting angle of loose ores. *Nonferrous Met.* **1983**, *35*, 24–29.
- Kejin, L.; Zhaoran, X.; Shihao, W. Discrete element-based simulation of silo storage unloading arch formation process and silo wall pressure distribution. *Trans. Chin. Soc. Agric. Eng.* **2018**, *34*, 277–285.
- Wensrich, C.M.; Katterfeld, A. Rolling friction as a technique for modelling particle shape in DEM. *Powder Technol.* **2012**, *217*, 409–417. [CrossRef]
- Iwashita, K.; Oda, M. Rolling resistance at contacts in simulation of shear band development by DEM. *J. Eng. Mech.* **1998**, *124*, 285–292. [CrossRef]
- Chen, S.; Xia, Z.; Feng, F.; Yin, D. Numerical study on strength and failure characteristics of rock samples with different hole defects. *Bull. Eng. Geol. Environ.* **2021**, *80*, 1523–1540. [CrossRef]
- Sun, H.; Zhao, L.S.; Wei, L.C.; Jia, J.Z.; Zhou, S.G. Numerical study of the influence of multiple parameters on hang-ups: Insight from a structural and mechanical characteristics analysis. *Rock Mech. Rock Eng.* **2024**. [CrossRef]
- Di Renzo, A.; Napolitano, E.S.; Di Maio, F.P. Coarse-grain DEM modelling in fluidized bed simulation: A Review. *Processes* **2021**, *9*, 279. [CrossRef]
- Bierwisch, C.; Kraft, T.; Riedel, H.; Moseler, M. Three-dimensional discrete element models for the granular statics and dynamics of powders in cavity filling. *J. Mech. Phys. Solids* **2009**, *57*, 10–31. [CrossRef]
- Radl, S.; Radeke, C.; Khinast, J.G.; Sundaresan, S. Parcel-based approach for the simulation of gas-particle flows. In Proceedings of the 8th International Conference on CFD in Oil & Gas, Metallurgical and Process Industries, Trondheim, Norway, 21–23 June 2011.
- Chen, X. Extension of classical Janssen loose mass pressure theory and its application in mining engineering. *Chin. J. Geotech. Eng.* **2010**, *32*, 315–319.
- Raúl, L.C.; Miguel, A.F.; Fernando, L. Experimental study of gravity flow under confined conditions. *Int. J. Rock Mech. Min. Sci.* **2014**, *67*, 164–169.
- Shi, C.; Zhang, Q.; Wang, S. Numerical simulation techniques and applications of granular flow (PFC5.0). *Rock Soil Mech.* **2018**, *39*, 36.
- Oda, M. Co-ordination number and its relation to shear strength of granular material. *Soils Found.* **1977**, *17*, 29–42. [CrossRef]
- Radjai, F.; Wolf, D.E.; Jean, M.; Moreau, J.J. Bimodal character of stress transmission in granular packings. *Phys. Rev. Lett.* **1998**, *80*, 61. [CrossRef]
- Majmudar, T.S.; Behringer, R.P. Contact force measurements and stress-induced anisotropy in granular materials. *Nature* **2005**, *435*, 1079–1082. [CrossRef]
- Azéma, E.; Radjai, F. Force chains and contact network topology in sheared packings of elongated particles. *Phys. Rev. E* **2012**, *85*, 031303. [CrossRef]
- Tempone, P.; Landro, M.; Fjær, E. 4D gravity response of compacting reservoirs: Analytical approach. *Geophysics.* **2012**, *77*, G45–G54. [CrossRef]
- Maiti, R.; Das, G.; Das, P.K. Experiments on eccentric granular drawing from a quasi-two-dimensional silo. *Powder Technol.* **2016**, *301*, 1054–1066. [CrossRef]

42. Cannavacciuolo, A.; Barletta, D.; Donsì, G.; Ferrari, G.; Poletto, M. Arch-Free flow in aerated silo discharge of cohesive powde *Powder Technol.* **2009**, *191*, 272–279. [CrossRef]
43. Han, G.; Gong, Q.; Zhou, S. Particle flow simulation analysis of micro-mechanism of soil arching effect of friction-type geotechni materials. *Rock Soil Mech.* **2013**, *34*, 1791–1798.

applied sciences

MDPI

rticle

Research on the Movement of Overlying Strata in Shallow Coal Seams with High Mining Heights and Ultralong Working Faces

uping Fu [1,2,*], Chuantian Li [1,2] and Yongliang He [1,2]

1 School of Engineering for Safety and Emergency Management, Taiyuan University of Science and Technology, Taiyuan 030024, China; hyl@tyust.edu.cn (Y.H.)
2 Intelligent Monitoring and Control of Coal Mine Dust Key Laboratory of Shanxi Province, Taiyuan University of Science and Technology, Taiyuan 030024, China
* Correspondence: 2011014@tyust.edu.cn

Abstract: To study the roof movement and ground pressure evolution characteristics of an ultralong working face in a shallow coal seam with a high mining height, the Shangwan Coal Mine in the Shendong mining area was used as the research background, and the physical and mechanical parameters of the surrounding rock were determined through rock mechanics experiments. A physical simulation model was built considering the 7 m mining height of the 12301 fully mechanized working face of the Shangwan Coal Mine to simulate and study the evolutions of the movement, fracture and collapse of the coal seam, direct roof, and basic roof and overlying strata during the mining process. The mechanical characteristics of the support, mechanism of roof collapse, and changes in the working resistance of the support were analysed and simulated. The research results indicate that when mining at a height of 7 m, the direct roof and basic roof strata collapse in layers; the basic roof strata collapse backwards, the rock block arrangement is more irregular, and the range of the basic roof that can form structural rock layers extends higher. After the basic roof rock fractures, it cannot form a masonry beam structure and can only form a cantilever beam structure. The periodic fracture of the cantilever beam causes periodic pressure on the working face. These research results are of great significance for planning the further mining of shallow coal seams with high mining heights and ultralong working faces in the Shendong mining area, as well as for improving the control of overlying strata.

Keywords: shallowly buried coal seams; high mining height fully mechanized mining; ultralong working face; physical simulation experiments; rock movement

1. Introduction

There are many shallow coal seams in northwest China, including Gansu, Ningxia, and Inner Mongolia. The burial depth of shallow coal seams is usually approximately 150 m, and such coal seams have typical characteristics, such as ultralong working faces, thin bedrock, and overlying thick loose sand layers [1–3]. The study of the mining pressure evolution at working faces with high mining heights is crucial for the efficient and safe mining of shallow and thick coal seams [4,5]. Based on the existing mining practice of fully mechanized shallow coal seam working faces with high mining heights in China, compared to the fully mechanized working faces of thin and moderately thick coal seams, the scale of goafs significantly increases [6]. After mining, the overlying roof rock layer fractures and fills a goaf, causing significant changes in the period, strength, and range of overlying rock movement and mining pressure in fully mechanized working faces with high mining heights [7]. Mining pressure problems caused by the high mining height and ultralong working face of shallow coal seams, such as roof collapse and roadway damage, seriously threaten the safety of personnel and equipment [8,9]. With increasing mining intensity and depth, the stress concentration and degree of damage intensify. Therefore, it is necessary to study the movement of overlying strata in shallow coal seams and the evolution of mining pressure distribution in high mining heights and ultralong working faces to ultimately achieve safe and efficient mining.

Many scholars have conducted extensive research on the development and evolution of mining pressure in shallow coal resources with different burial depths and working face lengths using theoretical analysis, physical simulation, and numerical simulation methods [10]. Li Yuanhui [11] used numerical and physical simulations to study the stress and failure characteristics of gently inclined thin ore bodies. As the dip angle of an ore body increases, the roof of the corresponding deeply buried mining area changes from experiencing stress release to experiencing stress concentration at different mining stages. When the dip angle of the ore body is 30°, shear failure is most likely to occur. Liu Chang [12] used a dual-mode parallel electrical method to dynamically monitor the overlying rock of a shallowly buried coal seam with a high mining face. Due to the influence of mining, cracks appeared in the near-surface loose layer above the working face. Zhang Libo [13] studied the spatial distribution characteristics and corresponding support design of the development and evolution of mining support pressure in the 8.2 m superhigh fully mechanized mining face with the 108 working face of the 12^{-2} of Jinjitan Mine as the background, revealing the distribution characteristics of the dynamic stress field and internal and external stress fields. Li Ang [14] explored the safety and feasibility of deep coal seam floor mining in the North China Coalfield. Taking the coal seam mining in the Pingdingshan Coal Mine as the engineering background, $FLAC^{3D}$ was used to simulate and analyse the influence of mining length, mining depth, and mining height on the depth of floor fractures. Ling Chunwei [15] noted that the hydraulic support in the mining faces of shallow coal seams with large mining outputs and strong rock strata around the stopes often resulted in damage. On the basis of basic experiments and physical similarity simulations, he studied the overlying rock fractures of shallow coal seams in the western mining area of China under the action of hydraulic coupling. Du Wengang [16] embedded distributed fibre optic and fibre optic grating sensors to study the deformation and internal stress evolution characteristics of overlying strata after shallow and thick coal seam mining and used them to monitor the deformation of overlying strata during coal mining. Zhang Cun [17] used drones, visible light cameras, and infrared cameras to determine the distribution characteristics and dynamic evolution of surface cracks. Based on the theory of rock movement and combined with monitoring results, a three-dimensional model of overlying rock fractures after mining in the 12,401 longwall working face was established. Liang Shun [18] studied the surface deformation during shallow coal seam mining and observed that as the working face depth increased, the surface deformation increased, reaching its maximum value near the coal seam. Zhang Jie [19] used theoretical analysis, numerical simulation, and similar simulation methods to study the evolution characteristics and formation mechanism of interlayer rock fractures during coal seam mining. Lan Tianwei [20] studied the mining pressure characteristics of shallow coal seams in extremely dense goafs and overlying coal–rock working faces and proposed distinguishing areas of the working face with similar mining stresses according to the advancing length of the working face, namely, the high-pressure zone, transition zone, and low-pressure zone. Niu Huiyong [21] reported that the gradual reduction in shallow coal seam resources, increase in coal seam mining depth, and increase in underground temperature have led to an increasing risk of residual coal fires each year.

The methods used in the study of thick and hard roof under the influence of mining mainly involve on-site measurement of mining pressure parameter data and analysis of mining pressure laws, similar simulation test research and inversion of overlying rock fracture structure morphology, numerical simulation calculation research and surrounding rock stress analysis, as well as construction mechanics models and theoretical analysis [22,23]. When using theoretical analysis for research, it mainly includes elastic–plastic mechanics, analysing thick and hard rock layers as rock slabs overlapped with the coal rock mass below, or the use of material mechanics to simplify the overlying rock as a rock beam model for research. Alternatively, based on structural mechanics, the stress analysis of rock blocks after the fracture of thick and hard rock layers can be conducted to study the criteria for structural instability after the fracture of thick and hard rock layers. By using similar simulation research, it is possible to analyse the morphology and evolution law of overlying rock fracture structures under the influence of mining from a macro perspective, and to more intuitively study the fracture characteristics of thick and hard rock layers under the influence of mining [24,25]. Numerical simulation calculation research is a research method that has emerged with the help of computer software. Based on failure criteria and models, it can efficiently and conveniently calculate and simulate the fracture mechanism of thick and hard rock layers under the influence of mining. The mining pressure data and parameters obtained from on-site engineering tests are the results of the fracture behaviour of thick and hard rock layers under the influence of mining. However, for the fracture characteristics of overlying rock structures, comprehensive analysis needs to be carried out in combination with the above theoretical research methods to verify the fracture situation of thick and hard roof slabs under the influence of mining.

The above research investigated the overlying rock failure in coal seams, the characteristics of crack development, the manifestation of shallow coal seam pressure, and effects of rock control measures. There is currently no systematic theory to guide the production of coal with high mining heights and ultralong working faces under shallow burial conditions. The movement of the overlying rock layers in the shallow coal seams of the Shangwan Coal Mine at ultralong and high mining heights was investigated. This article uses similar physical simulations to study the movement of rock layers due to shallow coal seam mining and establishes a theory of surrounding rock control for long working faces in shallow coal seams. The results of this modelling work can provide a theoretical reference for the subsidence of shallow coal seam strata during mining and has important theoretical and practical value for coal mine safety production, surface building protection, and rock movement research.

2. Geological Conditions of Coal Seams in the Shangwan Coal Mine

The Shangwan Coal Mine is located in the territory of the E'er'yijinhuoluo Banner, Inner Mongolia. The minable coal seams of the Shangwan Coal Mine include the #12 upper coal, #22 coal, and #31 coal seams from top to bottom. A total of 10 boreholes were surveyed in the 1^{-2} coal seams of the Shangwan Mine, including 114, b214, sb10, sb11, sb12, sb15, b212, b213, Sh5, and W28. The rock parameters of these 10 geological boreholes are shown in Table 1. The 1^{-2} coal seams in the Shangwan mining area have burial depths ranging from 84.32 to 165.04 m, with an average of 115.47 m. The thickness of the alluvial sand varies from 0.8 to 23.85 m, with an average thickness of 13.37 m. The thickness of the bedrock is 73.04~148.31 m, with an average thickness of 102.1 m. The direct roof lithology is mostly sandy clay, sandy mudstone, or siltstone, with a thickness range of 0.55–7.11 m and an average thickness of 2.52 m. The basic roof lithology is mostly coarse sandstone and fine sandstone, with a thickness of 0–15.11 m and an average of 9.24 m. The thickness of the 1^{-2} coal seams in the Shangwan mining area ranges from 3.8 to 9.25 m, with an average thickness of 7 m.

From bottom to top, sandy mudstone, fine-grained sandstone, siltstone, fine-grained sandstone, siltstone, medium-grained sandstone, and siltstone is present. The coal seam spacing is small, and the overlying rock layer has a greater strength. A stratigraphic column of the coal seams is shown in Figure 1.

Table 1. Characteristics of the 1^{-2} coal seams in the Shangwan coal mine.

Borehole	Burial Depth/m	Alluvium/m	Bedrock/m	Immediate Roof/m	Thickness/m	Main Roof/m	Thickness/m	Coal Thickness/m
114	118.83	8.2	110.63	Sandy mudstone	7.11	Coarse sandstone	9.25	4.68
B214	84.32	11.28	73.04	Sandy clay	0.55	Fine and siltstone	4.56	7.1
Sb10	125.62	11.7	113.92	Sandy mudstone	0.6	Coarse sandstone	9.89	6.86
Sb11	110.25	13.88	96.37	Sandy mudstone	1.22	/	0	7.5
Sb12	120.8	20.93	99.87	Mudstone, sandy mudstone	1.71	Fine sandstone	5.83	5.88
Sb15	97.41	10.33	87.08	Sandy clay	5.62	Coarse sandstone	13.89	3.8
b212	88.77	0.8	87.97	Fine sandstone	5.49	Middle sandstone	15.11	9.25
b213	98.82	23.85	74.97	Fine sandstone	1.25	Coarse sandstone	12.41	8.64
Sh5	165.04	16.73	148.31	Sandy mudstone	1.03	Coarse sandstone	9.55	6.63
W25	144.8	15.98	128.82	Sand clay	0.6	Middle sandstone	11.95	9.62
Average	115.47	13.37	102.1		2.52		9.24	7.00

Columnar	Name	Thickness	Lithology description
	Siltstone	3.78	Black, mainly dark coal, containing a small amount of vitrinite, bright coal. Silky luster, dull luster, more developed cracks.
	Medium sandstone	0.40	Gray, mainly quartz and feldspar.
	Siltstone	0.78	Dark grey, argillaceous cement, wavy bedding.
	Fine sandstone	3.56	Grey, argillaceous cement, well sorted, central thin siltstone.
	Siltstone	3.81	Dark gray, argillaceous cement, oblique bedding development, visible a small amount of plant fragments and fossils, sandwiched thin layers of medium sandstone.
	Fine sandstone	1.46	Dark grey, argillaceous cement, wavy bedding.
	Sandy mudstone	2.53	Dark gray, uneven fracture, a small number of plant leaf fossils can be seen, local development of slippery surface.
	Coal	4.75	Black, mainly dark coal, containing a small amount of vitrinite, bright coal, silk luster, dull luster, cracks more developed.

Figure 1. Coal seam stratigraphic column.

3. Similarity Simulation Experiment

3.1. Experimental Design

According to the prototype conditions, a large planar strain testing device with a length × height × thickness of 3000 mm × 3000 mm × 200 mm was selected for laboratory testing [26,27]. The surroundings and bottom plate of the test bench are constrained by 20# channel steel and a 25 mm thick organic glass plate, with the top of the model being the free end. The studied coal seam is buried at a depth of 118.76 m and has a thickness of 7 m. Based on the prototype conditions and experimental setup, the geometric similarity ratio is determined to be $C_l = 1/50$, the bulk density similarity ratio is $C_\gamma = 17/25 = 0.68$, the stress similarity ratio is $C_\alpha = C_l \times C_\gamma = 0.68/50 = 0.0136$, and the time similarity ratio is $C_t = \sqrt{C_l} = 0.1414$.

The model has a length of 3000 mm, with 200 mm wide coal pillars left on both sides of the model. The model can simulate a working face advancement length of 2600 mm, which can reflect a real working face advancement length of 130 m. The model is 200 mm thick and can simulate a working face length of 10 m. In the model, the thickness of the coal seam floor is 208.6 mm, the thickness of the coal seam is 140 mm, the thicknesses of the coal seam roof stratum and aeolian sand are 2375.2 mm, and the total height of the model is 2723.8 mm. Based on the coal seam thickness and geological conditions in the Shendong mining area, a planar similarity simulation test bench was used to simulate the development of mining pressure and the evolution of the overlying rock movement during the mining process of a full-thickness working face with a coal seam thickness of 7 m.

Based on the similarity theory, experiments are conducted using similar simulation materials to simulate the hinge structure morphology and migration evolution process of overlying fractured rock blocks during underground coal and rock excavation construction. Essentially, it is based on the principle of similarity, calculating and reducing the physical quantities of underground mine rock layers according to similarity ratios, preparing similar materials, and establishing models. Then, in the model of similar simulation research, simulating underground coal and rock excavation construction according to the pre-designed plan can provide a visual analysis of the evolution process and failure-instability-induced disaster mechanism of the hinged structure of overlying rock blocks under the influence of strong mining from a macro perspective of the mining site. The micro and internal damage situation cannot be displayed, and the model is a simplified model, which has certain differences from the on-site production situation. The similarity phenomenon refers to the proportional relationship between geometric-related physical quantities and their corresponding instantaneous changes being a certain value, and the existence of the same physical laws. We refer to the ratio of the same dimensional physical quantity at the corresponding point in a similar process as the similarity coefficient, also known as the similarity constant. The similarity constant exists in a certain combination relationship in similar phenomena. Substituting the same physical quantities in the similarity index into the same system can generate dimensionless combinations of various physical quantities. We call this combination the similarity criterion. Calculate the geometric similarity ratio based on the dimensions of the prototype's working face and experimental setup.

3.2. Model Materials and Proportions

Considering the discontinuity of rock materials and the measured mechanical strengths of rocks, a crack influence coefficient of 0.8 was considered [28,29]. Based on the mechanical parameter data of on-site rock testing on the working face and referencing other relevant literature on geological rock testing on the working face, uniaxial compression experiments were conducted on standard specimens made in the laboratory, and the optimal ratio of similar materials through repeated adjustments were obtained. Then, based on the similarity theory, the physical and mechanical parameters of each rock layer in the model were obtained, the material ratio was selected, and the material dosage was calculated. For details on the model parameters, please refer to Table 2.

Table 2. Main parameters of the roof and floor of the simulated mining coal seam and proportions of materials adopted in the model.

Layer	Lithology	Layer Thickness /m	Density /(kg/m^3)	Compressive Strength /MPa	Elastic Modulus /GPa	Similarity Ratio	Sand /kg	Cement /kg	Calcium Carbonate /kg	Gypsum /kg	Water /kg	Borax /g
1	Windblown sand	11.70	1700	/	/	/	238.7	/	/	/	/	/
2	Siltstone	8.60	2460	40.6	35	655	152.9	/	12.3	12.3	33.6	336
3	Sandy mudstone	5.10	2240	22.8	23	346	74.6	9.3	/	14	9.3	186
4	Coarse sandstone	10.28	2430	36.6	35	855	185.7	/	11.3	11.3	20.4	204
5	Fine sandstone	7.03	2500	44.6	32	955	131.9	7.2	/	7.2	14.3	143
6	Sandy mudstone	9.8	2240	22.8	23	337	143.3	/	13.4	31.4	25.6	512
7	Coarse sandstone	11.24	2430	36.6	35	855	203.1	/	12.4	12.4	22.3	223
8	Sandy mudstone	27.52	2240	22.8	23	337	402.4	/	37.7	88.0	71.9	1437
9	Coarse sandstone	9.89	2430	36.6	35	855	178.7	/	10.9	10.9	19.6	196
10	Sandy mudstone	6.17	2240	22.8	23	337	90.2	/	8.5	19.7	16.1	322
11	Fine sandstone	7.27	2500	44.6	32	955	136.4	7.4	/	7.4	14.8	148
12	Sandy mudstone	4.16	2240	22.8	23	337	60.8	/	5.7	13.3	10.9	217
13	Coal seam	5.5	1480	10.5	15	373	53.1	11.6	/	5	6.6	133
14	Siltstone	1.48	2460	40.6	35	655	26.3	/	2.1	2.1	4.2	42
15	Sandy mudstone	3.25	2240	22.8	23	337	47.5	/	4.5	10.4	8.5	170
16	Fine sandstone	5.70	2500	44.6	32	955	107	5.8	/	5.8	11.6	116

3.3. Model Testing

The Nikon DTM-531E total station was used to measure the coordinates of the measurement points arranged in the direct roof, basic roof, and overlying rock layers of the coal seam to obtain the displacement of the model. A CM-2B static strain gauge was used to collect the strain of the stress box pre-embedded inside the model. The stress box was made of a BE120-3AA resistance strain gauge, a custom-made copper ring, and a copper sheet. The calibration of the stress box was carried out before the test. The strain of the stress box measured by the strain gauge in the experiment was calculated via the calibration data. The stress box is shown in Figure 2, and the stress testing system is shown in Figure 3. A digital camera was used to record typical phenomena during the mining process.

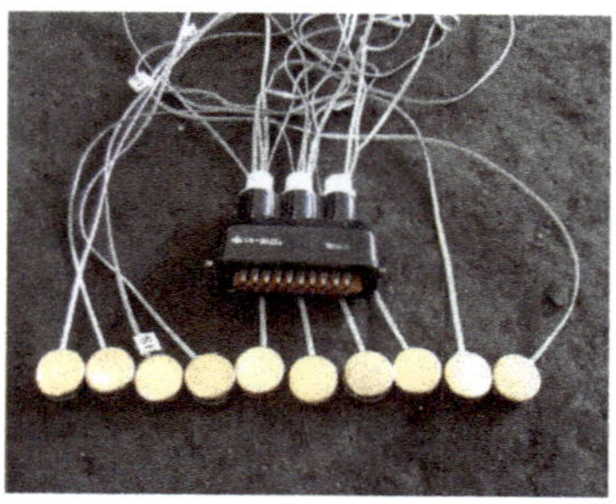

Figure 2. Pressure cell for measuring stress.

(1) Layout and testing methods of the rock movement measurement points

The displacement measurement points inside the rock layer can be conveniently used to monitor the movement of the rock layer, as shown in Figure 4. Thirteen rows of measurement points are arranged horizontally on the overlying strata of coal seam 1–2, with a spacing of 20 cm. The survey points were arranged 5 cm above the coal seam in the vertical direction. The first four rows had a spacing of 10 cm, and the last three rows had a spacing of 15 cm. A total of 96 displacement measurement points were arranged.

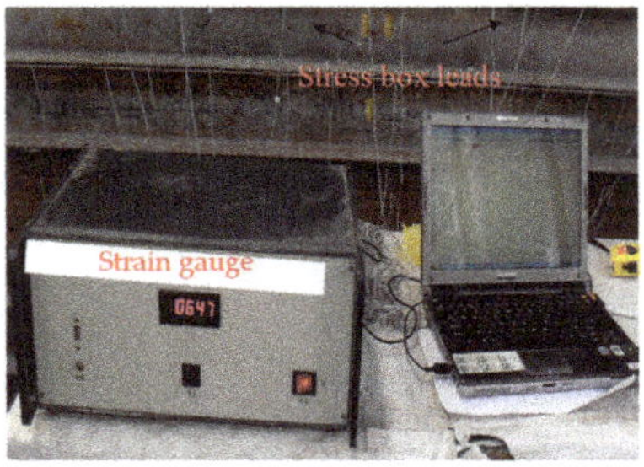

Figure 3. System for measuring stress.

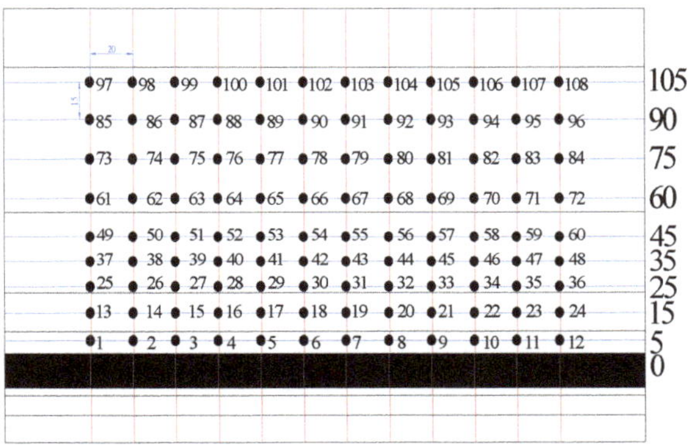

Figure 4. Schematic diagram of the displacement measurement points.

(2) Layout and testing methods of pressure measurement points

In the model experiment, to determine the distribution of the support pressure in front of the working face during the mining process, stress sensors were installed in conjunction with static and dynamic strain gauges, as well as a computer automatic data acquisition system for model stress measurement. Two rows of BE120-03AA (11) resistance strain gauges were buried with the stress sensors, 5 cm, 15 cm, and 25 cm from the simulated mining coal seam roof. The stress of the direct roof and basic roof during the mining process was measured. The spacing between points was 20 cm. The layout of the stress observation sensors for simulating the mining of coal seams 1–2 is shown in Figure 5.

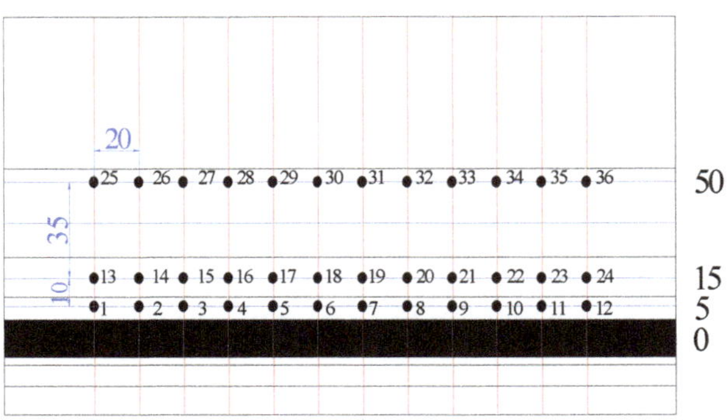

Figure 5. Schematic diagram of the stress measurement points.

4. Analysis of the Similar Experiment Results

4.1. Experimental Process Description

After the model was made and dried for 5 days, the glass panel was opened for displacement measurement point arrangement, as shown in Figure 6. The base point for the total station measurement was selected, the initial readings of each measurement point were taken, and the pressure box and strain gauge were connected, as shown in Figure 7.

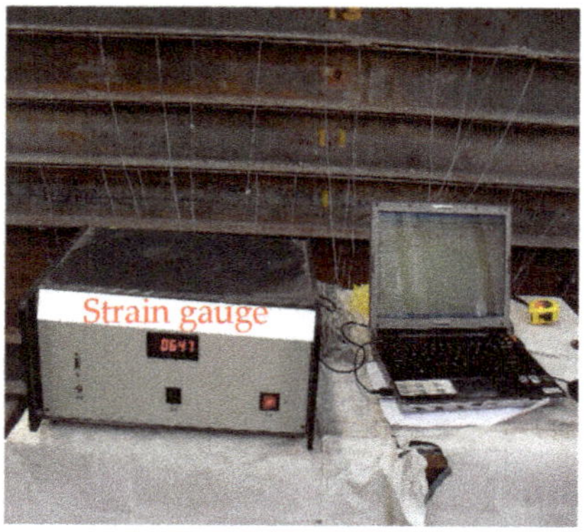

Figure 6. The connection of the stress boxes and testing instrument.

Figure 7. Diagram of the displacement measurement points.

A cut was made at a distance of 10 m from the edge of one side of the model, with a width of 8 m and a height of 5.5 m. A four-column hydraulic support was placed inside the cut. The underground work was carried out according to the 38th working system, with two-and-a-half shifts for coal cutting and half-shifts for maintenance. The coal cutting time for each knife cut in the actual scenario is approximately 1.3 h, and the advance distance is 0.8 m. Therefore, each shift cuts 6 coal seams and 15 coal seams per day. According to the time and geometric similarity ratio, the coal cutting time for each knife cut in the experiment was 11 min, and the advance distance was 0.8 m, with a total progress of 12 m per day. The relationship between the distance and cutting process is shown in Table 3.

Figure 8 shows the evolution process of the first collapse of the direct roof during the advancement of the working face. When the working face was advanced to 15.2 m (9th cut), the first and second layers separated. When the working face was advanced to 17.6 m (12th cut), directly above the first layer of collapse, there was a collapse thickness of 1 m. Separation occurs between the second and third layers. When the working face was advanced to 19.7 m (the 14th cut), the second layer collapsed, and the third layer collapsed in blocks.

The total thickness of the collapse was 3 m, and the remaining height of the goaf space was 4.5 m. When the working face was advanced to 20 m (15th cut), all the collapsed roof moved. The goaf was directly filled without causing collapse, with a maximum free space height of 3.6 m. The direct roof collapsed completely, and the basic overhanging length of the roof was 10 m. The direct collapse angle (on the side near the coal pillar) was 52°.

Table 3. Distances and cutting process.

Distances/m	Cutting Process	Distances/m	Cutting Process	Distances/m	Cutting Process
15.2	19	40	50	84.8	106
17.6	22	50.4	63	95.2	119
19.7	24	55.2	69	100	125
20	25	64.8	81	110.4	138
29.6	37	69.2	86.5	115.2	144
35.2	44	80	100		

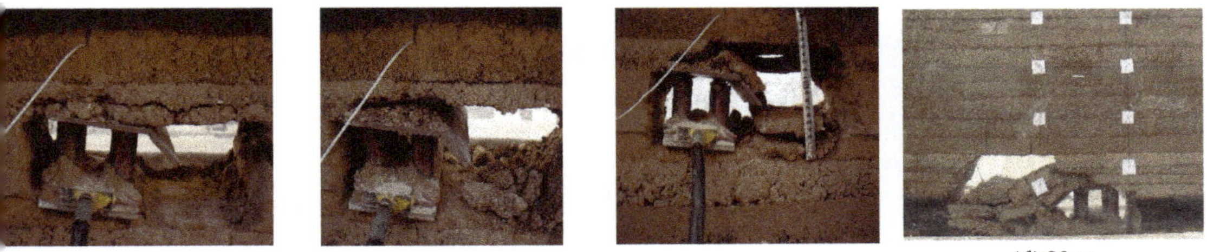

(**a**) 15.2 m (**b**) 17.6 m (**c**) 19.7 m (**d**) 20 m

Figure 8. Evolution process of the first collapse of the immediate roof.

Figure 9 shows the evolution process of the initial collapse of the basic roof during the advancement of the working face. When the working face was advanced to 29.6 m (the 27th cut), the first and second layers of the basic roof began to collapse. When the working face was advanced to 35.2 m (the 34th cut), the second layer of the structure formed by collapse experienced sliding instability. When the working face was advanced to 20 m, the third and fourth layers of the basic roof collapsed. At 50.4 m (53rd cut), the fifth layer of the basic roof collapsed, causing the overlying rock layer to collapse. The working face was under initial pressure, and when the working face was advanced to 50.4 m, a crack with a length of 20 m appeared at the top of the third layer at 8.5 m in front of the working face.

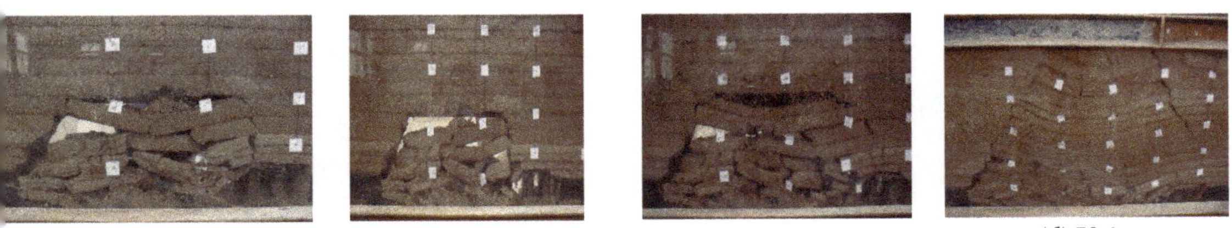

(**a**) 29.6 m (**b**) 35.2 m (**c**) 40 m (**d**) 50.4 m

Figure 9. Evolution process of the initial collapse of the basic roof.

When the working face was advanced to 55.2 m in Figure 10, the cracks in front of the working face extended upwards and downwards, penetrating directly the rock above the support. When the working face was advanced to 64.8 m, the cracks in front of the working face continued to extend. New fracturing cracks appeared directly above the support. When the working face was advanced to 69.6 m, with various cracks developing and the basic roof breaking, the overlying rock layer collapsed and caused the first periodic pressure on the working face. After being left for 12 h, cracks appeared at a height of 12.5 m above the fifth layer in front of the working face at a height of 15 m. When the working face was advanced to 75.2 m, the crack in the working face extended downwards to the third layer of the basic roof, with a length of 25 m. When the working face was advanced to 80 m, the cracks continued to expand downwards to the direct roof, reaching 50 m upwards. The subsidence of the top plate increased, and the second cycle of pressure developed.

(**a**) 55.2 m (**b**) 64.8 m

Figure 10. The crack in the working face during coal mining.

When the working face was advanced to 84.8 m, after the removal of the support, the roof rock layer was cut off along the back of the support, and the second cycle of pressure was completed. When the working face was advanced to 95.2 m, a crack with a length of 17.25 m above the coal wall formed, indicating the start of the third cycle of pressure. When the working face was advanced to 100 m, the third basic roof fracture caused periodic pressure with a step distance of 20 m. When the working face was advanced to 110.4 m, a crack approximately 20 m long appeared in the fourth and fifth layers above the coal wall. When the working face was advanced to 115.2 m, cracks expanded above the coal wall, the roof sank, and the fourth cycle of pressure was applied to the working face in Figure 11.

4.2. Roof Collapse Evolution

(1) Direct roof: First collapse step distance

When the working face was advanced to 15.2 m, there was separation between the first and second layers. When the working face was advanced to 17.6 m, the rock directly above the first layer collapsed, with a collapse thickness of 1 m. Separation occurred between the second and third layers. When the working face was advanced to 20 m, the entire roof collapsed. The goaf was directly filled without collapsing, with a maximum free space height of 3 m. The direct roof collapsed completely, and the basic overhanging length of the roof was 10 m. The direct collapse angle (on the side near the coal pillar) was 52°.

(2) Basic roof: First collapse step distance

When the working face was advanced to 29.6 m, the first and second layers of the basic roof began to collapse. After the second layer fractured, the two pieces interlocked to form a three-hinged arch structure. At 40 m, the third and fourth layers of the basic roof collapsed. At 50.4 m, the fifth layer of the basic roof collapsed, and the overlying rock layer collapsed accordingly. The working face was under initial pressure, with a step distance of 50.4 m in Table 4.

(3) Periodic loading step distance

After experiencing the initial pressure, as the working face continued to advance, the structure formed by the basic roof strata always underwent a change from stable to unstable and then back to stable, which was a cyclic process called periodic pressure [30,31]. During the excavation process of the model, the working face underwent four cycles of pressure, with pressure steps of 19.2 m, 10.4 m, 20 m, and 15.2 m.

(**a**) 69.2 m

(**b**) 80 m

(**c**) 84.8 m

(**d**) 95.2 m

(**e**) 110.4 m

(**f**) 100 m

(**g**) 115.2 m

Figure 11. Roof collapse at the working face.

Table 4. Statistics of the immediate roof and basic roof collapse intervals.

Collapse Interval	Pressure	Distance of the Working Face Advance under Pressure		Loading Step	
		Model/mm	Prototype/m	Model/mm	Prototype/m
1	First direct roof collapse	400	20	400	20
2	Basic roof initial pressure	1008	50.4	1008	50.4
3	First cycle of pressure	1392	69.6	384	19.2
4	Second cycle of pressure	1600	80	208	10.4
5	Third cycle of pressure	2000	100	400	20
6	Fourth cycle of pressure	2304	115.2	304	15.2
7	Mean cycle pressure step distance			324	16.2

4.3. Movement Pattern of the Overlying Rock Layers

A displacement measurement point was arranged in the middle of the direct roof, with the number of measurement points ranging from #1 to #12. A measurement point slowly sank after a hole was cut from the working face. As the working face was advanced, layer separation occurred in the roof, and the displacement of the measurement point increased sharply. When the working face was advanced to 20 m, displacement measurement point #1 collapsed directly. When the working face was advanced to 35.2 m, displacement measurement point #2 collapsed directly. When the working face was advanced to 50.4 m, displacement measurement point 3 directly collapsed. The corresponding displacement curves are shown in Figure 12.

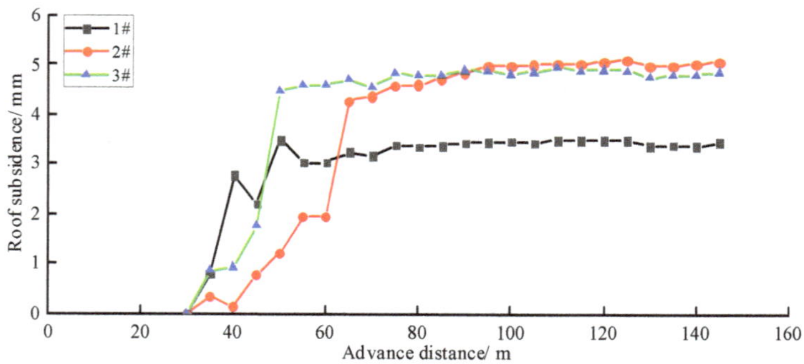

Figure 12. Immediate roof convergence with advance of the working face.

Through the analysis of the relationship curves between the settlement of displacement measurement points #1, #2, and #3, and the advancing distance of the working face, it can be concluded that before direct roof collapse, the direct roof generally experienced delamination and settlement, usually approximately 1–2.5 m. The amount of sinking of the collapsed roof was between 3.5 and 4.5 m. As the working face continued to advance, the amount of sinking of the direct roof under the pressure of the overlying rock layer gradually grew to approximately 5 m and then stabilized.

4.4. Basic Roof Subsidence and Collapse

Two layers of displacement measurement points in the middle and upper parts of the basic roof were arranged to measure the settlement of the basic roof [32,33]. The measurement points are numbered from #13 to #36. The variation curves of the subsidence of measurement points #13~15 and #25~27 with respect to the advancing distance of the working face are shown in Figure 13. The basic roof collapsed in layers. When the working face was advanced to 35 m, the lower layer measurement points collapsed when the basic roof first collapsed, while the upper layer measurement points collapsed when the basic roof collapsed only approximately 50 m. The settlement of the upper and lower displacement measurement points before collapse was between 1 and 1.5 m.

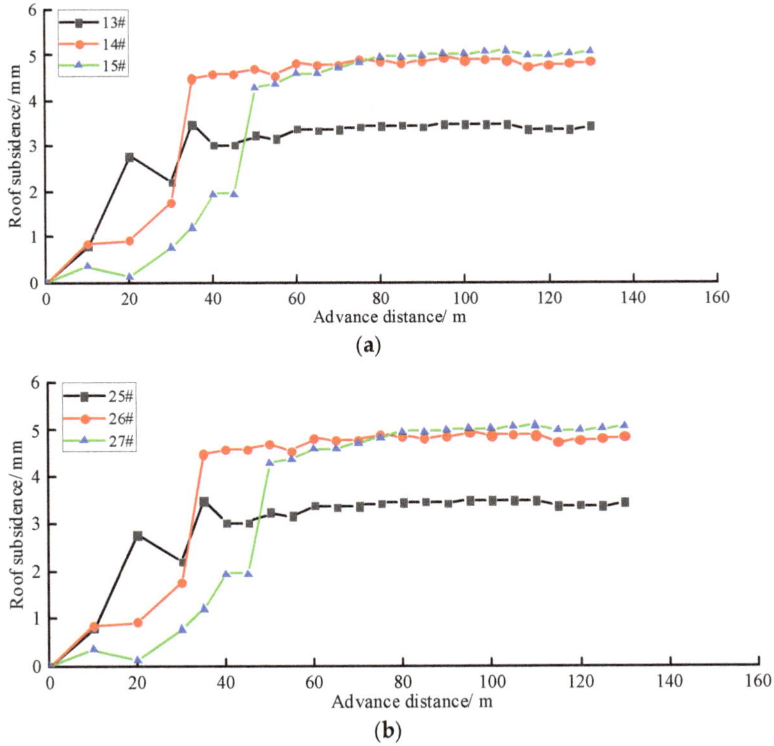

Figure 13. Basic roof convergence with advance of the working face.

4.5. Subsidence and Collapse of Rock Layers

There were five layers of displacement measurement points buried in the rock layer above the basic roof, with the first layer 100 mm from the measurement point above the basic roof and the other four layers spaced 150 mm apart. The measurement points at the third and sixth layers were blocked by channel steel, making it impossible to measure the data. The selected displacement measurement points were 1400 mm from the model boundary and are numbered #43, #55, #79, and #91. The relationship curve between the settlement of the measurement points and the advancing distance of the working face is shown in Figure 14. Before the working face was advanced to 65 m, the subsidence of the rock layer above the basic roof was very small. Later, under the pressure of the roof, tensile cracks appeared in the overlying rock layer in front of the working face, and the subsidence of the displacement measurement point gradually increased. When advancing to 80 m, the sinking speed of the displacement measurement point increased, and it entered the second cycle of pressure on the basic roof. The rock layer above the basic roof collapsed with the collapse of the basic roof and was finally compacted. The sinking amount of the displacement measurement point tended to stabilize [34].

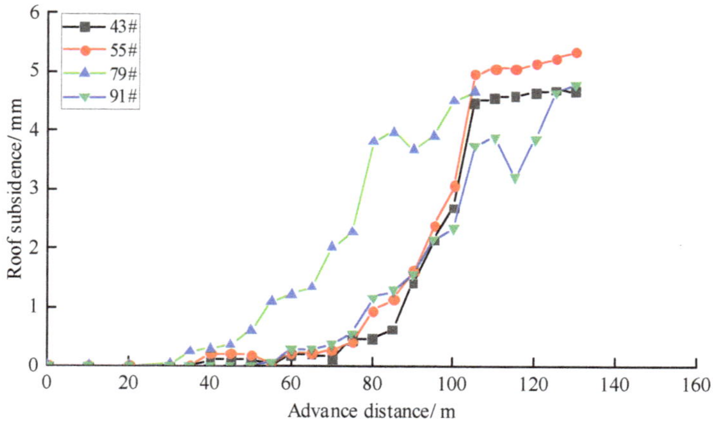

Figure 14. Variation in the convergence of the overlying strata with the advance of the working face.

4.6. Distribution Characteristics of Support Pressure

The maximum stress at the first-layer stress measurement point occurred an average distance of 12.5 m from the working face, with an average stress of 7.68, an average stress concentration coefficient of 2.64, and an average range of influence of the support pressure of 32.5 m. The maximum stress at the second-layer stress measurement point occurred 8.75 m from the working face on average, with an average stress of 7.62, an average stress concentration coefficient of 2.74, and an average range of influence of the support pressure of 32.5 m. The maximum stress at the third stress measurement point occurred an average distance of 5 m from the working face, with an average stress of 7.71, an average stress concentration coefficient of 2.9, and an average range of influence of the support pressure of 35 m. According to the stress measurement point data, as the layer of the roof rock at the stress measurement point increased, the maximum support pressure approached the working face, and the stress magnitude, stress concentration coefficient, and influence range did not substantially change. The maximum stress at the third-layer stress measurement point was 8.75 m from the working face on average, with an average stress of 7.67, an average stress concentration coefficient of 2.76, and an average range of influence of the support pressure of 33.3 m.

The overlying rock in the mining area experienced the influence of repeated mining on the previous adjacent working face and this working face before and after mining. After mining, the thick and hard rock layer in the goaf broke, and the adjacent fractured rock blocks formed a masonry rock beam structure under the interlocking effect. After the coal seam in the lower working face is mined out, an unloading space is formed in the overlying rock of the roof above the goaf. As the coal seam is gradually mined, when the unloading space range meets certain conditions, the overlying rock layer will collapse under the action of its own weight and overlying rock load [35,36]. The overlying rock collapse pattern is a transitive development trend from the bottom to the top of the coal seam roof, with obvious layering and grading characteristics. The activity of the overlying strata in the goaf is the result of the interaction between the stress in the mining area and the strength of the overlying strata. When the roof overlying strata collapse to the position where the ultimate bearing rock layer is located, their collapse height no longer develops. In the overlying strata in the goaf, the structural effect of "stress arch" is formed due to the self-organizing limitation of the overlying strata failure and movement. At the same time, the interaction between the fractured blocks after the collapse of the overlying rock in the mining area forms the "fractured arch"; the "fractured arch" in the overlying rock structure of the mining site is located within the range of the "stress arch" pressure relief zone. The stress arch mainly bears and transmits the load of the overlying rock layer; the hinge structure formed by the fracture of the overlying rock in the "fractured arch" mining area is the main load factor for the manifestation of strong mining pressure.

5. Discussion

Through the observation of mining pressure in the practice of mining working faces with high heights and ultralong lengths, as well as research on existing theories of mining pressure and rock layer control, it is shown that changes in the length of the working face will lead to changes in the roof pressure during the mining process, which indicates that the height of the roof rock layer movement and collapse in the mining area changes, and this change conforms to the pressure balance arch in the Proctor theory. The Proctor theory suggests that the roof strata above a certain span space (working face length) will experience arch collapse, and the height of the arch is related not only to the collapse but also to the strength of the rock. The specific relationship is shown in Formula (1). The height of the roof rock caving is

$$H_{M,L_g} = \frac{L_g}{2f} \tag{1}$$

where H_{M,L_g} is the falling height, m; L_g is the working face length, m; and f is Proctor's coefficient.

5.1. Theoretical Prediction of the Impact of Changes in Working Face Length on Support Strength

The collapse height of the roof rock layer under different working face lengths can be determined, and therefore, the mining pressure induced during the unloading period of the working face can be obtained as

$$q_2 = \frac{\gamma \bullet L_g}{2f} \tag{2}$$

where q_2 is the induced mining pressure in the normal working face, kPa, and γ is the rock bulk density, kN/m^3.

There are many coal seams with thicknesses greater than 6.3 m and a geological reserve of 1351.99 Mt in the Shendong mining area. The current problem faced by the Shendong mining area is what kind of coal mining method to adopt to achieve scientific, economic, safe, and efficient mining and to maximize the recovery rate of the coal resources. In general, there are three main mining methods for thick coal seams: layered mining, top coal caving mining, and full-height mining. The following work analyses and demonstrates the advantages and disadvantages of these three methods and their applicability to the Shendong mining area from different perspectives.

5.2. Analysis of the Applicability, Advantages, and Disadvantages of Layered Mining in Extrathick Coal Seams

(1) Layered mining is technically feasible

For the mining of thick coal seams, layered mining was mainly adopted in the 1970s and 1980s. Many experts and scholars have conducted comprehensive research on the technical issues in layered mining of thick coal seams, making it a mature and feasible technology. When mining thick coal seams in layers, metal mesh must be laid during the mining process of the first layer to form a false roof and facilitate the mining of the lower layer. The hanging mesh method for long-wall layered mining of thick coal seams was studied for different roof types, and good results were achieved via trial. A study was conducted on the movement of overlying strata in the layered mining of ultrathick coal seams. Due to the increase in total mining height, the movement of rock layers in the fractured zone is different from that in the curved subsidence zone. The rock layers in the curved subsidence zone still follow the sinking trend of masonry beams, while the rock layers in the fractured zone move in a stepped manner. The horizontal distance of the steps is suitable for the pressure step distance in the mining area. An experiment using a cemented roof to mine thick coal seams layer-by-layer showed that after the underground solidification period of 3 months, the time for the cemented roof material to reach its maximum strength is 28 days. Thus, it is best to use cover-type supports at the mining face.

(2) Layered mining has poor applicability in terms of safety, efficiency, production, and economy

For the 7 m thick coal seam in the Shendong mining area, if layered mining technology is adopted, the impact on the high production and efficiency of the mine and the economic efficiency of coal seam mining mainly includes the following aspects.

① Large amount of excavation work and maintenance costs for layered mining tunnels

The 7 m thick coal seam adopts a layered mining method, which requires at least two layers of mining. Therefore, the amount of excavation work for the tunnel needs to be doubled. Moreover, the roof of the lower layer roadway is prone to fragmentation due to the compaction of the collapsed gangue in the upper layer, resulting in an increase in maintenance costs for the roadway.

② The increasing number of moves in the layered mining face makes it difficult to achieve high productivity and efficiency.

For the 7 m thick coal seam, layered mining requires at least two layers of mining, so the layered mining face needs to be moved at least twice, and frequent movement of the face will lead to a waste of time and difficulty in achieving high production and efficiency.

③ Low recovery rate of layered mining

The practice of layered mining in thick coal seams has shown that for the Shendong 7 m thick coal seam, layered mining is generally divided into two layers, and a top coal layer of 0.5–1.0 m must be left between the layers to ensure the stability, safety, and control of the roof of the machine passage space during mining. Leaving the top coal will decrease the recovery rate of the coal seam.

④ Increased risk of spontaneous combustion of coal remaining in the goaf during layered mining

The layered mining process requires leaving a portion of the top coal to maintain the stability of the roof during lower layer mining. This portion of the coal is left in the goaf after the lower layer of the working face is mined, which poses a risk of spontaneous combustion. Therefore, the cost of fire prevention during the mine production process will definitely increase.

The above analysis shows that the use of layered mining in a 7 m thick coal seam is technically feasible, but it will increase coal mining costs, increase risks, and make it difficult to achieve high production and efficiency.

5.3. Analysis of the Applicability, Advantages, and Disadvantages of Fully Mechanized Caving Mining in Extrathick Coal Seams

The coal seams in the Shendong mining area are characterized by shallow burial, thin bedrock, moderate hardness, and high toughness. These conditions cause fully mechanized top coal caving mining to be ineffective; the recovery rate is low, and it is difficult to achieve high production and efficiency.

① Shallowly buried coal seam conditions (50–200 m) result in a poor roof coal caving ability

Under the conditions of shallowly buried coal seams, the support pressure is low, and the top coal cannot be fully broken under the support pressure. Therefore, if top coal caving is used to extract the extra-thick coal seam in the Shendong mining area, the block size of the falling top coal during mining is large, and the caving performance is poor.

② The conditions of the moderately hard and high-toughness coal seams in the Shendong mining area make it difficult to ensure the top coal release rate

The coal seams in the Shendong mining area are Jurassic coal seams with a uniaxial compressive strength of 20–30 MPa and high toughness. The crushing and caving characteristics of the top coal are closely related to its uniaxial strength. The difficulty in ensuring the top coal release rate leads to a lower recovery rate of the entire coal seam.

Considering the technical conditions and high production and efficiency of the Shendong extra-thick coal seam, fully mechanized top coal caving mining cannot adapt to the mining of the 7 m extra-thick coal seam.

5.4. Applicability Analysis of High Mining Height in Ultrathick Coal Seams

In recent years, the mining height in the Shendong mining area has increased from working face heights of 4.5 m to 5.5 m and 6.3 m, and the mining technology used to achieve high mining heights in this mining area is state-of-the-art within China and even worldwide.

(1) Large-mining-height technology can adapt to the mining of thick coal seams in the Shendong mining area

There have been many successful production practices in the Shendong mining area using high-mining-height technology, with rich technical experience. Moreover, the heights of the large working faces in the Shangwan and Bulianta mines reached 6.3 m. With the study of practical rock control and coal mining technology at a height of 6.3 m and scientific research and key technology development for a height of 7 m, it is possible to carry out high mining of the 7 m thick coal seam in the Shendong mining area while ensuring safety, high production, and efficiency.

(2) The Shendong mining area has an extremely thick coal seam, and high-mining-height technology is economically optimal

The main factors affecting the economic viability of the Shendong mining area's ultrathick coal seam, if high-mining-height technology is used, are as follows:

① The amount of excavation work for high-mining-height mining tunnels is small

Compared to layered mining and fully mechanized top coal caving mining, mining extra-thick coal seams with working face heights greater than 7 m requires less excavation and support at high mining heights. The tunnel system is simple, and excavation replacement is easier to coordinate, which is conducive to the high yield and efficient mining of extra-thick coal seams.

② Low frequency of movement during high-mining-height mining

For a coal seam with a thickness of 7 m, an area of coal seam is moved only once at full height, with the lowest cost.

③ High coal seam recovery rate during high-mining-height mining

The Shendong mining area uses a 7 m thick coal seam and hydraulic supports with appropriate support heights. High-mining-height mining technology can extract most of the resources from thick coal seams, and the recovery rate of the working face can reach over 95%.

Based on the above analysis, high-mining-height mining can achieve a high yield and efficient mining of 7 m thick coal seams both technically and economically.

6. Conclusions

(1) This study was based on the parameter information of rock pressure manifestation in the surrounding rock of the fully mechanized caving face with thick and hard roof under mining in Shangwan Coal Mine. The manifestation law of mining pressure induced by the fracture of thick and hard roof under mining was clarified. When mining at a height of 7 m, the initial collapse step distance of the direct roof is 20.8 m, and the initial collapse step distance of the basic roof is 44.8 m. During the mining simulation process, a total of six cycles of pressure are experienced, with an average step distance of 12.9 m.

(2) Based on the manifestation law of mining pressure induced by roof fracture under mining, this paper reveals the time course required for the development, penetration, and instability of thick and hard rock strata cracks in the working face under strong mining, and elaborates on the temporal effect and regional characteristics of the development and evolution of rock strata cracks under mining. Before direct roof collapse, the amount of direct roof subsidence is generally approximately 0.5–2.5 m, while after collapse, the amount of direct roof subsidence is 3.5–5 m. As the working face continues to advance, the amount of direct roof subsidence under the pressure of the overlying rock layer gradually increases to approximately 5–7 m and then stabilizes.

(3) Based on the fracture structure and evolution law of a thick and hard roof under strong mining, the energy storage mechanism of a thick and hard roof under strong mining and the main control factors causing disasters are elucidated. When mining at a height of 7 m, the advanced support pressure in front of the coal wall has a certain distribution. From the analysis of the data measured by the stress sensor in the experiment, the average distance between the maximum stresses of the first-layer stress measurement point and the working face is 9.25 m, the average stress is 7.45 MPa, the average stress concentration coefficient is 2.56, and the average range of influence of the support pressure is 30.75 m.

Author Contributions: Methodology, software, validation, C.L.; writing—review and editing, Y.F.; writing—original draft, Y.H. All authors have read and agreed to the published version of the manuscript.

Funding: This work was supported by the Taiyuan University of Science and Technology Scientific Research Initial Funding (20222112), Fundamental Research Program of Shanxi Province (202203021222184), the Reward Fund for Excellent Doctors Working in Shanxi Province (20232039).

Institutional Review Board Statement: Not applicable.

Informed Consent Statement: Not applicable.

Data Availability Statement: The data used to support the findings of this study are available from the corresponding author upon request. The data are not publicly available due to privacy.

Conflicts of Interest: The authors declare that they have no conflict of interest regarding the publication of this study.

References

1. He, X.; Zhao, Y.; Yang, K.; Zhang, C.; Han, P. Development and formation of ground fissures induced by an ultra large minin height longwall panel in Shendong mining area. *Bull. Eng. Geol. Environ.* **2021**, *80*, 7879–7898. [CrossRef]
2. Shi, X.; Zhu, S.; Zhang, W. Study on the mechanisms and prevention of water inrush events in a deeply buried high-pressure co seam—A case study of the Chensilou Coal Mine in China. *Arab. J. Geosci.* **2019**, *12*, 614. [CrossRef]
3. Li, L.; Wu, K.; Hu, Z.; Xu, Y.; Zhou, D. Analysis of developmental features and causes of the ground cracks induced by oversize working face mining in an aeolian sand area. *Environ. Earth Sci.* **2017**, *76*, 135. [CrossRef]
4. Ma, Z.; Zhang, D.; Cao, Y.; Yang, W.; Xu, B. Study of Key Technology of Gob-Side Entry Retention in a High Gas Outburst Co Seam in the Karst Mountain Area. *Energies* **2022**, *15*, 4161. [CrossRef]
5. Zhao, P.; Zhuo, R.; Li, S.; Lin, H.; Shu, C.; Wang, B.; Jia, Y.; Suo, L. Fractal characteristics of gas migration channels at differer mining heights. *Fuel* **2020**, *271*, 117479. [CrossRef]
6. Bustin, A.; Bustin, R. Contribution of non-coal facies to the total gas-in-place in Mannville coal measures, Central Alberta. *Int. Coal Geol.* **2016**, *154–155*, 69–81. [CrossRef]
7. Fan, H.; Wang, L.; Lu, Y.; Li, Z.; Li, W.; Wang, K. Height of water-conducting fractured zone in a coal seam overlain by thi bedrock and thick clay layer: A case study from the Sanyuan coal mine in North China. *Environ. Earth Sci.* **2020**, *79*, 12 [CrossRef]
8. Li, Z.; Zhang, H.; Jiang, Z.; Feng, G.; Cui, J.; Ma, J. Research on failure criteria and collapse height of roadway roof strata based o energy accumulation and dissipation characteristics. *Energy Sci. Eng.* **2021**, *9*, 2461–2473. [CrossRef]
9. Fu, Q.; Yang, J.; Song, H.; Wu, X.; Liu, Y.; Wei, X. Study on the Method of Pressure Relief and Energy Absorption for Protectin Roadway Under Thick and Hard Roof. *Rock Mech. Rock Eng.* **2023**, *56*, 7177–7196. [CrossRef]
10. Fu, J.; Wen, G.; Sun, H.; Yang, X. Study on the shear movement law of overlying strata by slice mining. *Energy Sci. Eng.* **2020**, 2335–2351. [CrossRef]

Li, Y.; Xiong, Z.; Li, K.; Yu, P.; Ding, Y.; Li, Z. Stress Evolution and Failure Characteristics of Overburden During Multi-Stope Mining for a Gently Inclined Thin Orebody, Mining. *Metall. Explor.* **2023**, *40*, 637–653. [CrossRef]
Liu, C.; Zhang, P.; Yao, D.; Ou, Y.; Tian, Y. Study on failure characteristics of overburden in extra thick coal seam mining. *Environ. Earth Sci.* **2022**, *81*, 479. [CrossRef]
Zhang, L.; Shen, W.; Li, X.; Wang, Y.; Qin, Q.; Lu, X.; Xue, T. Abutment Pressure Distribution Law and Support Analysis of Super Large Mining Height Face. *Int. J. Environ. Res. Public Health* **2023**, *20*, 227. [CrossRef] [PubMed]
Li, A.; Ding, X.; Yu, Z.; Wang, M.; Mu, Q.; Dai, Z.; Li, H.; Zhang, B.; Han, T. Prediction model of fracture depth and water inrush risk zoning in deep mining coal seam floor. *Environ. Earth Sci.* **2022**, *81*, 315. [CrossRef]
Ling, C.; Shi, X.; Wang, H.; Zhang, K.; Ren, Q.; Lv, Y. The Effect of Hydraulic Coupling on Mechanical Deformation Characteristics of Shallow Coal Seam in Western Mining Area. *Geofluids* **2022**, *2022*, 9727124. [CrossRef]
Du, W.; Chai, J.; Zhang, D.; Lei, W. Application of optical fiber sensing technology in similar model test of Shallow-buried and thick coal seam mining. *Measurement* **2021**, *181*, 109559. [CrossRef]
Zhang, C.; Zhao, Y.; He, X.; Guo, J.; Yan, Y. Space-sky-surface integrated monitoring system for overburden migration regularity in shallow-buried high-intensity mining. *Bull. Eng. Geol. Environ.* **2021**, *80*, 1403–1417. [CrossRef]
Liang, S.; Elsworth, D.; Li, X.; Yang, D. Topographic influence on stability for gas wells penetrating longwall mining areas. *Int. J. Coal Geol.* **2014**, *132*, 23–36. [CrossRef]
Zhang, J.; Wu, J.; Yang, T.; Yang, S.; He, Y.; Gao, S.; Peng, B. Analysis of Fracture Evolution Characteristics and Formation Mechanism of Inter-Layer Rock Under Different Mining Areas. *Rock Mech. Rock Eng.* **2024**, *57*, 3787–3811. [CrossRef]
Lan, T.; Liu, Y.; Yuan, Y.; Liu, H.; Liu, H.; Zhang, S.; Wang, S. Mine pressure behavior law of isolated island working face under extremely close goaf in shallow coal seam. *Sci. Rep.* **2023**, *13*, 20576. [CrossRef]
Niu, H.; Sun, S.; Sun, Q.; Wang, H. Study on Thermal Behavior Characteristics and Microstructure Evolution Mechanism of Residual Coal Combustion Process in Deep High Initial Temperature Environment. *Combust. Sci. Technol.* **2023**, 1–19. [CrossRef]
Li, H.; Li, X.; Fu, J.; Zhu, N.; Chen, D.; Wang, Y.; Ding, S. Experimental study on compressive behavior and failure characteristics of imitation steel fiber concrete under uniaxial load. *Constr. Build. Mater.* **2023**, *399*, 132599. [CrossRef]
Li, H.; Li, X.; Fu, J.; Gao, Z.; Chen, P.; Zhang, Z. Research on acoustic emission multi-parameter characteristics in the failure process of imitation steel fiber reinforced concrete. *Phys. Fluids* **2023**, *35*, 107109. [CrossRef]
Liu, H.; Li, X.; Yu, Z.; Tan, Y.; Ding, Y.; Chen, D.; Wang, T. Influence of hole diameter on mechanical properties and stability of granite rock surrounding tunnels. *Phys. Fluids* **2023**, *35*, 064121.
Li, X.; Chen, D.; Fu, J.; Liu, S.; Geng, X. Construction and application of fuzzy comprehensive evaluation model for rockburst based on microseismic monitoring. *Appl. Sci.* **2023**, *13*, 12013. [CrossRef]
Jin, Z.; Peng, T. Control of the Internal and External Staggered Distance of Coal Mining Face to the Water-Conducting Fissures in the Overlying Strata of the Near Coal. *Adv. Civ. Eng.* **2021**, *2021*, 1499675. [CrossRef]
Zhu, Y.; Zhang, Y.; Kong, J.; Yang, Q. Surface Movement and Deformation Law Caused by Different Coal Pillar Stagger Distances in Strip Filling IoT-Enabled Sustainable Mining. *Mob. Inf. Syst.* **2022**, *2022*, 2893172. [CrossRef]
Zhang, C.; Xu, J.; Yin, G.; Peng, S.; Li, Q.; Chen, Y. A Novel Large-Scale Multifunctional Apparatus to Study the Disaster Dynamics and Gas Flow Mechanism in Coal Mines. *Rock Mech. Rock Eng.* **2018**, *52*, 2829–2889. [CrossRef]
Liu, P.; Gao, L.; Zhang, P.; Wu, G.; Wang, Y.; Liu, P.; Kang, X.; Ma, Z.; Kong, D.; Han, S. Physical Similarity Simulation of Deformation and Failure Characteristics of Coal-Rock Rise under the Influence of Repeated Mining in Close Distance Coal Seams. *Energies* **2022**, *15*, 3503. [CrossRef]
Yang, H.; Liu, J.; Zhang, D.; Xiao, W.; Wang, X.; Wang, K. Response Characteristics of Coal Measure Strata Subjected to Hydraulic Fracturing: Insights from a Field Test. *Energy Fuels* **2021**, *35*, 19410–19422. [CrossRef]
Zhang, L.; Zhang, Z.; Zhang, R.; Gao, M.; Xie, J. The Ultrasonic P-Wave Velocity-Stress Relationship and Energy Evolution of Sandstone under Uniaxial Loading-Unloading Conditions. *Adv. Mater. Sci. Eng.* **2021**, *2021*, 9921716. [CrossRef]
Dai, W.; Wang, S.; Wang, S. Longwall Mining Automation—The Shearer Positioning Methods between the Longwall Automation Steering Committee and China University of Mining and Technology. *Appl. Sci.* **2023**, *13*, 12168. [CrossRef]
Chen, B.; Mei, H.; Li, Z.; Wang, Z.; Yu, Y.; Yu, H. Retrieving Three-Dimensional Large Surface Displacements in Coal Mining Areas by Combining SAR Pixel Offset Measurements with an Improved Mining Subsidence Model. *Remote Sens.* **2021**, *13*, 2541. [CrossRef]
Zhang, Z.; Liu, G.; Wang, X.; Lv, R.; Lin, J.; Barakos, G.; Chang, P. A fractal Langmuir adsorption equation on coal: Principle, methodology and implication. *Chem. Eng. J.* **2024**, *488*, 150869. [CrossRef]
Fu, J.; Chen, D.; Li, X.; Li, H.; Liu, S.; Li, C.; Zhang, J. Research on the technology of gob-side entry retaining by pouring support beside the roadway in three soft coal seam A case study. *Phys. Fluids* **2024**, *36*, 017123. [CrossRef]
Jiang, C.; Li, X.; Wang, F.; Wang, R. Comprehensive evaluation of coal burst risk using optimized linear weighted model. *Phys. Fluids* **2024**, *36*, 057113. [CrossRef]

applied sciences

MDPI

Article

The Thermodynamic Change Laws of CO_2-Coupled Fractured Roc

Fei Yu and Guangzhe Deng *

College of Energy Engineering, Xi'an University of Science and Technology, Xi'an 710054, China; 23103077016@stu.xust.edu.cn
* Correspondence: denggz@xust.edu.cn

Abstract: Under the background of the "dual carbon" target, exploring the pathway of efficient geological storage and high energy utilization of CO_2 is a hot issue in CO_2 emission reduction research. Under the coupling effect of high geopathic stress in deep rock layers and thermal stress generated during the geological sequestration of CO_2, CO_2 infiltration into coal rock changes the ambient temperature around the rock, while thermal diffusion effects cause damage to the rock and influence fracture expansion. In the present study, CO_2-water–rock coupling test system and characterization of fissure surface roughness were conducted to analyze the rock's mechanical properties, damage, and fracture evolution. Modeling of the equivalent fissure was employed to reveal the heat transfer mechanism between the rock matrix and CO_2. The results obtained illustrate that the rock samples coupled with CO_2 exhibited remarkable changes in mechanical properties. These changes include an increase in the number of pores, enhanced inter-pore connectivity, and a planar type of surface roughness in the fissures, ultimately resulting in an increase in conductivity. Conversely, the remaining rock samples displayed poor mechanical properties and surface fracture connectivity. As pressure decreased, the heat transfer coefficient decreased from 86.9 $W/m^2{\cdot}K$ to 57.5 $W/m^2{\cdot}K$, accompanied by a temperature drop from 33.6 °C to 30.6 °C, demonstrating a proportional relationship between pressure and the heat transfer coefficient. Furthermore, the flow rate gradually increased with the rise in CO_2 pressure, indicating denser flow lines with faster flow rates. At 15 MPa, CO_2 exhibits enhanced mobility.

Keywords: CO_2 geological storage; crack extension; heat transfer; CO_2 fracturing; thermal damage

1. Introduction

The depletion of resources in shallow coal beds has led to a gradual shift towards deep mines in coal mining. It is characterized by multiple energy sources and low seepage due to high-intensity mining in deep coal beds. High-stress, low-permeability coal rocks seriously restrict safe, efficient production in mines. The transformation of coal rocks has emerged as a fundamental scientific concern in addressing technical challenges encountered in coal mining operations [1]. The excellent mobility and high permeability of supercritical CO_2 play a crucial role in diminishing the mechanical properties of coal rocks, which enables it as a substitute for methane during CO_2 injection into the stratum for geological storage. This holds immense significance in achieving the objective of double carbon [2–5]. As early as the 1840s, researchers studied the application of CO_2 in mines and discovered its potential to induce alterations in the mechanical properties of coal rocks, which stabilizes reservoirs [6].

There are many studies on the effect of CO_2 on the mechanical properties of rocks. Wen et al. [7] researched CO_2 fracturing and permeability enhancement. Supercritical CO_2, exhibiting acidification and unblocking characteristics as well as excellent permeability, can enhance coal rock permeability and depressurization. CO_2 weakens the compressive strength and elastic modulus of rocks, and the weakening effect of supercritical CO_2 is more obvious compared to CO_2 [8]. Lyu et al. thought that the compressive strength and elasticity modulus of rocks decreased gradually with increased pressure [9]. Tang et al. [10]

found that the triaxial compressive and tensile strength of rocks decreases gradually with the increased adsorption time of supercritical CO_2. The elasticity modulus and Poisson's ratio of rocks increased gradually with increased CO_2 pressure and temperature [11]. Liu et al. [12] believed that supercritical CO_2 has different penetration enhancement effects on coal under different pressures. CO_2 pressure affects the fluidity and modification of the fluid in the coal matrix. Researchers have found that supercritical CO_2 significantly reduces the mechanical properties of coal rocks. Microcracks and fissures develop in rock mass under the interaction of CO_2 and rocks [13,14]. Liang et al. [15] studied the distribution characteristics of cracks in coal after CO_2 fracturing, revealing the extension mechanism of sandstone cracks during the fracturing process. Since water is one of the main factors affecting the mechanical properties of materials [16], the decreased fracture width or increased surface roughness makes the fluid more prone to decreased overpressure [17]. It weakens the discharge capacity of fractures. The mechanical mechanism of hydraulic fracturing for coal structure destruction is investigated, revealing that hydraulic fracturing can enhance the effectiveness of top coal pre-cracking [18]. Li et al. [19–21] investigated the mechanism of fracture morphology evolution in fractured coal under water and supercritical CO_2 fracturing conditions. CO_2 significantly affects coal rock wettability and fracture expansion.

The above studies mainly focus on rock tensile strength. CO_2 plays an important role in weakening rock mechanical properties and microcrack extension. However, the process of injecting CO_2 into the deep earth involves solid mechanics and physical fields as well as temperature and flow field changes triggered by the phase change process of CO_2. Additionally, the analysis of coal–rock structural modification should consider multi-field coupling characteristics such as heat–hydraulic–solid [22,23]. CO_2 exchanges heat with the rock matrix during fissure seepage, which changes the temperature field distribution of the fissured rock and affects the expansion of the fissure.

MTS tests were performed to determine the alteration patterns of rocks' mechanical properties after the rock samples were treated with CO_2-H_2O coupling in this study. Additionally, the changes in rock pore and fracture structure, as well as surface roughness, were quantified using SEM and 3D electron microscope scanning technology. A coupled heat–fluid–solid model was developed to invert the effect of CO_2 heat convection on the ambient temperature around fissured coal rocks, revealing the heat transfer between the rock matrix and CO_2 in the pores. The results can provide the theoretical basis for the safety evaluation of CO_2's deep earth storage and industrialized application in fracturing and permeability enhancement.

2. Materials and Methods

2.1. Experimental Device and Materials

Thermodynamic testing of rock samples was performed using MTS equipment developed by Central South University. This equipment, with a triaxial hydraulic control cabinet and triaxial seepage control cabinet, was used for the triaxial and seepage characterization of rocks under normal and high temperatures. The maximum load was 2600 KN, with a triaxial peripheral pressure of 140 MPa, a penetration pressure of 140 MPa, and a temperature from room temperature to 200 °C (Figure 1).

A CO_2-water–rock coupled experimental device developed by Xi'an University of Science and Technology was employed to prepare the rock samples exhibiting coupled thermodynamic behaviors (Figure 2). The seepage experimental device consists of a pressure chamber, a temperature control system, a coupled fracturing fluid supply device, and a fixed-speed pump. The device can provide a maximum stress of up to 40 MPa and regulate the temperature of the pressure chamber from room temperature to 100 °C. Additionally, the pump can provide constant pressure in the range of 0 to 40 MPa.

Figure 1. Experimental device of MTS seepage.

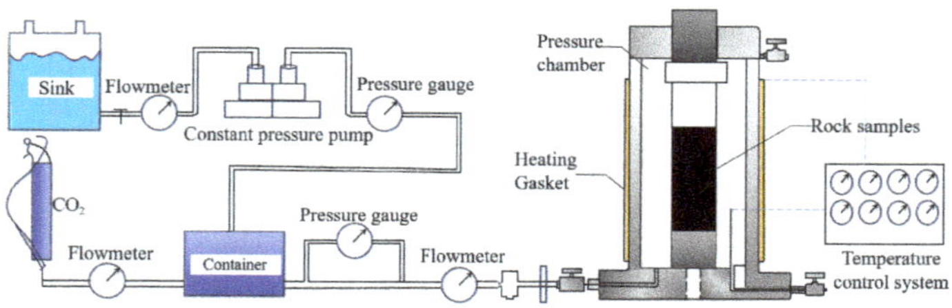

Figure 2. CO_2-water–rock coupled experimental device.

2.2. Experimental Methods

The specimens were taken from a low-permeability reservoir in the northern Shaanxi area of the Ordos Basin, which is the most concentrated area of low-permeability, high-stress coal rocks [24]. XRD analysis of the selected rock samples showed that the mass fraction of quartz in the sandstone was 30.15% and the mass fraction of carbonate minerals was 30.16%, including 25.16% of calcite and 5% of dolomite. Sandstone contained minor clay minerals, including 14.7% of illite and 3.4% of kaolinite.

Prior to experiments, standard rock samples were prepared (Φ 50 mm × 100 mm). The samples were classified as follows:

① Group 1 consisted of rock samples in their natural state. The samples were sealed and stored under atmospheric conditions and were labeled as FS-0.

② Group 2 involved water-saturated rock samples. The samples were immersed in water for 24 h and were labeled as water samples.

③ Group 3 comprised saturated CO_2-coupled samples. We placed the rock samples separately in a CO_2-water–rock coupled experimental device with liquid CO_2 concentrations of 0%, 4%, 6%, and 8% for full saturation, which were labeled as CW-0, CW-4, CW-6, and CW-8, as shown in Figure 3a. Then, coupling rock samples containing different concentrations of CO_2 were placed in a high-temperature and high-pressure reactor with temperatures of 25 °C, 35 °C, and 45 °C, which were labeled as B-1, B-2, and B-3; one natural rock sample was numbered F1, as shown in Figure 3b. Finally, we performed MTS seepage experiments under the same pore pressure and action time, as shown in Figure 3c, and the mechanical properties of four types of rock samples were analyzed. The mechanical parameters of the rock samples were obtained through uniaxial testing using the experimental device developed by Xi'an University of Science and Technology, as shown in Table 1.

Table 1. Mechanical parameters after CO_2-coal sample coupling action.

Sample	Compressive Strength/MPa	Tensile Strength/MPa	Elastic Modulus/GPa
CO_2-water–rock coupled rock samples	5.3	0.85	0.21
Clear water rock sample	9.8	1.52	0.38
Natural sample	16.5	2.98	0.57

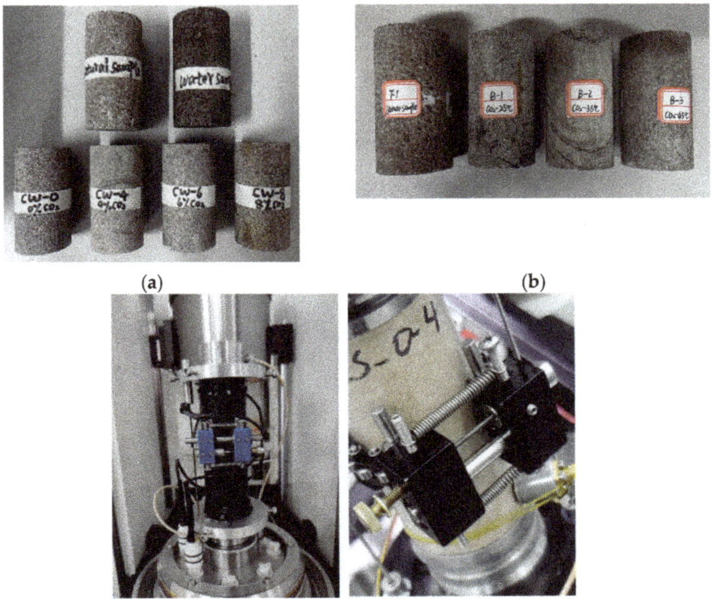

(a) (b) (c)

Figure 3. Rock samples under different conditions. (**a**) Rock samples in different states. (**b**) Rock samples prepared for the seepage test. (**c**) Installation of seepage test specimens.

3. Results

3.1. Experimental Results Thermodynamics of CO_2-Induced Rock Fracturing

The results of the CO_2-water–rock coupling seepage test demonstrate that the compressive strength and elastic modulus of samples subjected to immersion in water and coupled fracturing fluid with a CO_2 mass fraction of 8% exhibit varying degrees of reduction. Figure 4a indicates that the CW-8 sample exhibits a significantly weakened stress–strain diagram in comparison to the rock samples in other states. Figure 4b reveals that the compressive strength and elastic modulus of the rock samples exposed to CO_2-coupled fracturing fluid decreased greatly compared with those of the natural samples and the water-soaked samples; the larger the CO_2 mass fraction, the more obvious the decrease in compressive strength and the elastic modulus. The observed interaction between CO_2, water, and rock weakens the rock's resistance to mechanical damage. Among the coupled fracturing fluids, CW-8 had the most pronounced weakening effect on the strength of the rock samples. The results of the seepage test and uniaxial test are consistent.

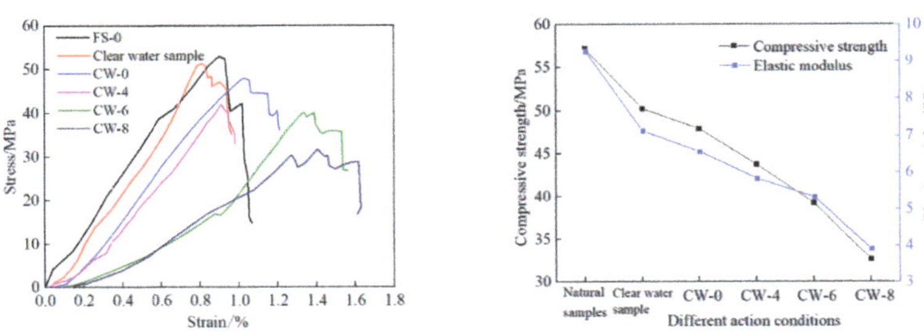

(a) (b)

Figure 4. Mechanical parameters of rock samples under different test conditions. (**a**) Stress–strain curves. (**b**) Variations in mechanical properties.

Therefore, this article examines the influence of temperature, stress, and time on the tensile strength of sandstone exposed to CW-8 fracturing fluid. Table 2 and Figure 5a,b show that during CO_2-water–rock coupling, the tensile strength of sandstone samples decreases at different temperatures when compared with the tensile strength of the natural rock sample F1. More specifically, the tensile strength was reduced by 2.63 MPa, 1.66 MPa, and 1.32 MPa, exhibiting a decrease of 57.19%, 36.17%, and 28.68% at 25 °C, 35 °C, and 45 °C, respectively. Figure 5a illustrates that CO_2-water–rock coupling results in rock samples reaching their peak load in the shortest duration, which is shorter than that of the natural samples. Moreover, it is observed that as the temperature increases, the time required to reach the peak load becomes longer. This is especially more pronounced at a temperature of 25 °C, with the shortest time to reach the peak being 58.2 s. As shown in Figure 5b, under these conditions, the tensile strength of rock samples experiences the largest decline and as the temperature rises; the reduction in tensile strength diminishes, and it consistently remains lower than the tensile strength of natural rock samples.

Table 2. Results of Brazilian splitting tests at different temperatures.

Rock Sample	Temperatures/°C	Pore Pressure/MPa	Active Time/h	Tensile Strength/MPa	Magnitude of Change/%
F1	/	7.0	10.0	4.59	69.85
B-1	25	7.0	10.0	1.96	57.19
B-2	35	7.0	10.0	2.93	36.17
B-3	45	7.0	10.0	3.27	28.68

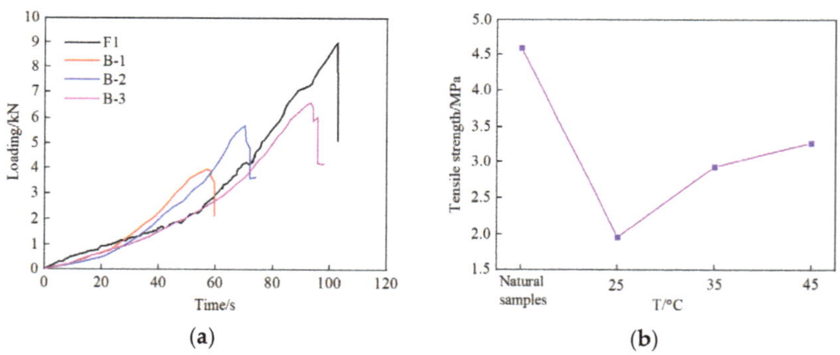

Figure 5. Effect of temperature on the tensile strength of sandstone. (**a**) Load–time characteristic curve. (**b**) Variations in tensile strength at different temperatures.

Accordingly, it is inferred that temperature affects the tensile strength of rock samples after CO_2-water–rock coupling. The higher the temperature, the lower the viscosity of CO_2-coupled fracturing fluid, and the higher the permeability coefficient. As the temperature rises, the coupled fracturing fluid exhibits greater wetting characteristics on rock samples, which reduces the strength of the rock mass. Additionally, the tensile strength exhibits a positive correlation with the temperature, which may be attributed to the lower solubility of CO_2 in hot water. Consequently, the PH of CO_2-coupled fracturing fluid increases and the acidity of the fluid weakens, thereby reducing the chemical damage to the rock mass.

Based on the distribution of tensile strength shown in Figure 6, the correlation between temperature and tensile strength after CO_2-water–rock coupling can be approximated using the following expression:

$$y = -0.0031x^2 + 0.286x - 3.2212. \quad (1)$$

Equation (1) indicates that the tensile strength of the rock sample gradually increases with increasing temperature.

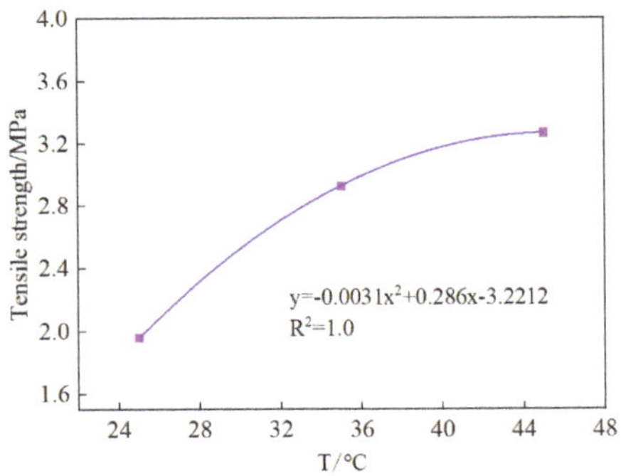

Figure 6. The distribution of tensile strength against temperature.

3.2. Characteristics of Fracture Evolution of Rock Samples before and after CO_2 Coupling

The coupling conditions were established using a CO_2 mass fraction of 8% in the fracturing fluid, a temperature of 25 °C, a pore pressure of 7 MPa, and a processing time of 10 h. Subsequently, scanning electron microscope (SEM) images of ×500 and ×2000 were captured after conducting the MTS procedure on the sandstone. These images were then compared with SEM images of rock samples in their natural state.

The results show that rock samples in their natural state have less distribution of fractures on the surface, and are primarily dominated by a single fracture. They also exhibit small pore diameters (red circles in the figure), low pore development, poor connectivity, a flatter surface, and mineral filling in both pores and fractures, as shown in Figure 7a,b.

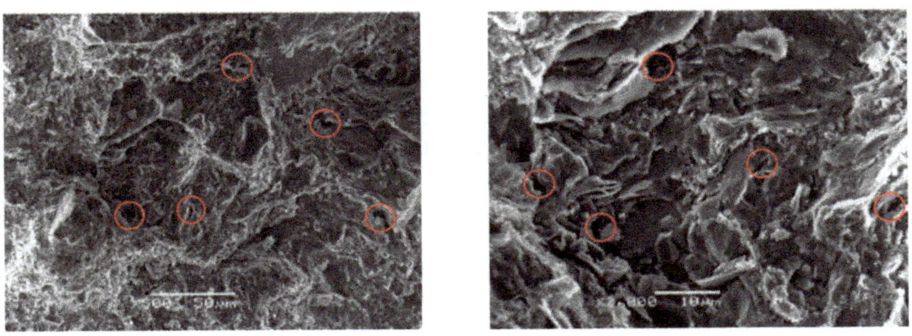

(**a**) (**b**)

Figure 7. SEM images of the rock sample in their natural state (the red circle indicates the hole). (**a**) ×500; (**b**) ×2000.

After CO_2 coupling, the rock samples were divided into uneven sheet-like structures through pores and cracks. There was a notable increase in the number of pores, and they became larger in size (red circles in the figure). The development degree of the pores significantly improved, with locally visible pores emerging and most of the pores being interconnected with each other. Figure 8a,b reveals that original cracks exhibit irregular changes, and the formation of new cracks resulted in interconnected patterns.

The observed changes in the fracture structure of rock samples before and after the CO_2 coupling effect may be attributed to the thermal and physical properties of CO_2 itself. When the CO_2-coupled fracturing fluid contacts the rock, it reduces the level of free energy in the system and strengthens the wettability of the rock. The presence of CO_2 fracturing fluid enhances the permeability of the rock mass. Furthermore, the convection heat transfer slightly increases the temperature of the rock mass, thereby reducing its density. Secondly, CO_2 transitions into the supercritical state after surpassing the critical temperature. In this supercritical state, CO_2 exhibits low surface tension and weakened intermolecular force. Finally, the combination of CO_2 and water results in the formation of carbonic acid, an acidic fluid that causes the erosion of the coal rocks in the reservoir and its cement. This process is accompanied by the deposition of new minerals and the transportation of debris particles, which changes the pore structure of coal rocks. The enhancement in CO_2 mobility within the fissure facilitates self-pressurization during the fissure expansion. This phenomenon accelerates rock seepage, promoting fissure expansion and effectively increasing the seepage capacity of low-permeability coal rock. It is worth noting that seepage is an affecting parameter during the phase transition of liquid CO_2 to a gaseous state.

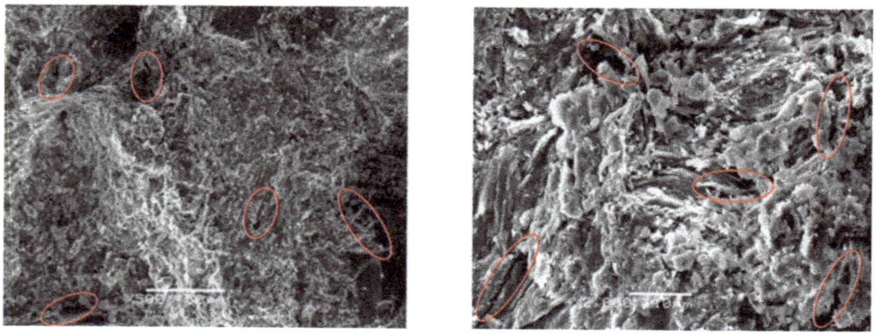

(a) (b)

Figure 8. SEM images of rock samples under MTS coupling (the red circle indicates the hole). (**a**) ×500; (**b**) ×2000.

3.3. Characterization of Fracture Surface Roughness of Rock Samples before and after CO_2 Coupling

To characterize the features of rock fracture surfaces before and after CO_2 coupling, a three-dimensional electron microscope scanner was employed for scanning, as depicted in Figure 9. The surface roughness of fractures reflects the micro-convex bodies' fluctuations on the rock fracture surface, which has a size effect and is a key indicator of fracture mechanics characteristics.

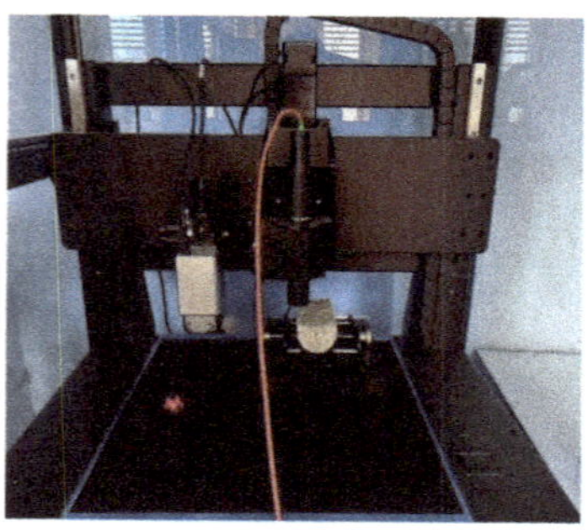

Figure 9. Three-dimensional electron microscope scanning instrument.

Utilizing the Joint Roughness Coefficient method (JRC), the surface of cracks undergoes digitization. In Figure 10a, the rough surface of fractures in the natural state of the rock sample exhibits a stepped shape. Due to significant differences in relief angles, conductivity is poor, resulting in a poor fracturing effect. The rough surface of fractures at the edge of the rock sample is uneven, with a maximum roughness of 0.82 mm. Figure 10b shows that the fracture roughness surface of MTS-coupled rock samples is planar, with the highest roughness reaching 1.13 mm. The fluctuation angle of the rock sample edge is relatively small, indicating good conductivity and a positive fracturing effect. Through comparison, it is evident that stepped rough surfaces exhibit significant size effects compared to planar rough surfaces.

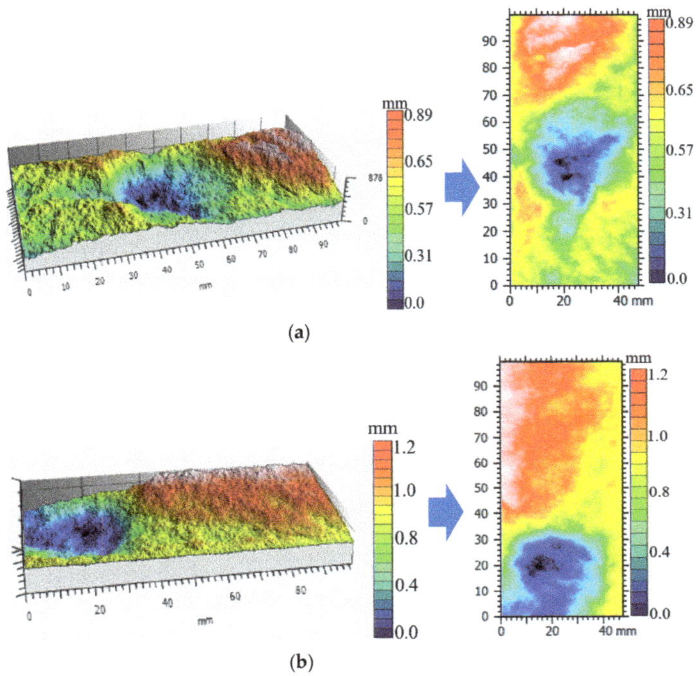

Figure 10. Characterization diagram of crack roughness: (**a**) natural samples and (**b**) rock samples under MTS coupling.

To quantify the roughness of surface cracks in rock samples, the roughness of the crack surface was quantified using an empirical expression that combines characterization statistical parameters with JRC [25].

$$JRC = 32.2 + 32.47 \log h_2, \tag{2}$$

$$h_2 = \left[\frac{1}{(n-1)(\Delta x)^2} \sum_{i=1}^{n} (h_{i+1} - h_i)^2 \right]^{\frac{1}{2}} \tag{3}$$

where x represents the root mean square of the fracture profile wall slope; n is the number of data points; and h denotes the axial distance between points $i + 1$ and i. The height of peak roughness can be obtained from the following expression:

$$\xi = h_{\max} - h_{\min} \tag{4}$$

where ξ represents the height of peak roughness, mm, and $h_{\max}$ and $h_{\min}$ are the maximum and minimum roughness heights, mm, respectively.

$$R_s = \sqrt{\frac{1}{n} \sum_{i=1}^{n} (h_i - h_a)^2} \tag{5}$$

$$R_n = \frac{1}{n} \sum_{i=1}^{n} |h_i - h_a| \tag{6}$$

where R_n and R_s represent the average and the root mean square of the roughness height, respectively, and h_i and h_a are the roughness height of point *i* and the average height of the elevation line, mm, respectively.

The roughness parameters of the fractured sample can be calculated using Equations (2)–(6), as shown in Table 3.

Table 3. Geometrical parameters and roughness of the fractured samples.

Sample	Peak Asperity Height ξ/mm	RS of Asperity Height R_s/mm	Mean Asperity Height R_n/mm	JRC
Natural sample	0.89	0.207	0.172	5.7
Coupled rock Sample	1.17	0.359	0.329	3.6

Table 3 indicates that compared to the sample under natural conditions, the CO_2-coupled rock sample has an average roughness 0.15 mm higher, with a difference of 0.16 mm in the root mean square of the roughness height. The JRC is less than that of the natural state samples, and the difference is small. Therefore, the rough surface of the coupled rock sample exhibits a planar shape. This indicates that the heat transfer effect generated by the rock matrix and CO_2 in the cracks is conducive to crack expansion, resulting in good fracture connectivity, strong conductivity, and a significant fracturing effect.

3.4. Simulation of CO_2 Heat Transfer in Pores

3.4.1. Pore Fracture CO_2 Phase Evolution

Based on the observations of high-temperature geothermal development, the surrounding rock is susceptible to cracking and subsidence under external loads, leading to the formation of a "holey" structure within the rock mass. In this structure, major fracture voids may form at intervals [26]. To facilitate the analysis, the complex fracture morphology was simplified using an equivalent fracture structure model where "holes" were used instead of "slits".

The pore flow CO_2 phase change was plotted based on NIST data [27]. It is found that the volume expansion of CO_2 affects temperature and pressure, while its temperature and pressure affect the phase changes. Furthermore, liquid CO_2 evaporates in the fracture hole [28]. It should be indicated that the triple-point temperature and pressure of CO_2 are 238.98 K (−58.9 °C) and 0.34 MPa, respectively. When CO_2 transforms from the liquid state to the supercritical state, its critical temperature and pressure are 304.2 K (31.14 °C) and 7.38 MPa, respectively. Figure 11 shows that supercritical CO_2 can quickly achieve the goal of energizing phase changes in fracture pores and promoting fissure propagation. However, the process of fracturing rock using liquid CO_2 reduces the temperature surrounding the rock body. Meanwhile, heat transfer takes place between liquid CO_2 and seepage within the rock body. This aligns with the characteristics of heat–fluid–solid coupling.

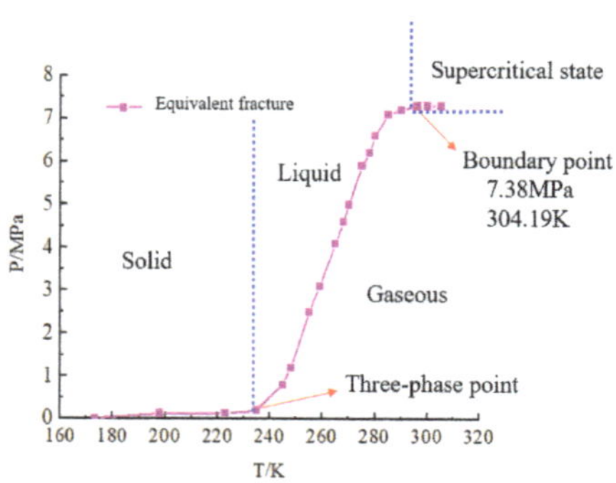

Figure 11. Phase change of CO_2 within the pore [24].

3.4.2. Fluid–Solid Coupling Model

The heat exchange between liquid CO_2 and the rock in the fractured rock fissure pore affects the temperature surrounding the rock mass [8]. Therefore, the CO_2 fluid and the fissure pore part of the rock mass are studied as the object of heat–fluid–solid coupling.

Based on the similar structural characteristics of the actual rock fracture curvature, the process of fracturing within the rock leads to the creation of a "porous" fracture structure within the rock mass [26]. In order to reveal the heat and mass transfer law between CO_2 in pores and rock mass, internal CO_2 fluid, pore, and external rock mass flow field models were established, as shown in Figure 12. The pore fissure (Φ1 mm) and the rock mass (2 m × 2.4 m × 1.8 m) are used as part of the thermal reservoir formed by CO_2 fracturing, as shown in Figure 13.

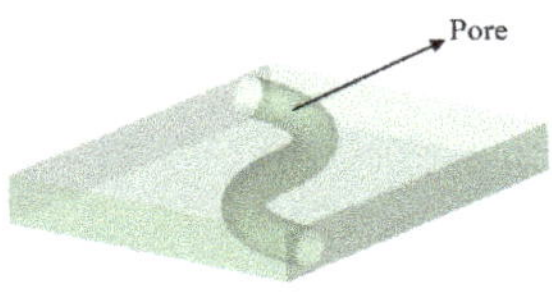

Figure 12. Pore model.

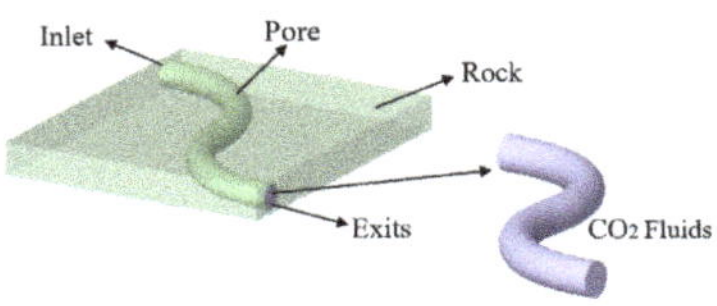

Figure 13. Fluid–solid coupling model.

3.4.3. Boundary Conditions

The boundary conditions were set as the pressure inlet, the inlet heat temperature was 31.04 °C (TSC-CO_2 $\geq$ 31.04 °C), and the total reflux temperature was set to 35 °C. The wall boundary conditions were set to thermal convection. To cover various scenarios, three inlet pressures of 15 MPa, 10 MPa, and 5 MPa were considered in the analysis.

3.4.4. Coupled Model Control Equations

The stress field is affected by the temperature and seepage, as well as mechanical and chemical damage during CO_2-coupled fracturing, which can cause rock deformation. The governing equation in the stress field can be expressed as follows [10]:

$$G(u_i + \frac{1}{1-2u}) - K(T - T_{tr}) - \alpha_p p_f = 0 \tag{7}$$

where $G = E^*/2(1+v)$ is the shear modulus; α_T denotes the thermal expansion coefficient of rock; K represents the bulk modulus; T is the rock temperature; T_{tr} denotes the rock temperature in its natural state; $\alpha_p = 1 - (K/K_s)$ represents the Biot number;K_s is the modulus of rock; and p_f denotes the fluid pressure within the pore.

During seepage of the liquid CO_2, heat transfer affects the temperature distribution within the rock mass. The total specific heat capacity of the rock mass is linearly correlated to its mass and the specific heat capacity of liquid CO_2. This can be mathematically expressed as follows [11]:

$$(\rho C)_t = \left(1 - \phi_m - \phi_f\right)\rho_s C_s + s_w \phi_f \rho_w C_w \tag{8}$$

where C_s and C_w are the specific heat capacity of the rock mass and CO_2, respectively.

During rock fracturing, deformation is related to the physical properties of the rock mass and its volumetric strain, which can be expressed in the form below [11]:

$$W_T = T\alpha_s K_v \frac{\partial \varepsilon_v}{\partial t} \tag{9}$$

where α_s is the thermal expansion coefficient of rock; K_v is the bulk modulus; ε_v denotes the volumetric strain; and T is the rock temperature.

The convective heat exchanged between liquid CO_2 and the rock mass is

$$Q_{tr} = \nabla \cdot \left[-(\frac{K_f k_w}{\mu_w} \nabla p_w \cdot \rho_w C_w) \Delta T \right] \tag{10}$$

where μ_w and p_w are the viscosity and pressure of the liquid CO_2, respectively.

Finally, the energy conservation law in the CO_2-coupled fracturing fluid can be summarized as follows:

$$\frac{\partial[(\rho C)_t \Delta T]}{\partial t} + T\alpha_s K \frac{\partial \varepsilon_v}{\partial t} + Q_{te} + \nabla\cdot(K_t \nabla T) + \Delta Q_m = Q_T. \tag{11}$$

The coupling between temperature and stress fields is initially manifested through heat exchange. Subsequently, it is obtained using fluid seepage and changes in the thermophysical properties of CO_2.

The mass conservation equation can be expressed as follows [11]:

$$\frac{\partial(\rho\mu)}{\partial t} + \nabla \cdot (\rho \mathrm{v}) = -\Delta \tag{12}$$

where ρ is the density of supercritical CO_2; μ is the rock porosity; t denotes time; Δ represents the seepage process source; and v is the fluid flow rate.

The temperature gradient between the liquid CO_2 and the rock induces convective heat transfer. According to Fourier's law, the heat transfer in the rock mass is proportional to the temperature difference between liquid CO_2 and the rock [11]:

$$q_c = -\lambda \Delta t \tag{13}$$

where λ and t are the thermal conductivity and temperature of liquid CO_2, respectively.

By coupling the stress and temperature field equations, the governing equations for heat transfer in CO_2-coupled fractured rocks can be expressed in the form below:

$$u_{ij} = (1 - \Delta Q_m)\alpha_p p_f + Q_{\mathrm{te}}. \tag{14}$$

The geometric model involving flow and rock mass was constructed and meshed, and the initial boundary conditions were set. The k-epsilon model was employed to approximate turbulent stresses.

3.5. Mass Transfer Analysis

When fracturing rock formations to create rock fracture voids through CO_2 fracturing fluids injected into the pipeline, the fracturing process results in increased seepage through the rock. This seepage affects conductive and convective heat transfer, causing spatial changes in CO_2 density. Figure 14a indicates that when the temperature is 320 K, the density of CO_2 is 200 kg/m^3, but when the temperature increases to 420 K, the corresponding density of CO_2 reduces to 105 kg/m^3. On the other hand, when the stress is 11.5 MPa, the CO_2 density is 105 kg/m^3, but when the stress increases to 18 MPa, the corresponding density increases to 205 kg/m^3. When the temperature and pressure of CO_2 exceed the critical values of 304 K and 8 MPa, its density exhibits a positive correlation with pressure, while Figure 14b indicates that the temperature exhibits a negligible impact on the density.

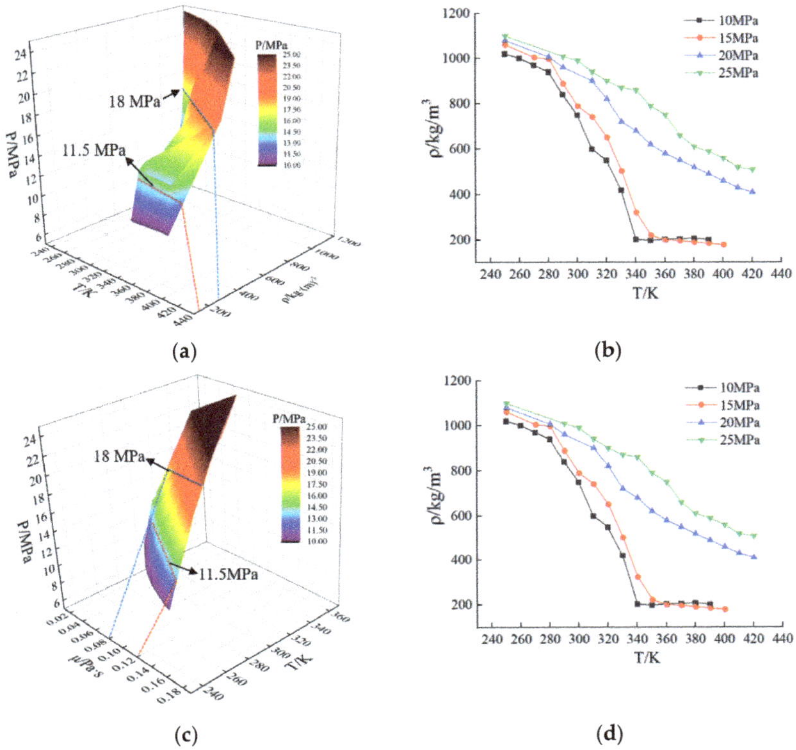

Figure 14. Thermal properties of supercritical CO_2 in the temperature range 240~440 K and pressure range 10~25 MPa. (**a**) Contours of density. (**b**) Density graphs. (**c**) Contours of viscosity. (**d**) Viscosity graphs.

The thermophysical properties of CO_2 change significantly near its critical point, which leads to rapid pressurization of CO_2 at the bottom of the wellbore. This accelerates the fracturing of the reservoir, intensifying the effects of rock seepage. Additionally, the thermal convective temperature gradient generated by liquid CO_2 contributes to the heat–fluid–solid-coupled heat-transfer effect within coal–rock fissures during seepage.

When CO_2 is used as the fracturing medium to fracture the rock body, it intensifies the role of rock seepage, and the CO_2 medium exchanges heat with the rock mass. When the temperature and pressure of CO_2 exceed their critical values of 304 K and 7.38 MPa, the CO_2 transits into the supercritical state and exhibits low surface tension, weak intermolecular forces, and strong mobility. Consequently, CO_2 rapidly diffuses in the pores, and seepage contributes to heat transfer.

Figure 14c shows that as the CO_2 temperature reaches 260 K, its viscosity is 0.10 μ/Pa·s. When the temperature increases to 320 K, the CO_2 viscosity reduces to 0.06 μ/Pa·s. On the other hand, the CO_2 viscosity is 0.08 μ/Pa·s at the pressure of 18 MPa. As the pressure reduces to 11.5 MPa, the corresponding viscosity increases to 0.12 μ/Pa·s. Figure 14d indicates that the viscosity exhibits a negative correlation with temperature, and the temperature is closely related to the flow rate. As a result, adjusting the flow rate can effectively increase the heat transfer coefficient while reducing the viscosity of the CO_2 flow.

The viscosity and pressure contours reveal that the change in pressure and temperature significantly affects the viscosity. As the fracture length expands during fracturing, the temperature of liquid CO_2 in the fracture of the rock gradually increases through convective and conductive heat transfer. Consequently, CO_2 gradually transforms into a low-viscosity supercritical state within the coupled fractured rock fractures, thereby increasing the resistance to seepage of supercritical CO_2 in the rock fractures. However, seepage is an effective parameter in heat–fluid–solid coupling. Therefore, it is of significant importance to analyze the flow rate variations to study heat–fluid–solid coupling.

3.6. Simulation Results

The temperature and heat transfer coefficient of supercritical fluids are affected by the heat exchange, thermophysical properties, pressure, and flow rate. According to the analysis, liquid CO_2 is very sensitive to the flow rate and pressure, and different flow rates and pressures result in significantly different heat transfer coefficients and temperatures.

Figure 15a shows that when liquid CO_2 is injected at a pressure of 15 MPa, heat transfer occurs to different degrees between the fluid and the rock mass. The heat transfer coefficient reaches a maximum value of 86.9 $W/m^2{\cdot}K$ at the edge of the middle part of the fluid, and the heat transfer coefficient reaches its maximum value when the temperature of the fluid approaches the critical temperature of 31.04 °C. Furthermore, it is found that the heat transfer coefficient decreases gradually at the outlet boundary. The maximum temperature at the fluid–rock interface is 33.6 °C, wherein CO_2 is in the critical state. Figure 15b indicates that the temperature of the fluid–solid interface gradually decreases from inlet to outlet and drops to 30.9 °C at the outlet of the lower right part.

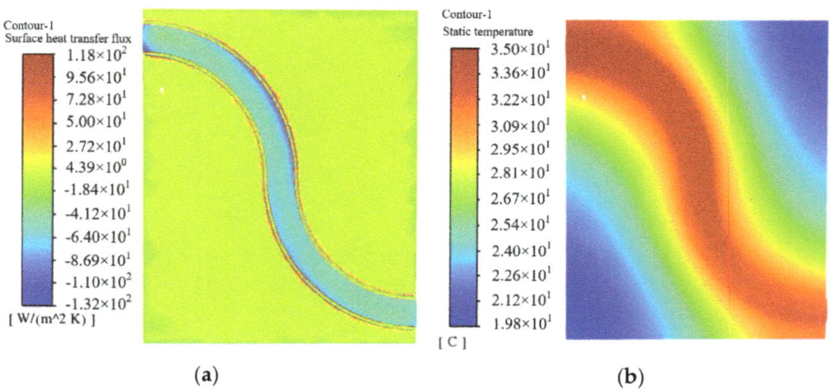

Figure 15. Contours of (**a**) heat transfer coefficient and (**b**) temperature for an injection pressure of 15 MPa.

Figure 16a shows the contours of temperature and the heat transfer coefficient when liquid CO_2 is injected at a pressure of 10 MPa. Compared with Figure 15, the heat transfer coefficient between the fluid and the rock mass decreases, with the maximum heat transfer coefficient at the middle edge of the fluid being 74.6 $W/m^2{\cdot}K$, indicating a reduction of 48.8 $W/m^2{\cdot}K$ at the outlet boundary. It is observed that as the pressure changes, the maximum temperature at the fluid–rock interface reaches 31.5 °C and the temperature decreases to 27.9 °C from inlet to outlet boundaries, as shown in Figure 16b.

Figure 17 illustrates the heat transfer and temperature contours for injection pressures of 5 MPa and 10 MPa. It is observed that the heat transfer coefficient between the fluid and the rock mass is significantly reduced compared with Figure 16. Figure 17a shows that the maximum heat transfer coefficient reaches 57.5 $W/m^2{\cdot}K$ and 28.1 $W/m^2{\cdot}K$ at the middle edge and outlet boundary, respectively. Furthermore, Figure 17b indicates that the temperature at the fluid–rock interface decreases from 32.1 °C at the inlet boundary to 24.7 °C at the outlet boundary.

The obtained results can be interpreted as follows: the temperature difference between liquid CO_2 and the rock mass is significant, resulting in a substantial heat exchange between the two. As the temperature decreases, the viscosity of CO_2 increases while its permeability decreases. During the phase transition of CO_2 from liquid to gas, the heat transfer between the fluid and the rock wall gradually decreases until it reaches an equilibrium state. This analysis aligns with the thermal dynamic relationship between rock tensile strength and temperature presented in Section 3.1.

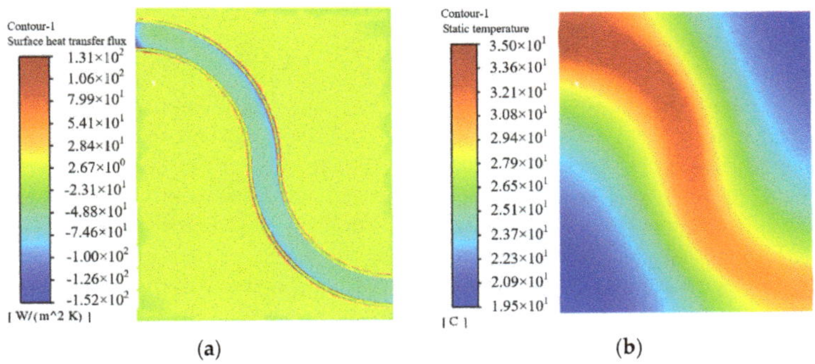

Figure 16. Contours of (**a**) heat transfer coefficient and (**b**) temperature for an injection pressure of 10 MPa.

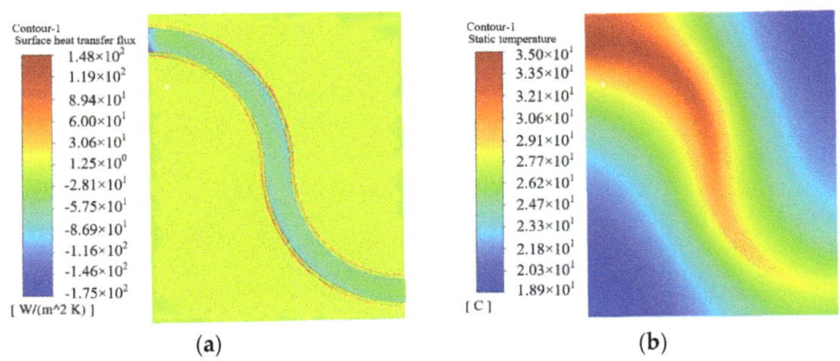

Figure 17. Contours of (**a**) heat transfer coefficient and (**b**) temperature for an injection pressure of 5 MPa.

The performed analysis is intended to establish a relationship between the heat transfer coefficient and temperature. As shown in Figure 18, the heat transfer coefficient of liquid CO_2 rapidly increases with the increase in fluid temperature. When the flow reaches its equilibrium state, the heat exchange gradually decreases as the temperature increases. This suggests that the heat transfer coefficient is directly related to the temperature in the early stage near the critical temperature of 31.04 °C. Meanwhile, the critical temperature significantly influences the heat transfer coefficient. This simulation demonstrates the heterogeneous characteristics of the thermal fluid–solid coupling process.

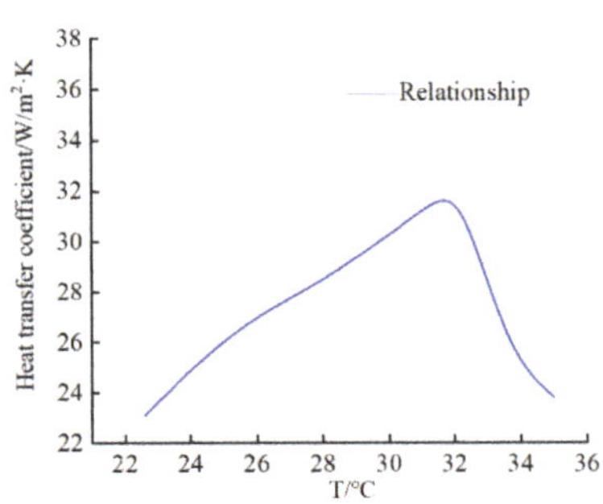

Figure 18. Distribution of the heat transfer coefficient against temperature.

Based on the performed heat transfer analysis, the fluid reaches high temperatures due to the extension of the crack length and exhibits high viscosity, which increases the resistance of liquid CO_2 in the crack. However, the variations in temperature and pressure are sensitive to the flow rate, and the appropriate flow rate can significantly enhance the heat exchange coefficient and reduce the flow resistance, which is enhanced for the less-viscous CO_2 fluid.

Figure 19 reveals that when liquid CO_2 is discharged from 5 MPa to 15 MPa, the faster the flow rate, the denser the flow line. Meanwhile, the flow volume increases gradually with increasing pressure. The initial pressure of the fluid decreases due to the influence of the permeability of the gas–water–rock coupling. The reason for this is that, with the continuous heat exchange between the fluid and the rock, the thermal expansion of the liquid CO_2 phase change is faster as the temperature and pressure increase. Ultimately, as gaseous CO_2 is discharged, there is an exponential increase in pressure and flow rate. Indicating that the sensitivity of flow velocity is directly proportional to the overall heat exchange coefficient, and the flow resistance gradually decreases with the increase in CO_2 temperature and pressure.

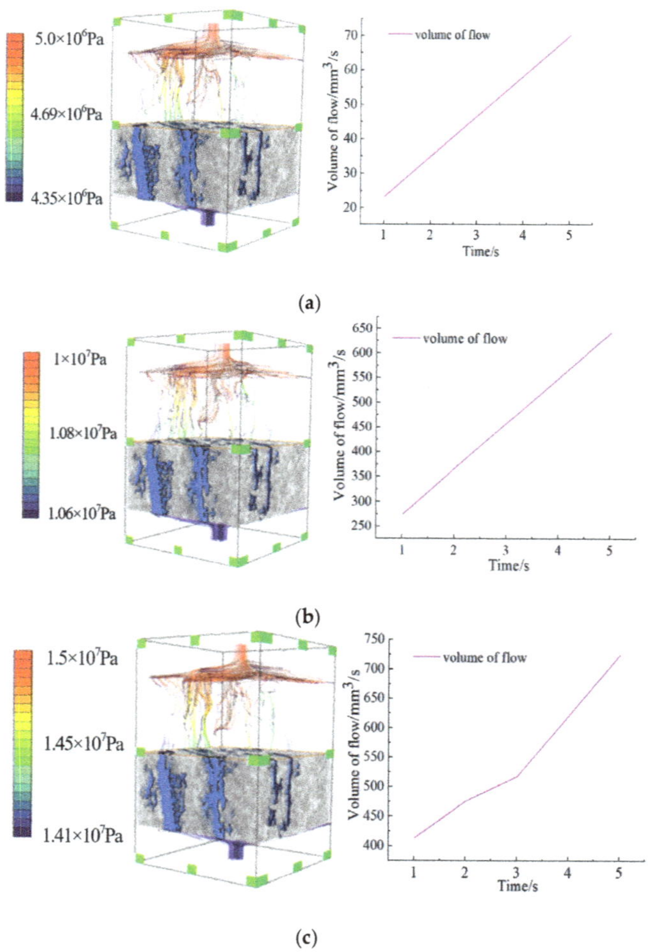

Figure 19. Flow rate variation diagram of CO_2 under different inlet pressure conditions: (**a**) CO_2-5 MPa; (**b**) CO_2-10 MPa; and (**c**) CO_2-15 MPa.

4. Discussion

This work investigated the impact of CO_2 coupling on the evolution of fractures and the roughness of rock samples, with a focus on the alteration process in the rocks' pore structure induced by acidic fluids generated from CO_2 and water. CO_2-induced rock fracturing involved obtaining the crack propagation law of fractured rocks and inverting the effect of CO_2 heat convection on the ambient temperature around fissured coal rocks, revealing the heat transfer between the rock matrix and CO_2 in the pores.

The results obtained illustrate that the interaction between CO_2, water, and rock weakens the rock's resistance to mechanical damage. It is inferred that temperature affects the tensile strength of rock samples after CO_2-water–rock coupling. The CO_2 exchanges heat with the rock matrix during fissure seepage, which changes the temperature field distribution of the fissured rock and affects the expansion of the fissure. The results can provide a theoretical basis for the safety evaluation of CO_2's deep earth storage and industrialized application in fracturing and permeability enhancement.

5. Conclusions

Based on the MTS tests and computational fluid dynamics analysis method, the mechanical evolution mechanism of CO_2-water–rock coupling in the fissure seepage process and the flow characteristics and the heat transfer mechanism of the CO_2 heat exchanges with the rock matrix are studied. Based on the obtained results and performed analyses, the main achievements can be summarized as follows:

As the temperature increased, the reduction in the tensile strength of rock samples became less pronounced, ranging from 2.61 to 1.30 MPa. The effect was attributed to the decreased viscosity of CO_2-coupled fracturing fluid. Additionally, wettability was enhanced and the strength of the rock samples was weakened.

The distribution of surface cracks on natural rock samples was relatively small, primarily composed of single cracks. The CO_2-coupled rock sample was divided into uneven sheet-like structures by pores and fractures. The nodal roughness coefficient of CO_2-coupled rock samples was smaller than that of the natural state sample, and the roughness surface was planar, indicating good fracture connectivity and significant permeability. The temperature, heat transfer coefficient, and flow rate changes between the rock wall and CO_2 were directly proportional to the CO_2 inlet pressure. The fluid flow line became denser, and the CO_2 volume flow rate gradually increased with increased pressure. The resistance to flow decreased gradually with increased CO_2 temperature and pressure.

Author Contributions: The first draft of the manuscript was written by F.Y. The concept of the manuscript was provided by G.D. All authors commented on the final version of the manuscript. All authors have read and agreed to the published version of the manuscript.

Funding: This study was supported by the National Natural Science Foundation of China (Grant No. 50149017) and the Fok Ying Tung Education Foundation Program (Grant No. 71076).

Institutional Review Board Statement: Not applicable.

Informed Consent Statement: Not applicable.

Data Availability Statement: The raw data supporting the conclusions of this article will be made available by the authors on request.

Conflicts of Interest: The authors declare that they have no conflicts of interest.

eferences

Yuan, Y.; Chen, Z.S.; Liang, X.K. Mechanism and application of carbon dioxide phase change blasting fracturing. *Coal Sci. Technol.* **2024**, *52*, 63–78.
Gao, F.; Tang, L.H.; Zhou, K.P. Mechanism analysis of liquid carbon dioxide phase transition for fracturing rock masses. *Energies.* **2018**, *11*, 2909. [CrossRef]
Xu, T.; Tian, H.; Zhu, H. China actively promotes CO_2 capture, utilization and storage research to achieve carbon peak and carbon neutrality. *Adv. Geo-Energy Res.* **2022**, *6*, 1–3.

4. Mwakipind, G.C.; Nyakilla, E.E.; Sanford, J. CO_2 sequestration and enhanced shale gas recovery by CO_2 injection: Numeri simulation method. *J. Energy Eng.* **2020**, *48*, 04022007. [CrossRef]
5. Xie, W.; Chen, S.; Wang, M. Progress and prospects of supercritical CO_2 application in the exploitation of shale gas reservo *Energy Fuels* **2021**, *35*, 18370–18384. [CrossRef]
6. Schoolerd, R. *The Use of Carbon Dioxide for Dislodging Coal in Mines*; Missouri University of Science and Technology: Rolla, M USA, 1944.
7. Wen, H.; Li, Z.B.; Wang, Z.P. Experiment on the liquid CO_2 fracturing process for increasing permeability and the characterist of crack propagation in coal seam. *J. China Coal Soc.* **2016**, *41*, 2793–2799.
8. Yin, H.; Zhou, J.P.; Xian, X.F. Experimental study of the effects of sub-and super-critical CO_2 saturation on the mechani characteristics of organic-rich shales. *Energy* **2017**, *132*, 84–95. [CrossRef]
9. Lyu, Q.; Long, X.P.; Ranjith, P.G. Experimental investigation on the mechanical properties of a low clay shale with differe adsorption times in sub-/super-critical CO_2. *Energy* **2018**, *147*, 1288–1298. [CrossRef]
10. Tang, J.R.; Lu, Y.Y.; Chen, Y.T. Experimental study of damage of shale mechanical properties under supercritical CO_2. *Rock S Mech.* **2018**, *39*, 797–802.
11. Ni, H.J.; Guo, X.; Ding, L. Experiment on mechanical properties of shale soaked with supercritical carbon dioxide. *J. China Un Pet. Ed. Nat. Sci.* **2019**, *43*, 77–84.
12. Liu, J.J.; Nie, Z.S.; Yu, B.Z. Analysis of the mechanism and influencing factors of supercritical carbon dioxide on coal permeabil enhancement. *Coal Sci. Technol.* **2023**, *51*, 204–216.
13. Zhou, J.; Yang, K.; Zhou, L. Microstructure and mechanical properties alterations in shale treated via CO_2/CO_2-water exposu *J. Pet. Sci. Eng.* **2021**, *196*, 108088. [CrossRef]
14. Zhang, K.; Sang, S.X.; Zhou, X.Z. Influence of supercritical CO_2H_2O-caprock interactions on the sealing capability of deep co seam caprocks related to CO_2 geological storage: A case study of the silty mudstone caprock of coal seam No.3 in the Qin sh Basin, China. *Int. J. Greenh. Gas Control.* **2021**, *106*, 103282.
15. Liang, W.G.; He, W.; Yan, J.W. Weakening and fracturing mechanism of mechanical properties of coal and rock caused supercritical CO_2. *J. China Coal Soc.* **2022**, *47*, 2557–2568.
16. Yang, K.; Wei, Z.; Dou, L.T. Research on dynamic tensile energy evolution and fractal characteristics of water-bearing coal sampl *J. China Coal Soc.* **2021**, *46*, 398–411.
17. Wang, J.H. *Mechanisms of Fluid Flow and Heat Transfer in Fractured Rock Mass and Its Implications to Heat Control for Geotherm Anomaly Mines*; China University of Mining and Technology: Xuzhou, China, 2022.
18. Deng, G.Z.; Wang, S.B.; Huang, B.X. Research on behavior character of crack development induced by hydraulic fracturing coal-rock mass. *Chin. J. Rock Mech. Eng.* **2004**, *23*, 3489–3493.
19. Ding, Y.; Li, S.G.; Zhu, B. Research on the feasibility of storage and estimation model of storage capacity of CO_2 in fissures of co mine old goaf. *Int. J. Min. Sci. Technol.* **2023**, *33*, 675–686. [CrossRef]
20. Chen, H.Q. Key problems and countermeasures in CO_2 flooding and storage. *Acta Geol. Sin.* **2023**, *97*, 858–872. [CrossRef]
21. Li, C.; Liang, W.G.; Hou, D.S. Morphology and formation mechanism of fractures in coal using hydraulic/ScCO_2 fracturing. *Ch J. Rock Mech. Eng.* **2020**, *39*, 761–772.
22. Li, W.X. *Finite Element Modelling Coupling of Fluid Flow and Heat Transfer in Fractured Rocs*; Beijing Jiaotong University: Beijin China, 2015.
23. Dong, H.Z.; Luo, R.H.; Zhang, L. Seepage and heat transfer model of a single fracture of rock and relevant parameter sensitivi analysis. *J. Hohai Univ. Nat. Sci.* **2013**, *41*, 42–47.
24. Deng, G.Z.; Liu, H. Comprehensive pre-splitting technology of coal seam hard dirt band based on gas-liquid-acid. *Sci. Techn Eng.* **2020**, *20*, 8916–8922.
25. Ha, S.J.; Choo, J.; Yun, T.S. Liquid CO_2 fracturing: Effect of fluid permeation on the breakdown pressure and cracking behavi *Rock Mech. Rock Eng.* **2018**, *51*, 3407–3420. [CrossRef]
26. Su, Z.; Wu, N.Y.; Zeng, Y.C. Research and development of enhanced geothermal system: A case of Fenton Hill in New Mexi (USA). *Prog. Geophys.* **2012**, *27*, 771–779.
27. Linstrom, P.J.; Mallard, W.G. The NIST chemistry WebBook: A chemical data resource on the internet. *J. Chem. Eng. Data* **20** *46*, 5. [CrossRef]
28. Span, R.; Wagner, W.A. New equation of state for carbon dioxide covering the fluid region from the triple-point temperature 1100K at pressure up to 80 MPa. *J. Phys. Chem. Ref. Data* **1996**, *25*, 1509–1596. [CrossRef]

applied sciences

MDPI

rticle

Development and Application of Unsealed Borehole Leakage Detection Device Based on Flow Method

ingqing Hao [1,2], Haidong Chen [1,2], Guangwei Xu [3,4], Yuqiang Yang [3,4,*], Xiangjun Chen [1,2], Zhaofeng Wang [1,2] and ang An [1,2]

1 State Key Laboratory Cultivation Base for Gas Geology and Gas Control, Henan Polytechnic University, Jiaozuo 454000, China
2 State Collaborative Innovation Center of Coal Work Safety and Clean-Efficiency Utilization, Henan Polytechnic University, Jiaozuo 454000, China
3 The First Exploration Team of Shandong Coal Geology Bureau, Jinan 250000, China
4 Shandong Engineering Research Center of Mine Gas Disaster Prevention, Jinan 250000, China
* Correspondence: yangyuqiang2024@163.com

Abstract: Poor sealing of gas extraction boreholes is one key to restrict gas extraction efficiency. In this paper, a novel borehole sealed quality detection device for the gas extraction of a coal mine is developed based on the theory of air leakage. By comparing the amount of gas extracted at different test points, it is possible to determine whether there is air leakage around the borehole, and the specific leakage position and leakage amount. Moreover, this device has the advantages of simple operation and a short test period. Based on the above analysis, a corresponding air leakage disposal method was proposed to handle the leaky boreholes. Field tests showed that the air–gas mixture flow in the test borehole was reduced by a factor of approximately 1.55 and the concentration of pure seam gas was increased approximately six times after the disposal of the air leaks. The combination of the leakage disposal method and the leakage detection device can accurately seal the borehole at the position of the leak, thus effectively ensuring the effectiveness of gas extraction from the borehole. The findings have important implications for improving the efficiency of gas extraction in coal mining operations.

Keywords: gas extraction; borehole sealing quality; unsealed borehole leakage detection; leakage disposal technique; extraction efficiency

1. Introduction

Coal mines in China generally have the occurrence characteristics of complex geological conditions, high gas content, and high gas pressure in coal seams, which make China one of the countries with serious coal and gas outburst accidents in the world [1–4]. Gas accidents account for more than 70 percent of major accidents in China's coal mines, according to statistics. Therefore, controlling the occurrence of gas accidents in coal mines is key to ensuring the safety of coal mine production. Years of coal mining practices have shown that gas extraction is an effective means of controlling coal mine gas disasters [5–10].

Ensuring the quality of borehole sealing is the key to improve the effectiveness of borehole gas extraction [11,12]. Currently, the commonly used sealing methods for drainage boreholes are the normal pressure sealing method and pressure sealing method. Normal pressure sealing methods mainly include polyurethane sealing, cement mortar sealing, and polyurethane–cement mortar joint sealing. Although this sealing method has a small investment, and simple equipment and process, the high polyurethane foaming ratio enables rapid sealing. But the cement mortar is naturally pumped without loading high pressure, the slurry penetration range is small, and the sealing effectiveness is not good for the surrounding rock when there are a large number of cracks [13–16]. Pressure sealing is mainly the bag type "two plugging and one injection" slurry (mucus) pressure sealing method. This sealing method achieves grouting of the borehole walls, which can effectively seal the cracks around the borehole well, and support the borehole after the slurry is consolidated. However, there are also problems such as high cost, complicated operation, and grout pressure being limited by the strength of the bag. Considering that the sealing quality of boreholes can be better guaranteed with the pressure sealing method under different conditions, most coal mines adopt the "two plugging and one injection" method of pressure sealing [17–19].

However, no matter which sealing method is used, air leakage will occur in the actual extraction process [20,21]. In general, the causes of unsealed borehole leakage are mainly reflected in the following aspects. First of all, when the sealing length is insufficient, the sealing section of the borehole may be in the range of the pressure relief area, which is an area of more developed cracks, permeability, and proneness to air leakage [22–26]. Secondly, when the sealing material is deformed or destroyed under the action of borehole deformation and its own gravity, there will be leakage from the sealing material itself or the contact surface between the material and the borehole wall [27,28]. Moreover, under the mutual superposition of double fracture area of the roadway fracture area and the fracture area around the borehole, the fractures around the borehole are developed and penetrated into each other, forming a macroscopic air leakage channel, thus affecting the sealing quality of the borehole [29].

The detection of leaky boreholes is a prerequisite for the disposal of air leaks and the improvement in the quality of borehole sealing. This has also been studied extensively by a number of scholars. Liu et al. [22], Chen et al. [26], Zhang et al. [30], and Zhang et al. [31] revealed the effect of sealing depth, extraction negative pressure, borehole diameter, and coal permeability on the quality of borehole sealing by establishing different borehole leakage models. Ba [32], Fan et al. [33], and Wang et al. [34] developed the air leakage detection device for gas extraction boreholes based on the distribution law and changes in gas concentration and extraction negative pressure in leaky boreholes, and the sealing quality of gas extraction boreholes was determined by detecting the gas sample parameters of different borehole depths. Qi et al. [35] and Cai et al. [36] injected tracer gas into the borehole, and then used the tracer gas concentration sensor to detect the tracer gas concentration in the extracted gas, so as to judge the sealing quality of the extraction borehole.

In summary, the air leakage detection of unsealed boreholes is crucial for enhancing the effectiveness of gas extraction. However, while the leakage detection methods and techniques proposed by scholars are effective, the detection period is generally long and the testing process is cumbersome. Therefore, there is an urgent need to develop new unsealed borehole leakage detection technologies to enable the rapid and efficient detection of air leakage around boreholes. In this paper, a flow-based technique was proposed for unsealed borehole leakage detection, which focuses on accurately determining the specific leakage position based on the flow variations at different locations in the borehole. On the one hand, this technique inherits the advantages of previous air leakage detection techniques. On the other hand, it makes up for their shortcomings. The results of this paper have positive implications for improving the gas extraction effectiveness of boreholes and ensuring gas extraction rates in mines.

2. Research and Development of Unsealed Borehole Leakage Detection Technology Based on Flow Method

2.1. Components of Air Leakage Detection Device

Most of the previous leakage detection is based on judging the concentration of seam gas extracted and negative pressure to determine whether the borehole is leaking. In this paper, a flow-based device was proposed for unsealed borehole leakage detection. The device (Figure 1) includes mainly a pressure limiting valve, pressure gauge, connecting rod, expandable capsule, sealing rubber plug, wet flow gauge, and other components. The specific leakage position and leakage amount can be judged based on the variation in the flow rate at different measurement points.

The specific operation of the leakage detection device is as follows.

(1) Remove the original connection device of the extraction pipe from the borehole, and the expanding capsule is then pushed to measurement point 1 using a connecting rod;

(2) Pass the outermost connection rod and the purple copper pipe (connected to wet flow gauge) through the double-hole rubber plug, and completely seal with the extraction pipe;
(3) After all components are connected, open the underway press-air system to inflate the expanding capsule to seal the section of the borehole to be tested for leakage;
(4) After the inflation is completed, open the extraction system valve, and observe the number of wet flow gauges;
(5) When the number of wet flow gauges is stable (the number remains unchanged within 3 min), turn off the extraction system valve, and deflate and depressurize the expanding capsule. Then, send the expanding capsule to the next measurement point for the next set of flow tests by mounting a link.

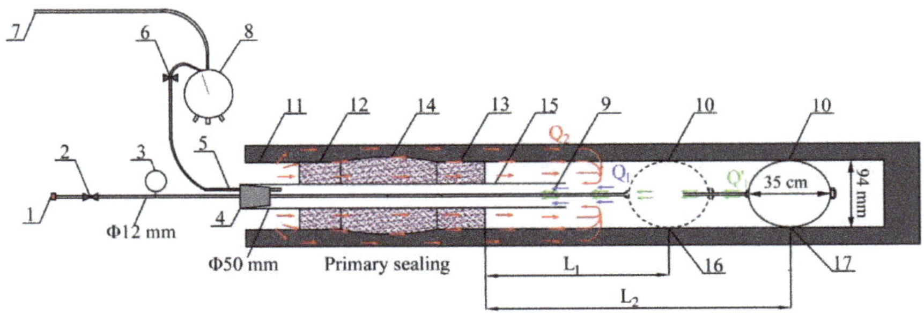

Figure 1. Air leakage detection device. 1—Press-air joint; 2—Pressure limiting valve; 3—Pressure gauge; 4—Rubber plug; 5—Purple copper pipe; 6—Extraction system valve; 7—Extraction system connector; 8—Wet flow gauge; 9—Connecting rod; 10—Expandable capsule; 11—Borehole; 12—Capsular bag 1; 13—Capsular bag 2; 14—Sealing material; 15—Extraction pipe; 16—Measurement point 1; 17—Measurement point 2 Q_1—Pure seam gas flow at measurement point 1 Q'_1—Pure seam gas flow at measurement point 2 Q_2—Air leakage around the borehole.

2.2. Selection of the Main Components of the Air Leakage Detection Device

The air leakage detection device developed is mainly to determine the leakage position around the borehole by detecting the difference of extraction flow before and after the leakage position. This section focuses on the selection process of the main components of the device on the ground, including the selection of expandable capsules, pressure limiting valve, and flow gauge.

2.2.1. Expandable Capsule Pressure-Bearing Test

As the main component of the air leakage detection device, the pressure-bearing capacity of the expandable capsule determines the usefulness and reliability of the device. The main material of the expandable capsule selected in this device is polyurethane resin; it has the characteristics of a light weight, corrosion resistance, and strong plasticity, which make the expandable capsule suitable for boreholes of any aperture and angle, and also improve the pressure-bearing capacity and service life of the capsule. Meanwhile, in order to avoid rupture of the expandable capsule during the operation of the device, it is necessary to simulate and investigate the maximum pressure that the expandable capsule can withstand in the borehole. The specific steps are as follows. After the expandable capsule is sent into the Φ94 mm PVC pipe (consistent with the diameter of the commonly downhole borehole), slowly inflate the expandable capsule to observe the final rupture pressure of the capsule. In order to prevent the contingency of the structure, repeat the steps three times, and the rupture pressures of the selected expandable capsule are tested to be 0.40 MPa, 0.43 MPa, and 0.45 MPa, respectively (Figure 2). According to the test results, the pressure in the expandable capsule should be controlled below 0.4 MPa.

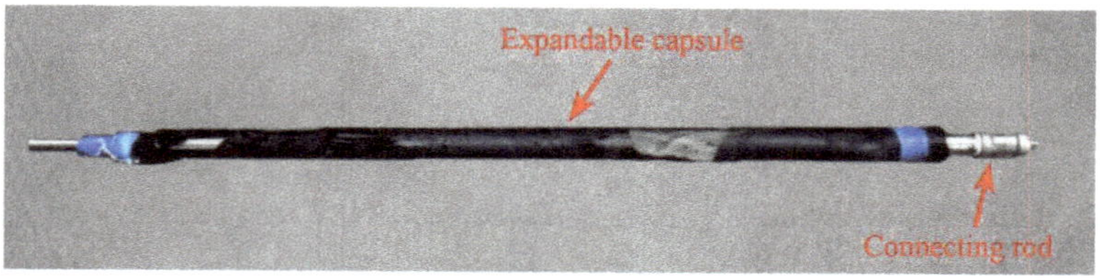

(**a**) Rupture pressure 0.4 MPa

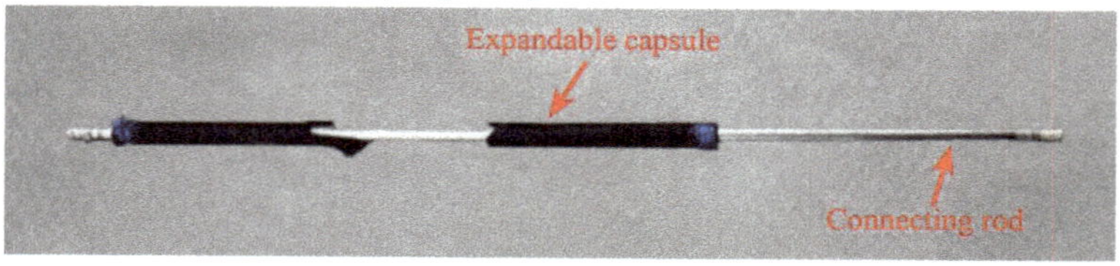

(**b**) Rupture pressure 0.43 MPa

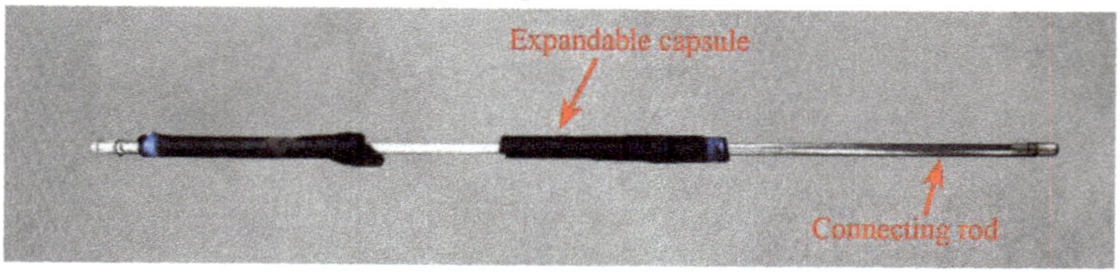

(**c**) Rupture pressure 0.45 MPa

Figure 2. Expandable capsule rupture pressure test.

2.2.2. Tightness Test of Expandable Capsule

In order to determine the tightness of the selected capsule, relevant tests were also carried out (Figure 3). The specific steps are as follows. Put the capsule into the middle of the Φ94 mm PVC pipe, and fill it with water with a pressure of 0.2 MPa~0.4 MPa to fully expand the capsule. The CH_4 gas with a pressure of 0.2 MPa then fills the PVC pipe sealed by the capsule. If the pressure inside the pipe remains constant for 20 min, it proves that the expandable capsule is well airtight. According to the test results, the tightness of the selected capsule is in accordance with the requirements.

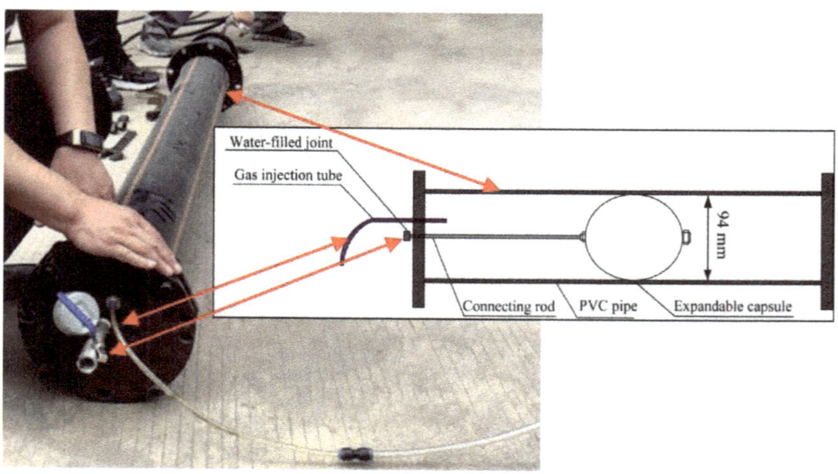

Figure 3. Tightness test of expandable capsule in Φ94 mm PVC pipe.

2.2.3. Selection of Pressure Limiting Valve

The expandable capsule is inflated through the underground press-air system, and the expansion of the capsule generally requires a gas charge in excess of 0.1 MPa. The pressure of the underground press-air system in coal mines is generally 0.6~0.8 MPa. Since the selected capsule can withstand a maximum pressure of 0.4 MPa, in order to prevent the capsule from rupturing directly after it is connected to the press-air system, a pressure limiting valve (Figure 4) needs to be installed between the press-air system and the connecting rod equipped with the expandable capsule. In order to keep a certain affluence coefficient, the final selection of the pressure limiting valve pressure is 0.25 MPa.

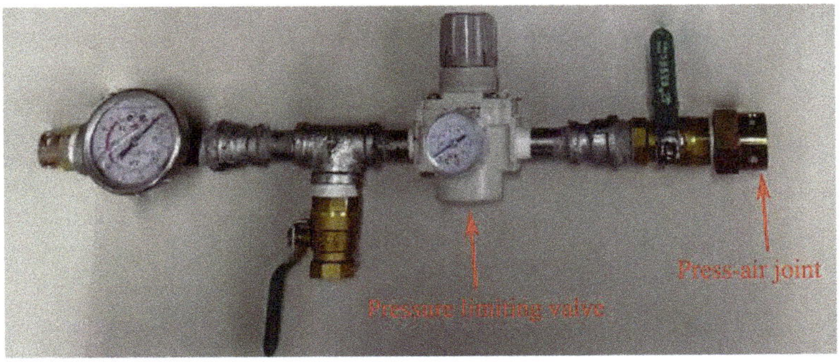

Figure 4. Installation position of pressure limiting valve.

2.2.4. Selection of Flow Gauge

This device determines the position of air leakage based on changes in the flow of air–gas mixture extracted from the borehole, so the choice of flow gauge is critical. Gas meters, rotors, and other flow gauges are mostly used to test gas flow in a barotropic environment. Considering that the flow gauges used in gas extraction are mostly under negative pressure conditions, after substantive investigations and selections, the wet flow gauge that can be used in the negative pressure environment of underground gas extraction is selected for this device. The specific model of the flow gauge can be chosen according to the actual working conditions of the coal mine.

2.3. Detection Principle of Leakage Detection Device

When the location of the expandable capsule is located at measurement point 1 in Figure 1, the first enclosed space is formed between the expandable capsule and the grout-sealed section, and the length of the first enclosed space is L_1. The flow measured at this time is called Q_{11}, and $Q_{11} = Q_1 + Q_2$, since Q_1 is the pure seam gas flow extracted by the first enclosed space; Q_2 is the amount of air leakage from the grout seal section. If the seal is intact, the flow measured in the first enclosed space is the pure seam gas flow Q_1.

When the location of the expandable capsule is located at measurement point 2 in Figure 1, the second enclosed space is formed between the expandable capsule and the grout-sealed section, and the length of the second enclosed space is L_2. The flow measured at this time is called Q_{22}, and $Q_{22} = Q'_1 + Q_2$. This is since Q'_1 is the seam gas flow extracted by the second enclosed space. If the seal is intact, the flow rate measured in the second enclosed space is the pure seam gas flow Q'_1.

The two measurements were subtracted to obtain

$$\Delta Q = Q_{22} - Q_{11} = (Q'_1 + Q_2) - (Q_1 + Q_2) = Q'_1 - Q_1 \quad (1)$$

where ΔQ is the flow corresponding to the length of the L_2-L_1, and thus the flow rate per unit length of borehole Q_E is obtained:

$$Q_E = (Q'_1 - Q_1)/(L_2 - L_1) \quad (2)$$

Then, $Q_1 = Q_E L_1$, and then obtain

$$Q_2 = Q_{11} - Q_1 = Q_{11} - Q_E L_1 \quad (3)$$

Based on the magnitude of Q_2, it is possible to determine not only whether there are air leaks in the sealed section of the borehole, but also to obtain a specific value for the amount of leakage.

According to the theory of air leakage around boreholes, if there is no leakage in the detected L_1 and L_2 sections, the increase in the flow rate between the first and second measurement points is only the pure seam gas extraction in the tested section ΔL (the length of L_2–L_1); the increment of ΔQ (the flow rate corresponding to the length of L_2–L_1) will be very small, as shown in Figure 5a. If there is a leakage in the test section ΔL, the increment of ΔQ will be larger when the expandable capsule is tested at measurement point 2, as shown in Figure 5b. Due to the complex and variable extraction conditions in coal mines, a complete absence of air leakage rarely exists. Generally, when the ratio of the ΔQ to the flow measured at the previous point is less than 5%, consider the measured borehole to be in a non-leakage state, and when this ratio is greater than 30%, the borehole is leaking.

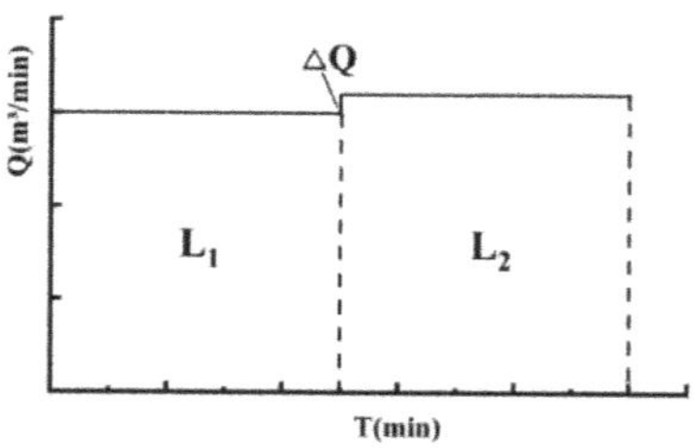

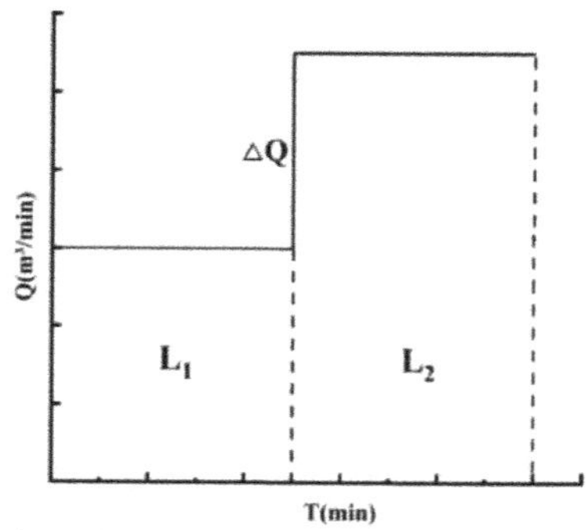

(**a**) Effect of non-leakage flow in ΔL section (**b**) Effect of air leakage flow in ΔL section

Figure 5. Leakage detection flow effect diagram.

3. Results and Discussions

3.1. Field Application of Air Leakage Detection Technique

The soundness of the developed air leakage detection device was verified through field tests. The field site was selected at the 12,316 working face of Wangjialing Coal Mine in Shanxi Province, China, and the test site was selected at 850 m of the return airway in the 12,316 working face, with a total of six boreholes named borehole 1–borehole 6 and a test borehole length of 40 m. According to previous studies [26] and combined with the parameters of Wangjialing mine coal formation, the plastic area extent of the return airway in the 12,316 working face is 6 m. Therefore, seal the 1#–6# boreholes by the method of "two plugging and one injection"; the initial sealing depth of the borehole is 6 m and the depth of the sealing section is 6 m. The borehole spacing is set to 5 m based on the extraction radius of the test face.

After the boreholes are constructed, the extraction system is connected and the negative pressure 15 kPa is debugged; then, detect the changes in the air–gas mixture flow from boreholes by a wet flow gauge under the condition of constant negative pressure every day, as shown in Figure 6.

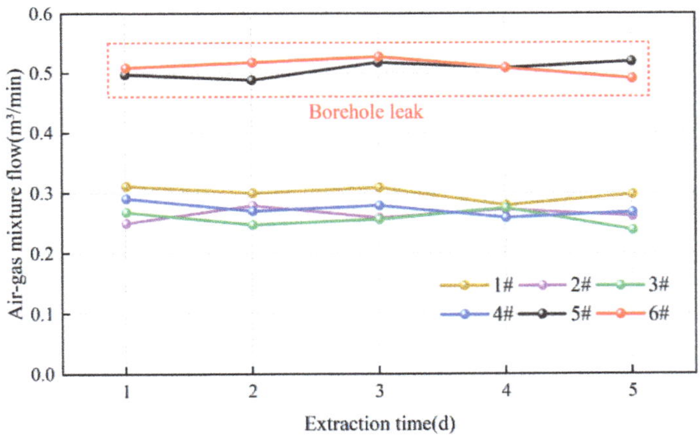

Figure 6. Air–gas mixture flow from 1#–6# boreholes.

According to Figure 6, the 5# and 6# boreholes extract significantly more mixed flow than the other boreholes compared to the 1#, 2#, 3#, and 4# boreholes in the previous five days. It is assumed that there is leakage around 5 # and 6 # boreholes.

For the detection of the 6# borehole, the expanding capsule was initially lowered to 7.5 m with almost no flow at a negative pressure of 15 kPa. When the capsule was placed at 8 m for the second detection, the flow of air–gas mixture extracted reached 0.2 m^3/min at a negative pressure of 15 kPa. When the capsule was placed at 8.5 m for the third detection, the flow of air–gas mixture extracted reached 0.4 m^3/min. However, when the capsule was placed at 9 m for the fourth detection, the flow of air–gas mixture extracted reached 0.4008 m^3/min. Therefore, the leakage position can be judged to be between 7.5 m and 8.5 m. The same method was used to detect the leakage position of the 5# borehole; it was concluded that the position of the air leaks of the 5# borehole was approximately between 8 and 9 m.

3.2. *Air Leakage Disposal Technique*

3.2.1. Principle of Air Leakage Disposal Technique

When using the air leakage detection device described above to detect a test borehole, it is necessary to provide secondary sealing to the borehole if there are air leaks inside. Currently, most approaches for addressing air leakage around boreholes are primarily based on the "three plugging and two injection" technique. This approach involves pre-judging potential air leakage locations based on "two plugging and one injection", while reserving a bag in advance. During drilling and mining operations, the bag can be activated to perform secondary sealing if the concentration of seam gas extracted decreases. This approach has the disadvantage that air leakage disposal cannot achieve precise positioning and higher sealing costs [37,38]. In view of the shortcomings, this paper proposes the Φ10 mm sealing capsular bag secondary sealing method, which directly sends the secondary sealing capsular bag to the front end of the air leakage position. This method achieves precise disposal of the leakage position and reduces the cost of sealing boreholes.

The specific usage is as follows. Firstly, the conventional "two plugging and one injection" sealing method is used to perform the primary sealing to the extraction borehole, and observe the changes in the flow rate and concentration; if there is an increase in the flow rate and a decrease in the concentration, the borehole is tentatively judged to be leaking. Then, use the air leakage detection device to detect the leakage, as shown in Figure 1, after accurately determining the leakage position; a 10 mm diameter extraction pipe with a capsular bag of sealing holes is passed through the Φ50 mm extraction pipe commonly used in the borehole to the front end of the leakage position for secondary grouting and sealing. After secondary sealing, the diameter of the extraction pipe was changed from the original 50 mm to 10 mm. The principle of the secondary sealing of a Φ10 mm sealing capsular bag is shown in Figure 7.

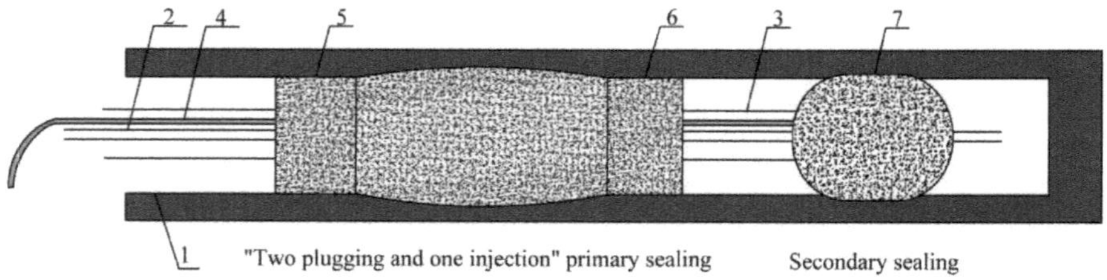

Figure 7. Principle of Φ10 mm sealing capsular bag secondary sealing. 1—Borehole; 2—Φ10 mm extraction pipe with capsular bag; 3—Φ50 mm extraction pipe; 4—Grouting tube; 5—Capsular bag 1; 6—Capsular bag 2; 7—Secondary sealing capsular bag.

3.2.2. Leakage Disposal Effect

After detecting the specific leakage locations of the 5# and 6# boreholes with the new air leakage detection device proposed in this study, the above-mentioned Φ10 mm sealing capsular bag secondary sealing method is used for leakage disposal. After detecting the air–gas mixture flow before and after the secondary sealing, the changes in the seam gas concentration in the mixture were detected by the optical gas detector. The specific detection process of the optical gas detector is as follows.

(1) Preparation: Check the detector chemical reactants, gas sample channel, and air tightness of the suction ball to confirm its good performance. The chemical reactants include the calcium chloride, silica gel, and sodium lime to make sure that they are not invalid. Then, check whether the gas sample channel and suction ball are leaking.
(2) Zeroing: Place the instrument in air, and pinch and release the suction ball several times to ensure that the zero scale of the micro-reading disk (decimal display disk) coincides with the pointer and that the selected black baseline coincides with the zero position of the dividing plate (integer display panel).
(3) Measurement: Extend the rubber tube connected to the air inlet to the air–gas mixture extracted, press the suction ball for more than 10 times, and ensure that the gas to be measured enters the gas chamber. Press the light source electric gate and observe the position of the black baseline and pointer from the eyepiece. Read out the integer number shown on the dividing plate and the decimal number shown on the micro-reading disk; the sum of the two is the measured gas concentration.

The seam gas concentration was tested for 8 days in 5# and 6# boreholes following the above testing process. The changes in the air–gas mixture flow and seam gas concentration in the mixture before and after the secondary sealing are shown in Figures 8 and 9.

From the 5th day, the Φ10 mm sealing capsular bag method was carried out to seal the 5# and 6# leaky boreholes for the secondary sealing. It can be seen from Figure 8 that before the disposal of the leakage around the 5# and 6# boreholes, the air–gas mixture flow was as high as 0.51 m^3/min. After the secondary sealing of 5# and 6# boreholes by the Φ10 mm sealing capsular bag method, the air–gas mixture flow decreased to 0.33 m^3/min and 0.29 m^3/min, respectively, which was decreased by a factor of approximately 1.55 and similar to other boreholes. The seam gas concentration in the air–gas mixture in 5# and 6# boreholes was improved by six times after the secondary sealing, as shown in Figure 9, which shows the significant increase in gas extraction efficiency.

Compared with the traditional method of "three plugging and two injection", the air leakage disposal method proposed in this paper avoids the blindness of the secondary sealing. It can also accurately seal boreholes at the position of leaks, which saves the cost of sealing boreholes in mines and effectively ensures the effectiveness of gas extraction from boreholes.

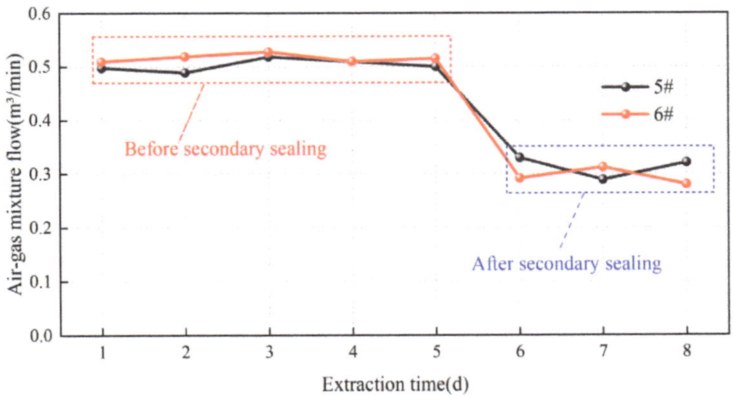

Figure 8. Air–gas mixture flow from 5# and 6# boreholes.

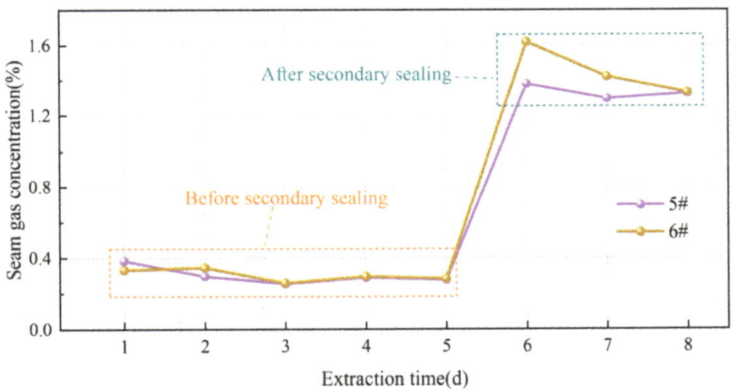

Figure 9. Seam gas concentration of 5# and 6# boreholes.

4. Conclusions

In this paper, a new unsealed extraction borehole leakage detection technique and a corresponding air leakage disposal method were proposed with the aim of improving gas extraction levels in mines, reducing gas emissions, and enhancing gas resource utilization. The following conclusions were drawn from this study.

(1) A new type of air leakage detection device for borehole gas extraction in coal mines has been developed to address the problem of a more cumbersome operation and longer testing period of the existing air leakage detection devices. By comparing the flow of gas extracted at different test points, it is possible to determine whether there is air leakage around the borehole, and the specific leakage position and leakage amount. This device has the advantages of simple operation and a short test period.

(2) In view of the shortcomings of the "three plugging and two injection" secondary sealing technique, this paper proposes the Φ10 mm sealing capsular bag secondary sealing method. Field tests showed that the air–gas mixture flow in the test borehole was reduced by a factor of approximately 1.55 and the concentration of pure seam gas was increased approximately six times after the disposal of the leaky position.

(3) The combination of the leakage disposal method and the leakage detection device can accurately seal the unsealed borehole at the position of the leak, thus avoiding the blindness of secondary sealing, reducing the cost of sealing boreholes, and effectively ensuring the effectiveness of gas extraction from the borehole.

Author Contributions: Conceptualization, Q.H. and H.C.; Methodology, Q.H., H.C. and X.C.; Software, Q.H. and K.A.; Formal analysis, Q.H. and Z.W.; Data curation, Y.Y. and G.X.; Validation, Y.Y. and G.X.; Supervision, H.C., Y.Y. and X.C.; Writing—original draft, Q.H. and H.C.; Writing—review and editing, Q.H., H.C. and Y.Y.; Project administration, Y.Y. and G.X.; Funding acquisition, H.C. All authors have read and agreed to the published version of the manuscript.

Funding: This work is financially supported by the Scientific and Technological Breakthrough Project of Henan Province (242102321030), the Henan Provincial Natural Science Foundation (242300421651 and 242300421012), the Innovation and Entrepreneurship Training Program for College Students (202210460028), the Fundamental Research Funds for the Universities of Henan Province (NSFRF230401), the Safety Discipline "double first-class" Creation Project of Henan Polytechnic University (AQ20230713 and AQ20230305), the Innovative Research Team in University of Henan Province (24IRTSTHN013), the Research Fund of Shandong Coalfield Geological Bureau (2022-004), and the Innovation Training Program of Henan Polytechnic University (202310460098).

Institutional Review Board Statement: Not applicable.

Informed Consent Statement: Not applicable.

Data Availability Statement: All data used to support the findings of this study are available from the corresponding author upon request.

Conflicts of Interest: The authors declare no conflict of interest.

References

1. Nie, B.; Sun, S.J. Thermal recovery of coalbed methane: Modeling of heat and mass transfer in wellbores. *Energy* **2023**, *263*, 1258[illegible] [CrossRef]
2. Yan, F.; Xu, J.; Peng, S.J.; Zou, Q.J.; Li, Q.G.; Long, K.; Zhao, Z.G. Effect of capacitance on physicochemical evolution characteristi[illegible] of bituminous coal treated by high-voltage electric pulses. *Powder Technol.* **2020**, *367*, 47–55. [CrossRef]
3. Kong, B.; Wang, E.Y.; Lu, W.; Li, Z.H. Application of electromagnetic radiation detection in high-temperature anomalous are[illegible] experiencing coalfield fires. *Energy* **2019**, *189*, 116144. [CrossRef]
4. Kang, Y.S.; Huangfu, Y.H.; Zhang, B.; He, Z.P.; Jiang, S.Y.; Ma, Y.Z. Gas oversaturation in deep coals and its implications for co[illegible] bed methane development: A case study in Linxing Block, Ordos Basin, China. *Front. Earth Sci.* **2023**, *10*, 1031493. [CrossRef]
5. Tang, S.L.; Zhai, H.B.; Tang, H.; Yang, F. Isothermal desorption hysteretic model for deep coalbed methane development. *Geoflui[illegible]* **2022**, *2022*, 5259115. [CrossRef]
6. Li, H.Y.; Lau, H.C.; Huang, S. China's coalbed methane development: A review of the challenges and opportunities in subsurfa[illegible] and surface engineering. *J. Pet. Sci. Eng.* **2018**, *166*, 621–635. [CrossRef]
7. Li, X.L.; Cao, Z.Y.; Xu, Y.L. Characteristics and trends of coal mine safety development. *Energy Sources Part A* **2020**, *12*, 1–[illegible] [CrossRef]
8. Tao, S.; Pan, Z.J.; Chen, S.D.; Tang, S.L. Coal seam porosity and fracture heterogeneity of marcolithotypes in the Fanzhuang bloc[illegible] southern Qinshui Basin, China. *J. Nat. Gas Sci. Eng.* **2019**, *66*, 148–158. [CrossRef]
9. Wan, L.M.; Hou, B.; Meng, H.; Chang, Z.; Muhadasi, Y.; Chen, M. Experimental investigation of fracture initiation position an[illegible] fluid viscosity effect in multi-layered coal strata. *J. Pet. Sci. Eng.* **2019**, *182*, 106310. [CrossRef]
10. Qu, Q.D.; Guo, H.; Khanal, M. Monitoring and analysis of ground movement from multi-seam mining. *Int. J. Rock Mech. Min. S[illegible]* **2021**, *148*, 104949. [CrossRef]
11. Men, X.Y.; Tao, S.; Liu, Z.X.; Tian, W.G.; Chen, S.D. Experimental study on gas mass transfer process in a heterogeneous co[illegible] reservoir. *Fuel Process. Technol.* **2021**, *216*, 106779. [CrossRef]
12. Kong, X.G.; Li, S.G.; Wang, E.Y.; Wang, X.; Zhou, Y.X.; Ji, P.F.; Wei, Z.Y. Experimental and numerical investigations on dynam[illegible] mechanical responses and failure process of gas-bearing coal under impact load. *Soil Dyn. Earthq. Eng.* **2021**, *142*, 10657[illegible] [CrossRef]
13. Tittarelli, F.; Mobili, A.; Giosuè, C.; Belli, A.; Bellezze, T. Corrosion behaviour of bare and galvanized steel in geopolymer an[illegible] Ordinary Portland Cement based mortars with the same strength class exposed to chlorides. *Corros. Sci.* **2018**, *134*, 64–7[illegible] [CrossRef]
14. Iacobescu, I.R.; Angelopoulos, N.G.; Jones, T.P.; Blanpain, B.; Pontikes, Y. Ladle metallurgy stainless steel slag as a raw material [illegible] Ordinary Portland Cement production: A possibility for industrial symbiosis. *J. Clean. Prod.* **2016**, *112*, 872–881. [CrossRef]
15. Fu, J.H.; Wang, D.K.; Li, X.L.; Wang, Z.M.; Shang, Z.J.; Jiang, Z.G.; Wang, X.B.; Gao, X. Experimental study on the cement-base[illegible] materials used in coal mine gas extraction for hole sealing. *ACS Omega* **2021**, *6*, 21094–21103. [CrossRef] [PubMed]
16. Mu, B.Y.; Li, M. Fabrication and characterization of polyurethane-grafted reduced graphene oxide as solid-solid phase chang[illegible] materials for solar energy conversion and storage. *Sol. Energy* **2019**, *188*, 230–238. [CrossRef]
17. Zhang, F.J.; Qin, Y.P.; Xu, H.; Wu, F.; Chu, X.Y.; Wang, K. Multi-angle analysis of the mechanism of polymer materials to impro[illegible] the sealing quality of boreholes. *Energy Sources Part A* **2022**, *44*, 3205–3222. [CrossRef]

Qin, W.; Xu, J.L. Dynamic secondary borehole-sealing method for gas drainage boreholes along the coal seam. *Geofluids* **2018**, *2018*, 1723019. [CrossRef]

Jing, M.; Ni, G.H.; Zhu, C.J.; Li, Z.; Wang, G.; Wang, Z.Y.; Huang, Q.M. Effect of new modified materials on the microscopic pore structure and hydration characteristics of sealing materials in coal seam boreholes. *Constr. Build. Mater.* **2023**, *377*, 131076. [CrossRef]

Zhang, Y.J.; Zou, Q.L.; Guo, L.D. Air-leakage model and sealing technique with sealing-isolation integration for gas-drainage boreholes in coal mines. *Process Saf. Environ. Prot.* **2020**, *140*, 258–272. [CrossRef]

Zhou, A.T.; Xu, Z.Y.; Wang, K.; Wang, Y.H.; An, J.Y.; Shi, Z. Coal mine gas migration model establishment and gas extraction technology field application research. *Fuel* **2023**, *349*, 128650. [CrossRef]

Liu, T.; Lin, B.Q.; Fu, X.H.; Zhu, C.J. Modeling air leakage around gas extraction boreholes in mining-disturbed coal seams. *Process Saf. Environ. Prot.* **2020**, *141*, 202–214. [CrossRef]

Zhang, C.; Jin, G.H.; Liu, C.; Li, S.G.; Xue, J.H.; Cheng, R.H.; Liu, H. Sealing performance of new solidified materials: Mechanical properties and stress sensitivity characterization of pores. *Adv. Polym. Technol.* **2020**, *2020*, 5397697. [CrossRef]

Wang, K.; Wang, L.; Ju, Y.; Dong, H.Z.; Zhao, W.; Du, C.G.; Guo, Y.Y.; Lou, Z.; Gao, H. Numerical study on the mechanism of air leakage in drainage boreholes: A fully coupled gas-air flow model considering elastic-plastic deformation of coal and its validation. *Process Saf. Environ. Prot.* **2022**, *158*, 134–145. [CrossRef]

Wu, W.; Wang, Z.F.; Jen, A. Effect of stresses and strains of roadway surrounding rocks on borehole airtightness. *J. Eng. Sci. Technol. Rev.* **2016**, *9*, 66–73.

Chen, H.D.; Chen, X.J.; Wang, Z.F.; Li, Z.Q.; An, F.H. Plastic zone range of a roadway considering the creep effect. *Sci. Rep.* **2020**, *10*, 20341. [CrossRef] [PubMed]

Xu, C.Y.; Zhang, H.L.; Kang, Y.L.; Zhang, J.Y.; Bai, Y.R.; Zhang, J.; You, Z.J. Physical plugging of lost circulation fractures at microscopic level. *Fuel* **2022**, *317*, 123477. [CrossRef]

Wang, Z.M.; Sun, Y.N.; Wang, Y.L.; Zhang, J.X.; Sun, Z.D. A coupled model of air leakage in gas drainage and an active support sealing method for improving drainage performance. *Fuel* **2019**, *237*, 1217–1227. [CrossRef]

Hong, Z.J.; Xiong, Z.Q.; Wang, C. Gas drainage sealing mechanism and optimization of key leakage parameters: Analysis, experiment, and field verification. *Arab. J. Geosci.* **2022**, *15*, 1431. [CrossRef]

Zhang, X.B.; Gao, J.L.; Jia, G.N.; Zhang, J.W. Study on the influence mechanism of air leakage on gas extraction in extraction boreholes. *Energy Explor. Exploit.* **2022**, *40*, 1344–1359. [CrossRef]

Zhang, J.X.; Liu, Y.W.; Ren, P.L.; Han, H.K.; Zhang, S. A fully multifield coupling model of gas extraction and air leakage for in-seam borehole. *Energy Rep.* **2021**, *7*, 1293–1305. [CrossRef]

Ba, Q.B. Research on air leakage mechanism and detection technology of coal mine gas drainage. *IOP Conf. Ser. Earth Environ. Sci.* **2020**, *446*, 052018. [CrossRef]

Fan, X.G.; Ma, Q.Q. Application test of hole sealing quality testing device in Wangpo Coal Mine. *Coal Mine Saf.* **2019**, *50*, 133–135. (In Chinese)

Wang, N.; Zhang, T.J.; Fan, J.D.; Deng, Z.S.; Ji, X. Testing technology and application of sealing hole quality in coal mine gas extraction. *Coal Technol.* **2019**, *38*, 87–89. (In Chinese)

Qi, Q.J.; Zhao, Y.X.; Lv, Y.C.; Jia, X.L.; Zhou, X.H. New process technology of "two blocks, one note and one row" for gas extraction. *Chin. J. Saf. Sci.* **2018**, *28*, 100–104. (In Chinese)

Cai, J.T.; Wu, J.S.; Yuan, S.Q.; Liu, Z.; Kong, D.S. Numerical analysis of multi-factors effects on the leakage and gas diffusion of gas drainage pipeline in underground coal mines. *Process Saf. Environ. Prot.* **2021**, *151*, 166–181. [CrossRef]

Yu, S.Y.; Su, X.B.; Song, J.X.; Wang, Q.; You, Z.J. The hole sealing technology of solid–liquid materials with three pluggings and two injections for gas extraction hole in the coal mine. *ACS Omega* **2022**, *7*, 43847–43855. [CrossRef]

Sun, X.Y.; Li, K.; Wang, X. Capsule-bag-type sealing technology for gas drainage boreholes and its application. *Geofluids* **2022**, *2022*, 1671859. [CrossRef]

applied sciences

MDPI

Article

Coal Dust and Methane as a Hazard in Coal Preparation Plants

Aleksander Lutyński [1] and Marcin Lutyński [2,*]

[1] KOMAG Institute of Mining Technology, Pszczyńska 37, 44-101 Gliwice, Poland; alutynski@komag.eu
[2] Department of Geoengineering and Natural Resources Extraction, Faculty of Mining, Safety Engineering and Industrial Automation, Silesian University of Technology, Akademicka 2, 44-100 Gliwice, Poland
* Correspondence: marcin.lutynski@polsl.pl

Abstract: This article presents the results of analysis of the hazards posed by coal mine dust and methane in the coal preparation plants of hard coal mines in Poland. It was shown how the number of workplaces in plants at risk of coal dust explosion and the highest permissible dust concentration changed in the period from 2003 to 2022 when compared with coal production. The methodology of assessing mine dust hazards was based on hazard ratios related to one million tons of hard coal enriched in preparation plants. As a result of the analysis, it was found that the explosion hazard index with zone 20 showed an increasing trend in the analyzed period, while the explosion hazard indices with zones 21 and 22 analyzed together and the maximum permissible dust concentration showed decreasing trends following a decrease in hard coal production. In the case of methane, no zone 0 explosion hazards were found, and there were only a few instances of zone 1 explosion hazards. However, it was determined that the explosion hazard index for zone 2 showed an increasing trend during the analyzed period, which is directly proportional to the coal produced and is a result of increasing depth of mining.

Keywords: coal dust; coal preparation plant; explosion hazard; coal production

1. Introduction

Coal dust and methane represent two of the most critical hazards in coal preparation plants, significantly impacting both worker safety and operational efficiency. Coal dust, a by-product of mechanical processes such as coal extraction, transportation, and beneficiation, poses severe risks due to its potential to cause catastrophic explosions under specific conditions [1,2]. Similarly, mine dust frequently encountered in these environments presents hazards ranging from injuries and fatalities to long-term occupational illnesses [3].

The generation of mine dust in coal preparation plants occurs during the mechanical extraction and transport of coal, as well as during the processing of coal and waste rock. Key factors influencing dust generation include:

- geological characteristics of the coal seam and the properties of the coal mined (location, mineral composition, hardness, degree of coalification, etc.),
- mining method, run-of-mine haulage method to the coal preparation plant (type of machinery and equipment for mining and loading, modes and means of transport),
- technology adopted for the preparation of coal in the preparation plant (beneficiation methods, type of machinery and equipment),
- Physical properties of the dust, such as particle size and dispersibility.

Suspended coal dust is transported by air currents, with some particles depositing on structures and machinery. Resuspension of this deposited dust can elevate atmospheric concentrations, posing persistent hazards to workers [4,5].

Coal dust recognized as a hazardous factor in a coal preparation plant originates from a coal seam and has a volatile content of more than 10% (particle size <1 mm). All currently mined hard coal seams in Poland contain more than 10% volatile matter in dry ash-free conditions. Such dust mixed with air in quantities of 50 to 1000 g/m^3 constitutes a mixture that, under the right conditions and as a result of initiation, can lead to an explosion.

Explosive dust atmospheres are divided into three zones based on the International standard IEC 60079-10-2 [6], which in the case of coal dust are classified as follows:

- zone 20—a place in which an explosive dust atmosphere, in the form of a cloud of dust in air, is present continuously, or for long periods or frequently,
- zone 21—a place in which an explosive dust atmosphere, in the form of a cloud of dust in air, is likely to occur in normal operation occasionally,
- zone 22—area in which an explosive dust atmosphere, in the form of a cloud of combustible dust in air, is not likely to occur in normal operation but, if it does occur, will persist for a short period only.

Prolonged exposure to coal dust at high concentrations poses significant health risks to workers. This is addressed in the Regulation of the Minister of Labour and Social Policy [7,8], which establishes the Maximum Allowable Concentration (MAC) for workplace dust exposure. The MAC represents the concentration level that, over an 8-h workday and a typical workweek, should not cause adverse health effects during a worker's lifetime or affect the health of future generations. This regulation underscores the importance of maintaining strict control over workplace dust concentrations to safeguard worker health.

The regulation specifies permissible dust concentrations based on the silica content in the dust. The Maximum Allowable Concentration (MAC) values cited in Table 1 are derived from [7], rather than [8], as the comparison focuses on results prior to 2018.

Table 1. Maximum allowable dust concentrations at workplaces, mg/m^3 [7].

Dust Fraction	Silica Content in Dust, %			
	<2	2 to 10	10 to 50	>50
Inhalable (<100 μm)	10	4	2	2
Respirable (<7 μm)	-	2	1	0.3

Adverse effects of coal dust mine on human health have been widely investigated [9–12] contributing to obstructive lung disease, pulmonary fibrosis, and lung cancer, in addition to the Coal Workers' Pneumoconiosis (CWP) and silicosis that receive greater attention. Therefore, fighting coal dust not only contributes to the safety in terms of explosion hazard but also contributes to the overall health of workers.

The second important hazard in coal preparation plant is methane. Methane is emitted during all dry processing operations, yet crushing and grinding common in coking coal processing plants (and to a lesser extent in thermal coal processing plants) significantly contributes to its release. Conveyor belts, especially at transfer points, also release methane from the material they transport. Other places where methane accumulates in coal processing plants are all sorts of bunkers for raw coal, concentrate, and occasionally coal tailings.

The increasing depth of coal mining, which has reached an average of 780 m in Poland and can exceed 1000 m in some cases, has led to a rise in methane emissions in processing plants. The average annual increase in mining depth for Poland is 7–8 m, and the relative methane content has grown from approximately 10 m^{3}CH$_4$/Mg in 2003 [13] to around 15 m^{3}CH$_4$/Mg in 2022 [14].

To assess the risk of Explosive Atmospheres, areas are categorized into three zones: zone 0 (where Explosive Atmospheres are continuous or frequent), zone 1 (where they may occur during normal operations), and zone 2 (where they are unlikely but may persist for short periods if they do occur).

The aim of this study is to analyze and compare the total coal production with coal dust explosion indicators and methane hazard at workplaces in coal preparation plants.

Data were collected from all of the polish coal preparation plants from 2003 to 2022. New indexes were derived based on the yearly production and occurrence of hazard zones, which allow an assessment of the trends in relation to the total coal production.

2. Coal Dust Control in Hard Coal Preparation Plants

Coal dust hazard and methane in hard coal preparation plants is controlled using administrative and legal regulations as well as technical methods. Administrative-legal regulations include the application of directives, laws, orders, recommendations, and regulations and organizational measures. A number of recommendations and orders are of a local nature, related to the conditions prevailing at a particular site. Technical methods aim at eliminating or at least minimizing the amount of dust and methane in the atmosphere surrounding workplaces to non-explosive levels and not exceeding the Maximum Allowable Concentrations (MAC). Technical methods involve dust collection or water spraying on areas where dust is generated and extensive ventilation in case of methane. The following are selected examples that illustrate solutions in this area applied in mechanical processing plants. One such solution is the MB-M-25A circulating dust collector. The results of dust collection tests carried out at a coal preparation plant using this dust collector are described in [15,16]. It was shown that the operation of the dust collector results in a significant decrease in the amount of dust precipitating on the surfaces around the coal preparation equipment. The largest decreases in the precipitation of the coal dust, more than twofold, were recorded in the vicinity of the screens.

As a result of the work carried out at ITG KOMAG, interesting design solutions have been developed that allow them to be used in many places in mining plants where a dust explosion hazard arises [17]. These devices are of four different types. The most popular one, which has been applied in several coal processing plants, is the UO type dust collector [18]. In this device, air mixed with coal dust is sucked in and introduced into the device through a vortex nozzle with a fan. Inside the dust collector, the coal dust introduced into it with the air is mixed with water. The high kinetic energy of the water droplets imparted by the vortex nozzle makes the dust removal highly efficient. Dust removal efficiency is estimated at 99% of the dust of the inhalable dust fraction and 97% of the respirable dust fraction [19].

Another interesting solution that has found application in coal processing plants is the PASAT-type sprinkler system [18]. This system has been installed in one mine above the transfer from the crushers to the belt conveyors, and in another mine above the transfer of the belt conveyor system. The PASAT system can be equipped with several sprinkler batteries, creating a curtain of water mist that dampens and loads the rising dust that falls onto the conveyor. The system uses compressed air, which reduces the use of water needed to reduce the dust concentration. The PASAT system is supplied with water directly from the mine's fire main. The sprinkler unit is equipped with filters, reduction and shut-off valves, and a UV25 type sterilizer. The system is activated by sensors detecting the movement of the conveyor belt. A view of the sprinkler system is shown in Figure 1. PASAT-type sprinkler systems demonstrate reliable operation and high dust reduction efficiency.

In the case of methane, technical methods of hazard mitigation involve placing measurement sensors in critical locations (wells, bunkers) that operate within the mine's automated methane monitoring system. The accumulation of methane at a concentration exceeding 1% in these locations triggers the shutdown of the electrical power supply in the hazardous area.

One measure to prevent the formation of an explosive atmosphere is to ventilate areas where such an atmosphere may occur. For instance, in some coal processing plants, the start-up of technological processes, including transportation, is preceded by the activation of ventilation fans.

A commonly used technical measure to eliminate potential ignition sources from explosion-hazardous areas is the use of electrically powered safety devices in these areas. These devices comply with ATEX Directives.

Figure 1. View of the PASAT sprinkler system.

3. Indexes for Coal Dust and Methane Hazards Estimation in Hard Coal Preparation Plants

The analysis of the coal dust hazard in hard coal mines covers the period from 2003 to 2022, which is marked by extremely dynamic organizational changes in the Polish hard coal mining industry. In 2003, there were 41 mines in operation [20], and each operated a preparation plant with a different range of coal beneficiation range. In 2022, there were 20 mines operating 30 mine sites, i.e., practically former mines [21]. There was also a significant decrease in coal production [22] as shown in Table 2 and in Figure 2.

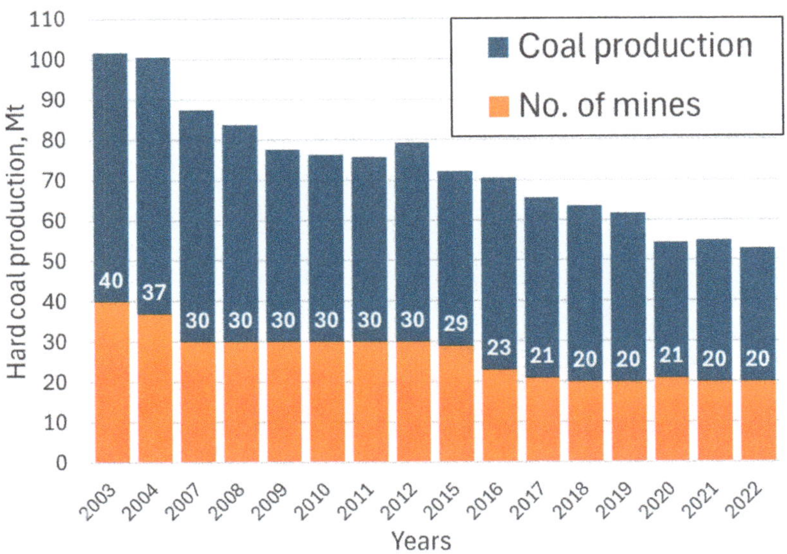

Figure 2. Hard coal production and number of mines in Poland 2003–2022 [22].

Information on hazards at workplaces in the coal preparation plants of Polish mines was obtained from annual reports on the state of natural and technical hazards in hard coal mining periodically issued by the Central Mining Institute [13,14,23–34]. Information on the number of workplaces in coal preparation plants operating in explosive dust atmospheres and the number of workplaces where permissible dust concentrations were exceeded in the years 2003 to 2022 are presented in Table 2.

Table 2. Hard coal production in Poland [35] versus number of workplaces in zone 20 and zones 21 + 22, zone 2 methane hazard and number of workplaces with exceeded maximum permissible dust concentrations (MACs) [13,14,23–34,36] in the years 2003 to 2022.

Year	Coal Production, Million Mg	No. of Workplaces in Zone 20	No. of Workplaces in Zone 21 + 22	No. of Workplaces in MAC	No. of Workplaces in Zone 2 Methane Explosive Atmosphere
2003	101.7	30	504	395	240
2004	100.5	30	587	303	325
2007	87.4	44	508	472	225
2008	83.7	10	458	541	274
2009	77.5	10	414	588	225
2010	76.2	12	400	516	232
2011	75.7	12	393	478	351
2012	79.2	15	410	516	279
2015	72.2	24	364	415	280
2016	70.4	21	309	458	251
2017	65.5	19	309	380	352
2018	63.4	17	308	315	217
2019	61.6	17	196	265	219
2020	54.4	19	190	203	237
2021	55.0	31	253	185	259
2022	52.8	19	157	177	282

No data available for years 2005, 2006, 2013, and 2014.

In order to observe trends related to changes of the abovementioned hazards in time and in relation to production, it was decided to develop indexes that could allow an estimation of hazards based on the available historical data.

Indexes were elaborated for workplaces with explosive dust atmospheres zone 20 (high risk zones) and lower risk zones as the sum of zones 21 and 22. The example mathematical expression for the index IEZ_{20} is the following:

$$IEZ_{20} = \left(\frac{N_{20}}{N_{20} + N_{21+22}} \cdot \frac{1}{P}\right) \cdot 100 \tag{1}$$

where IEZ_{20} is the explosion hazard index for workplaces at high explosion risk. This index measures the proportion of workplaces in explosive dust atmosphere zone 20 (high explosion risk) (N_{20}) relative to the total number of workplaces in explosive dust atmosphere zones 20, 21, and 22 (cumulative explosion risk), normalized by annual coal production (P). By multiplying by 100, the index expresses this proportion as a percentage, providing a standardized indicator to assess and compare the relative risk of zone 20 workplaces across different years or production scales. This metric highlights the specific contribution of high-risk areas to overall safety concerns in coal preparation plant operations. Similarly, index IEZ_{21+22} indicates workplaces where the number of explosion hazard risk is low and is derived in similar way as IEZ_{20}. In Figure 3 the total number of workplaces in explosive dust atmosphere zones (sum of zones 20, 21, and 22) was plotted against yearly coal production.

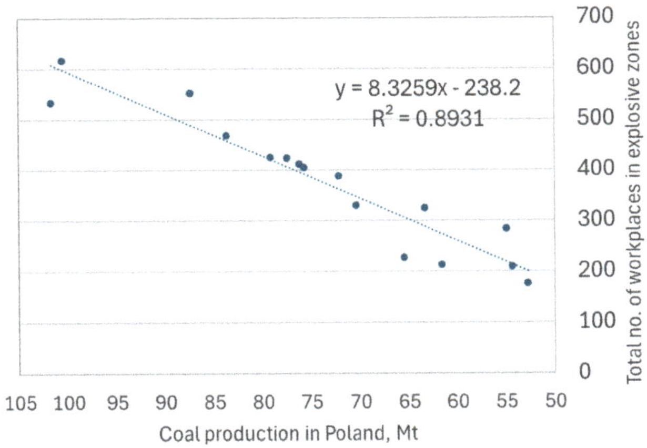

Figure 3. Total number of workplaces in explosive dust atmosphere zones 20, 21, and 22 plotted against total coal production in Poland in 2003–2022 [13,14,23–34,36].

The plot in Figure 3 demonstrates a strong correlation between the decline in coal production, driven by mine closures, and the reduction in workplaces located in explosion hazard zones within coal preparation plants. The Pearson correlation coefficient for this dataset is 0.945, indicating a very strong relationship. Despite the overall downward trend in the total number of workplaces in explosion hazard zones, the proportion of workplaces in zone 20 (high risk) has been increasing, even as coal production continues to decline (Figure 4).

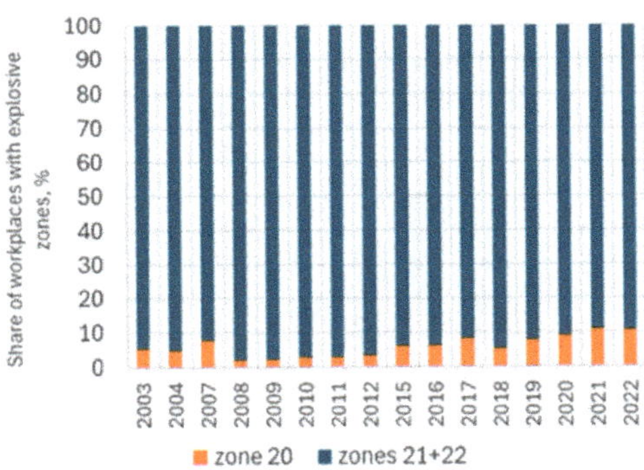

Figure 4. Share of workplaces in explosive zone 20 (high risk) and zones 21 + 22 (lower risk) in Poland in 2003–2022.

The dynamics of the change in calculated indexes IEL_{20} and IEL_{21+22} can be observed in Figure 5 where an upward trend is visible for both indexes. Despite the fact that the coal production is decreasing, the relative number of workplaces with dust Explosive Atmospheres is increasing. In fact, this trend is particularly visible in case of the number workplaces in zone 20 (high risk) since the Pearson coefficient shows a negative value of -0.592, indicating that the number is increasing despite the coal production shrinking. On the other hand, Pearson coefficient for zones 21 + 22 has a value of 0.591, indicating decrease of low risk zones with decreasing production trend.

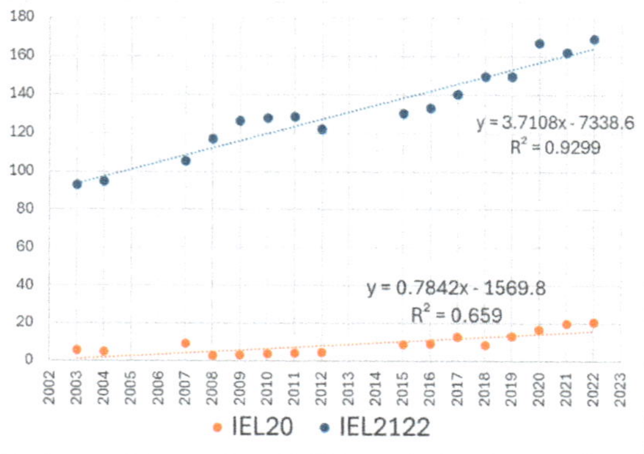

Figure 5. Calculated indexes IEL_{20} (high risk) and IEL_{21+22} (low risk), indicating dust Explosive Atmospheres at workplaces in hard coal preparation plants in Poland in relation to total coal production from 2003 to 2022.

To evaluate trends in the Maximum Allowable Concentration (MAC) for dust exposure in coal preparation plant workplaces, the MAC values were analyzed in relation to annual coal production. The proposed IMAC index is calculated by dividing the reported number of workplaces exceeding dust MAC by the respective coal production for each year. The calculated IMAC index for 2003–2022 is presented in Figure 6. In this case, no clear trend is observed, which is further reflected in the Pearson correlation coefficient of 0.145, indicating a weak relationship between production and reported number of workplaces in dust MAC.

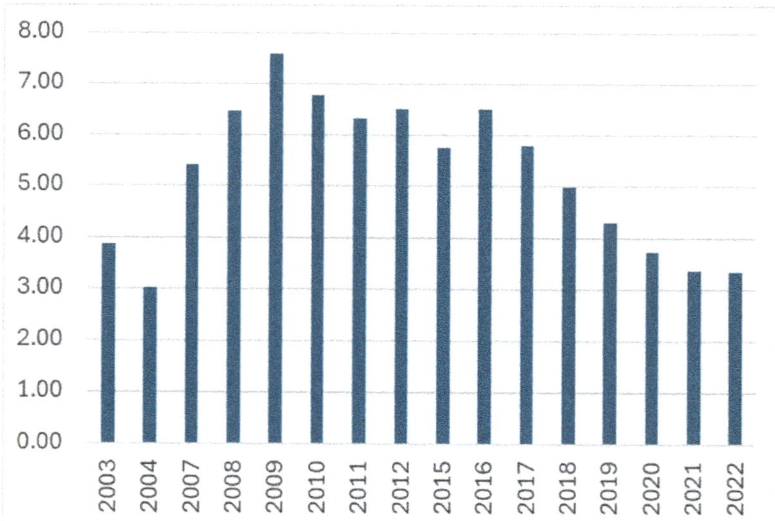

Figure 6. Changes in calculated IMAC index for 2003–2022.

To assess the risk of methane Explosive Atmospheres in the workplace in coal preparations plants, the number of workplaces in zone 2 was compared at first with annual production resulting in the IMC index. Analysis of the data shows that, during the analyzed period, no zone 0 methane explosion hazard was found at workplaces in the processing plant. Zone 1 occurred at 24 workplaces in 2007, at two workplaces in 2009, 2011, and 2012, and at one workplace in 2015, 2016, 2017, 2021, and 2022. Therefore, occurrences of zones 0 and 1 were neglected from the analysis. In the case of zone 2, there is a clear increasing trend showing index growth despite the decrease in coal production. This comparison does not provide any logical explanation as methane content in coal depends on the coal origin, mining depth rather than production rates. Therefore, it was decided to compare the number of workplaces in methane explosive atmosphere with data on the annual average methane content in coal provided in the reports of State Mining Authorities starting from 2000 to 2022 [21]. Results of the comparison are shown in Figure 7.

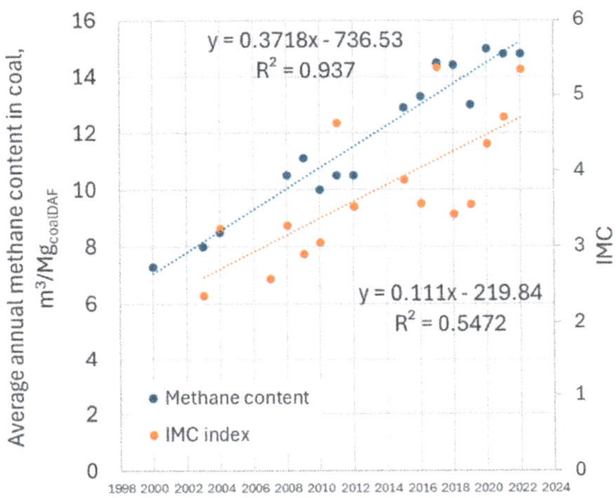

Figure 7. Changes in calculated IMC index and average annual methane content in coal from 2000 to 2022.

There is a clear correlation between increasing methane content in coal and the IMC index for coal preparation plant which is reflected in Pearson coefficient of 0.71 for the analyzed set of data.

4. Discussion

The analysis of coal dust and methane explosion hazards in coal preparation plants highlights significant trends and evolving risks over the two decades from 2003 to 2022. The period was characterized by substantial organizational changes within the Polish hard coal mining sector, including a sharp decline in the number of operational mines and coal production volumes. Despite these reductions, certain hazard indicators show concerning upward trends, reflecting complex relationships between workplace safety and operational adjustments.

The total number of workplaces in explosive dust atmosphere zones (zones 20, 21, and 22) exhibited a strong correlation with coal production, as evidenced by a Pearson coefficient of 0.945. This finding aligns with the general expectation that fewer mines and lower production volumes would reduce workplace hazards. However, the share of workplaces in high-risk zone 20 increased relative to the total number of workplaces in explosive zones, even as overall production declined. The calculated index IEZ_{20} demonstrated a negative correlation with production ($r = -0.592$), suggesting that high-risk areas may persist despite overall safety improvements. This is a surprising result that may indicate an obsolete infrastructure and inefficient safety measures.

The findings of this study align with the work of Dong et al. 2023 [37], who employed the Systems-Theoretic Accident Model and Process (STAMP) combined with the Rank-order Centroid (ROC) method to identify critical risk factors in non-coal mine explosion accidents. Dong et al.'s emphasis on systemic risks—such as inadequate licensing, confusion in safety management systems, and failures in supervision—might reflect some of the hassle in Polish coal preparation plants. Their approach offers a valuable framework for improving risk identification by integrating broader systemic considerations into hazard assessment methodologies. Niu et al. 2023 [38] highlight the critical role of human factors in explosion hazards. Their study shows that unsafe acts—such as habitual violations driven by mental state and other factors—are significant contributors to coal mine gas explosions. This aligns with our findings that, despite declining production, the proportion of workplaces in high-risk zones continues to grow. These insights underscore the need to address behavioral and organizational factors alongside physical hazards. Integrating frameworks like the Human Factor Analysis and Classification System (HFACS-GE) and Bayesian Networks (BNs) could help identify and mitigate unsafe acts, enhancing safety compliance in coal preparation plants.

On the other hand, technological advancements such as positive–negative pressure dust removal systems [39] can play a pivotal role in dust control. Their study showed that a positive- and negative-pressure composite dust removal system reduced average dust concentration in coal preparation plants by over 78%.

The IMAC index, representing workplaces exceeding the Maximum Allowable Concentration (MAC) for dust, did not show a clear trend over the analyzed period. The weak Pearson correlation coefficient of 0.145 suggests no consistent relationship between coal production and dust-related hazards. This may reflect variability in workplace conditions, safety measures, or monitoring practices, rather than a direct influence of production volumes. Fluctuations in IMAC index may be also attributed to changes in coal geology where access to coal with a higher silica content may be encountered. This is in line with a broader studies [9,40] in US mines, which show large variations in silica content depending on the strata and distance from longwall operation.

The IMC index, which evaluates workplaces exposed to methane in zone 2 Explosive Atmospheres, revealed an increasing trend despite declining coal production. This anomaly underscores the influence of factors such as methane content in coal seams that increases with mining depth, and geological conditions rather than production levels alone. A stronger correlation ($r = 0.71$) was observed between the IMC index and annual average methane content in coal, highlighting the need for targeted mitigation strategies that address these underlying factors.

The observed trends underscore the complexity of managing safety hazards in coal preparation plants. While declining production and mine closures have contributed to a reduction in overall risk, persistent or increasing risks in high-priority areas, such as zone 20 and methane hazards, warrant closer attention. These findings highlight the importance of robust safety protocols, continuous monitoring, and adaptive hazard management strategies to address dynamic risks effectively.

5. Conclusions and Further Recommendations

The comprehensive analysis of workplace hazards in Polish coal preparation plants between 2003 and 2022 highlights critical insights into the evolution of safety risks:

Dynamic Organizational Changes: The period saw significant structural shifts in the mining sector, with a sharp reduction in operational mines and coal production. Despite these changes, high-risk areas persist and, in some cases, have intensified.

Explosion Hazard Trends: The number of workplaces in explosive dust atmospheres decreased overall, correlating strongly with declining production. However, the rising share of zone 20 (high-risk areas) workplaces indicates persistent challenges in mitigating hazards in these critical zones.

Dust Exposure and MAC Compliance: The IMAC index showed no definitive trend, underscoring variability in safety practices and workplace conditions. This variability suggests that factors beyond production levels significantly impact dust exposure risks.

Methane Hazard Dynamics: An increasing trend in the IMC index, despite lower production levels, aligns with a rise in methane content in coal. This indicates that geological and operational factors, rather than production alone, play a significant role in methane hazard exposure.

Future Recommendations

Safety Measures for High-Risk Zones: Given the increasing proportion of zone 20 workplaces, targeted safety interventions, such as improved ventilation and advanced explosion suppression systems and dust suppression systems, should be prioritized for high-risk zones.

Enhanced Dust Monitoring and Control: The variability in IMAC trends suggests a need for standardized dust control protocols across all plants. Adoption of real-time dust monitoring technology and stricter enforcement of dust suppression measures could help mitigate risks.

Methane Hazard Mitigation: The correlation between methane content and IMC index calls for tailored strategies, such as pre-drainage of methane from coal seams and advanced gas detection systems.

Continuous Data Analysis and Index Development: Refining existing hazard indices, such as an IMC, to include additional variables (e.g., geological factors, equipment performance) can improve their predictive value. Establishing automated systems to monitor and analyze safety data in real time would enhance proactive risk management.

Training and Awareness: Regular training programs on hazard recognition and mitigation should be conducted for workers and supervisors. Emphasis on high-risk zones and methane hazards can reduce human error and enhance safety compliance.

Policy and Regulation Alignment: Updating safety regulations to reflect the latest technological advancements and emerging risks is essential. Collaboration between government bodies, industry leaders, and research institutions can ensure regulations are robust and relevant.

These recommendations aim to bolster safety practices and adapt to the evolving risks in coal preparation plants, ensuring worker protection and operational sustainability in the Polish hard coal industry.

Author Contributions: Conceptualization, A.L.; methodology, A.L.; validation, M.L.; formal analysis, M.L.; investigation, A.L.; resources, A.L. and M.L.; data curation, M.L.; writing—original draft preparation, A.L.; writing—review and editing, M.L. All authors have read and agreed to the published version of the manuscript.

Funding: This research received no external funding.

Institutional Review Board Statement: Not applicable.

Informed Consent Statement: Not applicable.

Data Availability Statement: Data derived from public domain resources.

Conflicts of Interest: The authors declare no conflicts of interest.

References

1. Chunmiao, Y.; Chang, L.; Gang, L. Coal Dust Explosion Prevention and Protection Based on Inherent Safety. *Procedia Eng.* **2011**, *26*, 1517–1525. [CrossRef]
2. Divers, E.F.; Cecala, A.B. *Dust Control in Coal Preparation and Mineral Processing Plants*; Bureau of Mines: Pittsburgh, PA, USA, 1990.
3. Attfield, M.; Castranova, V.; Hale, J.M.; Suarthana, E.; Thomas, K.C.; Wang, M.L. Coal Mine Dust Exposures and Associated Health Outcomes; a Review of Information Published since 1995. National Institute for Occupational Safety and Health. DHHS (NIOSH) Publication No. 2011–172. 2011. Available online: https://www.cdc.gov/niosh/docs/2011-172/pdfs/2011-172.pdf (accessed on 9 February 2024).
4. Jiang, J.; Wang, P.; Pei, Y.; Liu, R.; Liu, L.; He, Y. Preparation and Performance Analysis of a Coking Coal Dust Suppressant Spray. *Int. J. Coal Sci. Technol.* **2021**, *8*, 1003–1014. [CrossRef]
5. Prostański, D. Experimental Study of Coal Dust Deposition in Mine Workings with the Use of Empirical Models. *J. Sustain. Min.* **2015**, *14*, 108–114. [CrossRef]
6. *IEC 60079-10-2*; Explosive Atmospheres—Part 10-2: Classification of Areas—Explosive Dust Atmospheres. International Electrotechnical Commission (IEC): Geneva, Switzerland, 2015.

7. Rozporządzenie Ministra Pracy i Polityki Społecznej. Rozporządzenie Ministra Pracy i Polityki Społecznej z dnia 6 czerwca 20 r. w sprawie najwyższych dopuszczalnych stężeń i natężeń czynników szkodliwych dla zdrowia w środowisku pracy; Dz. 2014, poz. 817. Regulation of the Minister of Labor and Social Policy. Regulation of the Minister of Labor and Social Policy June 6, 2014, on the maximum allowable concentrations and intensities of harmful factors for health in the work environme Journal of Laws 2014, item 817.
8. Rozporządzenie Ministra Pracy i Polityki Społecznej. Rozporządzenie Ministra Pracy i Polityki Społecznej w sprawie najwyższy dopuszczalnych stężeń i natężeń czynników szkodliwych dla zdrowia w środowisku pracy; Dz. U. 2018, poz. 1286. (Regulati of the Minister of Labor and Social Policy. Regulation of the Minister of Labor and Social Policy on the maximum allowab concentrations and intensities of harmful factors for health in the work environment; Journal of Laws 2018, item 1286.).
9. Rahimi, E.; Shekarian, Y.; Shekarian, N.; Roghanchi, P. Investigation of Respirable Coal Mine Dust (RCMD) and Respirab Crystalline Silica (RCS) in the U.S. Underground and Surface Coal Mines. *Sci. Rep.* **2023**, *13*, 1767. [CrossRef] [PubMed]
10. Liu, T.; Liu, S. The Impacts of Coal Dust on Miners' Health: A Review. *Environ. Res.* **2020**, *190*, 109849. [CrossRef] [PubMed]
11. Go, L.H.T.; Cohen, R.A. Coal Workers' Pneumoconiosis and Other Mining-Related Lung Disease: New Manifestations of Illne in an Age-Old Occupation. *Clin. Chest Med.* **2020**, *41*, 687–696. [CrossRef] [PubMed]
12. Mocek, P.; Mocek, K. Health Effects of Long-Term Exposure to Industrial Dust in Preparation Plants of Hard Coal Mines. *Sci Marit. Univ. Szczec.* **2022**, *72*, 102–112. [CrossRef]
13. *Józef Raport Roczny o Stanie Podstawowych Zagrożeń Naturalnych i Technicznych w Górnictwie Węgla Kamiennego w Roku 2003*; Główn Instytut Górnictwa: Katowice, Poland, 2004; ISBN 978-83-65503-.
14. Makówka, J. *Raport Roczny o Stanie Podstawowych Zagrożeń Naturalnych i Technicznych w Górnictwie Węgla Kamiennego w Roku 20* Główny Instytut Górnictwa: Katowice, Poland, 2023; ISBN 978-83-65503-53-4.
15. Lutyński, A. *Zwalczanie Zagrożeń Wywoływanych Przez Metan, pył i Hałas w Zakładach Przeróbki Mechanicznej Kopalń Wę Kamiennego*; Innowacyjne Techniki i Technologie Mechanizacyjne; Instytut Techniki Górniczej KOMAG: Gliwice, Poland, 20 ISBN 978-83-65593-03-0.
16. Prostański, D.; Vargová, M. Installation Optimization of Air-and-Water Sprinklers at Belt Conveyor Transfer Points in the Aspe of Ventilation Air Dust Reduction Efficiency. EBSCOhost. Available online: https://openurl.ebsco.com/contentitem/gc 135672923?sid=ebsco:plink:crawler&id=ebsco:gcd:135672923 (accessed on 26 November 2024).
17. Lutyński, A.; Prostański, D. Zwalczanie Zagrożenia Wywoływanego Pyłem w Zakładach Przeróbki Mechanicznej Kopalń Węg Kamiennego. *Inżynieria Górnicza* **2017**, *3–4*, 44–51.
18. Bałaga, D.; Siegmund, M. Systemy Do Redukcji Zapylenia w Zak\ladach Przeróbczych. *Masz. Górnicze* **2015**, *33*, 8–15.
19. Jedziniak, M. Malogabarytowe Urządzenie Odpylające. *Masz. Górnicze* **2013**, *31*, 34–38.
20. Górniczy, W.U. *Stan Bezpieczeństwa i Higieny Pracy w Górnictwie*; State Mining Authority (Wyższy Urząd Górniczy): Katowi Poland, 2004.
21. Ocena Stanu Bezpieczeństwa Pracy. *Ratownictwa Górniczego Oraz Bezpieczeństwa Powszechnego w Związku z Działalnością Górnic Geologiczną w 2022 Roku*; State Mining Authority (Wyższy Urząd Górniczy): Katowice, Poland, 2023.
22. Coal Production and Consumption Statistics. Available online: https://ec.europa.eu/eurostat/statistics-explained/index.ph title=Coal_production_and_consumption_statistics (accessed on 22 January 2023).
23. Kabiesz, J. *Raport Roczny o Stanie Podstawowych Zagrożeń Naturalnych i Technicznych w Górnictwie Węgla Kamiennego w Roku 202* Główny Instytut Górnictwa: Katowice, Poland, 2021; ISBN 978-83-65503-32-9.
24. Kabiesz, J. *Raport Roczny o Stanie Podstawowych Zagrożeń Naturalnych i Technicznych w Górnictwie Węgla Kamiennego w Roku 201* Główny Instytut Górnictwa: Katowice, Poland, 2020; ISBN 978-83-65503-28-2.
25. Kabiesz, J. *Raport Roczny o Stanie Podstawowych Zagrożeń Naturalnych i Technicznych w Górnictwie Węgla Kamiennego w Roku 201* Główny Instytut Górnictwa: Katowice, Poland, 2019; ISBN 978-83-65503-16-9.
26. Kabiesz, J. *Raport Roczny o Stanie Podstawowych Zagrożeń Naturalnych i Technicznych w Górnictwie Węgla Kamiennego w Roku 201* Główny Instytut Górnictwa: Katowice, Poland, 2018; ISBN 978-83-65503-.
27. Kabiesz, J. *Raport Roczny o Stanie Podstawowych Zagrożeń Naturalnych i Technicznych w Górnictwie Węgla Kamiennego w Roku 201* Główny Instytut Górnictwa: Katowice, Poland, 2017; ISBN 978-83-65503-.
28. Kabiesz, J. *Raport Roczny o Stanie Podstawowych Zagrożeń Naturalnych i Technicznych w Górnictwie Węgla Kamiennego w Roku 201* Główny Instytut Górnictwa: Katowice, Poland, 2016; ISBN 978-83-65503-.
29. Kabiesz, J. *Raport Roczny o Stanie Podstawowych Zagrożeń Naturalnych i Technicznych w Górnictwie Węgla Kamiennego w Roku 201* Główny Instytut Górnictwa: Katowice, Poland, 2013; ISBN 978-83-65503-.
30. Kabiesz, J. *Raport Roczny o Stanie Podstawowych Zagrożeń Naturalnych i Technicznych w Górnictwie Węgla Kamiennego w Roku 201* Główny Instytut Górnictwa: Katowice, Poland, 2012; ISBN 978-83-65503-.
31. Kabiesz, J. *Raport Roczny o Stanie Podstawowych Zagrożeń Naturalnych i Technicznych w Górnictwie Węgla Kamiennego w Roku 201* Główny Instytut Górnictwa: Katowice, Poland, 2011; ISBN 978-83-65503-.
32. *Józef Raport Roczny o Stanie Podstawowych Zagrożeń Naturalnych i Technicznych w Górnictwie Węgla Kamiennego w Roku 2004*; Główn Instytut Górnictwa: Katowice, Poland, 2005; ISBN 978-83-65503-.
33. *Józef Raport Roczny o Stanie Podstawowych Zagrożeń Naturalnych i Technicznych w Górnictwie Węgla Kamiennego w Roku 2007*; Główn Instytut Górnictwa: Katowice, Poland, 2008; ISBN 978-83-65503-.

Józef Raport Roczny o Stanie Podstawowych Zagrożeń Naturalnych i Technicznych w Górnictwie Węgla Kamiennego w Roku 2009; Główny Instytut Górnictwa: Katowice, Poland, 2010; ISBN 978-83-65503-.
Coal Production and Consumption up in 2022—Products Eurostat News—Eurostat. Available online: https://ec.europa.eu/eurostat/web/products-eurostat-news/w/ddn-20230622-2 (accessed on 9 February 2024).
Eurostat. Europe 2020 Indicators—Climate Change and Energy. Available online: http://ec.europa.eu/eurostat/statistics-explained/index.php/Europe_2020_indicators_-_climate_change_and_energy (accessed on 23 October 2017).
Dong, X.; Yang, Z.; Guo, L.; Gao, Y. Assessment of the Explosion Accident Risk in Non-Coal Mining by Hasse Diagram Technique. *Processes* **2023**, *11*, 582. [CrossRef]
Niu, L.; Zhao, J.; Yang, J. Risk Assessment of Unsafe Acts in Coal Mine Gas Explosion Accidents Based on HFACS-GE and Bayesian Networks. *Processes* **2023**, *11*, 554. [CrossRef]
Liu, Q.; Zhu, Z.; Hua, Y.; Guo, L. Study on Dust Control Technology and Parameters for a Coal Washing Plant Screening Workshop Based on CFD Technology of Positive-Negative Pressure Dust Removal Systems. *Powder Technol.* **2024**, *448*, 120250. [CrossRef]
Sarver, E.; Keleş, Ç.; Afrouz, S.G. Particle Size and Mineralogy Distributions in Respirable Dust Samples from 25 US Underground Coal Mines. *Int. J. Coal Geol.* **2021**, *247*, 103851. [CrossRef]

applied sciences

MDPI

Article

Theoretical Analysis of Shaft Wall Damage and Failure Under Impacting of Ore-Rock Falling in Vertical Ore Pass

Qiangying Ma [1,†], Chi Ma [2,†], Jiaoqun Li [1,*], Zengxiang Lu [1,*] and Zhiguo Xia [1]

1 School of Mining Engineering, University of Science and Technology Liaoning, Anshan 114051, China; qiangying_ma@163.com (Q.M.); xzgyy88@163.com (Z.X.)
2 School of Civil and Resources Engineering, University of Science and Technology Beijing, Beijing 100083, China; machi@xs.ustb.edu.cn
* Correspondence: ljq_as@163.com (J.L.); zengxiang_lu@sohu.com (Z.L.)
† These authors contributed equally to this work.

Abstract: The impact of ore-rock blocks on the shaft wall of a vertical ore pass is a crucial cause of shaft wall damage and failure. Based on the structure and parameters of the ore pass in a case mine, the first collision's position of the ore-rock block with respect to the ore pass wall and the angle between the impacting direction of the ore-rock block and the horizontal plane before and after the collision are investigated via a kinematic analysis. The normal and tangential analysis models of ore rock impacting the shaft wall are established and analyzed based on contact mechanics. The results show that: (1) based on the kinematic analysis of ore rock moving in the ore pass and on the colliding condition of the ore-rock block the first time that it collides with the ore pass wall, the coordinates and angles of the collision are proposed; (2) the impacting process of ore rock is categorized into elastic compression, elastic–plastic compression, and rebound of the shaft wall material. The relationship between the normal impact force and the penetrating depth is determined, and the slipping distance of the ore-rock block along the shaft wall and the lost volume of the shaft material are established. (3) The wall material's normal, tangential, and total restitution coefficient is acquired. (4) The total lost volume during the collision is obtained through the analysis and solution of the model. (5) Based on the characteristics and parameters of the ore pass in the case mine, the influence of the impact velocity and angle of the ore-rock block on the restitution coefficient, maximum normal intrusion depth, maximum tangential displacement, and volume loss of the shaft wall are analyzed by using relevant formulas.

Keywords: ore pass wall deformation; ore rock impacting; contact mechanics; restitution coefficient; volume loss

1. Introduction

The ore pass system, widely used in underground mining worldwide, is a crucial development project of the mine where the ore or rock is lifted by a skip [1,2]. It facilitates the centralized downward transportation of ore and waste rock during multi-stage mining operations, it has a long service life, and it has a high production capacity [3,4]. However, the deformation and damage of the ore pass wall have emerged as significant issues impacting the efficiency of underground mining operations in metal mines. Impact and abrasion are the primary factors that cause such deformations and damage [5,6]. Ensuring the mine's safe and efficient production depends on the reliable realization of the chute system [7–9]. Nevertheless, complex engineering–geological conditions, coupled with harsh usage environments, have resulted in the deformation and frequent destruction of ore passes [10–13]. Therefore, it is imperative to research the damage characteristics of the ore pass wall while seeking reasonable and practical solutions to ensure the safe and efficient production of mines.

Numerous scholars and engineering technicians have extensively studied various aspects, including the structural design of the ore pass [14], supporting materials, reinforcement methods [9], movement behavior of the ore rock in an ore pass [15,16], deformation and failure mechanisms [17,18], as well as the rehabilitation strategies for damaged ore passes [19,20], of the ore pass and have achieved fruitful outcomes. For instance, Sjöberg J. et al. [21] used numerical models to identify the failure mechanisms of ore passes in the Kiirunavaara underground mine and developed a causative model to describe both failure mechanisms and failure development. Yang et al. [22] employed numerical simulation techniques to investigate force-chain changes among granular particles, revealing impact-tamping effects caused by the falling of ores/rocks on stored materials within vertical ore passes and their influence on ore pass walls. Luo et al. [23] explored the damage mechanisms affecting ore pass walls from an impulse perspective, developing a numerical analysis model based on a specific mine case which unveiled different degrees of impact-induced damage experienced by ore passes. Song et al. [24], based on the kinematic theory, analyzed the movement law and impact load of the ore-rock block in the ore pass at Chengchao iron mine in China, investigated the damage characteristics of the ore pass wall, and discussed the effect of the impact forces on the affected area. Lu et al. [25] analyzed the destruction mechanisms responsible for the damage to the ore pass wall from a kinetic energy perspective, highlighting that both impact and shear actions from moving ore-rock blocks are significant factors contributing to the deformation and failure of the wall. Yin Yue et al. [26] studied the ore-rock movement in a vertical ore pass and the characteristics of the damage caused by the impact with and the cutting of ore-rock blocks to the ore pass wall when the moving ore-rock blocks collided with the ore pass wall, and they constructed a damage model of the sharp and blunt surfaces of ore-rock blocks impacting the shaft wall based on contact mechanics. However, despite these studies, issues related to the deformation and destruction of ore passes still occur frequently, significantly impacting mining operations in terms of safety and efficiency.

For this reason, Qin et al. [27] studied the motion law of ore rock and the impact damage law using kinematic analysis and distinct element codes. They revealed the relationship among impact position, speed, degree of dispersion, and impact kinetic energy of ores in an ore pass. After that, Wu et al. [28] was able to gather the collapse information of an ore pass by using 3D visualization analysis. Ren et al. [29] established a laboratory bench of similarity simulations for the investigation of the damage mechanisms within the main ore pass. Most of these studies focus on the location and nature of ore-rock collisions with the ore pass wall, and only a few studies to date have focused on the ore pass wall's damage depth and lost volume after the ore-rock collision using the numerical simulation method.

Considering the Hertz contact theory [30], the elastoplastic body tangential contact theory [31–33], as well as the modified soil yield velocity formula [34], this paper, while taking into account the specific characteristics of the ore-rock blocks impacting the ore pass wall, establishes the colliding conditions of the ore-rock block colliding with the ore pass wall, proposes equations for calculating the three-dimensional coordinates of the collision point, its depression angle, and the direction angle of the ore-rock block's first impact with the ore pass wall. This study also presents a contact model between the ore-rock blocks and the ore pass wall under a range of impact forces in different directions to analyze the damage caused by the ore-rock blocks to the ore pass wall. Additionally, it investigates the calculation methods and the specific formulas for the normal restitution coefficient (NRC) and tangential restitution coefficient (TRC), as well as the effects of the impact velocity and ore-rock block sizes on NRC, TRC, and the loss area and volume of the ore pass wall.

2. Movement Characteristics of Ore-Rock Block in the Ore Pass

2.1. Background of the Case Mine

The Shunfeng iron mine, located in Northeast China, is considered in the context of this research. The footwall shaft system and a non-pillar sublevel caving mining method were employed to extract iron ore from this mine. The extracted ore was transported using LHDs equipped with a 4 m^3 bucket from the draw points to the mine pass, then it was collected on the main haulage level via a Granby car with a volume of 4 m^3 and conveyed to the main ore pass adjacent to the main shaft of the mine. Figure 1 illustrates the layout of the main ore pass in the Shunfeng iron mine. Table 1 presents key structural parameters of both the ore pass and the unloading station.

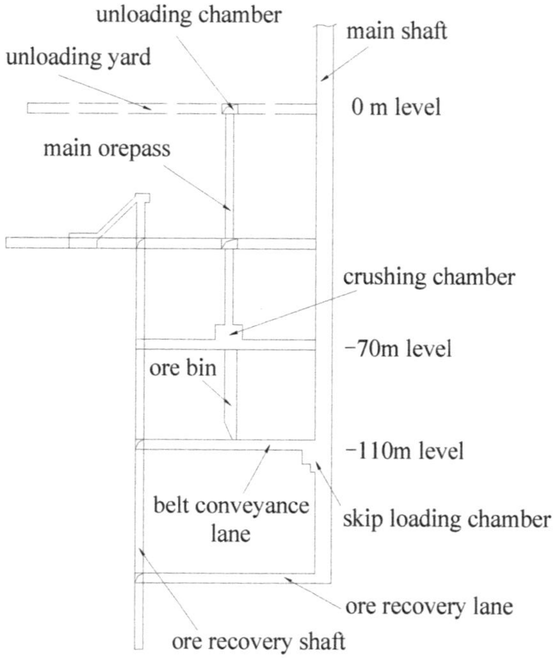

Figure 1. Layout of the main ore pass in the Shunfeng iron mine, a case mine located in Northeast China.

Table 1. Structural parameters of the main ore pass and of the unloading station.

Main Ore Pass			Unloading Chute		
Diameter/(m)	Height/(m)	Dip Angle/(°)	Height/(m)	Width/(m)	Dip Angle/(°)
5.0	70.0	90	4.0	1.2	45

Although the ore pass in this mine only has a height of 70 m, the deformation and deterioration of the ore pass wall cannot be avoided after several years of utilization.

2.2. Movement Analysis of Ore Rock in the Ore Pass

2.2.1. Basic Assumption

In order to know the spatial motion status of ore rock getting into the main ore pass, Figure 2, which is based on an actual mine (as shown in Figure 1), illustrates the spatial motion model of the ore-rock block as it enters the ore pass. When the ore-rock block moves through the unloading chute, it forms two angles along the horizontal and vertical planes, namely the depression angle α and the initial direction angle β. To analyze the characteristics related to the destruction of the ore pass wall, the following assumptions are established to study the motion patterns of ore-rock blocks in an ore pass.

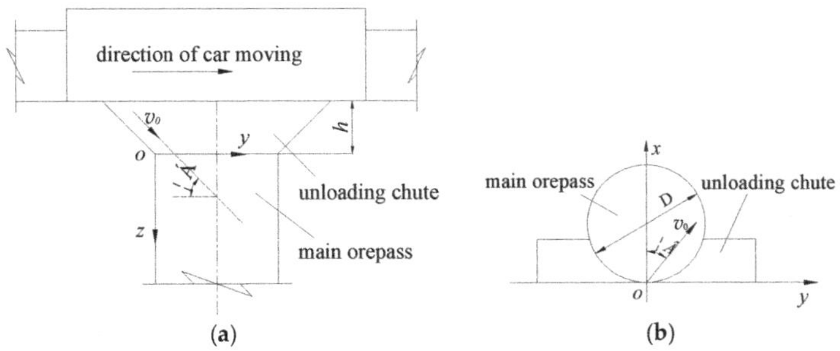

Figure 2. Spatial motion model of the ore-rock block entering the ore pass: (**a**) longitudinal projection along the length of the unloading chute; (**b**) a top view of figure (**a**).

(1) The ore-rock block is simplified into an elastoplastic sphere with a homogeneous and isotropic form, and the volume loss and rotation of the ore-rock block are not considered when the block falls in the ore pass.
(2) The impacting process does not consider the adhesive contact between the ore-rock block and the ore pass wall, i.e., the ore-rock block sticks to the ore pass wall.

2.2.2. Ore-Rock Block Moving in the Ore Pass

Due to the effect of the initial direction of the ore rock movement, the ore rock will collide with the ore pass wall [25]. When the velocity of the ore-rock block leaving the mine cart is ignored, the velocity of the ore-rock block entering the main ore pass, that is, the instantaneous velocity of the ore-rock block leaving point *O*, can be calculated through Equation (1).

$$v_0 = \sqrt{2gh\left(1 - \mu\frac{1}{\tan\alpha}\right)} \tag{1}$$

where α is the inclination of the chute floor relative to the horizontal plane, g is the gravitational acceleration, 9.8 m/s^2, h is the height of the inclined chute, m, and μ is the friction coefficient between the ore-rock block and the inclined chute floor.

Based on the kinematic principle and the motion model given in Figure 2, assuming v_0 is the instantaneous velocity of an ore-rock block entering the main ore pass from the unloading chute, α and β are the angles of the displacement vector of the ore-rock block relative to the *x-z* and the *x-y* plane, respectively. Then, according to the kinematic principle, the velocity and the spatial position of the ore-rock block at time *t* before its initial collision with the ore pass wall can be obtained through Equations (2) and (3):

$$\begin{cases} v_x = v_0 \cos\alpha \cdot \cos\beta \\ v_y = v_0 \cos\alpha \cdot \sin\beta \\ v_z = v_0 \sin\alpha + gt \end{cases} \tag{2}$$

$$\begin{cases} x_t = v_0 t \cdot \cos\alpha \cdot \cos\beta \\ y_t = v_0 t \cdot \cos\alpha \cdot \sin\beta \\ z_t = v_0 t \cdot \sin\alpha + \frac{1}{2}gt^2 \end{cases} \tag{3}$$

where α is the depression angle with a maximum value range of (0°, 90°), β is the initial direction angle ranging from −90° to 90°, x_t, y_t, and z_t denote the three-dimensional position coordinates of the block before the collision, and *t* is the duration of the movement preceding the impact. The remaining symbols are the same as above.

When the ore-rock block moves through the ore pass, there is a relationship between the diameter *D* of the ore pass and the displacement *x* of the ore-rock block in the horizontal direction, i.e., $D \geq x$. if $x = D$, the moving ore-rock block will come into contact with the ore pass wall, and, if $x > D$ (it is impossible to have $x > D$ due to the constraints imposed by the ore pass wall), the ore-rock block will collide with the ore pass wall. Therefore, $x \geq D$ constitutes the basic condition for the ore-rock block impact with the ore pass wall.

When both α and β are known values at initiation, D is the diameter of the main ore pass and the time t is:

$$t = \frac{D\cos\beta}{v_0\cos\alpha} \tag{4}$$

According to the above, we can obtain the coordinates of the collision point between the ore-rock block and the ore pass wall, as shown in Equation (5):

$$\begin{cases} x = D\cos^2\beta \\ y = \frac{1}{2}D\sin 2\beta \\ z = D\cos\beta\left\{\tan\alpha + \left[\frac{D\cos\beta}{4\text{h}\ \cos^2\alpha\left(1-\frac{\mu}{\tan\alpha}\right)}\right]\right\} \end{cases} \tag{5}$$

Consequently, Equation (5) provides the precise coordinates necessary to locate where the collision occurs. At this particular time, when the impact takes place between the ore-rock block and the ore pass wall, the depression angle α_1 and the direction angle β_1 of the ore-rock block coming into contact with the ore pass wall can be obtained as follows:

$$\begin{cases} \alpha_1 = \arctan\left(\frac{v_0\cdot\sin 2\alpha + gD\cdot cos\beta}{v_0\cdot\cos^2\alpha\cdot\cos\beta}\right) \\ \beta_1 = \arctan\left(\frac{\sin\beta}{\cos\beta}\right) \end{cases} \tag{6}$$

3. Theoretical Analysis of Ore Pass Wall Damage

3.1. Basic Assumptions

The following assumptions are proposed to simplify the calculation of the impact of the ore-rock mass with the ore pass wall, according to the Hertz contact theory:

(1) In the area range of the ore-rock block colliding with the ore pass wall, the ore pass wall is an elastic–plastic cylinder with infinite mass, homogeneity, and isotropy. The ore pass wall material is subject to the Von Mises criterion, and the strain hardening is not considered in the calculation process.
(2) The rock block is a uniform and isotropic elastoplastic sphere which moves only in the ore pass and experiences no mass loss in the collision process.
(3) In the model derivation process, it is assumed that the normal impact force and the tangential shear force are independent, and the coupling effect of normal and tangential force is ignored.

3.2. Theoretical Analysis of the Normal Impact

Affected by the initial movement direction, the horizontal velocity component of the ore-rock block acts vertically on the ore pass wall, with this being the main reason for ore pass wall damage from normal impact. The process of the ore rock impacting the shaft wall includes two stages. First, the ore rock comes into contact with the surface of the shaft wall and the material of the shaft wall becomes compressed until the deformation reaches its maximum, the kinetic and potential energy of the ore-rock block dissipates to 0, and the normal impact force on the shaft wall reaches its maximum. Subsequently, as a result of the restitution of the elasticity of the ore pass wall, the impact force on the shaft wall gradually decreases, and the ore-rock block bounces until it is completely separated from the surface of the ore pass wall.

3.2.1. Elastic Compression Stage

Based on the Hertz contact theory [32], the contact model of the ore rock impacting the ore pass wall was constructed, as shown in Figure 3a, and the velocity decomposition of both rock and ore impacting the shaft wall is shown in Figure 3b.

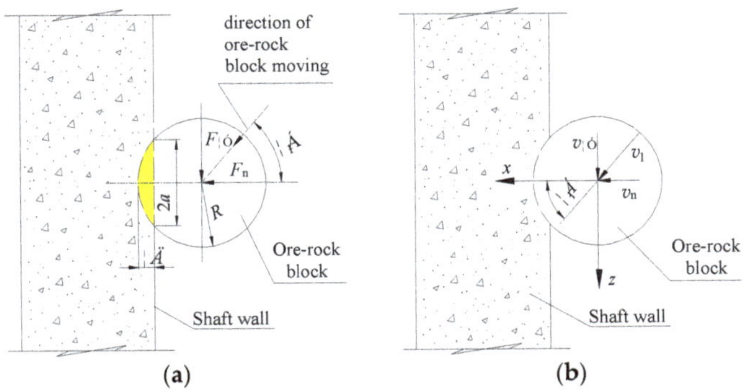

Figure 3. Contact model between the moving ore-rock block and the ore pass wall. (**a**) Calculation model of the moving ore-rock block impacting the ore pass wall and the yellow part is the deformation area of the ore pass wall; (**b**) decomposition of the velocity of the moving ore-rock block when it collides with the ore pass wall.

According to the Hertz contact theory, and based on the actual situation of the ore-rock block moving through the ore pass, the analytical solution to the elastic contact problem of the ore-rock block under normal contact pressure $F(r)$ is as follows [16,17]:

$$F(r) = \frac{3F_n}{2\pi a^2}\left[1 - \left(\frac{r}{a}\right)^2\right]^{\frac{1}{2}} \tag{7}$$

where F_n is the normal contact force in the elastic stage, N, a is the contact radius of the ore-rock block and ore pass wall, m, and r is also the contact radius, which is smaller or equal to a.

The normal contact force F_n and the contact radius a can be calculated as follows:

$$F_n = \frac{4E^*}{3} R^{\frac{1}{2}} \cdot \delta^{\frac{3}{2}} \tag{8}$$

$$a = \left(\frac{3F_n \cdot R}{4E^*}\right)^{\frac{1}{3}} \tag{9}$$

where δ is the depth of the ore-rock block intruding into the shaft wall, m, and R is the radius of the spherical ore-rock particles, m. The relationship among δ, R, and a can be described as in Equation (10). The term E^* from Equation (9) is the equivalent elastic modulus satisfied by Equation (11).

$$a^2 = R\delta \tag{10}$$

$$\frac{1}{E^*} = \frac{1 - \nu_1^2}{E_1} + \frac{1 - \nu_2^2}{E_2} \tag{11}$$

where E_1, ν_1 and E_2, ν_2 are the elastic modulus and Poisson's ratio of the ore pass wall and of the spherical ore-rock particles, respectively.

Jackson and Green [18] used the Von Mises yield criterion to derive the functional relationship between the initial yield intrusion depth δ_y and the initial yield contact force F_y at the elastic stage as follows [35]:

$$\begin{cases} \delta_y = \left(\frac{\pi \cdot C \cdot \sigma_y}{2E^*}\right)^2 \cdot R \\ F_y = \frac{4}{3}\left(\frac{R}{E^*}\right)^2 \cdot \left(\frac{\pi}{2} C \cdot \sigma_y\right)^3 \end{cases} \tag{12}$$

where σ_y is the yield limit stress of the shaft wall and $C = 1.295e^{0.736\nu_1}$.

The velocity of the ore block before colliding with the shaft wall is:

$$v_1 = \sqrt{{v_x}^2 + {v_y}^2 + {v_z}^2} \tag{13}$$

By combining Equations (2) and (4), the following can be obtained:

$$v_1 = \sqrt{{v_0}^2 + \frac{g^2 D^2 \cos^2\beta}{{v_0}^2 \cos^2\alpha} + 2 \cdot g \cdot D \cdot \cos\beta \cdot \tan\alpha} \tag{14}$$

Since the angle between the normal velocity v_n and v_1 of the ore-rock block moving is α_1, then:

$$v_n = v_1 \cdot \cos\alpha_1 \tag{15}$$

Assuming that v_c is the velocity at which the ore-rock block reaches the yield limit when it collides with the shaft wall, then, according to the functional relationship:

$$\frac{1}{2} m {v_n}^2 - \frac{1}{2} m {v_c}^2 = \frac{4}{3} E^* R^{\frac{1}{2}} \cdot \int_0^{\delta_y} \delta^{\frac{3}{2}} \cdot d\delta \tag{16}$$

By integrating Equation (16), we can obtain:

$$\frac{1}{2} m {v_n}^2 - \frac{1}{2} m {v_c}^2 = \frac{8}{15} E^* R^{\frac{1}{2}} \cdot {\delta_y}^{\frac{5}{2}} \tag{17}$$

It can be concluded that, at the end of the elastic compression stage, the velocity of the ore-rock block v_c is:

$$v_c = \left({v_n}^2 - \frac{16 E^* R^{\frac{1}{2}} \cdot {\delta_y}^{\frac{5}{2}}}{15 m} \right)^{\frac{1}{2}} \tag{18}$$

where m is the quality of the ore-rock block, kg, m equals $\frac{4}{3}\pi\rho R^3$, and ρ is the density of the ore-rock block, kg·m^{-3}.

When the average velocity of the ore-rock block colliding with the shaft wall is larger than v_c, plastic deformation occurs in the shaft wall material, resulting in deformation damage. Otherwise, only elastic deformation occurs, which has little influence on the ore pass wall. Therefore, it is imperative to calculate the critical velocity during the compression phase. Green [36] deduced the maximum strain energy accumulated by the collider before the plastic deformation of the sphere:

$$U_c = \frac{(\pi \cdot C \cdot \sigma_y)^5 R^3}{60 (E^*)^4} \tag{19}$$

Suppose $v_{c'}$ is the critical velocity between the elastic deformation and the plastic deformation of the wall which can be obtained from the perspective of the energy conversion:

$$\frac{1}{2} m ({v_c}')^2 = \frac{(\pi \cdot C \cdot \sigma_y)^5 R^3}{60 (E^*)^4} \tag{20}$$

By integrating Equation (12), the critical velocity of the plastic deformation of the shaft wall can be obtained from Equation (20) as follows:

$${v_c}' = \left(\frac{4\delta_y \cdot F_y}{5m} \right)^{\frac{1}{2}} \tag{21}$$

Specifically, $v_{c'}$ is the initial velocity that causes the wellbore to yield. When $v_{c'} \leq v_n$, only elastic deformation occurs in the wellbore which will not cause damage to the wellbore. When $v_{c'} > v_n$, plastic deformation occurs on the ore pass wall, resulting in impact damage.

3.2.2. Elastoplastic Compression Stage

When plastic deformation begins, the relation between the contact force and the depth does not conform to the Hertz contact model. At this stage, the Johnson contact theory is used for calculation. The average contact stress is related to the initial yield stress as follows [16]:

$$\sigma_n = \eta \sigma_y \tag{22}$$

where η is the dimensionless coefficient which, when plastic deformation occurs in the ore pass wall, equals 1.

The relationship between the normal impact force F_{ep} and the intrusion depth δ is as follows:

$$F_{ep} = F_y + \sigma_y \cdot \pi R \cdot \delta \tag{23}$$

In the elastoplastic impact stage, assuming that the maximum intrusion depth of the ore-rock block colliding with the shaft wall is δ_{max}, at this time, the velocity of the ore-rock block decreases to 0, and its kinetic energy is transformed into the strain energy required for plastic deformation. The relationship of the perspective of energy conversion in this process is as follows:

$$\frac{1}{2} m v_n^2 = \frac{8}{15} E^* R^{\frac{1}{2}} \cdot \delta_y^{\frac{5}{2}} + \int_{\delta_y}^{\delta_{max}} F_{ep} d\delta \tag{24}$$

Based on Equation (24), the maximum intrusion depth δ_{max} of the ore-rock block into the wall material can be calculated.

3.2.3. Rebound Stage

During this stage, the contact force between the shaft wall and the ore-rock block decreases from its maximum value to 0, and the ore-rock block bounces, producing a new falling movement in the ore pass. Its velocity increases continuously but is less than the normal velocity before the collision. The shaft wall cannot fully return to its original form after the plastic deformation; the relationship between the contact force and the intrusion depth is still consistent with the Hertz contact theory.

Assuming that the action of the elastic restoring force of the shaft wall materials begins from the maximum intrusion depth δ_{max}, that the permanent deformation caused by the plastic deformation of the wall surface is δ_{res}, and that the relative contact radius is R_{res}, then, using the Hertz contact theory, we can obtain [37]:

$$F_n' = (F_{ep})_{max} = \frac{4}{3} E^* R_{res}^{\frac{1}{2}} (\delta_{max} - \delta_{res})^{\frac{3}{2}} \tag{25}$$

When the ore block is no longer in physical contact with the shaft wall, the following relationship can be established from the perspective of the energy conversion:

$$\frac{1}{2} m v_n^2 = \int_{\delta_{max} - \delta_{res}}^{0} F'_n d\delta \tag{26}$$

By solving Equation (26), the velocity of the ore-rock block when it is no longer in physical contact with the shaft wall can be obtained:

$$v_n' = \sqrt{\frac{16E^*}{15m} (R_{res})^{\frac{1}{2}} \cdot (\delta_{max} - \delta_{res})^{\frac{5}{2}}} \tag{27}$$

where R_{res} and δ_{res} can be obtained as follows:

$$\begin{cases} \delta_{\text{res}} = \delta_{\text{max}} - \frac{3F_n'}{4E^* a_{\text{max}}} \\ R_{\text{res}} = \frac{4E^* (a_{\text{max}})^3}{3F_n'} \end{cases} \tag{28}$$

As it can be seen from Equation (10), the relationship between the maximum contact radius and the intrusion depth at this stage is as follows:

$$a_{\text{max}} = \sqrt{\delta_{\text{max}} \cdot R} \tag{29}$$

3.2.4. Normal Restitution Coefficient

Traditionally, the restitution coefficient can be calculated using the ratio of the ore-rock block velocity after colliding with the shaft wall to the ore-rock block velocity before the collision. The velocity of the ore-rock block after the collision with the shaft wall can be calculated by using Equation (27), and the velocity of the block before the collision can be calculated by using Equation (14). Thereby, the NRC e_n can be obtained as follows:

$$e_n = \begin{cases} 1 & (v_{\text{n}} \leq v_{\text{c}}) \\ \left| \frac{v_{\text{n}}'}{v_{\text{n}}} \right| & (v_{\text{n}} > v_{\text{c}}) \end{cases} \tag{30}$$

It can be seen from Equation (30) that the NRC is related to the yield velocity and to the normal velocity. When the normal impact velocity of the ore-rock block is less than the yield velocity, the velocity changes a little before and after the impact. When the impact velocity is greater than the yield velocity, the NRC is less than 1, the rebound velocity of the rock block is less than the normal velocity before the impact, and the impact damage of the shaft wall occurs.

3.3. *Theoretical Analysis of Tangential Shear*

3.3.1. Calculation of the Tangential Force

The collision between the ore-rock block and the ore pass wall is a process of oblique impact. The tangential force is controlled by the tangential velocity, the tangential deformation, the impact angle, and the friction coefficient, all of which are important parameters affecting the tangential movement of the ore-rock block. When the ore-rock block comes into contact with the shaft wall, if the adhesive contact between the two materials is not considered and only the slippage of the ore-rock block on the contact zone is considered, the formula for calculating the tangential force is as follows:

$$F_\tau = \mu F_{\text{n}}' \tag{31}$$

where μ is the friction coefficient between the rock block and the shaft wall and F_{n}' is the maximum normal contact force, N.

3.3.2. Calculation of the Tangential Restitution Coefficient

The TRC can be expressed as:

$$e_\tau = \frac{v_\tau'}{v_\tau} \tag{32}$$

where v_τ' and v_τ are the tangential velocity after and before the collision, respectively, m/s. As it can be seen from Figure 3, v_τ can be expressed as:

$$v_\tau = v_{\text{n}} \cdot \tan\alpha_1 \tag{33}$$

Suppose that the impulse ratio f between the normal and the tangential direction is:

$$f = \frac{P_\tau}{P_{\text{n}}} \tag{34}$$

According to Newton's second law, the normal impulses P_n and P_τ can be expressed as the velocities before and after the impact:

$$\begin{cases} P_n = m(v_n + v_n') \\ P_\tau = m(v_\tau - v_\tau') \end{cases} \quad (35)$$

The TRC can be obtained by solving Equations (32)–(35) and combining the outcome with Equation (30), as follows:

$$e_\tau = 1 - \frac{f(1 + e_n)}{\tan \alpha_1} \quad (36)$$

Considering that the ore-rock block will slip slightly during the collision with the shaft wall, the impulse ratio f in Equation (36) can be replaced by μ [38,39], and the theoretical solution to e_τ is as follows:

$$e_\tau = 1 - \frac{\mu(1 + e_n)}{\tan \alpha_1} \quad (37)$$

According to the NRC and TRC represented by the formula, the total restitution coefficient e of the ore-rock block during the collision with the shaft wall can be obtained as follows:

$$e = \sqrt{e_n{}^2 \cdot \cos^2 \alpha_1 + e_\tau{}^2 \cdot \sin^2 \alpha_1} \quad (38)$$

Therefore, it can be concluded that the total translational rebound velocity of the ore-rock block after colliding with the shaft wall is:

$$v_1{}' = e \cdot v_1 \quad (39)$$

4. Loss of the Shaft Wall Materials

4.1. Basic Model

When the impact load of the ore-rock block is larger than the ultimate yield strength of the wall material, the plastic deformation of the wall occurs, resulting in impact damage to the wall in the normal direction and in shear damage in the tangential direction. Therefore, it is essential to calculate the loss of the shaft wall material to evaluate the stability of the ore pass.

To simplify the calculation of the loss of the shaft wall material, according to the basic assumptions of the Hertz contact theory, when the loss of the per unit volume of the shaft wall material is regarded as the product of the maximum residual intrusion depth, the vertical area, and the distance of the ore-rock block sliding along the pass wall, a calculation model of the lost volume of the shaft wall is obtained, as shown in Figure 4a.

4.2. Lost Volume Calculation

According to Figure 4b, the area S_{ABCA} of the shaded part in the figure is:

$$\begin{aligned} S_{ABCA} &= 2\left(\frac{\pi R^2 \theta}{2\pi} - \frac{1}{2}a \cdot (R - \delta_{res})\right) \\ &= R^2 \cdot \arcsin\left(\frac{\delta_{res}}{R}\right)^{\frac{1}{2}} - (\delta_{res} \cdot R)^{\frac{1}{2}} \cdot (R - \delta_{res}) \end{aligned} \quad (40)$$

where $a^2 = R \times \delta_{res}$.

When the ore-rock block collides with the ore pass wall and produces the slip distance δ_τ, for the slip distance $d\delta_\tau$, the lost volume per unit of the wall material dQ is:

$$dQ = S_{ABCA} \cdot d\delta_\tau \quad (41)$$

When the ore-rock block moves in the tangential direction, part of the kinetic energy is lost due to the influence of the friction. According to the energy-changing relationship during the sliding process of the ore-rock block, it can be obtained:

$$\frac{1}{2}m \cdot {v_\tau}^2 - \frac{1}{2}m \cdot (v'_\tau)^2 = \mu F_n{}' \cdot \delta_{\tau\max} \tag{42}$$

where $\delta_{\tau\max}$ is the maximum slip distance in the tangential direction during the collision process, m. The other symbols are the same as above.

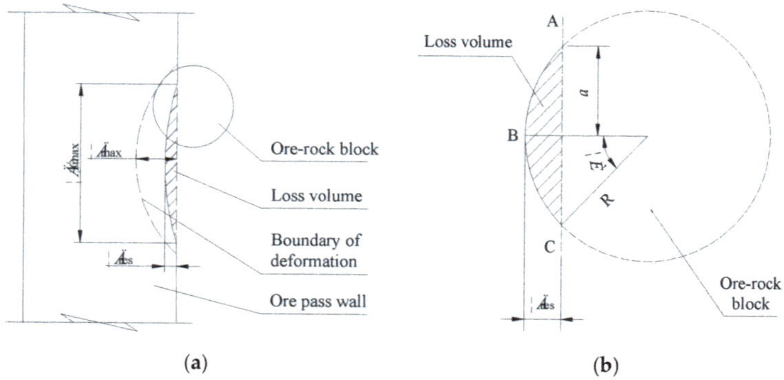

Figure 4. Calculation model of the lost volume of the ore pass wall. The shaded part in the figure is the lost volume of the ore pass wall: (**a**) profile along the center line of the ore pass, located at the point of the maximum intrusion depth of the ore-rock block; (**b**) section perpendicular to the center line of the ore pass, located at the point of maximum intrusion depth of the ore-rock block. The A.C. line in this figure represents the original surface of the ore pass wall and B is the point of maximum intrusion depth on the ore pass wall.

By integrating Equation (41), it can be concluded that, when the slip distance reaches the maximum $\delta_{\tau\max}$, the total volume of wall loss during the collision is:

$$Q = \delta_{\tau\max} \cdot R^2 \cdot \arcsin\left(\frac{\delta_{res}}{R}\right)^{\frac{1}{2}} - (\delta_{res} R)^{\frac{1}{2}} \cdot (R - \delta_{res}) \tag{43}$$

5. Analysis of Factors Affecting the Degree of Shaft Wall Loss

5.1. Influence of Impact Velocity and Angle on Restitution Coefficient of the Ore Pass Wall

When the impact load of the ore-rock block is larger than the ultimate yield strength of the wall material, the plastic deformation of the wall occurs, resulting in impact damage. Figure 5 shows the influence of v_1 variations on NRC and TRC.

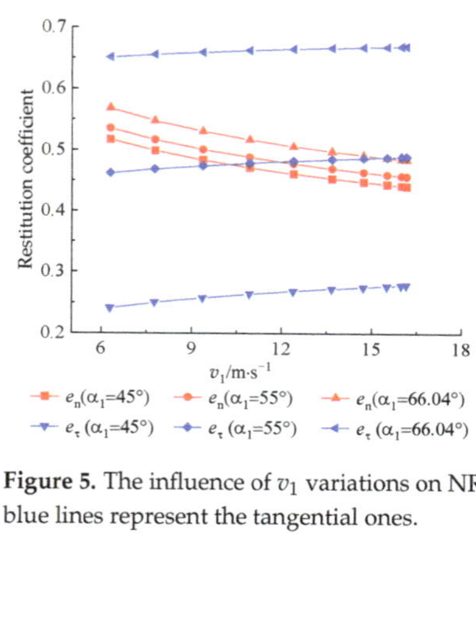

Figure 5. The influence of v_1 variations on NRC and TRC. The red lines represent the NRCs, and the blue lines represent the tangential ones.

It can be seen from Figure 5 that, when the ore-rock block collides with the ore pass wall, the NRC and the TRC of the shaft wall materials show an opposite growth characteristic relative to the change in the impact angle and velocity. Under different impact angles, with the increase in impact velocity v_1, the NRC decreases, and the TRC increases. Meanwhile, the NRC and the TRC are less sensitive to the impact velocity, and the TRC is more affected by the impact angle.

At different impact velocities, the NRC and the TRC all reach the maximum value when the impact angle is 66.04°, but they attain the minimum value when the angle is 45°. When the velocity is 6.26 m/s and the impact angle is 66.04°, the NRC and the TRC are 0.57 and 0.65, respectively, and, when the angle is 45°, they are 0.51 and 0.24, respectively. When the velocity is 16.12 m/s and the angle is 66.04°, the NRC and the TRC are 0.48 and 0.67, respectively, and 0.44 and 0.28, respectively, when the impacting angle is 45°.

5.2. Influence of Impact Velocity and Ore-Rock Block Sizes on the Maximum Normal Intrusion Depth

In the normal direction of the shaft wall, the change in the maximum intrusion depth will significantly affect the deformation of the shaft wall. When the impact angle is 45°, the restitution coefficient before and after the normal impact of the ore-rock block on the shaft wall is the smallest, and more energy is lost during the collision between the ore-rock block and the shaft wall. Through the theoretical model developed here, the maximum normal intrusion depth under different ore-rock block radii and impact velocities is calculated when the impact angle is 45°, as shown in Figure 6.

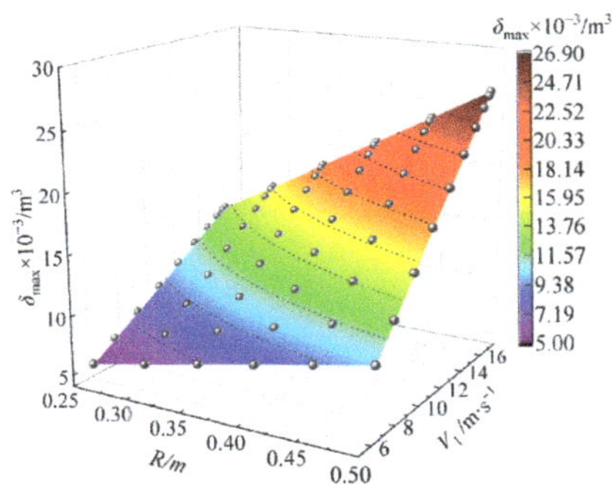

Figure 6. Influence of the ore-rock block radius R and of the impact velocity v_1 on the intrusion depth.

It can be seen from Figure 6 that the impact velocity and radius of the ore-rock blocks significantly influence the impact effect on the shaft wall. When the impact velocity is small, the intrusion depth is small, and, when the velocity increases, the intrusion depth also significantly increases. When the impact velocity is constant, the intrusion depth increases with the increase in the radius of the ore-rock block. Under the condition that the velocity is 6.26 m/s, the minimum intrusion depth is 5.07×10^{-3} m when the radius of the ore-rock block is 0.25 m, which is 10.14×10^{-3} m when the radius is 0.50 m. Under the condition that the velocity is 16.12 m/s, the minimum intrusion depth is 13.4×10^{-3} m when the radius is 0.25 m, and, when the radius is 0.5 m, the intrusion depth reaches its maximum, 26.80×10^{-3} m.

5.3. Influence of Impact Velocity and Ore-Rock Block Sizes on Tangential Maximum Displacement

Based on theoretical calculations, it is found that the volume loss of the shaft wall is more significant when the impact angle is 66.04°. Therefore, the law of tangential displacement changing with the radius and impact velocity of ore-rock blocks is studied based on the impact angle of 66.04°, and the results are shown in Figure 7.

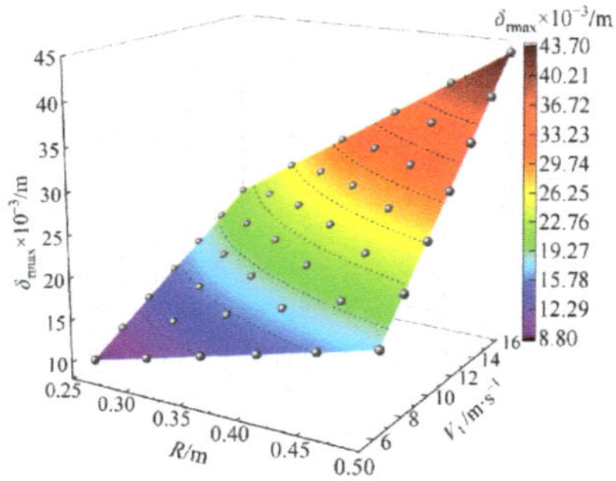

Figure 7. Influence of the rock block radius R change and of the impact velocity v_1 on the tangential displacement.

It can be seen from Figure 7 that the radius of the ore-rock block, and especially the impact velocity, significantly affects the tangential displacement. When the ore-rock block has a constant radius, the tangential displacement increases with the impact velocity. Under the condition that the velocity is 6.26 m/s, the minimum tangential displacement is 8.83×10^{-3} m when the radius of the ore-rock block is 0.25 m, which is 17.65×10^{-3} m when the radius is 0.50 m. Under the condition that the velocity is 16.12 m/s, the minimum intrusion depth is 21.82×10^{-3} m when the radius is 0.25 m, and, when the radius is 0.5 m, the tangential displacement reaches its maximum value of 26.80×10^{-3} m.

5.4. Influence of Impact Velocity and Ore-Rock Block Sizes on Volume Loss of the Shaft Wall

The larger the impact angle, the larger the tangential velocity before the collision, the larger the tangential force acting on the wall, the more significant the cutting and shearing effect, and the more serious the loss and deformation of the ore pass wall. Therefore, with an impact angle of 66.04°, as described in Section 5.1, the influence of impact velocity and ore-rock block sizes on the volume loss of the shaft wall is analyzed, and the results are shown in Figure 8.

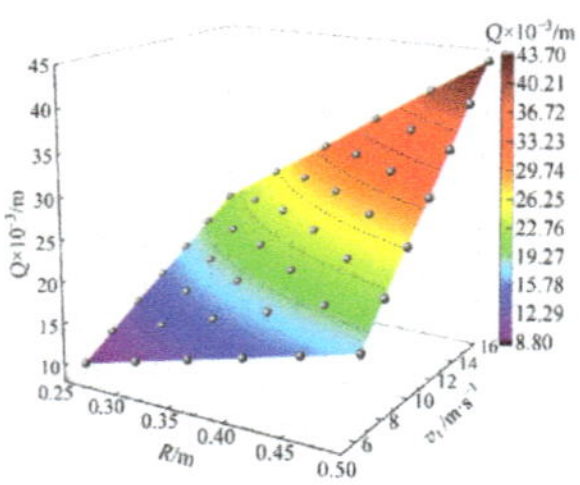

Figure 8. Influence of the rock block radius R change and of the impact velocity v_1 on the volume loss of the shaft wall.

It can be seen from Figure 8 that the volume loss of the ore pass wall following the impact with the colliding ore-rock block characterized by larger radii is much larger than that ensuing the collision with the ore blocks characterized by smaller radii and velocities. The larger the radius and the greater velocity of the ore-rock block, the more serious the damage and the deformation experienced by the shaft wall. When the radius of the ore-rock block is constant, the shaft wall's volume loss increases with the impact velocity. Under the condition that the velocity is 6.26 m/s, the minimum volume loss of the shaft wall is 2.49×10^{-3} m when the radius of the ore-rock block is 0.25 m, which is 19.88×10^{-3} m when the radius is 0.50 m. Under the condition that the velocity is 16.12 m/s, the minimum volume loss is 11.23×10^{-3} m^3 when the radius is 0.25 m, and, when the radius is 0.5 m, the volume loss reaches its maximum value of 89.85×10^{-3} m^3.

5.5. Experimental Verification of the Impact Damage of the Ore Pass Wall

There are great risks in measuring the volume loss of the shaft wall in a real ore pass, and it is difficult to judge the impact position and damage degree of a certain ore-rock block on the ore pass wall. After deducing the volume loss of wall material theoretically, it is necessary to verify the rationality of the theoretical deduction by conducting a physical similarity test. Yue, Y. et al. [26] conducted experiments using the sharp and blunt surfaces of the ore-rock blocks colliding with the ore pass wall and investigated the characteristics of the damage to the ore pass wall resulting from the impact with and the cutting of ore-rock blocks.

In the experiment, two shapes of magnetite quartzite blocks were selected as the impacting materials, including the sharp and blunt surface blocks, with a quality of 0.2 kg and a radius of approximately 25 mm. In order to obtain the impact positions, a series of experiments simulating collisions between two-shape ore-rock blocks and the ore pass wall were conducted under the conditions of a chute angle α of 50°, 55°, and 60°. Figure 9a shows the effect of two types of ore-rock blocks colliding with the ore pass wall.

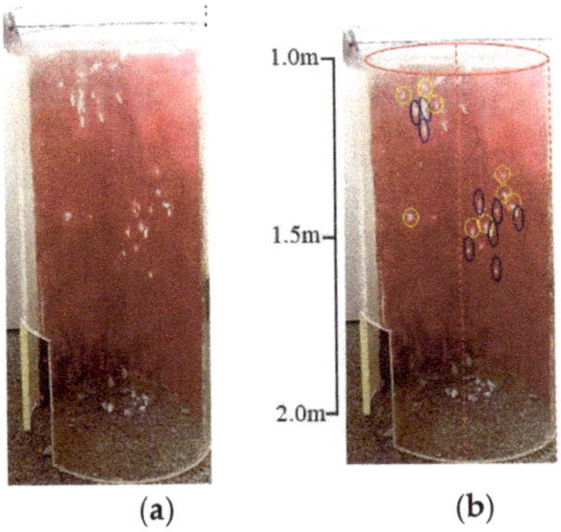

Figure 9. Characteristics of the damage to the ore pass wall resulting from impact with and cutting of ore-rock block: (**a**) characteristics of the damage to the ore pass wall according to Yue Y et al. [26]; (**b**) analysis of the damage characteristics of the ore pass wall—the yellow circles show the impact traces on the ore pass wall caused by the collision with the ore-rock blocks and the black ellipses show the damage caused by the impact with a sharp surface. The height meter shows the lower half of the ore pass model.

In order to verify the reliability of the formula presented in this paper, a PS50 three-dimensional surface topography scanner produced by NANOVEA Inc. (Irvine, CA, USA) was used to measure the intrusion depth δ of the ore pass wall caused by different shapes of ore-rock blocks colliding with the ore pass wall. Meanwhile, according to the parameters of the ore pass model [26], the maximum intrusion depth δ_{res} was calculated theoretically using the formula developed in this paper; the mechanical parameters of the ore pass model and the ore-rock blocks with the sharp and blunt surfaces are shown in Table 2. The results from the comparison of the experiment and the theoretical calculation are shown in Table 3.

As illustrated in Table 3, the depression angle α and the direction angle β affect the maximum intrusion depth of the ore-rock block into the ore pass wall. When the chute inclination α, i.e., the depression angle, was 50° and the direction angle β was 5°, the maximum intrusion depth obtained from the experiment was 30 μm; when α was 60° and β was 27°, the maximum intrusion depth was 12 μm. However, the maximum intrusion depths under these cases were 31 μm and 11 μm with relative errors of 2.82% and 8.09%, respectively. The relative errors derived from the comparison of the results calculated theoretically via the formula proposed in this paper and those obtained from the similarity simulation experiment were 2.33% to 8.09%. It also shows that the results of the two methods are consistent, and the theoretical calculation can reflect the degree of impact damage of the ore pass wall.

Table 2. Mechanical parameters of the ore pass model and of the ore-rock block.

Materials	Elastic Modulus/(Gpa)	Poisson's Ratio	Cohesion/(Mpa)	Density/(kg·m^{-3})	Friction Coefficient	Material Yield Stress/(Mpa)
Ore pass wall	23.5	0.27	2.6	2500	0.3	210
Ore-rock block	52.5	0.31	—	3200	—	—

Note: "—" means 'it does not exist'.

Table 3. Results from the comparison of the experiment and the theoretical calculation.

No. of Impacting Point	α/(°)	β/(°)	Experimental Intrusion Depth δ_{res}/μm		Calculating Intrusion Depth δ_{res}/μm	Relative Error/%
			Ore-Rock Block with a Sharp Surface	Ore-Rock Block with a Blunt Surface		
1	50	5	—	30	31	2.82
2	50	8	—	29	30	3.79
3	50	10	—	28	29	5.04
4	50	14	29	—	28	4.83
5	50	17	27	—	26	4.17
6	50	20	23	—	24	3.71
7	55	5	—	26	27	3.95
8	55	9	—	25	26	3.97
9	55	12	—	24	25	3.52
10	55	13	25	—	24	2.44
11	55	20	21	—	20	3.31
12	55	23	19	—	18	4.76
13	60	6	—	22	23	2.48
14	60	8	—	21	22	5.00
15	60	12	22	—	21	6.13
16	60	16	—	18	19	4.05
17	60	18	18	—	18	2.33
18	60	27	12	—	11	8.09
19	50	5	—	30	31	2.82
20	50	8	—	29	30	3.79

Note: "—" means 'it does not exist'.

Therefore, the theoretical model developed in this paper can be used to predict the degree of damage to the ore pass wall caused by a collision with ore-rock blocks.

It should be noted that the value of β in Table 3 was obtained by inverting the kinematic formula given in this paper according to the location of the collision between the ore-rock block and the shaft wall. The main reason behind this decision is that the ore-rock block used in the experiment was irregular in shape, and it was difficult to control the orientation of the irregular ore-rock block after it left the unloading chute, so it was difficult to set the accurate direction angle of the ore-rock block moving into the ore pass during the experiment.

6. Discussion and Suggestions

Based on contact mechanics and tribology, Yue Y. et al. [26] analyzed the calculation method of the volume loss experienced by the ore pass wall following the collision with the sharp and blunt face of the rock block. Only the damage and failure of the shaft wall under normal load were considered, and the combined action of both normal and tangential load on the rock block was not considered. The boundary between the elastic deformation and the plastic deformation of the wall materials is unclear.

The research results of this paper show that the quality of the ore-rock block, the velocity and angle of the ore-rock block impact with the ore pass wall, the structure parameters of the ore pass, and the physical and mechanical parameters of the ore pass wall are the main factors affecting the impact-induced damage to the shaft wall. This is consistent with the findings of Esmaieli et al. [13,14] and Qin et al. [27]. Therefore, it is essential to take appropriate measures to prevent damage to the ore pass wall caused by the ore-rock impact in the context of actual production whenever feasible.

(1) The ore pass should be constructed using good rock mass quality or its walls should be subjected to effective reinforcement measures to increase the resistance of the shaft wall to the damage and destruction caused by the impact of rock movement.
(2) Reasonable ore pass structure parameters should be selected, changing the movement state of the ore after entering the main ore pass and reducing the speed of the collision between the ore-rock block and the ore pass wall.
(3) During the design phase of the ore pass, minimizing the height of the main ore pass is advisable to reduce the impact energy of the falling ore-rock block whenever feasible.
(4) Installing a screen at the entrance of the ore pass restricts the size of the ore-rock blocks entering the main ore pass. Meanwhile, this is beneficial to reduce speed of ores entering the ore pass.

7. Conclusions

According to the movement characteristics of the ore-rock block in the vertical ore pass, the representation equation of the three-dimensional coordinates and the impact time of the ore-rock block moving in the ore pass are obtained. The angle between the movement direction of the ore-rock block and the shaft wall is proposed before and after the ore-rock block collides with the shaft wall.

Based on the Hertz contact theory, a contact model of ore-rock blocks colliding with the ore pass wall is developed, and the process of the ore-rock blocks coming into physical contact with the shaft wall is categorized into three stages: elastic compression, elastic–plastic compression, and rebound of the ore pass wall material. The relationship between the normal impact force and the depth of the ore-rock block intruding into the shaft wall, as well as between the maximum contact radius and the depth of penetration, is determined. The NRC of the wall material is also acquired.

Based on the tangential contact mechanics, the tangential cutting model along the shaft wall is developed. The tangential slipping distance of the ore-rock block along the shaft wall, the TRC, and the lost area and volume of the shaft wall material are derived based on the established model when the ore-rock block collides with the ore pass wall.

The calculation model for the volume loss of the shaft wall material during the collision between the ore-rock block and the shaft wall is formulated, and the total lost volume during the collision is obtained through the analysis and resolution of the model.

The reliability of the maximum intrusion depth of the ore-rock block into the ore pass wall, calculated using the formula proposed in this paper, is verified through simulation experiments using similar materials. The relative errors of the two depth results show that they are consistent, and the theoretical calculation can reflect the degree of impact damage exhibited by the ore pass wall.

According to the specific characteristics and parameters of the ore pass in the case mine, the influence of the impact velocity and angle of collision of the ore-rock block on the restitution coefficient of the shaft wall material, the maximum normal intrusion depth of the ore-rock block into the ore pass wall, the maximum tangential displacement of ore-rock block along the shaft wall, and the volume loss of the shaft wall when the ore-rock block collides with the ore pass wall are analyzed by using the relevant formulas obtained in this paper, and the corresponding rules are obtained.

Author Contributions: Investigation, formal analysis, data curation, visualization, writing—original draft, Q.M.; investigation, resources, formal analysis, data curation, writing—original draft, C.M.; methodology, formal analysis, writing—review and editing, project administration, J.L.; conceptualization, methodology, writing—review and editing, supervision, project administration, funding acquisition, Z.L.; data curation, formal analysis, writing—review and editing Z.X. All authors have read and agreed to the published version of the manuscript.

Funding: The research was supported by the National Natural Science Foundation of China (Grant No. 51774176).

Data Availability Statement: The detailed data are available upon request. The data are not publicly available due to privacy.

Conflicts of Interest: The authors declare no conflicts of interest.

References

1. Sredniawa, W.; Skawina, B.; Rapp, J.; Shekhar, G.; Gunillasson, J. Longevity chart for planning production and the renovati of orepasses. In *Caving 2022: Proceedings of the Fifth International Conference on Block and Sublevel Caving, Adelaide, Australia, August–1 September 2022*; Potvin, Y., Ed.; Australian Centre for Geomechanics: Perth, Australia, 2022; pp. 369–380. [CrossRef
2. Flyability. What Is an Ore Pass in Mining? 2024. Available online: https://www.flyability.com/blog/ore-pass (accessed on April 2024).
3. Salmi, E.F.; Phan, T.; Sellers, E.J.; Stacey, T.R. A review on the geotechnical design and optimisation of ultra-long ore passes deep mass mining. *Environ. Earth. Sci.* **2024**, *83*, 301. [CrossRef]
4. Hadjigeorgiou, J.; Lessard, J.F. Strategies for restoring material flow in ore and waste pass systems. *Int. J. Min. Reclam. Envir* **2010**, *24*, 267–282. [CrossRef]
5. Sredniawa, W.; Skawina, B.; Garcia, J.R.; Rapp, J. Analyzing degradation of ore pass walls in Kiirunavaara mine using scanni data. *Front. Built. Environ.* **2023**, *9*, 1250671. [CrossRef]
6. Zhao, Y.; Ye, H.; Lei, T.; Wang, C.; Wang, Q.; Long, M. Theoretical study of damage characteristics on ore pass wall based on erosion-wearing theory. *Yanshilixue Yu Gongcheng Xuebao/Chin. J. Rock Mech. Eng.* **2017**, *36*, 4002–4007. [CrossRef]
7. Vo, T.; Yang, H.W.; Russell, A.R. Cohesion and suction induced hang-up in ore passes. *Int. J. Rock Mech. Min. Sci.* **2016**, *87*, 113–1 [CrossRef]
8. Skawina, B.; Greberg, J.; Salama, A.; Gustafson, A. The effects of orepass loss on loading, hauling, and dumping operations a production rates in a sublevel caving mine. *J. S. Afr. I. Min. Metall* **2018**, *118*, 409–418. [CrossRef]
9. Dukes, G.B.; Oort, J.V.; Heerden, F.D.V. Rockpass stability and rehabilitation at AngloGold Ashanti's Tau Lekoa Mine. *J. S. Afr Min. Metall* **2006**, *106*, 5–10.
10. Hart, R. Case study of the rockpass system at Kloof No. 3 Shaft. *J. S. Afr. I. Min. Metall* **2006**, *106*, 1–4.
11. Maree, J.A. Orepass best practices at South Deep. *J. S. Afr. I. Min. Metall* **2011**, *111*, 257–272.
12. Hadjigeorgiou, J.; Stacey, T.R. The absence of strategy in orepass planning, design, and management. *J. S. Afr. I. Min. Met* **2013**, *113*, 795–801.
13. Esmaieli, K.; Hadjigeorgiou, J. Selecting Ore Pass-finger Raise Configurations in Underground Mines. *Rock Mech. Rock E* **2011**, *44*, 291–303. [CrossRef]
14. Esmaieli, K.; Hadjigeorgiou, J.; Grenon, M. Stability Analysis of the 19a ore pass at Brunswick Mine using a two-stage numeri modeling approach. Rock Mech. *Rock Eng.* **2013**, *46*, 1323–1338. [CrossRef]
15. Sato, A.; Tang, H. Analysis of ore pass hang-ups in long vertical ore passes by 3-D DEM. *Int. J. Min. Eng. Miner Proce* **2020**, *9*, 1–11.
16. Xia, Z.G.; Deng, Z.; Lu, Z.X.; Ma, C.L. Study on the macro-fine mechanical behavior of ore flow based on the discrete eleme method. *Appl. Sci.* **2024**, *14*, 3457. [CrossRef]
17. Li, C.; Qin, H. A new ore pass repair method based on cavity auto scanning laser system accurate detection technology. *Int. Min. Miner Eng.* **2016**, *7*, 37–50. [CrossRef]
18. Stacey, T.R.; Wesseloo, J.; Bell, G. Predicting the stability of rock passes from the geological structure. *J. S. Afr. Inst. Min. Met* **2005**, *105*, 803–808.
19. Gardner, L.J.; Fernandes, N.D. Ore pass rehabilitation—Case studies from Impala Platinum Limited. *J. S. Afr. Inst. Min. Met* **2006**, *106*, 17–23.
20. Chen, H.; Yu, S.B.; Wang, Z.X.; Yuan, Y. A new plugging technology and its application for the extensively collapsed ore pass the non-empty condition. *Energies* **2018**, *11*, 1599. [CrossRef]
21. Sjöberg, J.; Lundman, P.; Nordlund, E.; Quinteiro, C. Stability analysis of ore passes in the Kiirunavaara mine. In Proceedings the 10th ISRM Congress, OnePetro, Sandton, South Africa, 7–12 September 2003; pp. 1093–1098.
22. Yang, Y.J.; Deng, Z.; Lu, Z.X. Effects of ore-rock falling velocity on the stored materials and the force on the shaft wall in a vertic orepass. *Mech. Adv. Mater. Struc.* **2022**, *30*, 3455–3462. [CrossRef]
23. Luo, Z.Q.; Chen, J.; Xie, C.Y.; Wang, W.; Liu, X.M. Mechanism of impact-induced damage of main chute and its experiment validation. *Yantu Lixue/Rock Soil Mech.* **2015**, *36*, 1744–1751. [CrossRef]
24. Song, W.D.; Wang, H.Y.; Wang, X.; Du, J.H. Theoretical analysis and test of impact load due to ore dumping in chute. *Yan Lixue/Rock Soil Mech.* **2011**, *32*, 326–332+340.

- Lu, Z.X.; Wu, X.X.; Ma, C.; Yin, Y. Influences of initial movement of ore or rock block on the rule of impacting on orepass wall. *Metal Mine* **2020**, *531*, 60–64.
- Yue, Y.; Chi, M.; Zengxiang, L.; Zhiqiang, Z. Deformation and damage of orepass wall under impact and cutting. *World Min. Surf. Undergr.* **2020**, *72*, 205–210.
- Qin, H.; Li, C.; Ma, H.; Wang, Y. Study on impact damage laws of ore pass vased on particle flow. *J. Saf. Sci. Technol.* **2015**, *11*, 20–26.
- Wu, Q.; Luo, Z.; Huang, J.; Qin, Y. 3D visualization analysis of deep main chute falling based on CMS monitoring. *J. Saf. Sci. Technol.* **2016**, *12*, 148–153.
- Ren, Z.; Ma, H.; Wang, S.; Zeng, M.; Jin, L. Experimental study on damage laws of ore pass based on similarity simulation. *J. Saf. Sci. Technol.* **2016**, *12*, 98–102.
- Popov, V.L. *Contact Mechanics and Friction: Physical Principles and Applications*, 2nd ed.; Springer: Berlin/Heidelberg, Germany, 2017; ISBN -13: 9783662530801.
- Johnson, K.L. *Contact Mechanics*; Cambridge University Press: Cambridge, UK, 1985. [CrossRef]
- Landau, L.D.; Pitaevskii, L.P.; Kosevich, A.M.; Lifshitz, E.M. *Theory of Elasticity: Volume 7 (Course of Theoretical Physics)*, 3rd ed.; Kindle Edition; Butterworth-Heinemann: Oxford, UK, 2012; ISBN -13: 978-0750626330.
- He, S.; Li, X.; Wu, Y. Research on yield property of soil under rock-fall impact. *Yanshilixue Yu Gongcheng Xuebao/Chin. J. Rock Mech. Eng.* **2008**, *27*, 2973–2977.
- Mangwandi, C.; Cheong, Y.S.; Adams, M.J.; Hounslow, M.J.; Salman, A.D. The coefficient of restitution of different representative types of granules. *Chem. Eng. Sci.* **2007**, *62*, 437–450. [CrossRef]
- Jackson, R.L.; Green, I. A finite element study of elasto-plastic hemispherical contact against a rigid flat. *J. Tribol. Apr.* **2005**, *127*, 343–354. [CrossRef]
- Green, I. Poisson ratio effects and critical values in spherical and cylindrical Hertzian contacts. *Int. J. Appl. Mech. Eng.* **2005**, *10*, 451–462.
- Jackson, R.L.; Green, I.; Marghitu, D.B. Predicting the coefficient of restitution of impacting elastic-perfectly plastic spheres. *Nonlinear Dyn.* **2010**, *60*, 217–229. [CrossRef]
- Wu, C.Y.; Thornton, C.; Li, L.Y. Coefficients of restitution for elastoplastic oblique impacts. *Adv. Powder Technol.* **2003**, *14*, 435–448. [CrossRef]
- Kharaz, A.H.; Gorham, D.A.; Salman, A.D. An experimental study of the elastic rebound of spheres. *Powder Technol.* **2001**, *120*, 281–291. [CrossRef]

applied sciences

MDPI

Article

Magnetic–Thermal Coupling-Based Study on the Temperature Characteristics of Flameproof Cable Boxes for Coal Mines at Various Failure Modes

Hongkui Zhang [1,2,3,*], Zhaoxuan Cui [1] and Yipeng Lan [1,*]

[1] Department of Electrical Engineering, Shenyang University of Technology, Shenyang 110178, China; sygyczx@smail.sut.edu.cn
[2] CCTEG Shenyang Research Institute, Fushun 110178, China
[3] Fushun CCTEG Inspection Center Co., Ltd., Fushun 110178, China
* Correspondence: hongkui_2019@smail.sut.edu.cn (H.Z.); lanyipengg@163.com (Y.L.)

Abstract: Temperature monitoring is the main indicator to ensure the safe operation of the coal mine explosion-proof box. In this study, a three-way flameproof cable box was considered for coal mines. A three-dimensional transient temperature field simulation model was constructed for magnetic–thermal coupling and subsequently used to conduct temperature field simulation for the failure of the flameproof cable box for coal mines. An experimental platform was developed to test the temperature characteristics of failure flameproof cable boxes for coal mines and obtain temperature curves at various failure currents. Using the experimental results of temperature characteristics and temperature field simulation outcomes, we analyzed the time responses and spatial distribution characteristics of the temperature rise of flameproof cable boxes for coal mines. The error between simulation and experimental results does not exceed 1 K, and the upper limit of time to ensure safe operation under different failure modes is determined. Thus, the acquired temperature distribution characteristics in a failure mode can be used as a reference for the design, inspection, and status warning of flameproof cable boxes for coal mines.

Keywords: coal mine; flameproof cable box; failure operating mode; magnetic–thermal coupling; temperature characteristics

1. Introduction

Coal is a major fossil energy resource and is extensively used worldwide [1]. The underground environment in underground mining is usually dark and damp. The mining environment contains flammable and explosive gases and dust. Therefore, underground coal mines must impose strict safety requirements on electrical equipment. In an explosive underground atmosphere containing gases and dust, the Coal Mine Flameproof Cable Box is used for connecting and branching power cables in AC power supply networks. Explosion-proof cable boxes are typically used in coal mines to ensure that the external explosive environment is protected from electrical sparks, combustion, and internal explosions [2]. However, closed structures do not dissipate heat efficiently. Prolonged overloading and short-circuiting can increase the temperature of internal connecting wires, terminals, and other components [3]. High temperatures caused by operating in fault mode are the main cause of explosion-proof cable box failures in coal mines. Furthermore, operating the box at high temperatures reduces mechanical strength, causes structural damage, and reduces insulation capability. Thus, production accidents such as phase-to-phase short circuits, electrical fires, and gas explosions are highly likely to occur [4,5].

The temperature characteristics of electrical equipment have been extensively investigated. For example, Armand van Deursen et al. established a cable and cable joint temperature computing model based on the multi-physics field. Using this model, they not only analyzed the effects of thermal conductivity, temperature, and wind speed on the temperature distribution characteristics of the cable but also studied the relationship between the current propagation speed and the cable temperature [6–8]. Kim et al. developed a three-dimensional (3D) temperature field simulation model for electric motors to analyze the influence of the ambient temperature, wind velocity, and radiators on temperature fields; understand the relationship between temperature and silicon steel sheet losses; and develop a permanent magnet temperature evaluation method [9–12]. By simulating and experimentally validating temperature rises of medium voltage switchgear used for power transmission and distribution, Fjeld et al. proposed a novel temperature simulation method that can be implemented by adjusting heat transfer coefficients [13,14]. Wilkinson et al. proposed a simplified computational fluid dynamics model to investigate the heat dissipating capacity of a transformer and the temperature distribution characteristics of windings in the transformer [15–17]. Existing research on temperature characteristics is focused on temperature distribution characteristics, the influence of environmental factors on temperature, the relationship of temperature and losses, and the development of a permanent magnet temperature evaluation method in normal operating conditions. Limited research has been conducted to address the failure of explosion-proof electrical equipment in coal mines. In this scenario, it is crucial to study the temperature characteristics of magnetic coupling-based explosion-proof cable boxes for coal mines, especially in the failure mode.

2. Mathematical Model

Heat conduction or heat transfer is caused by the thermal motion of microscopic particles such as molecules, atoms, and free electrons when relative displacement does not occur between the various components of a flameproof cable box for coal mines during its working hours. A temperature gradient inside the flameproof cable box for coal mines can result in internal energy transfer, and the energy transfer rate can be expressed using the following formula:

$$q = -kA\frac{\partial T}{\partial n} \tag{1}$$

where $\frac{\partial T}{\partial n}$ refers to a temperature gradient in the normal direction of area A, k refers to a heat conductivity coefficient (unit: W/(m K)) that depends on the material and temperature of the selected box, and q refers to a heat flow density. Figure 1 schematic diagram describes Equation (1).

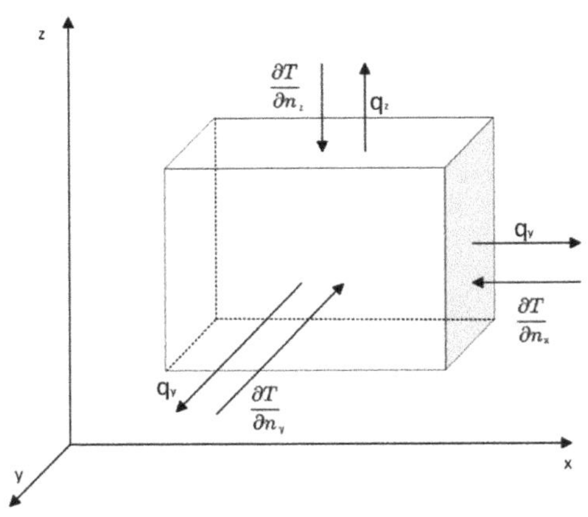

Figure 1. Schematic diagram.

A differential equation of heat conduction is derived according to the law of energy conservation and Fourier's law. Isotropous heat-conducting objects can be expressed as follows in a rectangular coordinate system:

$$\rho C\frac{\partial T}{\partial t}=\frac{\partial}{\partial x}\left(\lambda\frac{\partial T}{\partial x}\right)+\frac{\partial}{\partial y}\left(\lambda\frac{\partial T}{\partial y}\right)+\frac{\partial}{\partial z}\left(\lambda\frac{\partial T}{\partial z}\right)+\phi \tag{2}$$

where $\rho C\frac{\partial T}{\partial t}$ is the thermodynamic energy increment of a micro-body, $\frac{\partial}{\partial x}\left(\lambda\frac{\partial T}{\partial x}\right)+\frac{\partial}{\partial y}\left(\lambda\frac{\partial T}{\partial y}\right)+\frac{\partial}{\partial z}\left(\lambda\frac{\partial T}{\partial z}\right)$ represents a net heat flow imported into the micro-body, and ϕ is the formation heat of a heat source in the micro-body.

A differential equation set of convective heat transfer is constituted by a continuity equation, a momentum differential equation, and an energy differential equation. These equations are derived based on the law of mass conservation, the law of momentum conservation, and the law of energy conservation. The continuity equation and the momentum differential equation have been derived from fluid mechanics. Assumptions for derivations are as follows: (1) D, (2) incompressible Newtonian fluids, (3) constant-property, (4) no internal heat sources, (5) neglecting viscous dissipation, (6) neglecting variations in kinetic and potential energy caused by the net force that is incurred by unbalanced shearing force on microelements in different directions, and (7) neglecting radiation heat transfer between the fluid and the wall surface. The conclusion of inference is expressed as follows:

$$\frac{\partial T}{\partial t}+\frac{\partial T}{\partial x}+\frac{\partial T}{\partial y}=\frac{\lambda}{\rho C_p}\left(\frac{\partial^2 T}{\partial x^2}+\frac{\partial^2 T}{\partial y^2}\right) \tag{3}$$

where $\frac{\partial T}{\partial t}$ is an unsteady term, $\frac{\partial T}{\partial x}+\frac{\partial T}{\partial y}$ is a convective term, and the equation on the right of the equal sign is a heat conduction term.

The metallic conductor produces an eddy current because of the existence of an induced electric field when magnetic–thermal coupling simulation is conducted for the temperature rise of the flameproof cable box for coal mines. Moreover, the power loss of the box consists of the electric current and eddy current of the conductor. For simplifying calculations, the effects of space charge and displacement charge are ignored, and the relative permeability of the conductor and enclosure was assumed to be a constant and electrical conductivity to be only associated with the temperature. According to Maxwell's equations, the frequency domain of electromagnetic fields in the area to be solved was analyzed. The corresponding governing equation is expressed as follows:

$$J_z=\frac{1}{\mu_t}\left[\frac{\partial^2 A_z}{\partial x^2}+\frac{\partial^2 A_z}{\partial y^2}\right]-j\omega\rho_t(T_t)A_z \tag{4}$$

where J_z and A_z refer to the electric current density vector and magnetic vector potential along the z-axis, respectively. Here, μ denotes magnetic conductivity, $\rho(T)$ denotes time-varying electrical resistivity, and t indicates the enclosure or conductor.

Here, the power loss of the enclosure or conductor in the unit length is denoted as Q_t, and $\rho_t(T_t)$ represents the influence of the temperature on electrical resistivity. Next, Q_t and $\rho_t(T_t)$ can be expressed as follows:

$$Q_t=\int\rho_t(T_t)J_t^2 dS_t \tag{5}$$

$$\rho_t(T_t)=\rho_{tT0}[1+0.004(T_t-T_0)] \tag{6}$$

where S_t refers to cross-sectional area, T_0 is the ambient temperature, and ρ_{tT0} is the electrical resistivity at T_0.

3. Temperature Characteristics Simulation

This section takes the three-way explosion-proof cable box for coal mines as the research object, selects the Dell Precision T7920, Dell Technologies, China and finite element simulation and analysis software for simulation, introduces its rated parameters, composition, and structure, describes the process of creating a 1:1 3D model by the computer-aided design software and using it for thermodynamic simulation software to solve the problem, and performs numerical solutions to define the material properties and the hybrid meshing methods. In addition, the current load is applied to the explosion-proof cable box for the coal mine to simulate the temperature distribution characteristics under different fault currents, and the temperature changes and temperature rise under different fault currents (180 A, 240 A, 300 A, and 360 A) are demonstrated, and the corresponding temperature distribution cloud diagrams are given.

3.1. Simulation Models

A three-way flameproof cable box for coal mines was the research object. The rated current and rated voltage were 60 A and 1140 V, respectively. The box consisted of an explosion-resistant enclosure, wiring terminals, an insulating base, and cable entry. By using computer-assisted design software, a 3D model was developed on a 1:1 scale for the box and subsequently imported into thermodynamic simulation software for numerical solution [18,19]. Figure 2 details the 3D model of such flameproof cable boxes for coal mines.

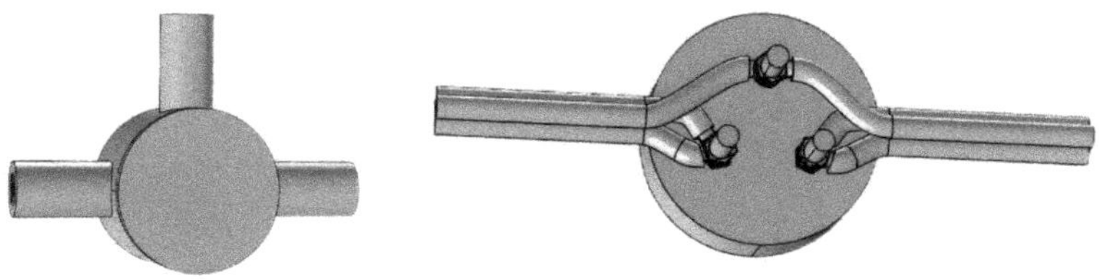

(**a**) Holistic mode (**b**) Internal model

Figure 2. Temperature characteristics simulation models for the flameproof cable box for coal mines.

The numerical solutions are as follows. Because the computational domain is replaced with a series of discrete points, the relationship of variables at these points was established by a governing equation. The differential equation was discretized into an algebraic equation. Variables at the discrete points could be solved [20]. Figure 3 depicts the flow diagram of numerical calculation. The governing equations of the computational domain are generally mathematical formulations of physical laws, and various conservation laws can be expressed using partial differential equations. However, the analytical solutions of a partial differential equation cannot be obtained in many cases. The partial differential equation should be first discretized and subsequently solved by a numerical method to acquire its approximate solution.

The material attribute was defined based on the material of flameproof cable boxes for coal mines. In this study, the enclosure was fabricated from cast iron. The specific heat capacity and heat conductivity coefficient were 460 J/(kg·K) and 66 W/(m·K), respectively. The insulating base was ceramic and had a specific heat capacity of 850 J/(kg·K) and a heat conductivity coefficient of 28 W/(m·K). The wiring terminals and connecting wires were prepared from copper, with a specific heat capacity of 381 J/(kg·K) and a heat conductivity coefficient of 388 W/(m·K). The simulation utilizes material homogeneity assumptions, isotropy assumptions, temperature boundary condition assumptions, electromagnetic boundary condition assumptions, and ideal geometry assumptions. Hybrid meshing was conducted to generate the meshes of the flameproof cable box model. Considering simulation requirements and the arithmetic speed of the computer, high mesh generation precision was selected for guidelines and wiring terminals, whereas mesh generation precision was set to be low for other parts [21–24]. Figure 4 depicts a schematic of mesh generation. The changes in key electromagnetic and thermal physical quantities after each mesh refinement of the coal mine blast box model were compared. The relative errors of magnetic field strength, eddy current density, and thermal physical quantities were calculated. The relative errors of the key physical quantities are less than the pre-set threshold (3%), and this meshing meets the above requirements and is considered to be mesh-independent.

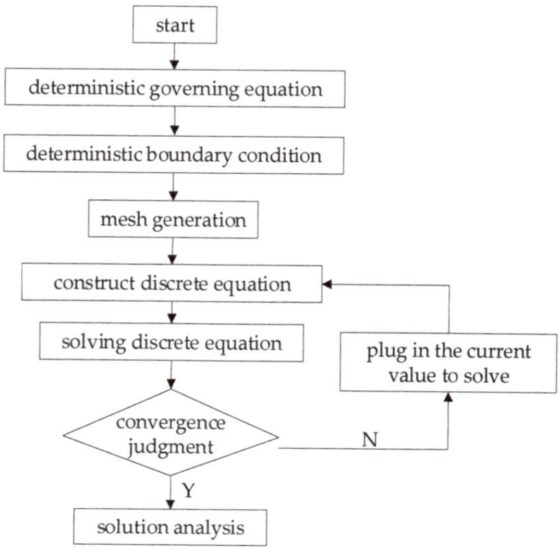

Figure 3. Flowchart of numerical calculation.

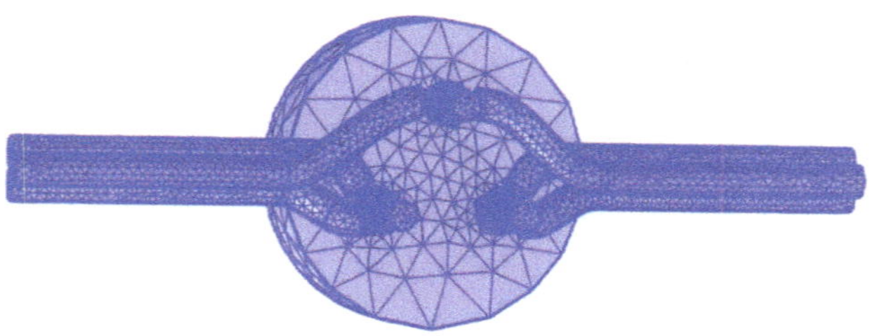

Figure 4. Schematic of mesh generation.

3.2. Failure Simulation

Current loads were applied to the flameproof cable box for coal mines to simulate the temperature distribution characteristics at various failure currents. During the failure simulation of the box, the failure current was configured at 180 A, 240 A, 300 A, or 360 A, and the ambient temperature was set to 15 °C. Because the maximum temperature rise suitable for the safe running of the flameproof cable box for coal mines was 65 K, the temperature characteristics simulation ceased when the temperature rise reached this value in the simulation. In this case, temperature distribution cloud maps were obtained, as presented in Figures 5–8. Specifically, Figure 4 is a temperature distribution cloud map for the flameproof cable box at 180 A. This figure reveals that the temperature reached 80.7 °C at 1065 s when the failure current was set at 180 A, and the temperature rise in this case was approximately 65.7 K. Figure 5 indicates the temperature distribution of the box when the failure current was set at 240 A. The temperature and temperature rise were up to 81.1 °C and 66.1 K, approximately, at 365 s. Figure 6 depicts the temperature distribution at 300 A, where the temperature and temperature rise, respectively, reached 81.3 °C and approximately 66.3 K at 210 s at such an operating condition. Finally, the temperature distribution of the box with a failure current of 360 A is plotted in Figure 7, which reveals that temperature and temperature rise, respectively, were 81.0 °C and 66 K at 135 s.

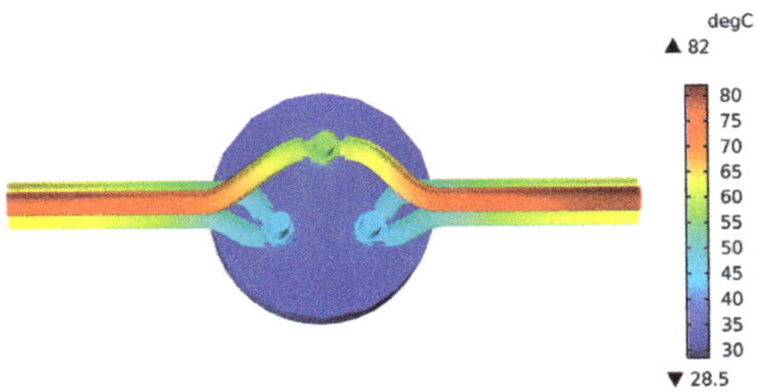

Figure 5. Temperature distribution cloud map of the flameproof cable box for coal mines at 180 A.

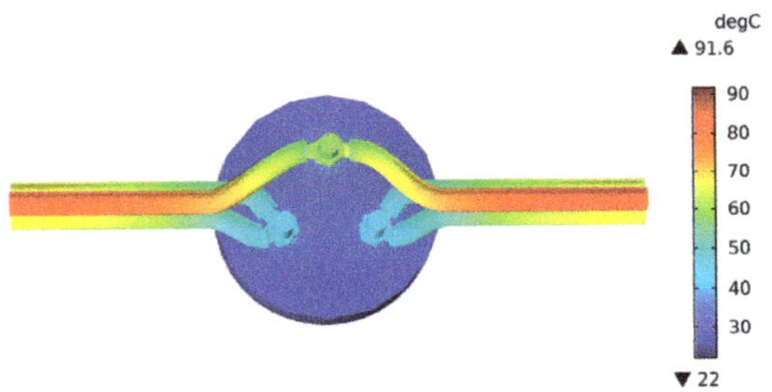

Figure 6. Temperature distribution cloud map of the flameproof cable box for coal mines at 240 A.

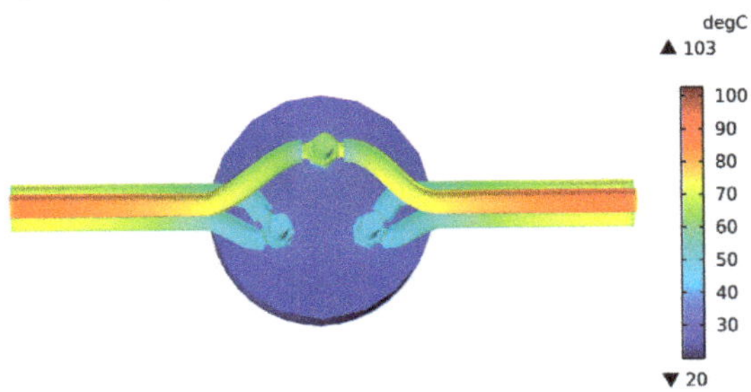

Figure 7. Temperature distribution cloud map of the flameproof cable box for coal mines at 300 A.

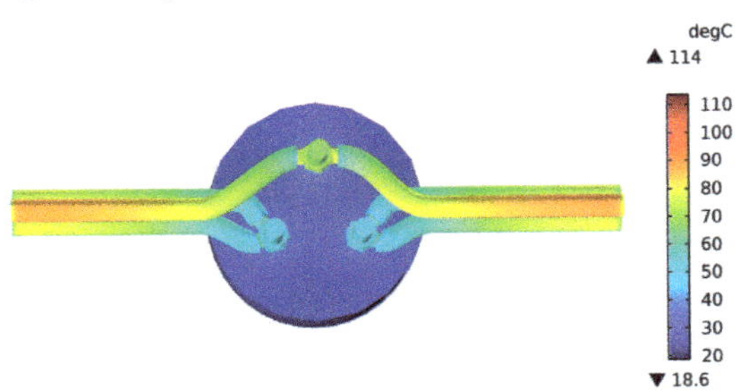

Figure 8. Temperature distribution cloud map of the flameproof cable box for coal mines at 360 A.

4. Testing Under Failure Conditions

This section mainly focuses on the temperature characterization test of explosion-proof cable junction boxes for coal mines. The conditions for testing in a closed environment, including avoiding abnormal heat exchange and controlling wind speed, are described in detail, and the equipment used for testing, such as high current generators, thermocouples, temperature recorders, industrial computers, and thermal imaging cameras, are introduced. Through experiments and simulations on the temperature of the test point under different fault currents (180 A, 240 A, 300 A, and 360 A) and comparing the results of the two, the temperature change of the junction box is analyzed, which verifies the validity of the simulation model and the reasonableness of the related experiments.

4.1. Test Experimental

Temperature characteristics testing was conducted in an enclosed environment specific to the flameproof cable box for coal mines. Figure 9 depicts the experimental flow chart. During the testing, the effects of unusual extraneous heating or cooling were avoided [25,26]. Furthermore, the wind velocity within 1 m from the box should not exceed 0.1 m/s, and external air flows were not introduced [27,28]. A high-current generator was used to supply power in the temperature characteristics testing for the box at failure conditions. A thermocouple and a temperature rise recorder were used to collect and convert temperature data. An industrial computer was adopted to acquire the temperature and experimental current values of the box at failure conditions. Furthermore, a thermal imaging camera was used to detect the temperature of test points during temperature characteristics testing.

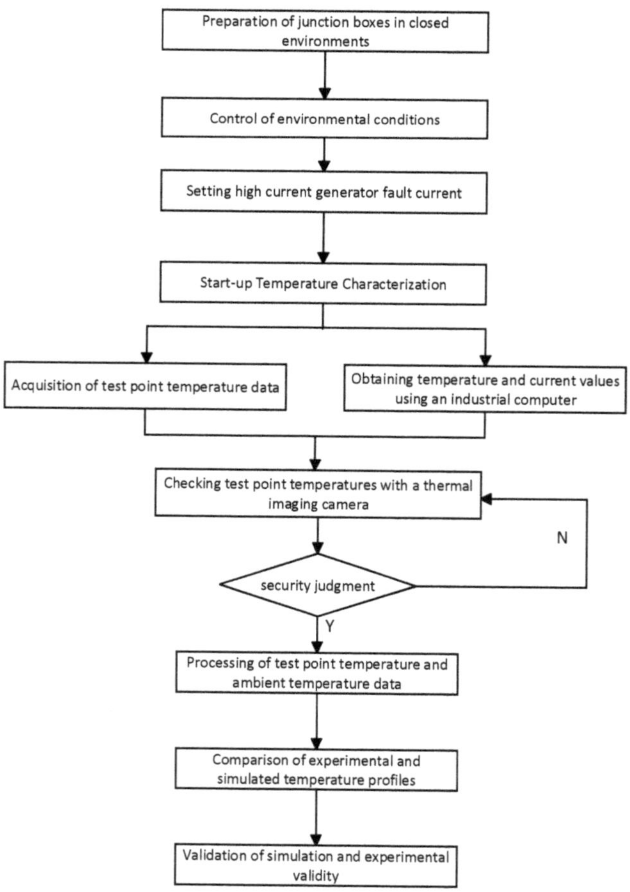

Figure 9. Flow chart of the experiment.

Figure 10 shows the structure of the box temperature test system. The experimental test system of coal mine explosion-proof box uses the WZP-231 temperature sensor from Guangzhou Wenlai Instrumentation Co., Guangzhou, China.

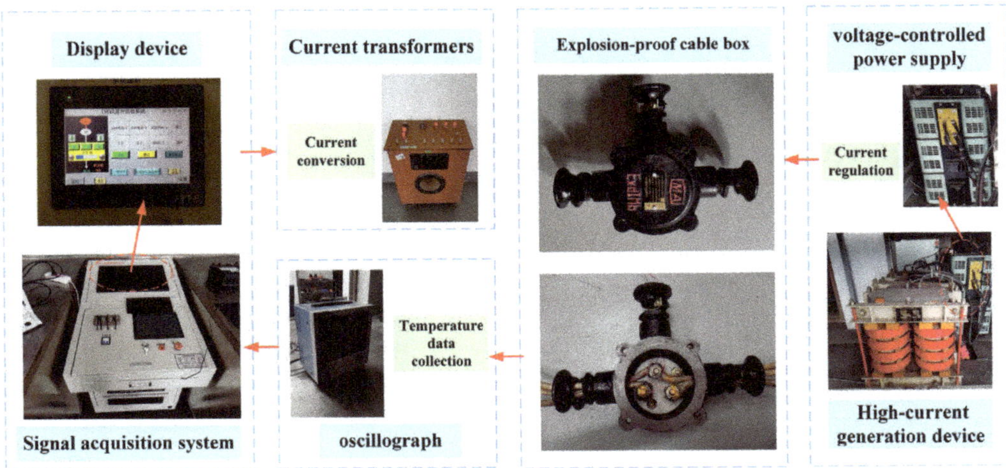

Figure 10. Temperature characteristics testing system structure of the flameproof cable box for coal mines.

Wiring terminal A, entry, wiring terminal B, and internal guidelines were used as test points based on a thermal imaging map produced by temperature characteristics experiments for the box and denoted as Test Points A, B, C, and D. Moreover, Figure 11 depicts the thermal imaging map. By taping, the thermocouple was fixed on the test points mentioned to test the temperature. Figure 12 depicts the positions of these test points. The temperature at test points of the box was obtained from the thermocouple and documented at an interval of 5 s at the temperature rise recorder since energization. During the temperature characteristics testing, the temperature of the box was tested at various failure currents of 180 A, 240 A, 300 A, and 360 A, and the testing stopped when the temperature rise reached its upper limit for safe running (i.e., 65 K). Next, time-varying temperature curves were obtained for the flameproof cable box for coal mines. We processed the temperature data of the test points and ambient temperature data. Thus, relation curves of the temperature rise and time were acquired for the box.

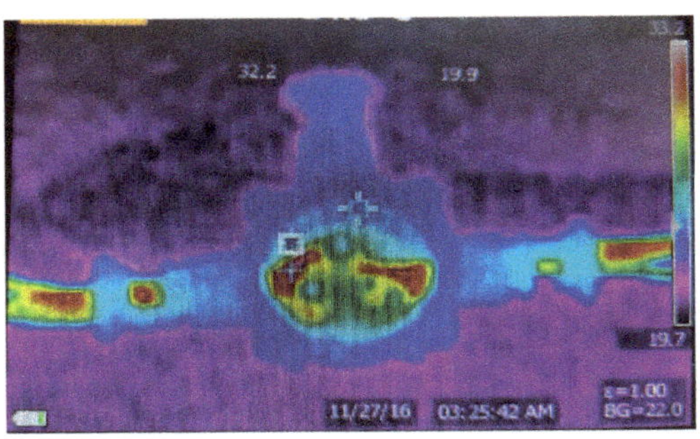

Figure 11. Thermal imaging map of the flameproof cable box for coal mines.

Figure 12. Schematic of the test points for the flameproof cable box for coal mines.

4.2. Discussion and Results Analysis

Specific to the flameproof cable box with a failure current at 180 A, Figures 13 and 14 depict the experimental and simulated temperature versus time curves, respectively. The two curves have the same trend of temperature rise; in this measurement time region, the temperatures at test points B and D increase rapidly, and the temperatures obtained at test points A and C remain consistent. As can be seen from the simulation results, the temperature at test point D is the first to reach the limiting temperature rise value at 1065 s, which is 80.831 °C, and the ambient temperature is 15 °C, with a corresponding temperature rise of about 65.831 K. The temperature at test point D is the first to reach the limiting temperature rise value. In the experiment, the temperature of measurement point D is 82.8 °C, the ambient temperature is 17.7 °C, and the corresponding temperature rise is about 65.1 K. The error between the two is 1.12%, which is in line with the standard.

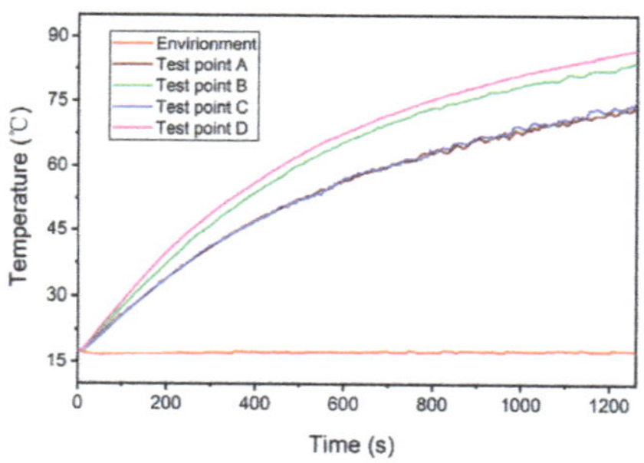

Figure 13. Temperature characteristic curve of explosion-proof cable box for coal mine under experiment at 180 A current.

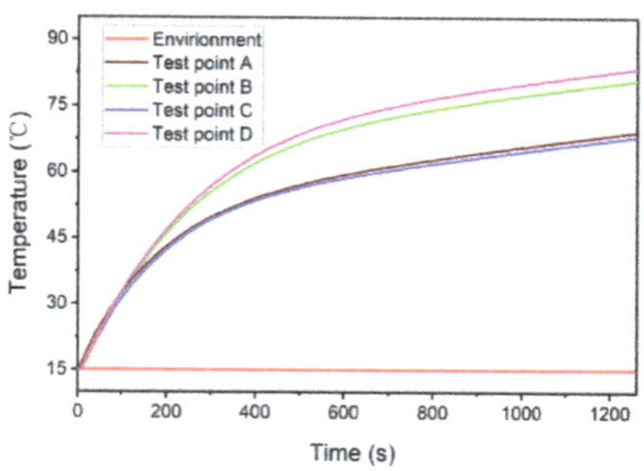

Figure 14. Temperature characteristic curve of coal mine explosion-proof cable box at 180 A under simulation.

Figures 15 and 16 present the temperature rise curves of the box at 240 A current under experiment and simulation, respectively. The results of temperature rise measured by simulation and experiment are similar, and the temperature reaches the limit value of temperature rise at the test point D first, the temperature limit of the box at 240 A current. They indicate that the temperature at Test Points B and D increased rapidly; the temperature at Test Point C is slightly above that at Test Point A. As can be seen from the simulation results, the temperature at test point D is the first to reach the limiting temperature rise value at 365 s, which is 80.234 °C, and the ambient temperature is 15 °C, with a corresponding temperature rise of about 65.234 K. In the experiment, the temperature of measurement point D is 85.5 °C, the ambient temperature is 19.9 °C, and the corresponding temperature rise is about 65.6 K. The error between the two is 0.06%, which is in line with the standard.

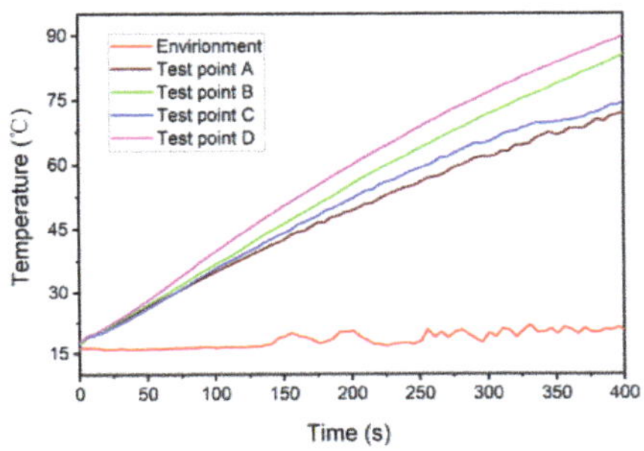

Figure 15. Temperature characteristic curve of explosion-proof cable box for coal mine under experiment at 240 A current.

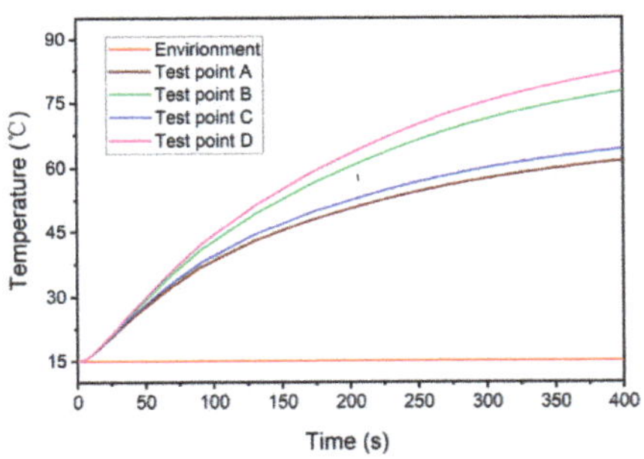

Figure 16. Temperature characteristic curve of coal mine explosion-proof cable box at 240 A under simulation.

Figures 17 and 18 illustrate the time-varying temperature curves of the box when the failure current was set to 300 A under experiment and simulation, respectively. The figures revealed that the temperature at Test Points B and D increased rapidly; the temperature at Test Point C was moderately higher than that at Test Point A, and the temperature at Test Point D was 84.0 °C at 210 s under experiment. In this case, the ambient temperature was 17.7 °C, and the corresponding temperature rise was approximately 66.3 K. The temperature at Test Point D was 81.898 °C at 210 s under simulation. In this case, the ambient temperature was 15 °C, and the corresponding temperature rise was approximately 66.898 K. The error between the two is 0.09%, and the two temperature rise results are similar.

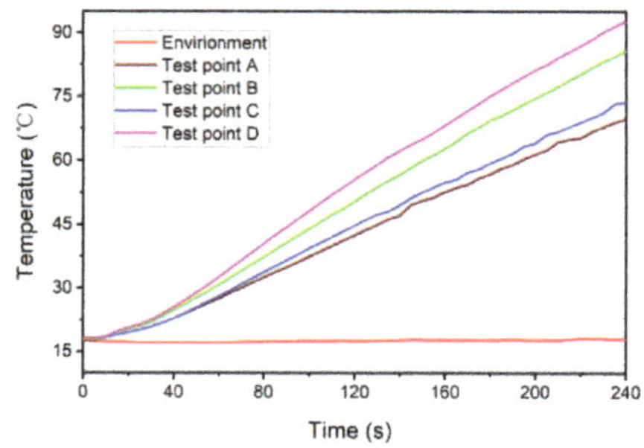

Figure 17. Temperature characteristic curve of explosion-proof cable box for coal mine under experiment at 300 A current.

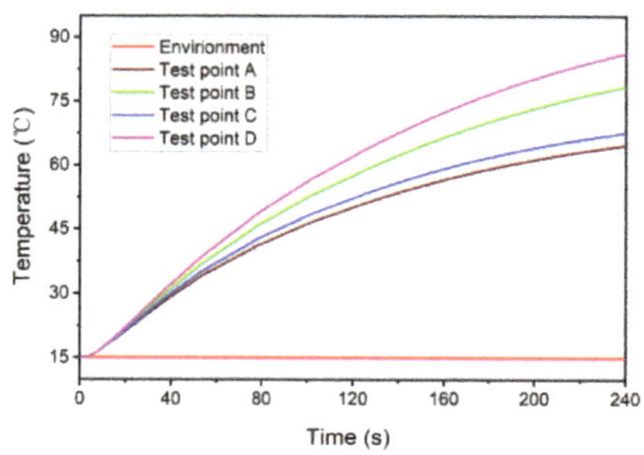

Figure 18. Temperature characteristic curve of coal mine explosion-proof cable box at 300 A under simulation.

Figures 19 and 20 plot the time-varying temperature and temperature rise curves under the failure current set to 360 A for the box under experiment and simulation, respectively. Similarly, the temperature at Test Points B and D increased rapidly, the temperature at Test Point C was higher than that at Test Point A, and the temperature of Test Point D at 135 s was 86.7 °C under experiment. The ambient temperature at 86.7 °C was 21.7 °C, and the corresponding temperature rise was approximately 65.0 K. The temperature of Test Point D at 135 s was 80.584 °C under simulation. The ambient temperature at 86.7 °C was 15 °C, and the corresponding temperature rise was approximately 65.584 K. The error between the two is 0.09%, and the two temperature rise results are similar.

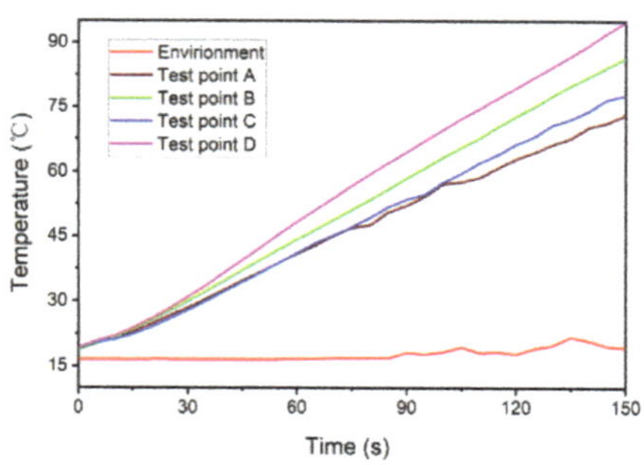

Figure 19. Temperature characteristic curve of explosion-proof cable box for coal mine under experiment at 360 A current.

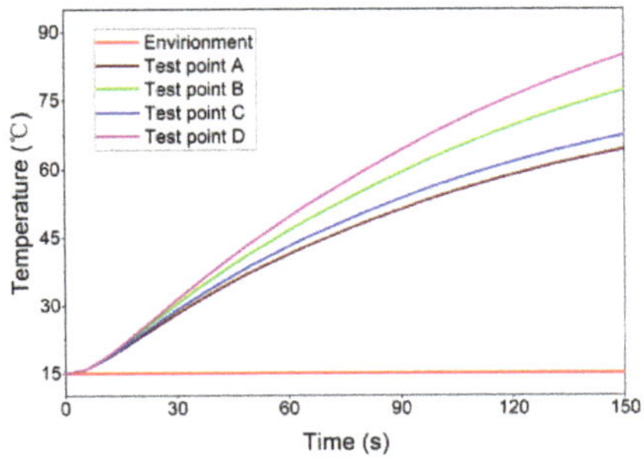

Figure 20. Temperature characteristic curve of coal mine explosion-proof cable box at 3600 A under simulation.

The simulation results of the temperature characteristics of the flameproof cable box for coal mines were compared with the experimental outcomes. Figure 21 depicts the relevant comparison results. This figure reveals that the difference between simulation and experimental results is not greater than 1 K, which proves the validity of the simulation model. At various failure modes, the upper time limit to ensure the safe running of the flameproof cable box for coal mines was separately configured to be 1065 s, 365 s, 210 s, and 135 s. The relationship between respective failure current and time is depicted in the failure characteristic curves in Figure 22. These characteristic curves were inverse-time, which conformed with the characteristic curve variation tendency of explosion-proof electrical equipment used in coal mines. Therefore, the validity of simulation and experiments was proved.

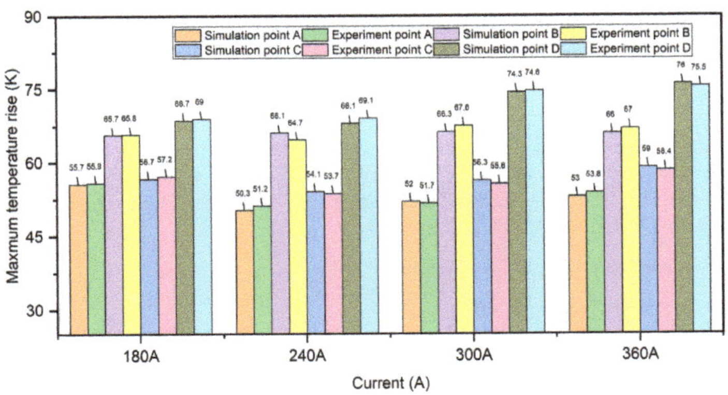

Figure 21. Comparison of experimental and simulation curves at failure modes.

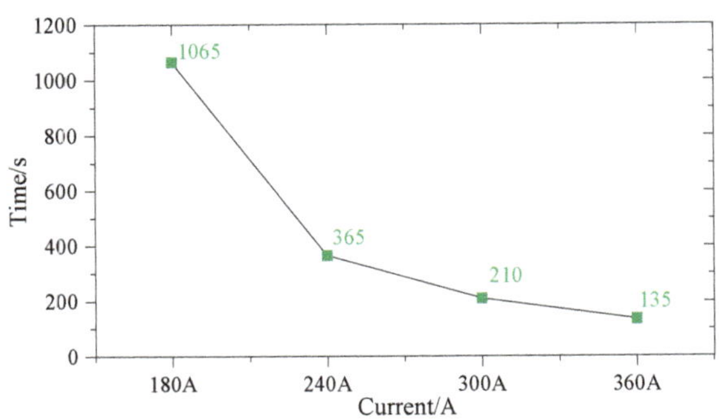

Figure 22. Characteristic curves at various failure currents.

Due to the experimental process, the temperature of the environment will increase due to the heat dissipation of the coal mine explosion-proof box. The actual ambient temperature will be slightly higher than 15 °C; the temperature of the simulation is constant, and the trend of the temperature rise between the simulation and the experiment is consistent.

5. Conclusions

To analyze the mathematical model developed for the flameproof cable box for coal mines, a temperature characteristics simulation model was established for such boxes to investigate their temperature distribution characteristics at failure modes. Additionally, relevant simulation results were compared with their experimental outcomes. The contributions of this paper can be summarized as follows:

(1) Effects of an eddy current loss and a heat conductivity coefficient on the temperature rise were considered when the simulation model was used, and simulation results obtained at various failure modes were compared with corresponding experimental outcomes. Both the simulation model and the selected method are valid.
(2) Temperature distribution characteristics and the time limit for ensuring safe running were obtained at various failure modes by combining global 3D temperature field simulation with the temperature rise measurement at multiple points inside the flameproof cable box for coal mines.
(3) For the flameproof cable box for coal mines, the temperature at the failure modes rapidly increased; the temperature rise tendencies at diverse test points remained constant; and characteristic curves at various failure currents were inverse-time. These results can provide effective values for production design, inspection, and temperature monitoring.

Author Contributions: Resources, H.Z.; methodology, Y.L.; writing—original draft, H.Z.; writing—review and editing, H.Z. and Z.C.; funding acquisition, H.Z.; project administration, H.Z.; investigation, H.Z., Z.C. and Y.L.; supervision, Y.L. and H.Z.; data curation, H.Z.; software, H.Z. and Z.C.; validation, Y.L.; visualization, H.Z. All authors have read and agreed to the published version of the manuscript.

Funding: This study was funded by the National key research and development program of China: 2022YFF0605300, This research was funded by the National Natural Science Foundation of China: 52174228 and Key Laboratory of Special Machine and High Voltage Apparatus (Shenyang University of Technology), Ministry of Education: KFKT202106.

Institutional Review Board Statement: Not applicable.

Informed Consent Statement: Not applicable.

Data Availability Statement: The data presented in this study are available on request from the corresponding author (privacy).

Acknowledgments: We would like to thank the anonymous reviewers for their helpful remarks. The authors would like to acknowledge the support for this study provided by the National Key Research and Development Program of China (Grant No. 2022YFF0605300), the National Natural Science Foundation of China (Grant No. 52174228), and the Key Laboratory of Special Machine and High Voltage Apparatus (Shenyang University of Technology), Ministry of Education (Grant No. KFKT202106).

Conflicts of Interest: Author Hongkui Zhang was employed by the company Fushun CCTEG Inspection Center Co., Ltd. The remaining authors declare that the research was conducted in the absence of any commercial or financial relationships that could be construed as a potential conflict of interest. The authors declare no conflict of interest.

ferences

Gao, K.; Meng, Z.; Jiang, K.; Zhang, H.; Zeng, Q. Shearer Height Adjustment Based on Mechanical-Electrical-Hydraulic Cosimulation. *IEEE Access* **2020**, *8*, 222064–222076. [CrossRef]

Gao, Y.; Zhang, X.; Zhang, T.; Li, Z. A Graph Convolutional Encoder-Decoder Model for Methane Concentration Forecasting in Coal Mines. *IEEE Access* **2023**, *11*, 72665–72678. [CrossRef]

Zhou, W.; Zhao, B.; Liu, J.; Chen, Z.; Tang, X.; Yu, Z.; Wu, J.; Zeng, R. Systematic Analysis and Characterization of Extreme Failure for IGCT in MMC-HVdc System—Part I: Device Structure, Explosion Characteristics, and Optimization. *IEEE Trans. Power Electron.* **2022**, *37*, 8076–8086. [CrossRef]

Yang, W.; Zhang, X.; Ma, H.; Zhang, G.-M. Geometrically Driven Underground Camera Modeling and Calibration with Coplanarity Constraints for a Boom-Type Roadheader. *IEEE Trans. Ind. Electron.* **2021**, *68*, 8919–8929. [CrossRef]

Yang, W.; Zhang, X.; Ma, H.; Zhang, G. Laser Beams-Based Localization Methods for Boom-Type Roadheader Using Underground Camera Non-Uniform Blur Model. *IEEE Access* **2020**, *8*, 190327–190341. [CrossRef]

van Deursen, A.; Wouters, P.; Steennis, F. Influence of Temperature on Wave Propagation in Low-Voltage Distribution Cables. *IEEE Trans. Dielectr. Electr. Insul.* **2021**, *28*, 1785–1792. [CrossRef]

Zhang, L. Temperature Analysis Based on Multi-Coupling Field and Ampacity Optimization Calculation of Shore Power Cable Considering Tide Effect. *IEEE Access* **2020**, *8*, 119785–119794. [CrossRef]

Chen, X.; Yu, J.; Yu, L.; Zhou, H. Numerical analysis of thermo-electric field for AC XLPE cable in DC operation based on conduction current measurement. *IEEE Access* **2019**, *7*, 8226–8234. [CrossRef]

Jung, H.-S.; Kim, H.; Sul, S.-K.; Berry, D.J. Permanent Magnet Temperature Estimation in a Mass-Produced Traction Motor for an Electric Vehicle. *IEEE Trans. Transp. Electrif.* **2022**, *8*, 1863–1873. [CrossRef]

Yang, G.; Liu, Y.; Sun, Q.; Qiao, J.; Wang, T. Numerical Simulation on the Flow and Temperature Field in the High-Power Motor. *IEEE Access* **2023**, *11*, 82830–82837. [CrossRef]

Li, Y.; Liu, G.; Cao, W.; Zhang, F. Study on Calculation Model of High Speed Machine Silicon Steel Sheet Electromagnetic Characteristics Considering Magnetic, Stress and Temperature Condition. *CES Trans. Electr. Mach. Syst.* **2022**, *6*, 288–297. [CrossRef]

Wu, S.; Hao, D.; Tong, W. Temperature Field Analysis of Mine Flameproof Outer Rotor Permanent Magnet Synchronous Motor with Different Cooling Schemes. *CES Trans. Electr. Mach. Syst.* **2022**, *6*, 162–169. [CrossRef]

Fjeld, E. Limitations in the Application of Computer Simulations of Temperature Rise of Completely New Designs of MV Metal Enclosed Switchgear. *IEEE Trans. Power Deliv.* **2021**, *36*, 2740–2747. [CrossRef]

Zheng, H.; Li, Y.; Luo, X.; Zhang, E.; Jing, J. Investigation on Hot Spot Temperature of Resin-Impregnated Paper High-Voltage Bushing Based Upon Dielectric Loss. *IEEE Trans. Dielectr. Electr. Insul.* **2023**, *30*, 1884–1892. [CrossRef]

Zhao, S.; Liu, Q.; Wilkinson, M.; Wilson, G.; Wang, Z. A Reduced Radiator Model for Simplification of ONAN Transformer CFD Simulation. *IEEE Trans. Power Deliv.* **2022**, *37*, 4007–4018. [CrossRef]

Raeisian, L.; Niazmand, H.; Ebrahimnia-Bajestan, E.; Werle, P. Thermal management of a distribution transformer: An optimization study of the cooling system using CFD and response surface methodology. *Int. J. Elect. Power* **2019**, *104*, 443–455. [CrossRef]

Li, H.; Liu, Y.; Zhuang, X.; Xiao, H.; Fan, X.; Wang, J.; Li, X.; Jiang, T. Test and Analysis on Extended Temperature Rise of 110 kV Transformer Based on Distributed Temperature Sensing. *IEEE Trans. Power Deliv.* **2023**, *38*, 1030–1041. [CrossRef]

Santisteban, A.; Delgado, F.; Ortiz, A.; Fernández, I.; Renedo, C.J.; Ortiz, F. Numerical analysis of the hot-spot temperature of a power transformer with alternative dielectric liquids. *IEEE Trans. Dielectr. Electr. Insul.* **2017**, *24*, 3226–3235. [CrossRef]

Zhao, B.W.; Jian, Y.F.; Jin, Y.T.; Li, Y.; Guo, H. Numerical calculation and analysis of foil winding loss and temperature rise of dry transformer. *Transformer* **2020**, *57*, 1–5.

Cui, Z.; Lin, X.; Zhong, J.; Yao, Y.; Jianyuan, X. Study on the Temperature Rise Characteristics of C4F7N/CO2 Mixed Gas Ultra High Voltage Bus. *Trans. China Electrotech. Soc.* **2023**, *38*, 2491–2499.

Zhou, H.B.; Zhang, Z.X.; Li, W.T.; Zhang, Q.J.; Du, Z.B.; Wang, J.M. Calculation and analysis of winding hot spot temperature of dry transformer under different load. *Transformer* **2022**, *59*, 1–5.

Chen, Z.; Cai, W.; Cai, L.; Cen, G.; Wang, H. Simulation and Experimental Analysis of the Temperature Field of Jet Flow of Aircraft Based on CFD Theory. *IEEE Access* **2020**, *8*, 55881–55892. [CrossRef]

Zhang, Q.J.; Wang, H.M.; Du, Z.B.; Jing, C.Y.; Wang, Y.F.; Wang, J.M. Numerical analysis of three-dimensional temperature field and temperature rise of dry transformer windings. *Tansformer* **2021**, *58*, 39–43.

Poudel, A.; Kim, S.; Cho, B.H.; Kim, J. Temperature Effects on the Natural Frequencies of Composite Girders. *Appl. Sci.* **2024**, *14*, 1175. [CrossRef]

Li, C.; Feng, Q.; Jia, M.; Gao, L.; Jia, P.; Fang, Z.; Lu, H.; Xiong, J. Accurate Real-Time Temperature Measurement Method in Ultra-High Temperature Rotational Environments for Aero Engines/Turbines. *IEEE Sens. J.* **2022**, *22*, 6482–6490. [CrossRef]

Yu, T.; Ren, C.; Jia, Y.; Li, J.; Zhang, J.; Xu, Y.; Yan, B.; Zhang, M.; Qiao, L.; Wang, T.; et al. Photovoltaic Panel Temperature Monitoring and Prediction by Raman Distributed Temperature Sensor With Fuzzy Temperature Difference Threshold Method. *IEEE Sens. J.* **2021**, *21*, 373–380. [CrossRef]

27. Chen, M.; Wang, H.; Pan, D.; Wang, X.; Blaabjerg, F. Thermal Characterization of Silicon Carbide MOSFET Module Suita for High-Temperature Computationally Efficient Thermal-Profile Prediction. *IEEE J. Emerg. Sel. Top. Power Electron.* **202** 3947–3958. [CrossRef]
28. Chai, X.; Ning, P.; Cao, H.; Zheng, D.; Li, H.; Huang, Y.; Kang, Y. Online Junction Temperature Measurement of Double-sic Cooling IGBT Power Module through On-state Voltage with High Current. *Chin. J. Electr. Eng.* **2022**, *8*, 104–112. [CrossRef]

applied sciences

MDPI

rticle

eepage Response of Fractured Sandstone to the True Triaxial Gas–Solid Coupling Effect

ngwen Gao [1,2,3,4], Tong Zhang [1,2,3,4,*], Xiang Yu [1,2,3,4], Yanfang Li [1,2,3,4] and Xin Yang [1,2,3,4]

1 National Local Joint Engineering Research Center for Safe and Precise Coal Mining, Anhui University of Science and Technology, Huainan 232001, China; 2023100100@aust.edu.cn (X.Y.)
2 Institute of Energy, Hefei Comprehensive National Science Center, Hefei 230031, China
3 School of Energy and Safety, Anhui University of Science and Technology, Huainan 232001, China
4 State Key Laboratory of Mining Response and Disaster Prevention and Control in Deep Coal Mines, Anhui University of Science and Technology, Huainan 232001, China
* Correspondence: 2018013@aust.edu.cn

Abstract

The fluid flow in underground reservoirs is directly related to resource recovery and hazard prevention. In this study, the evolution of fractured sandstone deformation and permeability under an in situ stress influence is investigated using the true triaxial percolation system. The results show that the strain of fractured sandstone increases logarithmically with the increase in axial stress. The evolution of axial strain is dominated by the maximum principal stress, and the minimum principal stress and the intermediate principal stress affect the strain amplitude. The fracture morphology of low-permeability sandstone affects permeability and strain evolution. Small fractures are more sensitive to the increase in the maximum principal stress, and the response in principal strain to the increase in principal stress is obvious in large fractures. There is a negative exponential relationship between pore pressure and the conductivity of fractures. When pore pressure is 0.3 MPa, the conductivity is the highest; meanwhile, when pore pressure is 1.8 MPa, the conductivity is the lowest. The decreasing range of the conductivity increases with the increase in fracture size. In situ stress significantly affects the evolution of principal strain and related permeability. The permeability decreases with an increase in the minimum and middle principal stresses. Under low pore pressure (0.3–0.6 MPa), the permeability decreases with an increase in the principal stress; meanwhile, under high pore pressure (0.6–1.8 MPa), permeability changes slightly with an increase in the principal stress. The findings provide reference to the engineering practice of underground mining.

Keywords: fractured sandstone; permeability; true triaxial stress; deformation

1. Introduction

The world's 4000 trillion cubic meters of unconventional resources are the potential clean energy of the future. Understanding fluid flow and migration in unconventional geology is beneficial to the mining and safety of natural resources. The geology structure composed of fractures and pores is directly related to the mechanical properties, permeability characteristics, and stress sensitivity, and it contributes to the fluid flow in resource reservoirs. The conductivity capacity of sandstone, characterized by permeability, is influenced by in situ-dependent pore fracture. The strain and permeability of fractured sandstone is significant to resource recovery. Numerous studies were conducted, and abundant achievements were obtained in this field [1–4].

To simulate the real geological environment, uniaxial and true triaxial tests were carried out with a consideration of fluid flow, ground stress, and geological conditions. In terms of uniaxial compression, Liu et al. [5] carried out uniaxial compression experiments with frozen weakly sandstone under different loading and unloading rates, and they revealed the failure mechanism of weakly frozen sandstone. Vasyliev et al. [6] used four experimental features to mathematically model the cracks in cylindrical rock samples and compared and verified the design strength with experimental data obtained from uniaxial compression. Abbas et al. [7] conducted uniaxial compression tests on sandstone–shale-sandstone composites, and based on the obtained mechanical properties and stress–strain relationships, it was pointed out that shale interlayers exert significant effects on the mechanical properties of the composites. Jiang et al. [8] used PFC2D particle-flow numerical simulation software to simulate the mechanical properties and damage evolution process of broken rock, and they discussed the multi-parameter stage characteristics and precursor information of acoustic emission under uniaxial compression. In terms of true triaxial compression, Ishibashi et al. [9,10] discussed the energy balance in the hydraulic shear process of granite fractures and pointed out the key role of the surrounding mass in the storage of elastic potential energy. Yu et al. [11] constructed a prediction model of sandstone permeability under the coupling effect of real triaxial stress field and pore pressure, and they pointed out that the permeability of sandstone is proportional to the pore pressure and inversely proportional to the principal stress. Niu et al. [12] studied the deformation, damage evolution, seepage characteristics, and energy response of columnar jointed basalt based on true triaxial compression experiments, and they pointed out that the permeability and axial strain of the sample have an exponential relationship. Lu et al. [13] obtained permeability under various true triaxial stresses using a newly developed multifunctional true triaxial geophysical (TTG) instrument, and a theoretical model was proposed to describe the permeability characteristics of composite coal rock under a real triaxial stress condition. Liu et al. [14] established a new mathematical model to predict the influence of shale permeability, considering true triaxial stress.

Solid deformation and fluid migration occurred during underground activity, and the coupling relationship was investigated through a simulation, a theoretical analysis, and an experimental test. And the characteristic of seepage, fracture propagation, and the strain response and the corresponding relationship was obtained [15–25]. Among them, Zhang et al. [26] conducted true triaxial tests on virtual granite samples using a properly calibrated DEM model to explore the grain-scale mechanism and energy release of the σ_2 effects. Zhang et al. [27] numerically investigated the mechanics and failure characteristics of fractured rock samples under true triaxial compression, considering the effects of true triaxial loading parameters and pre-existing fracture geometry parameters. To investigate the damage evolution and stress dependency, Gao et al. [28] conducted a series of true triaxial cyclic loading tests on the collected intact and jointed marbles, irreversible strains and dissipated energy were calculated as damage variables, and progressive failure was analyzed. Wang et al. [29] studied the effects of true triaxial unloading on the deformation and failure characteristics of dacite, and they analyzed the strength and deformation behaviors caused by unloading σ_3 based on the results of stress–strain curves. During the true triaxial experiment, the mechanical and conductivity parameter response to loading and unloading behavior was analyzed based on the acoustic emission (AE) methods of LVDT et al. [30–36]. Fracture as a conductivity medium significantly influences the mechanical behavior and conductivity capacity, while a rare study of the fractured rock response considered the influence of the loading path.

The behavior of geological reservoirs under disturbed stress conditions has been extensively investigated using a true triaxial experiment system. The true triaxial stress–

strain and seepage characteristics in fractured sandstone significantly contribute to the resource extraction in tight resource reservoirs. In this study, the seepage, strain, and permeability response to the variation in in situ stress was analyzed in fractured sandstone. The correlations among permeability, pore pressure, stress, and strain were analyzed. A new method for predicting the permeability of fractured sandstone has been developed. The findings provide significant insight into the engineering practice.

2. Experimental Setup

2.1. Material Preparation

The experiment samples were collected from the Yanchang Formation 7–2, an ultra-low-permeability reservoir. The sandstone samples were machined into dimensions of 100 mm × 100 mm × 200 mm using the numerical control cutting equipment (Figure 1). Prefabricated fractures were created through splitting tests performed on a universal testing machine. The samples were sealed with NANDA 704 silicone rubber (Xiangyang High-Temp Resistant New Material Technology Co., Ltd., Xiangyang, China), which was applied evenly over the sandstone surface. This type of silicone rubber exhibits strong adhesion, high strength, non-corrosiveness, excellent sealing performance, and aging resistance, with a maximum curing depth of 3–5 mm. The coated specimens were allowed to cure for more than 24 h.

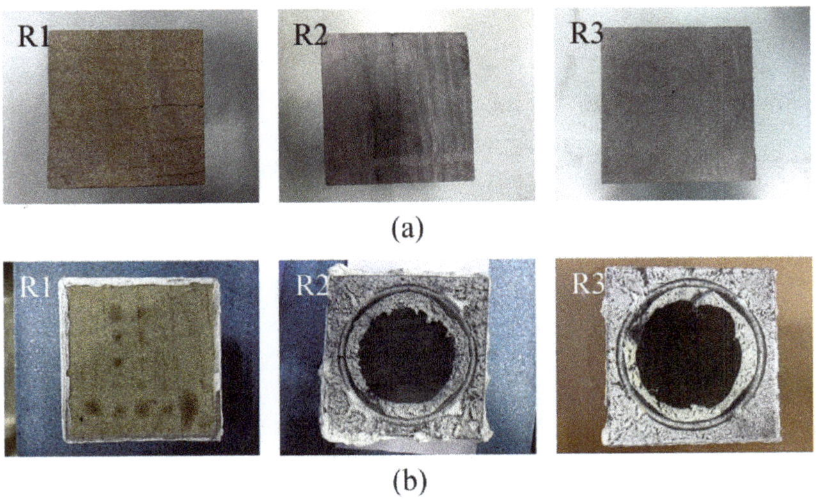

Figure 1. Prefabricated cracks before and after experiment: (**a**) before experiment; (**b**) after experiment.

2.2. True Triaxial Gas–Solid Coupling Rock Mass Seepage Test System

The true triaxial gas-solid coupling system developed by Anhui University of Science and Technology was used to carry out the test. This test system can independently control the three-dimensional stress through the servo system $\sigma_1, \sigma_2, \sigma_3$ (Figure 2). The experimental device can be used to study the deformation, failure and gas seepage of coal and rock mass under the coupling of true triaxial stress and gas pressure in a closed environment. The experimental system mainly includes hydraulic servo control system and monitoring control system. The main technical parameters of true triaxial gas–solid coupling rock mass seepage test system is shown in Table 1.

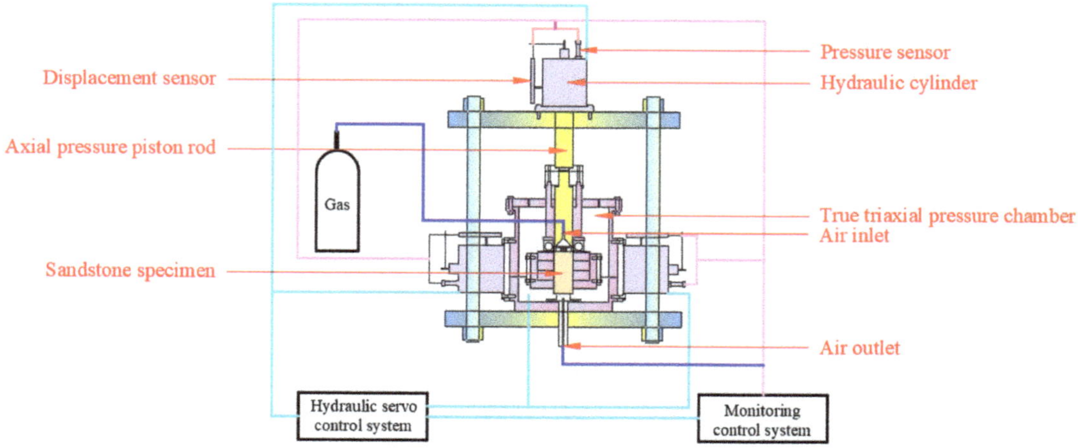

Figure 2. True triaxial coupling experiment apparatus.

Table 1. True triaxial gas–solid coupling rock mass seepage test system.

Item	Technical Parameter
Maximum axial pressure (σ_1)	70 MPa
Maximum lateral pressure (σ_2)	35 MPa
Maximum lateral pressure (σ_3)	10 MPa
Maximum gas pressure	6 MPa
Maximum axial displacement	50 mm
Maximum lateral displacement (unilateral)	30 mm
Sample size	100 mm × 100 mm × 200 mm
Force value test accuracy	±1% of indicated value
Force value control accuracy	±0.5% of the indicated value
Displacement test accuracy	±1% of indicated value
Axial and lateral loading control mode	Force control and displacement control
Front and rear loading control mode	Force control
Overall stiffness of plant	Greater than 10 GN/m

2.3. Experimental Scheme

To simulate the in situ stress condition, the experiment scheme was designed as below: firstly, gradually apply $\sigma_1 = \sigma_2 = \sigma_3$ to a predetermined value of 2 MPa under hydrostatic pressure, and then apply 1 MPa carbon dioxide gas. After the gas is fully adsorbed and balanced, the gas pressure is changed to measure the permeability of the specimen under each gas pressure. After the measurement is completed, the gas permeability at each hydrostatic pressure is measured by applying $\sigma_1 = \sigma_2 = \sigma_3$ to a predetermined value of 4, 6, and 8 MPa. Since the compressive strength of the specimen is much greater than 32 MPa, the stress state of the specimen can be changed by adding and releasing pressure. Then hold $\sigma_1 = 8$ MPa unchanged and decrease $\sigma_2 = \sigma_3$ simultaneously to 2 MPa. Decrease $\sigma_1 = 4$ MPa and determine the permeability of the test piece. Holding $\sigma_2 = \sigma_3 = 2$ MPa unchanged, increase σ_1 to 8 MPa, $\sigma_2 = \sigma_3 = 4$ MPa, measure the permeability of the test piece, and then maintain $\sigma_2 = \sigma_3 = 4$ MPa unchanged, increasing successively $\sigma_1 = 12$, 16, 20, 24, 28, and 32 MPa, measuring the permeability of the test piece throughout. Subsequently, reduce $\sigma_1 = 30$ MPa, increase $\sigma_2 = \sigma_3 = 6$ MPa, and measure the permeability of the test piece concurrently. Then, maintaining $\sigma_2 = \sigma_3 = 6$ MPa unchanged, decrease σ_1 successively to 24, 18, and 12 MPa, and determine the gas permeability. Subsequently, increase $\sigma_1 = 16$ MPa, $\sigma_2 = \sigma_3 = 8$ MPa, measure the permeability of the test piece, and then hold $\sigma_2 = \sigma_3 = 8$ MPa unchanged, increasing σ_1 successively to 24 and 32 MPa, and measure the permeability of the test piece. Finally, decrease the permeability of the test piece by setting σ_1 = 6 MPa, $\sigma_2 = \sigma_3$ = 2 MPa, and then increase σ_1 to 8, 10, 12, 14, and 16 MPa. The permeability of the test piece at each level of σ_1 was measured. The permeability changes in specimens at pressure of 0.3, 0.6, 0.9, 1.2, 1.5 and 1.8 MPa were measured successively. The scheme is shown in Figure 3.

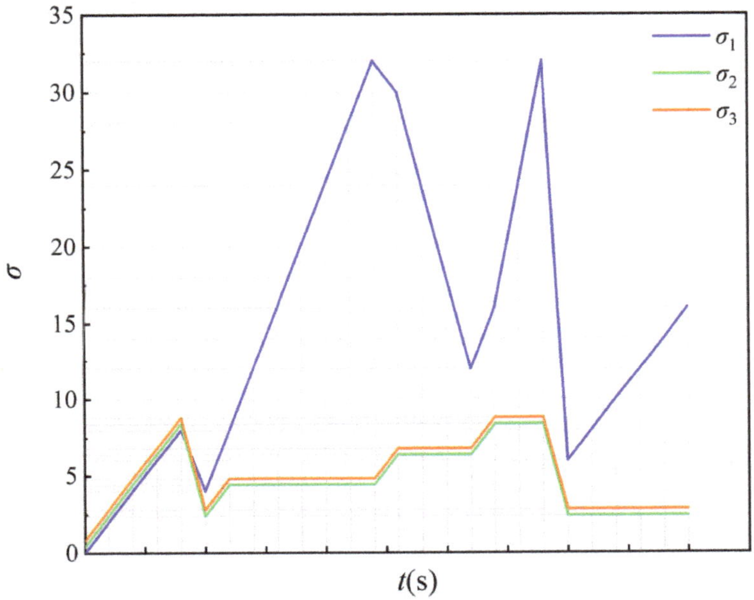

Figure 3. Schematic diagram of stress loading.

3. Experimental Results and Discussion

3.1. Characteristics of Stress–Strain

Figure 4 presents the stress–strain curves of three sandstone samples (R1, R2, R3) under true triaxial loading. The results demonstrate that an increase in lateral stresses (σ_2, σ_3) enhances the yield strength of the rock. Under lower lateral stress, the rocks approach their strength limits as the axial stress increases, a trend particularly pronounced in sample R2. Based on the stress–strain relationships, no sample failure occurred as the axial stress was raised from 2 MPa to 32 MPa. The deformation process mainly involved compaction and elastic stages, with higher lateral stresses leading to more extended elastic deformation.

Axial strain generally increases with rising axial stress. However, due to inherent differences in initial rock properties, the stress responses varied among the samples. R1 and R2 exhibited distinct fractures, with R2 containing notably longer interconnected fractures, whereas R3 remain intact. This structural difference explains why R1 and R2 showed stronger responses to changes in σ_2 and σ_3 (especially R2), while R3 displayed minimal variation. The contrasting behaviors of R1, R2, and R3 underscore the critical influence of fracture structures on stress–strain evolution.

In summary, rock strength increases under higher triaxial stress levels. The evolution of axial strain is primarily controlled by the maximum principal stress, while the magnitudes of the intermediate and minimum principal stresses affect the strain amplitude.

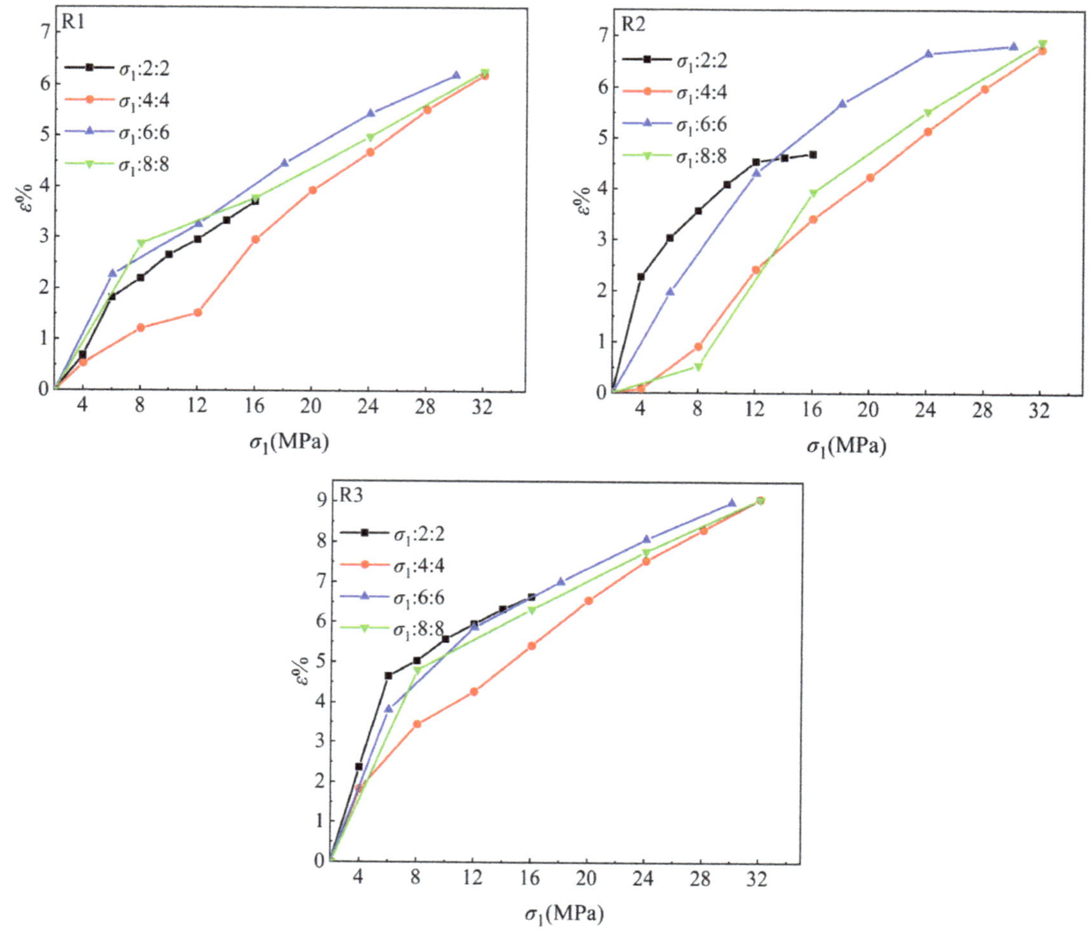

Figure 4. Axial stress–strain relationship of true triaxial seepage test.

3.2. The Influence of Pore Pressure on Permeability

Figure 5 illustrates the evolution of permeability under different pore pressures at a constant triaxial stress ratio. The results indicate a negative correlation between permeability and pore pressure. The characteristics of permeability evolution during pore pressure variations are mainly influenced by differences in fracture structures of the rock samples and the applied triaxial stress ratios.

With respect to fracture structures of rock cores, it was observed that under the same triaxial stress ratio, higher stress levels lead to a more pronounced permeability response to pore pressure changes in R2 sample, followed by R1, with R3 exhibiting the least sensitivity. This finding is consistent with the stress–strain results in Section 3.1. Taking the triaxial stress ratio of 1:1:1 with $\sigma_1 = \sigma_2 = \sigma_3 = 2$ MPa as an example, experimental results demonstrate that under identical stress conditions, when pore pressure increased from 0.3 MPa to 1.8 MPa, the permeability of samples R2, R1, and R3 decreased by 89.81%, 88.24%, and 75%, respectively.

Regarding the influence of triaxial stress ratios, results indicate that the stress ratio σ_1:σ_2:σ_3 increased from 1:1:1 to 6:1:1, while permeability experienced dynamic decreases and slight increase in response to rising pore pressure. In detail, for sample R1, at stress ratios of 1:1:1 and 2:1:1, although the triaxial stress ratio is the same, with the increase in triaxial stress, the permeability decreases gradually under the same pore pressure, and σ_1:σ_2:σ_3 = 1:1:1, the permeability decreases more than σ_1:σ_2:σ_3 = 2:1:1. It illustrates that under increasing triaxial stress, original pore and cracks are gradually closed by three-dimensional compression, with essentially no crack propagation or new crack generation at this stage. And at higher stress ratios, 3:1:1, 4:1:1, 5:1:1, 6:1:1, 7:1:1 and 8:1:1, permeability varies with the stress ratio. Under low triaxial stress, permeability increases gradually with the increase in triaxial stress. When the triaxial stress reaches 24:8:8, 32:8:8, and 30:6:6, the permeability suddenly decreases, but it is still greater than the initial permeability under the corresponding stress ratio. This implies that micro cracks are generated in the fractured sandstone under high stress ratio, resulting in the increase in connected pores and thus enhanced permeability. As the triaxial stress increases from 24:8:8 to 30:6:6, both newly generated microcracks and pre-existing fractures are compressed and closed, resulting in an overall permeability reduction.

For sample R2, as the stress ratio increases from 1:1:1 to 8:1:1, permeability decreases gradually with the increase in triaxial stress, though the extent of decrease is relatively small. Therefore, the change in permeability is very different from R1. For sample R3, the permeability shows the same change trend as sample R2, and both show a gradual decrease with the increase in triaxial stress, but the change degree of permeability is greater than in sample R2. This can be attributed to the generation of new fractures in R3 under increasing triaxial stress, where the extent of fracture development exceeds the degree of compression-induced closure.

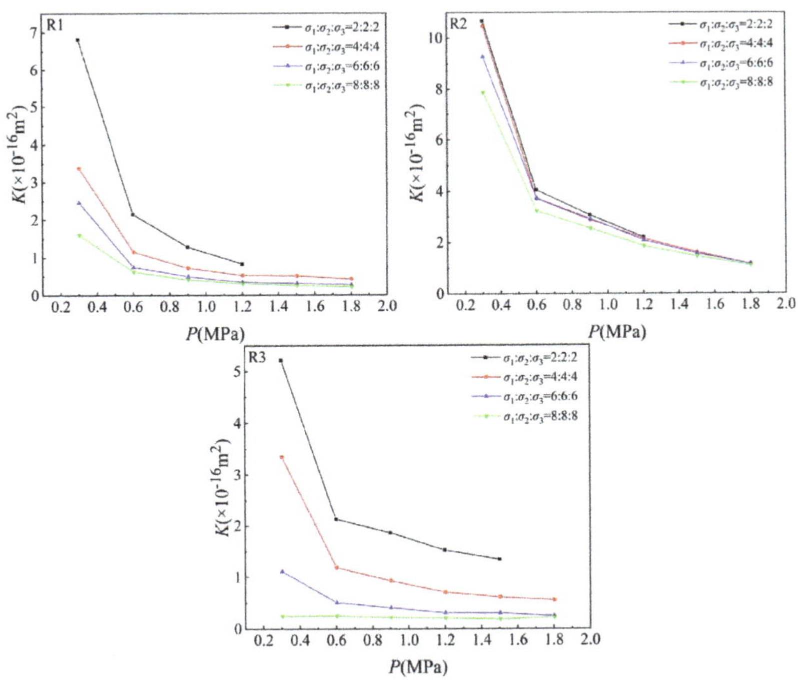

(a) σ_1:σ_2:σ_3 = 1:1:1

Figure 5. *Cont.*

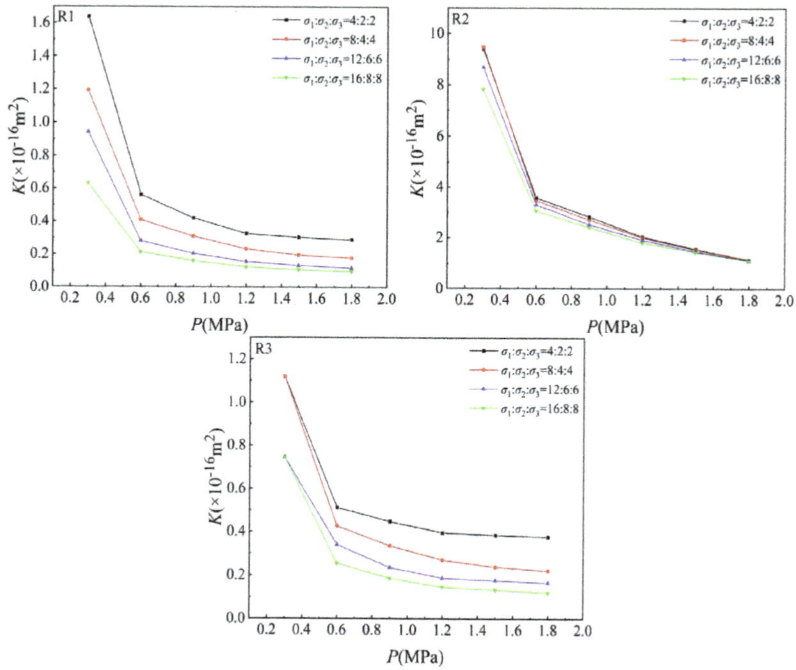

(**b**) σ_1:σ_2:σ_3 = 2:1:1

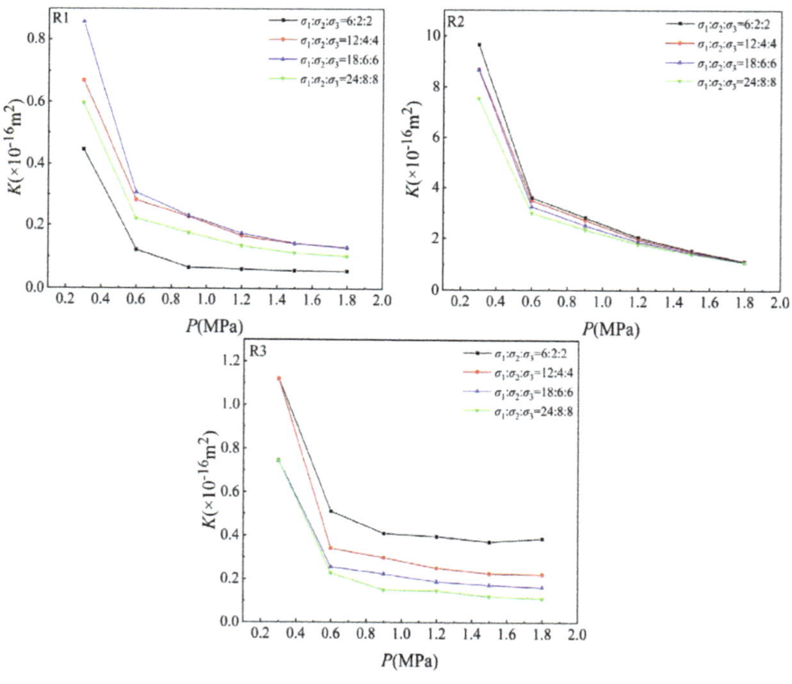

(**c**) σ_1:σ_2:σ_3 = 3:1:1

Figure 5. *Cont.*

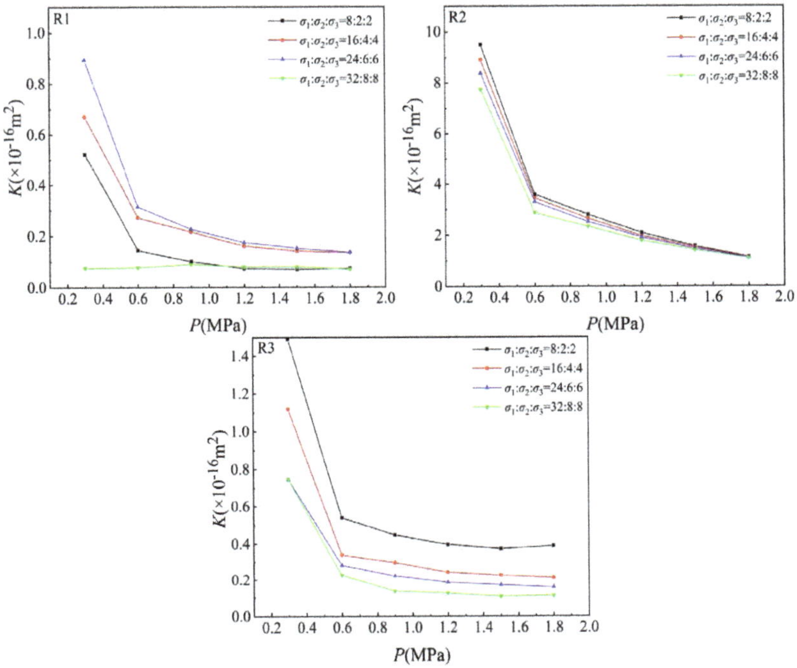

(**d**) $\sigma_1:\sigma_2:\sigma_3$ = 4:1:1

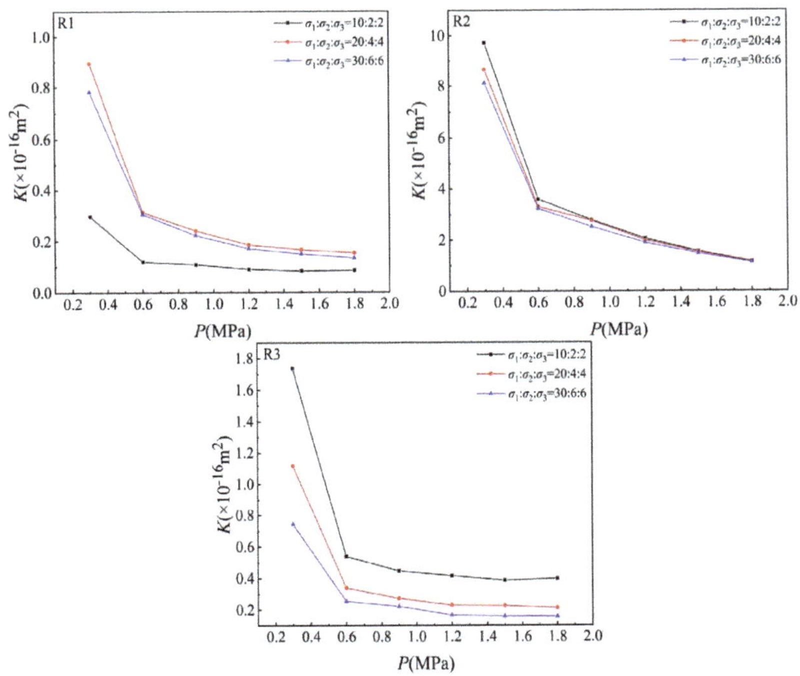

(**e**) $\sigma_1:\sigma_2:\sigma_3$ = 5:1:1

Figure 5. *Cont.*

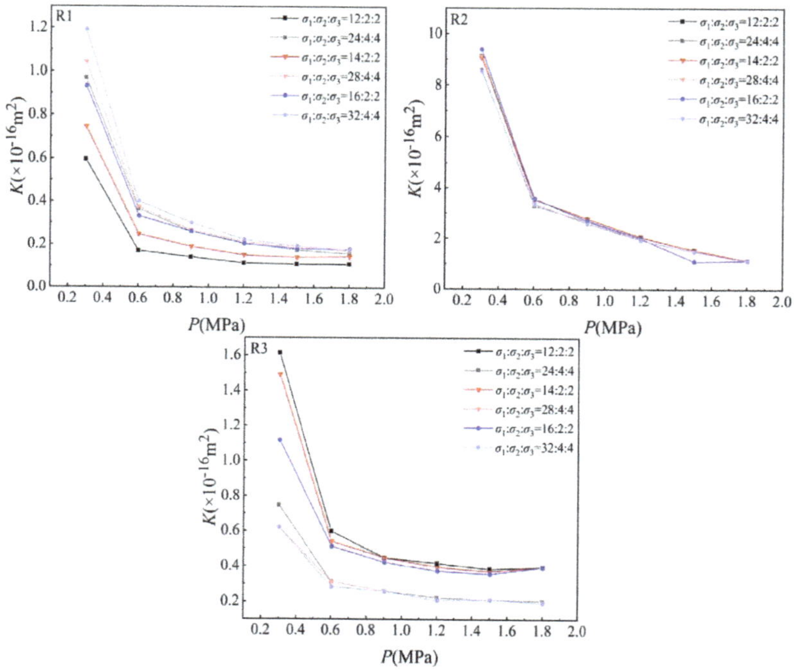

(**f**) $\sigma_1:\sigma_2:\sigma_3$ = 6:1:1/7:1:1:1/8:1:1

Figure 5. Variation in permeability with pore pressure under the same stress ratio.

3.3. Influence of the Fracture on Permeability

Fracture conductivity, characterized by permeability during the true triaxial loading process, was identified as corresponding to small, medium, and large fractures in samples R1, R3, and R2, respectively. As the maximum principal stress increased, slight and dynamic permeability responses were observed in samples with large and small fractures. The phenomenon is different from the previous work [37,38], which reported that dynamic response typically occurs in fractures with large apertures. In these experiments, the maximum principal stress was applied parallel to the fracture orientation. Under this configuration, the increase in maximum principal stress only marginally influences the fluid flow channel through stress-dependent strain. In addition, the permeability response was weakened with the increase in the in situ stress. Fractures within the ultra-low-permeability sandstone significantly influence pore pressure distribution. Non-Dary flow behavior occurred at high pore pressure of 0.6–1.8 MPa, which can be attributed to the special fracture structures. This observation contrasts with findings from previous work [39,40]. Thus, the novelty of this study lies in identifying the facture contribution under different in situ stresses, particularly considering the influence of the maximum principal stress.

3.4. Influence of In Situ Stress on Permeability

The influence of in situ stress on the evolution of permeability was shown in Figure 6. The permeability was negatively correlated with the increase in in situ stress. In detail, for sample R1, under the stress condition $\sigma_1:\sigma_2:\sigma_3$ = x:2:2, as the axial stress increases from 4 MPa to 8 MPa, the permeability decreases gradually, but the rate of decline decreases slowly. When the axial stress further increases from 8 MPa to 16 MPa, permeability begins to increase gradually with the increase in axial stress, and the permeability is always between the permeability change curves corresponding to the axial stress of 8 MPa and

4 MPa. Similarly, under the stress condition σ_1:σ_2:σ_3 = x:4:4, the axial stress increases from 8 MPa to 12 MPa, while permeability decreases with the increase in axial stress. However, when the axial stress continues to increases from 12 MPa to 32 MPa, permeability shows an increasing trend. In contrast, under higher confining stress conditions σ_1:σ_2:σ_3 = x:6:6 and x:8:8, permeability remains largely unchanged with increasing axial stress. Permeability shows different changes with the increase in axial stress, which is mainly because of the fact that when the triaxial stress is relatively small (σ_1:σ_2:σ_3 = x:2:2 and x:4:4), the prefabricated fractures in the sandstone specimen, coupled with the high strength of the sandstone itself, basically do not develop pore fractures under low axial stress. These pore fractures close with the increase in low horizontal axial stress, resulting in the decrease in sandstone permeability. However, with further increase in axial stress, these prefabricated fractures begin to expand under a high level of axial stress, which increases the pore fractures in the sandstone and increases the permeability of the sandstone, when the triaxial stress is relatively large (σ_1:σ_2:σ_3 = x:6:6 and x:8:8). With the increase in axial stress, the expansion of primary fractures and the development of new fractures in sandstone specimens are approximately consistent with the fracture rate of sandstone compression closure. It is observed that permeability basically does not change with the increase in axial stress.

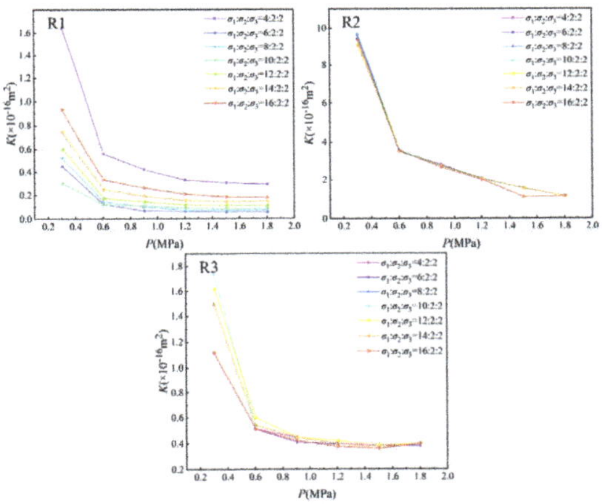

(a) σ_1:σ_2:σ_3 = x:2:2

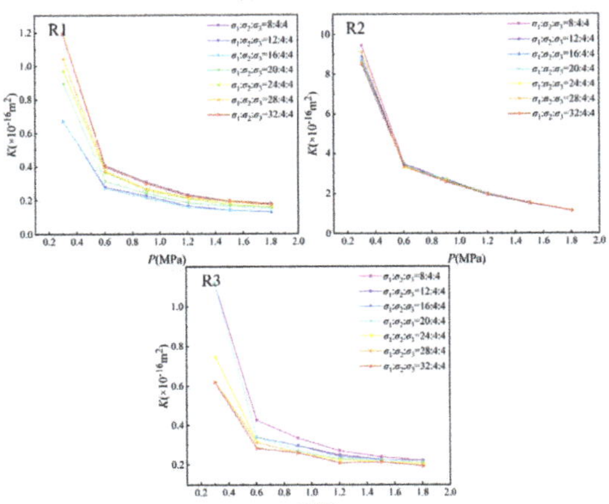

(b) σ_1:σ_2:σ_3 = x:4:4

Figure 6. *Cont.*

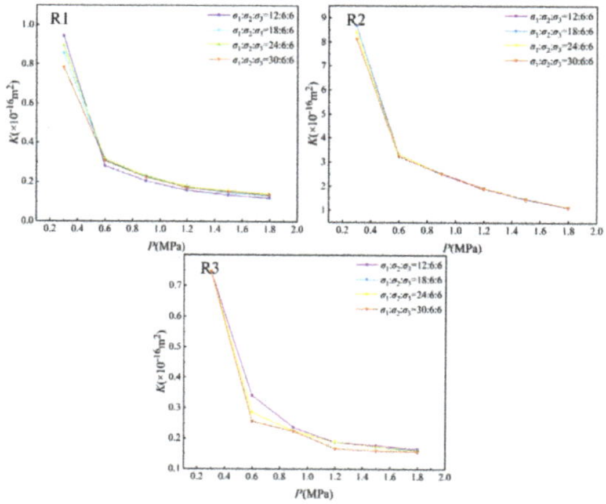

(**c**) σ_1:σ_2:σ_3 = x:6:6

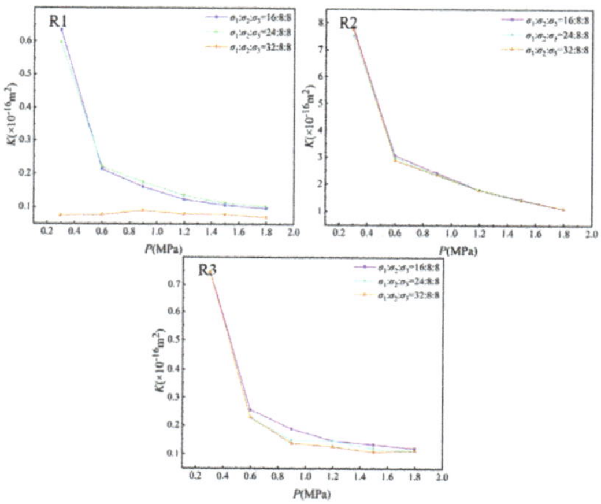

(**d**) σ_1:σ_2:σ_3 = x:8:8

Figure 6. Variation in permeability with pore pressure during the increase in axial stress.

For sample R2, the permeability remains largely unchanged with the increase in axial stress under different triaxial stress ratios. For sample R3, when σ_1:σ_2:σ_3 = x:2:2, the permeability shows little variation as the axial stress increases from 4 MPa to 6 MPa. However, when the axial stress increases from 6 MPa to 10 MPa and further to 16 MPa, the permeability increases gradually. When σ_1:σ_2:σ_3 = x:4:4, the permeability decreases gradually with the increase in axial stress. When σ_1:σ_2:σ_3 = x:6:6 and x:8:8, the permeability stabilizes and shows no significant change with axial stress. The analysis shows that under a very low triaxial stress ratio (σ_1:σ_2:σ_3 = x:2:2), multiple fractures were generated in the sandstone during the pre-fracturing process, and some tip fractures further expand and develop in the initial triaxial loading stage, resulting in an increase in pore-fracture connectivity and a gradual rise in permeability. However, when the triaxial stress ratio reaches σ_1:σ_2:σ_3 = x:4:4, the high strength of sandstone prevents fracture tips from reaching the critical stress required for propagation. Under these conditions, the rate of fracture closure due to compression exceeds the rate of partial fracture extension. As a result, permeability gradually decreases with increasing axial stress. With the further increase in triaxial stress ratio (σ_1:σ_2:σ_3 = x:6:6 and x:8:8), the applied stress approaches the fracture strength of the sandstone. At this time, the fracture closure rate is almost equal to the expansion rate of the new fracture. Consequently, permeability remains largely unchanged with increasing axial stress.

3.5. Relationship of Permeability and Axial Strain

The permeability evolution of the fractured sandstone is directly related to the increase in stress-dependent strain, as shown in Figure 7. A logarithmic relationship was observed between stain and stress in situ, while an exponential was identified between permeability and in situ stress. In detail, permeability initially decreases rapidly and then more gradually as pore pressure increases. The maximum permeability of 9×10^{-16} m^2 presented at 0.3 MPa, while the minimum permeability of 0.1×10^{-16} m^2 presented at 1.8 MPa. Gas conductivity decreased with the occurrence of turbulence flow due to the interaction of dynamic flow and limited connected channel. Permeability decreases with the increase minimum and intermediate principal stress. This decreasing trend in response to principal stress is pronounced at low pore pressure, while only a slight permeability response is observed at higher pore pressures ranging from 0.6 to 1.8 MPa.

3.6. Establishment of Stress-Dependent Permeability Theoretical Modeling

The influence of pore connectivity and fracture conductivity on the conductivity of intact rock was considered. In this study, the effective permeability k_e was developed, which consists of two parts, namely k_p of pore structure within rock matrix and k_f of fracture. According to Darcy's law,

$$Qr = A \cdot \frac{ke}{\mu} \cdot \frac{\Delta P}{L} = Qf + Qp = A \cdot \frac{kf}{\mu} \cdot \frac{\Delta P}{L} + A \cdot \frac{kp}{\mu} \cdot \frac{\Delta P}{L} \tag{1}$$

where Q_r, Q_f, and Q_p are the flow rates of whole rock, fracture and matrix pores, respectively, k_e, k_p and k_f are the corresponding permeability, ΔP is the fluid pressure difference between the upstream and downstream, A is the cross-section area, L is the length of the flow path, and μ denotes the fluid viscosity.

Permeability values k_f and k_p can be correlated with fracture porosity and pore porosity according to the modified Kozeny–Carman model and Poiseuille equation, which are controlled by the following formula:

$$kf = \frac{a^2\phi_f^3}{12} \tag{2}$$

$$kp = \frac{\phi_p^3}{c\tau^2 S^2} \tag{3}$$

where a is rock matrix size, φ_f is the fracture porosity, φ_p is the pore porosity of the rock matrix, c, τ and S denote the Kozeny constant, tortuosity of the tube, and tube surface area per unit volume.

Partial differentiation of effective permeability k_e to effective stress σ_e:

$$\frac{\partial k_e}{\partial \sigma_e} = \frac{\partial\left(k_f + k_p\right)}{\partial \sigma_e} = \frac{a^2\phi_f^2}{4} \cdot \frac{\partial \phi_f}{\partial \sigma_e} + \frac{3\phi_p^2}{c\tau^2 S^2} \cdot \frac{\partial \phi_p}{\partial \sigma_e} \tag{4}$$

Rock compressibility C_f and pore medium compressibility C_p are defined as [41]

$$C_f = -\frac{1}{\phi_f} \cdot \frac{\partial \phi_f}{\partial \sigma_e} \tag{5}$$

$$C_p = -\frac{1}{\phi_p} \cdot \frac{\partial \phi_p}{\partial \sigma_e} \quad (6)$$

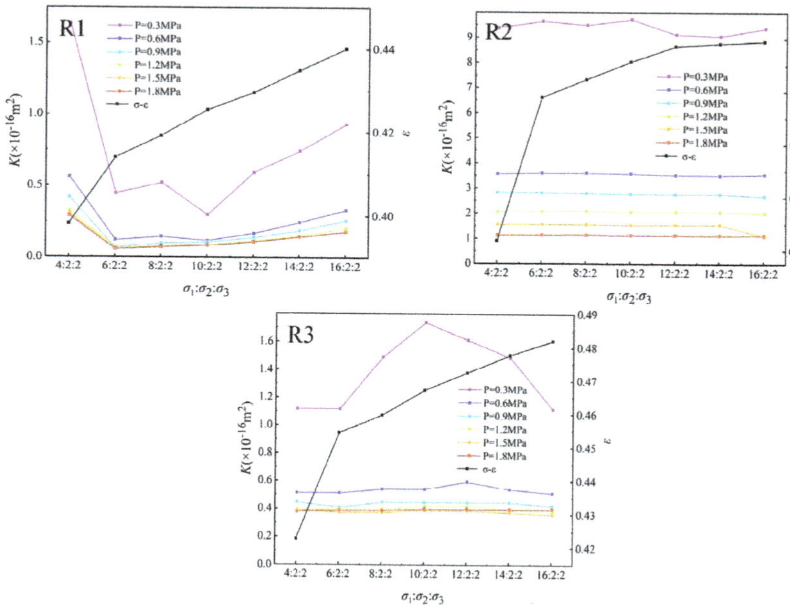

(**a**) σ_1:σ_2:σ_3 = x:2:2

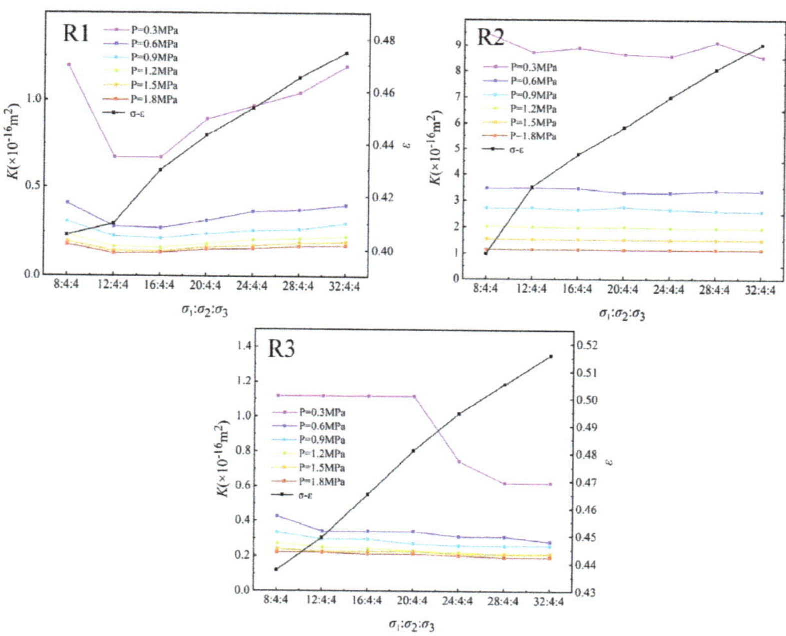

(**b**) σ_1:σ_2:σ_3 = x:4:4

Figure 7. *Cont.*

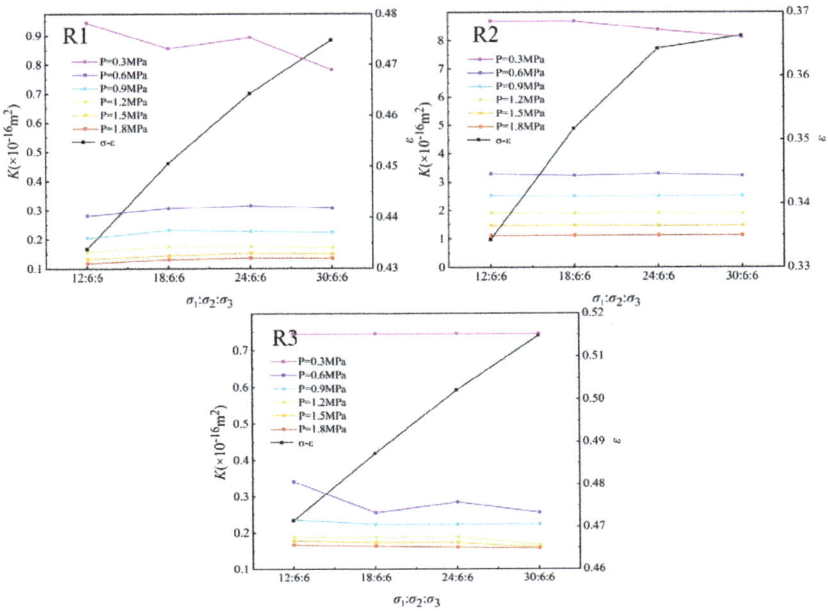

(c) $\sigma_1:\sigma_2:\sigma_3$ = x:6:6

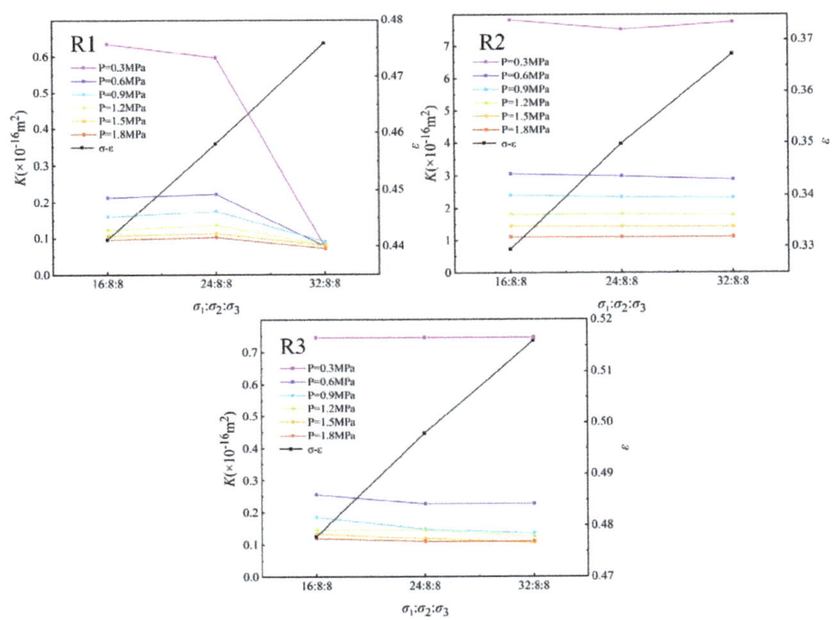

(d) $\sigma_1:\sigma_2:\sigma_3$ = x:8:8

Figure 7. Variation in permeability and axial strain.

Taking Equations (6) and (7) into Equation (5), the effective permeability was deduced [42]:

$$k_e = k_{f0} \cdot e^{-3C_f(\sigma_e-\sigma_{e0})} + \left(k_{e0} - k_{f0}\right) \cdot e^{-3C_p(\sigma_e-\sigma_{e0})} \quad (7)$$

Considering that compressibility varies with the applied effective stress, the stress-related fracture/pore compressibility can be calculated as follows:

$$\overline{C_f} = \frac{C_{f0}}{\alpha_f(\sigma_e - \sigma_{e0})}\left(1 - e^{-\alpha_f(\sigma_e - \sigma_{e0})}\right) \tag{8}$$

$$\overline{C_p} = \frac{C_{p0}}{\alpha_p(\sigma_e - \sigma_{e0})}\left(1 - e^{-\alpha_p(\sigma_e - \sigma_{e0})}\right) \tag{9}$$

where C_{f0} and C_{p0} are the initial fracture and pore compressibility, α_f and α_p are the declining rate of fracture or pore compressibility with the increase in the effective stress. The first-order Taylor formula expansion is carried out for Equations (9) and (10):

$$\overline{C_f} = \frac{C_{f0}}{1 + \alpha_f(\sigma_e - \sigma_{e0})} \tag{10}$$

$$\overline{C_p} = \frac{C_{p0}}{1 + \alpha_p(\sigma_e - \sigma_{e0})} \tag{11}$$

Taking Equations (11) and (12) into Equation (8), the equivalent effective model was established:

$$k_e = k_{f0} \cdot e^{-3\overline{C_f}(\sigma_e - \sigma_{e0})} + \left(k_{e0} - k_{f0}\right) \cdot e^{-3\overline{C_p}(\sigma_e - \sigma_{e0})} \tag{12}$$

The validity of developed model was verified based on the experiment data, as shown in Figure 8. The coefficient of correlation for R2 was controlled at 0.9554, and the result demonstrates the developed stress-dependent permeability model can accurately predict the evolution of the pore-fracture rock fluid conductivity. The logarithmic strain-stress relationship observed in experiments (Figure 4) aligns with the exponential permeability–stress dependency predicted by the model (Equation (12)).

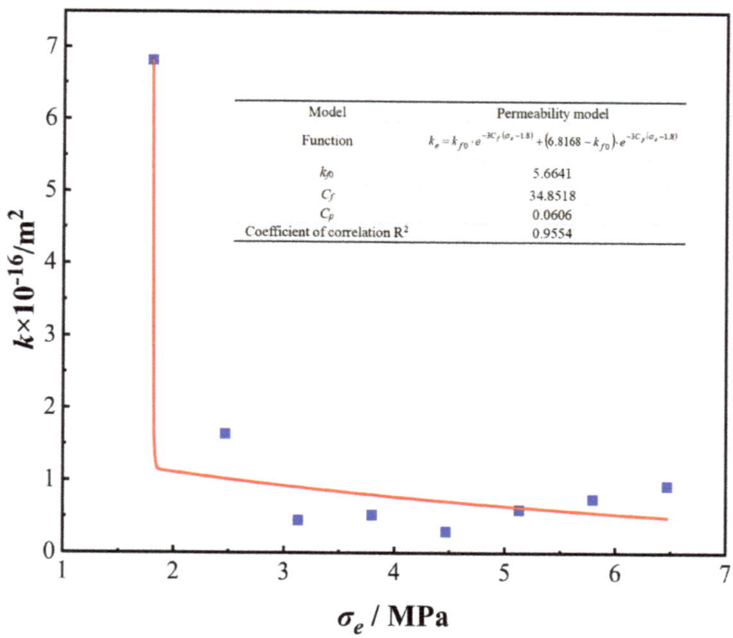

Figure 8. Comparison of the model predicted and experimental data.

4. Conclusions

(1) The morphology of fractures in low-permeability sandstone significantly influences the evolution of both permeability and strain. Among different fracture sizes, small fractures exhibit the greatest sensitivity to increases in the maximum principal stress, followed by medium and large fractures. In contrast, the most pronounced principal strain response to increasing principal stress occurs in large fractures, with medium and small fractures showing successively lower responses. Overall, fracture conductivity generally decreases, with minor variations observed specifically in small fractures.

(2) Hydrodynamic pressure influences the fracture conductivity, and an exponential correlation was observed. The permeability of sandstone is negatively correlated to hydrodynamic pressure; the maximum permeability of 9×10^{-16} m^2 presented at 0.3 MPa, and minimum permeability of 0.1×10^{-16} m^2 presented at 1.8 MPa. The decrease in amplitude increased with the increase in fracture size. The gas conductivity decreased with the occurrence of turbulence flow due to the interaction of dynamic flow and limited connected channel.

(3) In situ stress significantly influences the evolution of both principal strain and pore pressure-dependent permeability. Permeability decreases with increasing minimum and intermediate principal stresses. This decreasing trend in response to principal stress is pronounced at low pore pressures, while only a slight permeability response is observed at higher pore pressures ranging from 0.6 to 1.8 MPa in situ.

Author Contributions: Research design and original draft, Y.G.; supervision and editing, T.Z.; technical support, X.Y. (Xiang Yu); software development, Y.L.; data collection, X.Y. (Xin Yang). All authors have read and agreed to the published version of the manuscript.

Funding: This paper was supported by the National Natural Science Foundation (42477201), China; National key research and development plan (2021YFC28000903-05), China; the National Science and Technology Major Project (2024ZD1700201); the National Key Laboratory of Deep Coal Mining and Environmental Protection Fund (HNKY2024ZD201) China. Excellent Youth Project of Anhui Province (2022AH030086), China; the Anhui Provincial General Fund Project (2308085ME154), China; the Anhui Provincial Key Research and Development Program Fund Project (2023z04020001), China.

Institutional Review Board Statement: Not applicable.

Informed Consent Statement: Not applicable.

Data Availability Statement: The data presented in this study are available on request from the corresponding author. The data are not publicly available due to privacy considerations and ongoing research.

Conflicts of Interest: The authors declare no conflicts of interest.

References

Dang, F.Q.; Li, S.Y.; Dong, H.; Wang, Z.J.; Zhu, J.Z. Experimental study on the oil recovery characteristics of CO_2 energetic fracturing in ultralow-permeability sandstone reservoirs utilizing nuclear magnetic resonance. *Fuel* **2024**, *366*, 131370. [CrossRef]

Tang, M.; Zhang, T.; Ma, Y.K.; Hao, D.Y.; Yang, X.; Li, Y.F. Experimental study on fracture effect on the multiphase flow in ultra-low permeability sandstone based on LF-NMR. *Geoenergy Sci. Eng.* **2023**, *222*, 211399. [CrossRef]

Zhang, T.; Tang, M.; Ma, Y.K.; Zhu, G.P.; Zhang, Q.H.; Wu, J.; Xie, Z.Z. Experimental study on CO_2/Water flooding mechanism and oil recovery in ultralow—Permeability sandstone with online LF-NMR. *Energy* **2022**, *252*, 123948. [CrossRef]

Zhang, T.; Tang, M.; Yuan, L.; Zhu, G.P.; Zhang, C.; Liu, Y.; Li, Y.F.; Wang, W.; Yang, X. Experimental study on stress-dependent multiphase flow in ultra-low permeability sandstone during CO_2 flooding based on LF-NMR. *Energy* **2023**, *278*, 127874. [CrossRef]

Liu, J.Q.; Lyu, X.Z.; Liu, Y.; Zhang, P.P. Energy evolution and macro-micro failure mechanisms of frozen weakly cemented sandstone under uniaxial cyclic loading and unloading. *Cold Reg. Sci. Technol.* **2023**, *214*, 103947. [CrossRef]

6. Vasyliev, L.; Malich, M.; Vasyliev, D.; Katan, V.; Rizo, Z. Improving a technique to calculate strength of cylindrical rock samples terms of uniaxial compression. *Min. Miner. Deposits* **2023**, *17*, 43–50. [CrossRef]
7. Abbas, H.A.; Mohamed, Z.; Kudus, S.A. Deformation behaviour, crack initiation and crack damage of weathered compos sandstone-shale by using the ultrasonic wave and the acoustic emission under uniaxial compressive stress. *Int. J. Rock. Me Min. Sci.* **2023**, *170*, 105497. [CrossRef]
8. Jiang, X.Y.; Wu, D.; Zhao, Z.X.; Yang, Y.; Tian, J.W.; Wang, F.L. Experimental and numerical study on the damage evolution a acoustic emission multi-parameter responses of single flaw sandstone under uniaxial compression. *Theor. Appl. Fract. Mech.* **20** *133*, 104535. [CrossRef]
9. Ishibashi, T.; Asanuma, H.; Mukuhira, Y.; Watanabe, N. Laboratory hydraulic shearing of granitic fractures with surface roughn under stress states of EGS: Permeability changes and energy balance. *Int. J. Rock. Mech. Min. Sci.* **2023**, *170*, 105512. [CrossRe
10. Ishibashi, T.; Yamaya, Y.; Asanuma, H. Experimental insights into coupled hydraulic, mechanical, and electrical behavi of granite fractures: Implications for indirect estimation of crustal permeability changes. *Int. J. Rock. Mech. Min. Sci.* **20** *188*, 106060. [CrossRef]
11. Yu, B.C.; Zhao, H.G.; Tian, J.B.; Liu, C.; Song, Z.L.; Liu, Y.B.; Li, M.H. Modeling study of sandstone permeability under tr triaxial stress based on backpropagation neural network, genetic programming, and multiple regression analysis. *J. Nat. Gas. S Eng.* **2021**, *86*, 103742. [CrossRef]
12. Niu, Z.H.; Zhu, Z.D.; Que, X.C.; Xie, X.H.; Jin, K. Experimental study on damage evolution process, seepage characteristics, a energy response of columnar jointed basalt under true triaxial cyclic loading. *Geoenergy Sci. Eng.* **2024**, *237*, 212774. [CrossRe
13. Lu, J.; Yin, G.Z.; Deng, B.Z.; Zhang, W.Z.; Li, M.H.; Chai, X.W.; Liu, C.; Liu, Y.B. Permeability characteristics of layered compos coal-rock under true triaxial stress conditions. *J. Nat. Gas. Sci. Eng.* **2019**, *66*, 60–76. [CrossRef]
14. Liu, C.; Yin, G.Z.; Li, M.H.; Deng, B.Z.; Song, Z.L.; Liu, Y.B.; Yin, S.Y. Shale permeability model considering bedding effect und true triaxial stress conditions. *J. Nat. Gas. Sci. Eng.* **2019**, *68*, 102908. [CrossRef]
15. Duan, X.L.; Wang, W.; Liu, S.F.; Cao, Y.J.; Zheng, Z.; Zhu, Q.Z. Experimental investigation on mechanical behavior, ener evolution and gas permeability of anisotropic phyllite subjected to triaxial compression and cyclic loading. *Geomech. Ener Environ.* **2023**, *35*, 100483. [CrossRef]
16. Liu, J.C.; Jia, W.S.; Duan, H.F.; Li, X.; Xia, K.W. Experimental study on the deformation and failure characteristics of cement sto under true triaxial stress state. *Constr. Build. Mater.* **2024**, *449*, 138397. [CrossRef]
17. Zhang, J.Z.; Long, Y.D.; Zhang, T.; Zhou, X.P. A true triaxial experiment investigation of the mechanical and deformation failu behaviors of flawed granite after exposure to high-temperature treatment. *Eng. Fract. Mech.* **2024**, *306*, 110273. [CrossRef]
18. Zhou, H.Y.; Liu, Z.B.; Liu, F.J.; Shao, J.F.; Li, G.L. Anisotropic strength, deformation and failure of gneiss granite under high stre and temperature coupled true triaxial compression. *J. Rock. Mech. Geotech. Eng.* **2024**, *16*, 860–876. [CrossRef]
19. Meng, F.B.; Shi, L.; Hall, S.; Baud, P.; Wong, T.F. Onset of pore collapse and dilatancy in porous sandstone under true triaxi compression: Experimental observation and micromechanical modeling. *Int. J. Rock. Mech. Min. Sci.* **2025**, *186*, 105983. [CrossRe
20. Zhang, Q.H.; Meng, X.R.; Zhao, G.M.; Liu, C.Y.; Liu, Z.X.; Wu, X.K. Energy dissipation and damage mechanism of rocks und true triaxial graded perturbations. *Soil. Dyn. Earthq. Eng.* **2025**, *190*, 109192. [CrossRef]
21. Singh, C.; Gupta, P.K. Numerical analysis of failure mechanics of concrete under true dynamic triaxial loading using a four-pha meso-model. *Constr. Build. Mater.* **2024**, *450*, 138661. [CrossRef]
22. Wibisono, D.Y.; Gutierrez, M.; Majumder, D. Experimental investigation of tunnel damage and spalling in brittle rock using true-triaxial cell. *Int. J. Rock. Mech. Min. Sci.* **2024**, *182*, 105884. [CrossRef]
23. Couture, C.; Besuelle, P. A true triaxial experimental study on porous Vosges sandstone: From strain localization precursors failure using full-field measurements. *Int. J. Rock. Mech. Min. Sci.* **2022**, *153*, 105031. [CrossRef]
24. Lin, H.F.; Ji, P.F.; Kong, X.G.; Li, S.G.; Long, H.; Xiao, T.; Li, B. Experimental study on the influence of gas pressure on CH adsorption-desorption-seepage and deformation characteristics of coal in the whole process under triaxial stress. *Fuel* **202** *333*, 126513. [CrossRef]
25. Gan, L.; Liu, Y.; Xu, F.; Ma, H.Y.; Xu, W.C. Experimental investigation of the seepage characteristics of a single fracture in limesto with different roughness and seepage fluids. *J. Hydrol.* **2023**, *622*, 129699. [CrossRef]
26. Zhang, S.H.; Wu, S.C.; Duan, K. Study on the deformation and strength characteristics of hard rock under true triaxial stress sta using bonded-particle model. *Comput. Geotechnol.* **2019**, *112*, 1–16. [CrossRef]
27. Zhang, Y.J.; Liu, S.J.; Kou, M.M.; Wang, Z.Q. Mechanical and failure characteristics of fissured marble specimens under tr triaxial compression: Insights from 3-D numerical simulations. *Comput. Geotechnol.* **2020**, *127*, 103785. [CrossRef]
28. Gao, Y.H.; Feng, X.T. Study on damage evolution of intact and jointed marble subjected to cyclic true triaxial loading. *Eng. Fra Mech.* **2019**, *215*, 224–234. [CrossRef]
29. Wang, S.S.; Xu, W.Y.; Yan, L.; Feng, X.T.; Xie, W.C.; Chen, H.J. Experimental investigation and failure mechanism analysis f dacite under true triaxial unloading conditions. *Eng. Geol.* **2020**, *264*, 105407. [CrossRef]

Liu, C.Y.; Pan, C.; Zhao, G.M.; Xu, W.S.; Meng, X.R.; Liu, W.J.; Cheng, X. Experimental study on the failure mechanism of roadway surrounding rock under true triaxial loading: Insights from acoustic emission (AE) characteristics. *Measurement* **2025**, *246*, 116727. [CrossRef]
Zhang, Q.H.; Meng, X.R.; Zhao, G.M.; Li, Y.M.; Xu, W.S.; Liu, C.Y.; Liu, Z.X.; Wu, X.K.; Qin, Z.H.; Wang, K. Damage evolution and acoustic emission characteristics of sandstone under different true triaxial cyclic loading and unloading modes. *Eng. Fail. Anal.* **2025**, *167*, 108947. [CrossRef]
Li, B.L.; Shi, W.H.; Zhang, H.J.; Long, Y. Failure behaviors and acoustic emission characteristics of flawed granite of different grain sizes: Insights from true triaxial experiments with a free face. *Theor. Appl. Fract. Mech.* **2024**, *131*, 104394. [CrossRef]
Zhou, Y.S.; Zhang, S.C.; Ma, X.F.; Zhang, X.H.; Zhang, S.P. Hydraulic fracture morphology and conductivity of continental shale under the true-triaxial stress conditions. *Fuel* **2023**, *352*, 129056. [CrossRef]
Zhang, Z.; Li, R.H.; Pan, P.Z.; Qi, J.H.; Su, G.S.; Zheng, H. Shear failure behaviors and degradation mechanical model of rock mass under true triaxial multi-level loading and unloading shear tests. *Int. J. Min. Sci. Technol.* **2024**, *34*, 1385–1408. [CrossRef]
Huang, J.; Hu, Q.T.; Qin, C.Z.; Song, Z.L.; Wang, X.D. Pre-peak acoustic emission characteristics of tight sandstone failure under true triaxial stress. *J. Nat. Gas. Sci. Eng.* **2022**, *102*, 104576. [CrossRef]
Gao, H.; Xie, H.P.; Zhang, Z.T.; Lu, J.; Zhang, D.M.; Zhang, R.; Wu, M.Y. True triaxial energy evolution characteristics and failure mechanism of deep rock subjected to mining-induced stress. *Int. J. Rock. Mech. Min. Sci.* **2024**, *176*, 105724. [CrossRef]
Jiang, C.B.; Yang, Y.; Wei, W.H.; Duan, M.K.; Yu, T. A new stress-damage-flow coupling model and the damage characterization of raw coal under loading and unloading conditions. *Int. J. Rock. Mech. Min. Sci.* **2021**, *138*, 104601. [CrossRef]
Fan, L.; Liu, S.M. Evaluation of permeability damage for stressed coal with cyclic loading: An experimental study. *Int. J. Coal Geol.* **2019**, *216*, 103338. [CrossRef]
Yang, S.; Wen, G.C.; Yan, F.Z.; Li, H.L.; Liu, Y.B.; Wu, W.B. Swelling characteristics and permeability evolution of anthracite coal containing expansive clay under water-saturated conditions. *Fuel* **2020**, *279*, 118501. [CrossRef]
Xue, Y.; Ranjith, P.G.; Gao, F.; Zhang, D.C.; Cheng, H.M.; Chong, Z.H.; Hou, P. Mechanical behaviour and permeability evolution of gas-containing coal from unloading confining pressure tests. *J. Nat. Gas. Sci. Eng.* **2017**, *40*, 336–346. [CrossRef]
Jiang, T.W.; Yao, W.; Sun, X.W.; Qi, C.Y.; Li, X.; Xia, K.W.; Zhang, J.; Nasseri, M.H.B. Evolution of anisotropic permeability of fractured sandstones subjected to true-triaxial stresses during reservoir depletion. *J. Petrol. Sci. Eng.* **2021**, *200*, 108251. [CrossRef]
Gao, H.; Zhang, D.M.; Lu, J.; Yin, G.Z.; Wu, M.Y. Experimental study on infuence of intermediate principal stress on the permeability of sandstone. *Transport. Porous Med.* **2020**, *135*, 753–778. [CrossRef]

MDPI AG
Grosspeteranlage 5
4052 Basel
Switzerland
Tel.: +41 61 683 77 34

Applied Sciences Editorial Office
E-mail: applsci@mdpi.com
www.mdpi.com/journal/applsci